ORBITAL APPROACH TO THE ELECTRONIC STRUCTURE OF SOLIDS

Orbital Approach to the Electronic Structure of Solids

ENRIC CANADELL
Institut de Ciència de Materials de Barcelona (CSIC)

MARIE-LIESSE DOUBLET
CNRS – University of Montpellier

CHRISTOPHE IUNG
University of Montpellier

OXFORD
UNIVERSITY PRESS

Great Clarendon Street, Oxford, OX2 6DP,
United Kingdom

Oxford University Press is a department of the University of Oxford.
It furthers the University's objective of excellence in research, scholarship,
and education by publishing worldwide. Oxford is a registered trade mark of
Oxford University Press in the UK and in certain other countries

© E. Canadell, M.-L. Doublet & C. Iung 2012

The moral rights of the authors have been asserted

First published 2012
First published in paperback 2016

All rights reserved. No part of this publication may be reproduced, stored in
a retrieval system, or transmitted, in any form or by any means, without the
prior permission in writing of Oxford University Press, or as expressly permitted
by law, by licence or under terms agreed with the appropriate reprographics
rights organization. Enquiries concerning reproduction outside the scope of the
above should be sent to the Rights Department, Oxford University Press, at the
address above

You must not circulate this work in any other form
and you must impose this same condition on any acquirer

Published in the United States of America by Oxford University Press
198 Madison Avenue, New York, NY 10016, United States of America

British Library Cataloguing in Publication Data

Data available

Library of Congress Cataloging in Publication Data

Data available

ISBN 978–0–19–953493–7 (Hbk.)
ISBN 978–0–19–876705–3 (Pbk.)

Links to third party websites are provided by Oxford in good faith and
for information only. Oxford disclaims any responsibility for the materials
contained in any third party website referenced in this work.

Preface

Understanding the electronic structure of the materials on which he/she is working may not be an essential need for an experimental scientist but certainly can make his/her everyday work easier and more intellectually pleasing. The electronic structure is the most obvious and useful link between the structure and properties of any solid. Thus, understanding how the electronic structure of a given material can be assembled (and thus how it can be altered) from that of the chemically significant building blocks from which it is made up is a simple yet very suggestive approach to the main goal of any materials science researcher: the design and preparation of materials with controlled properties. Whether the new materials suggested in this way can be actually prepared or not is something that depends, among other things, on the preparative skills and art of the scientist. This is why knowledge of the electronic structure may not be essential. However, it can make the quest much more rational and straightforward, or it can direct the attention to something which otherwise could seem bizarre.

The impressive increase in computing power and the development of highly performing simulation codes for solids in recent years has provided chemists, physicists, and materials science researchers with very efficient tools to access the details of the electronic structure of practically any periodic solid. However, this does not necessarily mean that we can understand the electronic structure of any solid in a precise yet simple way. Indeed this is what is needed to truly master the link between the structure and properties of the solids of interest. The development of efficient computational *and* conceptual tools is the only way towards a fruitful interaction between theoretical and experimental approaches with the intention of developing a sound understanding in this field. Materials science being an essentially interdisciplinary field, the training of scientists in the area is very much dependent on the physical or chemical orientation of their curriculum. Nevertheless, understanding the structure–properties correlation needs both physical and chemical concepts, which are usually taught using quite different languages. The reason why the writing of this book has been undertaken is the observation that, to the best of our knowledge, none of the materials science books available at present extensively use a blend of band theory, the appropriate *physical* approach to the understanding of the structure and properties of many solids, and orbital interaction arguments, which is a transparent and *chemically* very insightful concept. We believe that this kind of interdisciplinary approach may be extremely enlightening.

There is certainly nothing novel in saying that knowledge of electronic structure is one of the more effective ways of making significant advances in materials science. J. Goodenough was among the first to systematically use concepts of electronic structure closely linked to structural details in looking for trends and predicting what materials could exhibit a certain physical property. This work had, and still has, a lasting influence on materials science. Pioneered by R. Hoffmann, J. K. Burdett, and M.-H. Whangbo in the 1980s, the introduction to materials science of the ideas of orbital interaction, which had been so useful in rationalising the structure and reactivity of molecules, was a major breakthrough. It soon became clear that the step-by-step building up of many of the tools used within the context of the band theory of solids, such as band structure, density of states, Fermi surface, etc., based on orbital interaction ideas, provided an invaluable yet intuitive and easy-to-handle tool with which to analyse the results of quantitative calculations or to rationalise experimental observations. Structural and transport properties, the origin of different phase transitions and structural modulations, the nature of scanning tunnelling and atomic force microscopy images of complex materials, etc. were successfully rationalised on the basis of this type of approach. Very detailed structural information is encoded within orbital-interaction-type arguments so that through this approach it is relatively easy to link the effect of possible structural modifications into say the band structure or the Fermi surface, etc. and, consequently, to anticipate how these changes could alter the stability, conductivity or related properties of a given structure.

With these developments in mind, around 1990 we thought that it would be timely to introduce these ideas into the curricula of chemistry, physics or materials science courses at the postgraduate or final-year undergraduate levels. This idea materialised as a course on the orbital approach to the electronic structure of solids given at Université de Paris-Sud Orsay, which was quite successful and was repeated for a number of years. It was also introduced at other French institutions such as the Ecole Normale Supérieure de Cachan, Université de Montpellier, and Université de Pau, as well as in several international events. Based on this experience a book entitled *Description orbitalaire de la structure électroniques des solides* by C. Iung and E. Canadell, covering the general principles and applications of such approach to one-dimensional solids was published in French by Ediscience International in 1997. Over the years many colleagues prompted us to complete this work by writing a new book fully covering the course, but academic and professional duties continuously delayed this project. The present book is a natural follow-up of the initial French publication in which we have generalised the content to cover two- and three-dimensional solids and added some new material.

The book contains 12 chapters, the first two being a sort of prelude. The first is a very brief overview of the free electron theory of solids with the purpose of introducing some very basic physical notions, which we will use throughout the book. In the second chapter we present a short overview of the basic notions currently used to understand the electronic structure of molecules, emphasising the symmetry and orbital interaction arguments. One of the purposes of this chapter is to show that the molecular orbital theory used for molecules and the band theory used for periodic solids are really simple variations of the same idea due to the discrete or periodic nature of the systems. The essential machinery of the band theory of solids and its orbital interaction analysis is developed in Chapter 3. Most of the formal tools that will be used throughout the book are explained there using the simplest periodic system we can think of: the infinite chain of hydrogen atoms. This keeps the formal developments simple and allows us to treat the same system in different ways so that the reader may be aware of different ways to approach a given problem. The fourth chapter is devoted to the ubiquitous Peierls distortions of solids. This is an important phenomenon exhibited by many solids and has strong consequences for transport and other properties. Chapters 5, 7, and 8 are essentially different applications of the ideas developed in the third and fourth chapters to organic and inorganic one-dimensional solids. Chapter 6 is a brief introduction to the handling of symmetry when studying the electronic structure of solids. The use of symmetry in band theory is an elegant yet not always simple matter, which cannot be developed at length in a book like the present one. However, we have discussed some useful and quite basic aspects of symmetry in this chapter. Up to the end of Chapter 8 the work is restricted to one-dimensional systems. Chapters 9–11 generalise the approach to two- and three-dimensional solids. In Chapter 9 the basic theoretical notions are generalised for systems of any dimensionality and some model systems are considered. The increase in dimensionality and structural complexity soon leads to the need to consider many orbitals and several directions of the Brillouin zone. The analysis of the results (or the qualitative building up of the electronic structure) may become too cumbersome, so that a simpler analytical tool must be devised. The simpler and more useful tool devised for this purpose is the density of states (DOS). The object of Chapter 10 is to present several ways to analyse this useful construct from the viewpoint of orbital interaction analysis using real examples. Chapter 11 deals with low-dimensional solids and the analysis of the Fermi surface, an extremely useful concept which, when appropriately decoded, contains much information about the transport and structural properties of metallic systems. In this chapter we will show that the essential aspects of the Fermi surface of a given metal may be obtained in a relatively simple way using the orbital interaction approach. The procedure will be illustrated by considering several classes of low-dimensional materials, which have given rise to considerable debate in the literature. Most of the present book uses a one-electron view of the electronic structure of solids. Although this is a perfectly legitimate option for a very wide range of materials and for the purposes of this book, it must be clearly stated that an explicit consideration of electronic repulsion is indispensable to understand certain classes of solids such as systems exhibiting magnetic properties. Discussion of this problem at a level consistent with the detailed approach of this book would have markedly increased its length and has not been considered realistic. However, we have included a final chapter in which the essentials of how the inclusion of electronic repulsion can modify the conclusions of a one-electron approach are outlined.

This is essentially a teaching book and consequently we have included a series of exercises so that readers may check their progress from time to time. Exercises that do not need to be considered on a first reading are marked with an asterisk. Answers to the exercises are provided, although sometimes they are deliberately only sketched. Since this is not a research book we have not made any attempt to present a detailed list of references. We generally mention some books or publications that may be helpful for readers interested in expanding their coverage of the subject. For the real examples discussed in the text we always make reference to the original publications reporting the structure of the system. In that way, readers interested in carrying out actual calculations for the system can prepare their inputs. In general we also provide reference to one or two papers in which the electronic structure is discussed. Because of the nature of the book we have always chosen those with a strong pedagogic orientation. We apologise for not mentioning the many excellent papers available for most of the systems considered.

This book would have been very different (and certainly less satisfying) without the input of the many students who attended our lectures. We are deeply indebted to them; their comments and questions have provided the impetus for the continuous polishing and revising of many aspects of this book. In addition we have benefited from the comments of many friends and colleagues who have read parts of the book, both the French and English versions. This book also owes much to the many discussions that took place before the actual writing with T. R. Hughbanks (Texas A & M University), M.-H. Whangbo (North Carolina State University), and the late J. K. Burdett (University of Chicago), and to Y. Jean (Palaiseau), and F. Volatron (Orsay) for pushing us to write the initial French version. We thank A. García for his help in implementing the tight-binding programs and F. Boyrie for his invaluable help in the LaTeX compilations. We also thank C. Raynaud and E. Clot for useful discussions about the methodological part of the book. We are grateful to Dunod Éditions for permission to use material from the French edition in the present work. We warmly thank Sonke Adlung, our editor at Oxford University Press, and his team (Lynsey Livingston, April Warman, and Clare Charles) for their continuous support, help and infinite patience with three authors who were continuously delaying the writing of the book. Last, but not least, we deeply thank our families for patiently enduring the writing of this book.

Enric Canadell,
Marie-Liesse Doublet,
and Christophe Iung

Bellaterra, Montpellier,
February 2011

Contents

1 **Elementary introduction to the transport properties of solids** 1
   1.1 Free electron model 1
      1.1.1 One-dimensional system 2
      1.1.2 Generalisation to a three-dimensional system 9
   1.2 Conductivity of real solids 10
      1.2.1 Factors influencing the conductivity 10
      1.2.2 Band structure of real solids 11
      1.2.3 Metallic behaviour 11
      1.2.4 Semiconducting and insulating behaviour 12
      1.2.5 Number of carriers 13

2 **Electronic structure of molecules: use of symmetry** 14
   2.1 Molecular orbital theory 14
      2.1.1 Born–Oppenheimer approximation 15
      2.1.2 One-electron approximation 15
      2.1.3 LCAO approximation 15
      2.1.4 Secular equations and secular determinant 16
      2.1.5 Basic features of the Hückel and extended Hückel methods 17
      2.1.6 Symmetry properties of the molecular orbitals 18
   2.2 A short review of the theory of symmetry point groups 19
      2.2.1 Different symmetry point groups 19
      2.2.2 Classes 21
      2.2.3 Basis for an irreducible representation 22
   2.3 Application to the study of the π system of regular cyclobutadiene 25
      2.3.1 Decomposition of the $\Gamma(p_z)$ basis 26
      2.3.2 Determination of the basis elements for different irreducible representations 27
      2.3.3 Molecular orbital diagram of the π system of regular cyclobutadiene 30
   2.4 Transition metal complexes 30
      2.4.1 Ligands and formal oxidation state 31
      2.4.2 The ML_6 octahedral complex 33
      2.4.3 Distortions of a complex 39

3 Electronic structure of one-dimensional systems: basic notions — 44

- 3.1 Bloch and crystal orbitals — 45
  - 3.1.1 Bloch orbitals — 46
  - 3.1.2 Crystal orbitals — 49
- 3.2 Electronic structure of the model chain H_n — 51
  - 3.2.1 Representation of the CO(Γ) and CO(X) functions — 51
  - 3.2.2 Energy of the crystal orbitals in the Hückel approach — 52
  - 3.2.3 Band structure — 54
  - 3.2.4 Basis for an energy level $E(\pm \vec{k})$ — 55
  - 3.2.5 Fermi level of the H_n chain — 57
- 3.3 Electronic structure of the dimerised model chain $(H_2)_{n'}$ — 58
  - 3.3.1 Formal determination of the band structure — 58
  - 3.3.2 Qualitative determination of the band structure — 61
- 3.4 Comparison of the regular H_n and dimerised $(H_2)_{n'}$ chains — 63
  - 3.4.1 Comparison of the band structures of the regular H_n chain generated by either a simple or a double unit cell — 63
  - 3.4.2 Dimerisation in the H_n chain: notion of distortion in a periodic system — 67

4 First-order Peierls distortions in periodic 1D systems — 72

- 4.1 Analysis of the model system $(H^{0.5+})_n$ — 72
  - 4.1.1 Effect of a tetramerisation on the Fermi level — 73
  - 4.1.2 Effect of a tetramerisation on the states near the Fermi level — 74
  - 4.1.3 Effect of a tetramerisation on the band structure — 76
- 4.2 Analysis of first-order Peierls distortion in terms of a charge density wave — 77
- 4.3 Nesting vector — 81
- 4.4 Commensurate and incommensurate distortions — 81
  - 4.4.1 Commensurate distortion — 81
  - 4.4.2 Incommensurate distortion — 83
  - 4.4.3 Comparison — 83
- 4.5 Conclusions — 83

5 Application to *trans*-polyacetylene — 85

- 5.1 Electronic structure of ethylene — 86
- 5.2 Main aspects of the band structure for *trans*-polyacetylene — 87
- 5.3 Detailed analysis of the band structure of *trans*-polyacetylene — 88
- 5.4 Determination of the band structure of *trans*-polyacetylene using the fragment formalism — 89
  - 5.4.1 Calculation of the band structure by means of the Hückel approach — 91
  - 5.4.2 Qualitative determination of the band structure — 92
- 5.5 Band gap opening at the Fermi level in *trans*-polyacetylene — 93

Contents   xi

6 Handling the symmetry in 1D compounds — 96

- 6.1 Analysis of the A_n system — 96
  - 6.1.1 Analysis of the cyclic A_n system — 96
  - 6.1.2 Analysis of the linear A_n system — 101
  - 6.1.3 Notion of group of a k point — 104
- 6.2 Application to the determination of the band structure for the A_n linear system, where A is an atom — 104
  - 6.2.1 Group of the different k points — 105
  - 6.2.2 Symmetry of the different Bloch orbitals — 105
  - 6.2.3 Bands associated with σ-type overlaps — 107
  - 6.2.4 Complete band structure — 108
- 6.3 Band structure of the hypothetical $(NaCl)_n$ chain — 109
  - 6.3.1 Group of the different k points — 110
  - 6.3.2 Bands associated with σ-type overlaps — 110
  - 6.3.3 Complete band structure — 112
- 6.4 Consequences of the existence of a glide plane — 113
  - 6.4.1 Using point group symmetry properties in *trans*-polyacetylene — 113
  - 6.4.2 Complete space group (non-symmorphic) of *trans*-polyacetylene — 115
  - 6.4.3 Crystal orbitals of *trans*-polyacetylene by means of the non-symmorphic space group $G = T_n \otimes C_{2h} \otimes \{E, g_\sigma\}$ — 117
  - 6.4.4 Concluding remarks — 119
- 6.5 Work plan for the study of a 1D system — 120

7 Application to polyacene — 122

- 7.1 Band structure near the Fermi level — 123
  - 7.1.1 Unit cell definition — 123
  - 7.1.2 Symmetry analysis of the chain — 123
  - 7.1.3 Appropriate fragment orbitals — 123
  - 7.1.4 Crystal orbitals at the Γ and X points — 124
  - 7.1.5 π-type band structure of polyacene — 126
- 7.2 Distortions in polyacene — 128
  - 7.2.1 Disappearance of the σ_{xy} symmetry plane — 128
  - 7.2.2 Disappearance of the σ_{yz} symmetry plane — 128
- 7.3 General remarks concerning Peierls distortions — 130
  - 7.3.1 First-order Peierls distortions — 130
  - 7.3.2 Second-order Peierls distortions — 131

8 Electronic structure of selected inorganic chains — 133

- 8.1 KCP — 133
  - 8.1.1 Band structure of the eclipsed chain $[Pt(CN)_4]^{(2-\delta)-}$ — 134
  - 8.1.2 Band structure of KCP (staggered chain) — 139
  - 8.1.3 Conclusions — 142
- 8.2 $(ML_4L')_n$ chains — 143
  - 8.2.1 Symmetry — 143

		8.2.2	Choice of the fragment orbitals to generate the Bloch orbitals	143
		8.2.3	Analysis of the Bloch orbitals at the Γ and X points	144
		8.2.4	Symmetry of the Bloch orbitals	144
		8.2.5	Band structure	145
		8.2.6	Study of the $(ReCl_4N)_n$ chain	147
		8.2.7	Electronic structure of the $(Pt(NH_2Et)_4Cl^{2+})_n$ chain	149
	8.3	Suggested studies		153

9 Electronic structure of 2D and 3D systems — 157

- 9.1 Basic concepts — 157
  - 9.1.1 Direct and reciprocal lattices — 157
  - 9.1.2 Bloch and crystal orbitals — 159
  - 9.1.3 Brillouin zone — 161
  - 9.1.4 Symmetry and the Brillouin zone — 162
- 9.2 Analysis of the electronic structure of 2D model systems — 166
  - 9.2.1 The square lattice $^2_\infty[H_n]$ system — 166
  - 9.2.2 The square lattice $^2_\infty[A_n]$ system — 169
  - 9.2.3 π-type band structure of hexagonal graphene layers — 173

10 Density of states — 181

- 10.1 Calculation and analysis of the density of states — 181
  - 10.1.1 Density of states — 181
  - 10.1.2 Projected density of states — 183
  - 10.1.3 Crystal orbital overlap population — 185
- 10.2 Combined use of DOS and COOP: electronic structure of the MPS_3 layered phases — 186
- 10.3 Step-by-step determination of the density of states: the $(Pt(NH_3)_4Cl)^{2+}$ chain — 188
- 10.4 Density of states and fragment molecular orbital interaction analysis: application to the $[(C_5H_5)M]$ chains — 193
- 10.5 Transition metal diborides with the AlB_2 structure type: a 3D case study — 196

11 Fermi surface and low-dimensional metals — 203

- 11.1 Notion of Fermi surface — 204
- 11.2 Nesting vector and electronic instabilities in low-dimensional metals — 207
- 11.3 Monoclinic TaS_3 versus $NbSe_3$ — 210
  - 11.3.1 Crystal structure and electron counting — 211
  - 11.3.2 Qualitative band structure — 212
  - 11.3.3 Qualitative Fermi surface: differences between $NbSe_3$ and TaS_3 — 214
- 11.4 Molybdenum bronzes — 215

Contents   xiii

        11.4.1   Octahedral distortions and t_{2g} level splitting in MoO_6 octahedra   216
        11.4.2 MoO_5 chain with corner-sharing octahedra: counting of $2p$ oxygen antibonding contributions   217
        11.4.3 $A_{0.33}MoO_3$ (A = K, Rb, Cs, Tl) 2D red bronzes: metallic or insulating?   219
        11.4.4 $A_{0.3}MoO_3$ (A = K, Rb, Tl) blue bronzes: 2D solids with pseudo-1D behaviour   224
        11.4.5   Looking for 1D systems where there seem to be none: the concept of hidden nesting   227
   11.5 Low-dimensional molecular conductors   232
        11.5.1   An archetypal molecular metal: $(TMTSF)_2PF_6$ 234
        11.5.2   Chemically modifying the electronic structure of molecular conductors   235
        11.5.3   Structurally complex materials with simple band structures   238
        11.5.4   A case study: 1D vs 2D character of the carriers in some α phases of BEDT-TTF   242
        11.5.5   Electronic structure and folding: how to relate the band structure and Fermi surface of different salts of the same family   247

12 Electron repulsion   256

   12.1 From the Hückel model to the Hubbard model   256
        12.1.1   The delocalised picture of H_2 256
        12.1.2   The localised picture of H_2 260
        12.1.3   From the molecule to the solid state   266
        12.1.4   Application to one-band systems   269
   12.2 Mean-field approaches   273
        12.2.1   The many-body problem   273
        12.2.2   The Hartree–Fock method   274
        12.2.3   Density functional theory   281
   12.3 Conclusion   287

Solutions for exercises   291

Appendix: Character tables   342
Index   345

Elementary introduction to the transport properties of solids

One of the main goals of this book is to build a bridge between the electronic structure of a periodic solid and its physical properties and, more particularly, its conducting properties. Consequently, the first thing that we must learn is what characterises a metal, a semiconductor, and an insulator. In this chapter we will use the simplest possible approach that provides some understanding of the nature of a metallic system: the *free electron model*. Even if this model is in many aspects quite simplistic, it provides a simple and essentially correct view of what makes a system a good or a bad conductor. In passing, a brief consideration of the free electron model will allow us to introduce many of the key concepts that we will use throughout this book.

1.1 Free electron model

For a system to behave as a metal, i.e. to be a good conductor, it must possess valence electrons not tightly bound to the nuclei. The free electron model, initially developed by Sommerfeld, is based on the assumption that a metal can be viewed as a series of electrons which move freely over a network of fixed A^+ cations (see Fig. 1.1). [1]

The potential felt by each electron *is assumed to be nil* in the solid, but equal to a positive and large value $(+V_0)$ outside. The electron is thus confined within the metallic piece and the different forces felt by the electron in the metal (attractive or repulsive) are neglected. Since we are interested in the collective properties of the bulk material and, in general, we will not be interested by the properties at the surface, we will adopt boundary conditions, making it

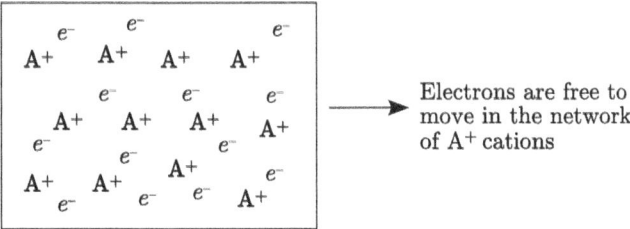

Electrons are free to move in the network of A^+ cations

Fig. 1.1
Free electron model: the electrons are represented as e^-.

possible to neglect the effects at the borders. Now that this approach is clearly stated we will proceed to the determination of the wavefunctions describing the behaviour of an electron in this *ideal* metal. For simplicity, we will initially consider a one-dimensional system and later we will generalise the results to a three-dimensional system.

1.1.1 One-dimensional system

Let us consider an electron constrained to move through a linear segment of length L.

Eigenvectors of the system

Let us write the Schrödinger equation for this one-dimensional system:

$$-\frac{\hbar^2}{2m}\frac{d^2\Psi(x)}{dx^2} + V(x)\Psi(x) = E\Psi(x) \tag{1.1}$$

where $\Psi(x)$ is the wavefunction describing an electron at point M with coordinate x, E is the energy of the electron and $V(x)$ is the potential at the M point. In the free electron model the potential is nil for any point of the system and we must therefore solve the following equation:

$$\forall x \in \left]-\frac{L}{2}, \frac{L}{2}\right] \quad -\frac{\hbar^2}{2m}\frac{d^2\Psi(x)}{dx^2} = E\Psi(x) \tag{1.2}$$

for which the general solution in the interval $]-L/2, L/2]$ is given by:

$$\Psi(x) = C_+ \exp(ikx) + C_- \exp(-ikx) \tag{1.3}$$

where $k = \frac{\sqrt{2mE}}{\hbar}$ and $E \geq 0$. C_+ and C_- are two constants such that the $\Psi(x)$ function is not nil in the $]-L/2, L/2]$ interval. In this way we obtain a linear combination of two waves associated with opposite wave vectors $\vec{k}$ with projections k and $-k$ on the Ox axis.

Let us now impose boundary conditions to be able to characterise the wavefunction for any point M.[1] Different approaches are possible (see Exercise (1.1) at the end of this chapter for another possible approach). Here we will adopt one that makes equivalent the linear system (see Fig. 1.2) and the cyclic system with perimeter L resulting from the condensation of the points with abscissas $+L/2$ and $-L/2$ (Fig. 1.3).[2, 3]

Such boundary conditions, initially proposed by Born and von Karman, of course ignore the effect of the borders because they do not exist in the cyclic system. This model is only valid if the system is very large. The point M in

[1] By imposing these boundary conditions we recognise the existence of the borders and impose the condition that the wavefunction describing the electron within and outside the metal is continuous and may be derived. Whatever the boundary conditions imposed, they lead to a quantisation of the energy and, consequently, of the k vector.

Fig. 1.2
One-dimensional system of length L assumed to be very large. The position of the M point is represented by a vector $\vec{r}_m$ and an abscissa x ($x \in]-L/2, L/2]$).

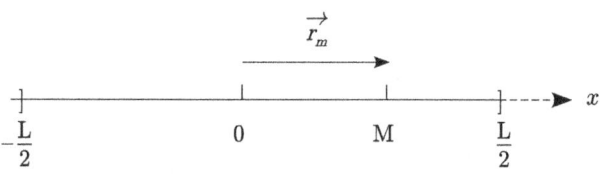

Free electron model

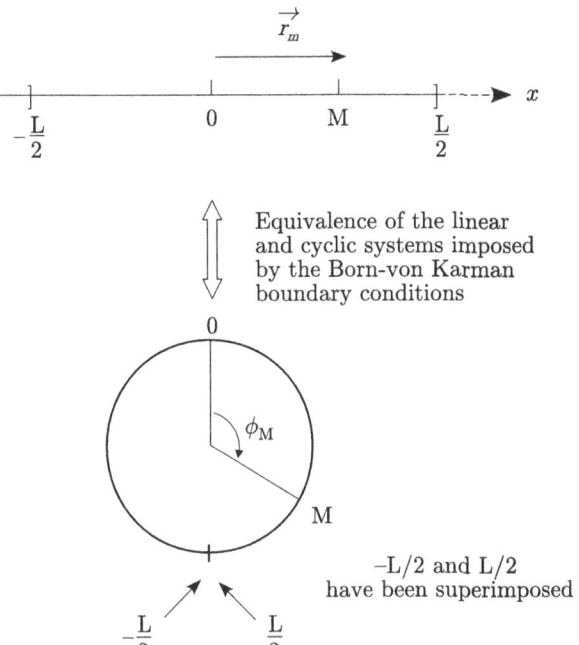

Fig. 1.3
Equivalence between a very large linear system and a cyclic system.

the linear representation is equivalent to the point M characterised by the angle ϕ_M in the cyclic representation:

$$\phi_M = \frac{x}{L} 2\pi \quad ; \quad \phi_M \in\,]-\pi, \pi\,] \tag{1.4}$$

The wavefunction of the electron in the cyclic representation as well as its derivative must verify the following boundary conditions:

$$\Psi(-\pi) = \Psi(\pi) \quad \text{and} \quad \Psi'(-\pi) = \Psi'(\pi) \tag{1.5}$$

where Ψ' refers to the derivative of the Ψ function.

Given the equivalence between the two systems, the wavefunction of the linear system must verify the periodicity condition:

$$\Psi\left(-\frac{L}{2}\right) = \Psi\left(\frac{L}{2}\right) \quad \text{and} \quad \Psi'\left(-\frac{L}{2}\right) = \Psi'\left(\frac{L}{2}\right) \tag{1.6}$$

We will only keep the eigenstates fulfilling the boundary conditions of eqn (1.6), i.e. those that verify:

$$C_+ \exp\left(-ik\frac{L}{2}\right) + C_- \exp\left(ik\frac{L}{2}\right) = C_+ \exp\left(ik\frac{L}{2}\right) + C_- \exp\left(-ik\frac{L}{2}\right) \tag{1.7}$$

$$kC_+ \exp\left(-ik\frac{L}{2}\right) - kC_- \exp\left(ik\frac{L}{2}\right) = kC_+ \exp\left(ik\frac{L}{2}\right) - kC_- \exp\left(-ik\frac{L}{2}\right) \tag{1.8}$$

Rearranging these formulas we obtain:

$$2i \sin\left(k\frac{L}{2}\right)(C_+ - C_-) = 0$$

$$\text{and} \quad 2ik \sin\left(k\frac{L}{2}\right)(C_+ + C_-) = 0 \quad (1.9)$$

However, the quantities $(C_+ - C_-)$ and $(C_+ + C_-)$ are not simultaneously nil except if the C_+ and C_- constants are both nil, which is impossible (eqn (1.3)). Consequently, the term $\sin(k\frac{L}{2})$ must be nil.

$$\sin(k\frac{L}{2}) = 0 \Rightarrow k\frac{L}{2} = n\pi \quad \text{if } n \text{ is an integer.} \quad (1.10)$$

As a result, k is quantised and must be multiple of $\frac{2\pi}{L}$. The solutions verifying the Schrödinger equation (eqn (1.2)) as well as the boundary conditions of eqn (1.6) are labelled $\Psi_k(x)$ and are given by:

$$\Psi_k(x) = N \exp(ikx) \quad \text{where } N \text{ is a normalisation constant} \quad (1.11)$$

$$\text{with} \quad k = \frac{2n\pi}{L} \quad \text{where } n \text{ is an integer} \quad (1.12)$$

In other words, k may adopt the values $0, \pm\frac{2\pi}{L}, \pm\frac{4\pi}{L}, \ldots$

Any linear combination of the degenerate wavefunctions $\Psi_k(x)$ and $\Psi_{-k}(x)$ is also a solution of the Schrödinger equation (1.2) and obeys the boundary conditions of eqn (1.6), because no restriction is imposed on the C_+ and C_- coefficients.

As shown in eqn (1.3), the quantisation of k also imposes quantisation on the energy. We will label $E(k)$ the energy associated with the waves $\Psi_k(x)$ and $\Psi_{-k}(x)$:

$$E(k) = \frac{(\hbar k)^2}{2m} \quad (1.13)$$

These functions may be written in the following vectorial notation:

$$\Psi_{\vec{k}}(\vec{r}_m) = N \exp(i\vec{k} \cdot \vec{r}_m) \quad \text{if} \quad \vec{k} = k\vec{i}_x \quad \text{and} \quad \vec{r}_m = x\vec{i}_x \quad (1.14)$$

The wavefunction $\Psi_{\vec{k}}(\vec{r}_m)$ is the wave associated with an electron with a momentum $\vec{p}$ equal to $\hbar\vec{k}$.[1] Let us note that the states $\Psi_{\vec{k}}(\vec{r}_m)$ and $\Psi_{-\vec{k}}(\vec{r}_m)$ are degenerate.

Now we must normalise the $\Psi_{\vec{k}}(\vec{r}_m)$ wavefunctions in the $]-L/2, L/2]$ interval to determine the constant N:

$$\int_{-\frac{L}{2}}^{\frac{L}{2}} |\Psi_{\vec{k}}(x)|^2 dx = 1 \Rightarrow N = \sqrt{\frac{1}{L}} \quad (1.15)$$

The wavefunction associated with an electron which must stay within a fragment of length L fulfilling the Born–von Karman boundary conditions is characterised by the wave vector $\vec{k}$ equal to $k\vec{i}_x$:

$$\Psi_{\vec{k}}(x) = \sqrt{\frac{1}{L}} \exp(ikx) = \sqrt{\frac{1}{L}} \exp(i\vec{k} \cdot \vec{r}_m) \quad (1.16)$$

Free electron model

The allowed values for k $\left(k = 0, \pm\frac{2\pi}{L}, \pm\frac{4\pi}{L}, \ldots\right)$ may be represented on an axis as shown below:

$$\ldots \quad -\frac{6\pi}{L} \quad -\frac{4\pi}{L} \quad -\frac{2\pi}{L} \quad 0 \quad \frac{2\pi}{L} \quad \frac{4\pi}{L} \quad \frac{6\pi}{L} \quad \ldots \longrightarrow k$$

The energy $E(\vec{k})$ associated with the function $\Psi_{\vec{k}}(x)$ is given by eqn (1.13).

Fermi level

Since electron–electron repulsions are neglected, electrons fill the lowest energy levels, with two electrons per allowed state in the ground state at $T = 0$ K. We will refer to the Fermi level, ε_f, the highest energy level filled in the ground state at $T = 0$ K. This highest filled level is characterised by the $\pm \vec{k}_f$ wave vectors and its energy ε_f is given by the formula:

$$\varepsilon_f = \frac{(\hbar k_f)^2}{2m} \quad (1.17)$$

As an example, let us assume that the system possesses seven electrons. In Fig. 1.4 we have plotted the curve of $E(k)$ (eqn (1.13)) as well as the allowed energy values $\left(E(k_0 = 0), E\left(k_1 = \pm\frac{2\pi}{L}\right), E\left(k_2 = \pm\frac{4\pi}{L}\right), \ldots\right)$. Given that all energy levels are doubly degenerate, except the lowest level ($k = 0$), the seven electrons fill the three lowest energy levels as shown in Fig. 1.4.

More generally, for a one-dimensional system possessing a large number of electrons N_e, at $T = 0$ K, the lowest level will be filled with two electrons and all other levels will be filled with four electrons up to the Fermi level, ε_f, which is characterised by the values of $\pm k_f$

$$\pm k_f = \pm \left(\frac{2\pi}{L}\right) \text{Int}\left(\frac{N_e + 1}{4}\right) \quad (1.18)$$

where $\text{Int}\left(\frac{N_e+1}{4}\right)$ denotes the integer part of $\frac{N_e+1}{4}$.

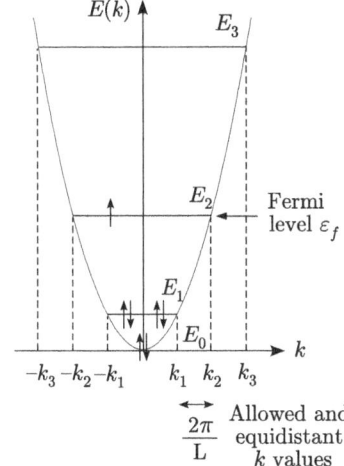

Fig. 1.4
Allowed energies for a one-dimensional system fulfilling the Born–von Karman boundary conditions. For clarity, the different energy levels have been drawn quite far from each other. In reality, they are very close since L is large ($2\pi/L$ is small).

Band structure

Since the system is very large, two adjacent allowed k values are very close to each other. Consequently, the energy spectrum is very dense and practically continuous. Thus, we are dealing with an *energy band* of allowed energy levels (see Fig. 1.5). Since the spectrum of allowed k values is practically continuous, we must distinguish the k values associated with filled levels at $T = 0$ K from those associated with empty k levels (Fig. 1.6).

Fermi–Dirac statistics

At a given finite temperature T, the probability $f(E)dE$ that a state with energy E is filled follows a Fermi–Dirac distribution $f(E)$ given by:

$$f(E) = \frac{1}{1 + \exp\left(\frac{E - \varepsilon_f}{k_B T}\right)} \quad (1.19)$$

where k_B is the Boltzmann constant ($k_B = 1.38066 \, 10^{-23}$ J.K^{-1}).

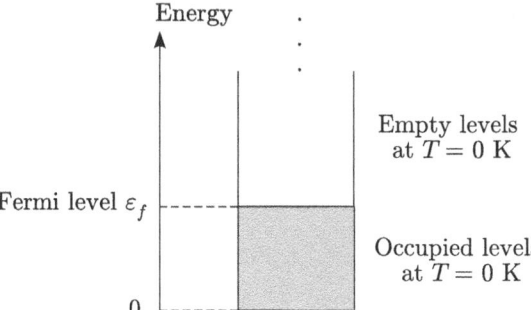

Fig. 1.5
Band of allowed energies in the free electron model.

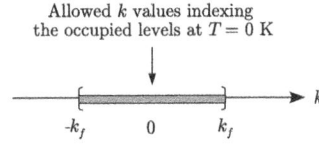

Fig. 1.6
k values associated with filled states at $T = 0$ K (in black).

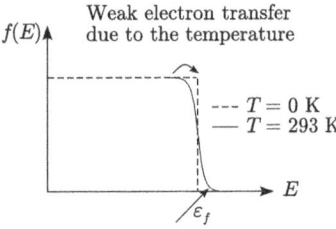

Fig. 1.7
Fermi–Dirac distribution.

The curve of $f(E)$ is a step function, which becomes increasingly rounded near the Fermi level as the temperature increases.

Given that the value of $k_B T$ at room temperature is 25 meV, only the states in the vicinity of the Fermi level are affected by this thermal excitation. In Fig. 1.7 we have schematically shown the temperature effect on the population of the allowed energy levels: because of the thermal excitation, some electrons that were below the Fermi level at $T = 0$ K lie slightly above at a finite temperature. Obviously, as the temperature increases the number of electrons affected by this transfer to higher energy states also increases.

Figure 1.8 shows the k values associated with states populated at a given finite temperature. The temperature leads to a transfer of electrons lying below the Fermi level at $T = 0$ K towards states which are above the Fermi level, characterised by k vectors with norm higher than k_f.

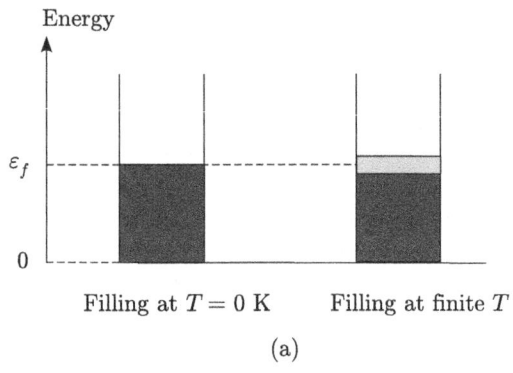

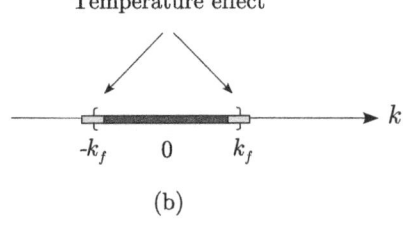

Fig. 1.8
(a) Population of a band at $T = 0$ K and at $T \neq 0$ K. (b) k values associated with states that are filled (in black) or partially filled (in grey) for $T \neq 0$ K.

Ohm's law

Since the wave vector $\vec{k}$ is proportional to the momentum vector $\vec{p}$, Fig. 1.8b shows that the system possesses as many electrons moving towards the right as electrons moving towards the left. What happens if, at time $t = 0$, the system is placed under the influence of a constant and uniform electric field $\vec{E}$ directed towards the left of the Ox axis? Every electron then feels an electric force $\vec{F}$ equal to $-e\vec{E}$ (e being the absolute value of the electron charge). Using the fundamental relation of dynamics for the electron characterised at the initial moment ($t = 0$) by a wave vector $\vec{k}(t = 0)$ and thus, by a momentum $\vec{p}(t = 0)$ equal to $\hbar \vec{k}(t = 0)$, we obtain:[2]

$$\vec{F} = \frac{d\vec{p}}{dt} = \hbar \frac{d\vec{k}}{dt} = -e\vec{E} \qquad (1.20)$$

Integration of this differential equation leads to:

$$\vec{p}(t) = \vec{p}(t=0) - e\vec{E}t \quad \text{and} \quad \vec{k}(t) = \vec{k}(t=0) - \frac{e\vec{E}t}{\hbar} \qquad (1.21)$$

The electrons in this model thus feel a constant acceleration towards the right. This result is not realistic since the electrons will decrease their speed as a consequence of the interaction with the lattice of A^+ cations (see Fig. 1.1) whose existence has been neglected so far.

Because of the collisions of the electrons with the underlying network, after a certain time τ these electrons will, on average, no longer be accelerated. In other words, the acceleration due to the electric field is cancelled after a time τ by the collisions of the electrons with the cations of the network. Clearly, after a certain time τ, the system reaches a stationary state in which every electron keeps the momentum acquired. Consequently, the electron initially described by a $\vec{p}$ momentum will possess a $\vec{p}$(stationary) momentum given by:

$$\vec{p}(\text{stationary}) = \vec{p}(t=0) + \delta \vec{p} \quad \text{with} \quad \delta \vec{p} = -e\vec{E}\tau \qquad (1.22)$$

Once the stationary state is reached, the electron associated with a wave vector $\vec{k}$ before the application of an electric field is characterised by a new $\vec{k}$(stationary) equal to $\vec{k} + \delta \vec{k}$, given by the equation:

$$\vec{k}(\text{stationary}) = \vec{k}(t=0) + \delta \vec{k} \quad \text{with} \quad \delta \vec{k} = -\frac{e\vec{E}\tau}{\hbar} \qquad (1.23)$$

Thus the electric field leads to the increase of every wave vector $\vec{k}$ by the same amount $\delta \vec{k}$. At $T = 0$ K, once the stationary state is reached, the filled states are characterised by $\vec{k}$ vectors whose projection lies in the $[-k_f + \delta k, k_f + \delta k]$ interval (see Fig. 1.9a), where δk is the projection of the $\delta \vec{k}$ vector.

Figures 1.9b and 1.9c show the schematic band structures, where the states associated with a vector $\vec{k}$ whose projection on the axis Ox is positive ($k > 0$) are separated from those characterised by a negative projection ($k < 0$). At $T = 0$ K, when the electric field is nil, there are as many states associated with positive values of k as states associated with negative values (see Fig. 1.9b). In contrast, application of the electric field leads to a depopulation

---
[2] Even if working in the framework of quantum mechanics, it is possible for this particular case to use the fundamental equation of classical mechanics.

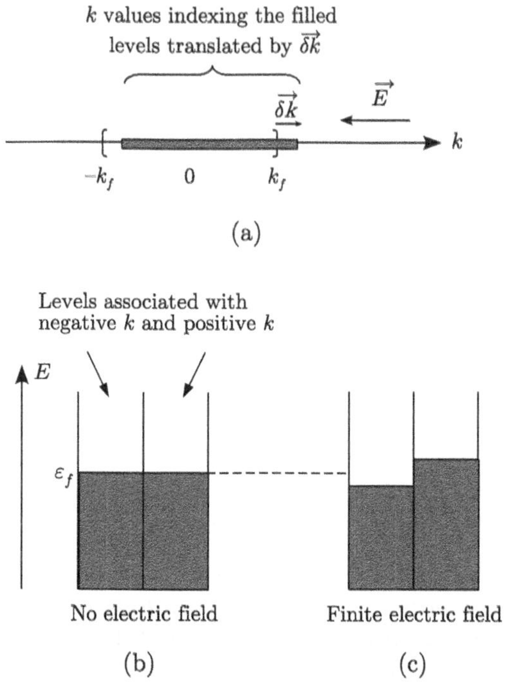

Fig. 1.9
(a) Effect of an electric field on the wave vectors characterising the filled states once the stationary state is reached at $T = 0$ K. (b) Population of the states at $T = 0$ K when the electric field is nil. (c) Population of the states at $T = 0$ K when an electric field in the negative direction of the Ox axis is applied.

of states near the Fermi level with a negative value of k, while populating states slightly above the Fermi levels characterised by a positive value of k (see Fig. 1.9c). Once the stationary state is attained, the system possesses more electrons moving towards the right than moving towards the left so that, globally, the system conducts current. *This scheme shows that a system may only transport current if it has allowed states just above the Fermi level ε_f.*

Once the stationary state is attained every electron has acquired an identical supplementary momentum $\delta\vec{p}$, associated with an increase in speed, $\delta\vec{v}$, which is given by the formula:

$$\delta\vec{v} = \frac{\delta\vec{p}}{m} = -\frac{e\tau\vec{E}}{m} \qquad (1.24)$$

As a result, the electric field is at the origin of the existence of a current density $\vec{j}$ through the system:

$$\vec{j} = \frac{i}{S}\vec{i}_x = \frac{dq}{Sdt}\vec{i}_x = \frac{n(-e)S\delta\vec{v}dt}{Sdt} = \frac{ne^2\tau}{m}\vec{E} = \sigma\vec{E} = \frac{1}{\rho}\vec{E} \qquad (1.25)$$

where

- i is the intensity crossing the system with section S,
- $\vec{i}_x$ is a normalised horizontal vector,
- dq is the charge crossing section S during the time dt,
- $(-e)$ is the charge of the electron,
- n is the number of free electrons per volume unit with a charge $-e$, i.e. the number of carriers per volume unit,
- σ is the conductivity of the material and ρ is the resistivity (usually in units of Ohm.cm).

Equation (1.25) is simply Ohm's law, and makes clear that the conductivity of a material depends on two factors: the number of carriers of the system, i.e. the number of free electrons, and the value of τ, which becomes larger as the coupling of the electronic motion with the positively charged network decreases.

> The free electron model correctly accounts for Ohm's law as far as the role of the underlying positive network is taken into account. The interaction between the electrons and the positive network is at the origin of the slowing down of the electronic motion through the system and, consequently, of the resistivity of the material.

1.1.2 Generalisation to a three-dimensional system

Let us now consider a metal with cubic shape and volume L^3 (with $x \in]-L/2, L/2]$, $y \in]-L/2, L/2]$, $z \in]-L/2, L/2]$). In the free electron model the electrons are free to move in the solid, which is a three-dimensional (3D) network of cations A^+. The potential within the solid is assumed to be nil and the eigenstates describing the behaviour of an electron are plane waves:

$$\Psi_{\vec{k}}(\vec{r}_m) = \sqrt{\frac{1}{L^3}} \exp(i\vec{k}\cdot\vec{r}_m) \tag{1.26}$$

where $\vec{k} = k_x \vec{i}_x + k_y \vec{i}_y + k_z \vec{i}_z$, $\vec{r}_m = x\vec{i}_x + y\vec{i}_y + z\vec{i}_z$ and $(\vec{i}_x, \vec{i}_y, \vec{i}_z)$ is an orthonormal basis, if the Born–von Karman boundary conditions are imposed along the three directions of space:

$$\Psi\left(-\frac{L}{2}, 0, 0\right) = \Psi\left(\frac{L}{2}, 0, 0\right) \text{ and } \frac{\partial \Psi}{\partial x}\left(-\frac{L}{2}, 0, 0\right) = \frac{\partial \Psi}{\partial x}\left(\frac{L}{2}, 0, 0\right) \tag{1.27}$$

$$\Psi\left(0, -\frac{L}{2}, 0\right) = \Psi\left(0, \frac{L}{2}, 0\right) \text{ and } \frac{\partial \Psi}{\partial y}\left(0, -\frac{L}{2}, 0\right) = \frac{\partial \Psi}{\partial y}\left(0, \frac{L}{2}, 0\right) \tag{1.28}$$

$$\Psi\left(0, 0, -\frac{L}{2}\right) = \Psi\left(0, 0, \frac{L}{2}\right) \text{ and } \frac{\partial \Psi}{\partial z}\left(0, 0, -\frac{L}{2}\right) = \frac{\partial \Psi}{\partial z}\left(0, 0, \frac{L}{2}\right) \tag{1.29}$$

These conditions are fulfilled if the k_x, k_y, and k_z components are quantified in the following way:

$$k_x = \frac{2n_x \pi}{L} \quad ; \quad k_y = \frac{2n_y \pi}{L} \quad ; \quad k_z = \frac{2n_z \pi}{L} \tag{1.30}$$

where n_x, n_y, and n_z are integers.

Every eigenstate is characterised by three quantum numbers (k_x, k_y, k_z). In eqn (1.26), $\vec{k}$ refers to the wave vector of the plane wave $\Psi_{\vec{k}}(\vec{r}_m)$ with which a momentum vector $\vec{p}$ equal to $\hbar\vec{k}$ can be associated. The energy $E(k)$ associated with this plane wave is given by:

$$E(\vec{k}) = \frac{\hbar^2 k^2}{2m} = \frac{p^2}{2m} \quad (1.31)$$

where k and p represent respectively the norm of the vectors $\vec{k}$ and $\vec{p}$. According to eqn (1.31), the energy $E(\vec{k})$ is given by:

$$E(\vec{k}) = \frac{\hbar^2 (k_x^2 + k_y^2 + k_z^2)}{2m} \quad (1.32)$$

In the free electron model a 3D system is treated in exactly the same way as the 1D system except for the fact that three quantum numbers are needed to characterise a state. The Fermi level can be obtained by populating the lowest energy states with two electrons. The $\vec{k}$ vectors, with components k_x, k_y, k_z characterising the Fermi level, fulfill the equation:

$$\varepsilon_f = \frac{\hbar^2 (k_x^2 + k_y^2 + k_z^2)}{2m} \quad (1.33)$$

Consequently, the endpoint of any $\vec{k}$ vector characterising a state at the Fermi level is found on a sphere with radius $k_f = \frac{\sqrt{2m\varepsilon_f}}{\hbar}$ (see Fig. 1.10).

We define the *Fermi surface* as the surface containing the endpoints of the k_f vectors characterising the states whose energy is equal to ε_f. In the free electron model the Fermi surface of a metal is the surface of a sphere with radius k_f (see Fig. 1.10). The study of a 3D system according to the free electron model is completely equivalent to that of a 1D system. From the formal point of view, the problem is a little bit more complex because the degeneracy of each level is higher. However, the reasoning is completely equivalent. When an electric field $\vec{E}$ is applied, every electron increases its speed by exactly the same amount, $\delta\vec{v}$ (eqn (1.24)). This is at the origin of the density current going through the system (eqn (1.25)). Even if for simplicity we have discussed Ohm's law using a 1D approach, we could equally have done so using a 3D free electron model.

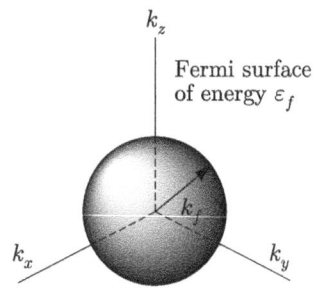

Fig. 1.10
Fermi surface according to the free electron model.

1.2 Conductivity of real solids

1.2.1 Factors influencing the conductivity

The free electron model provides a simplified but essentially correct description of the requirements for a system to conduct electricity. To begin with, a conductor needs to have electrons that are capable of participating in the conducting process, i.e. which are not strongly bound to the underlying network of nuclei. In addition, the process of electrical conduction needs the participation of states just above the Fermi level, ε_f (see Fig. 1.9c). Thus, good conductors usually have *a large density of allowed states very close to the Fermi level* so that a considerable number of electrons can participate in the conduction process. In addition, since conductivity is better when the interaction between the electrons and the underlying network is weak, it is clear that it depends on the nature of the atoms from which this network is built as well as the temperature and the purity of the material. For instance, when temperature is raised there is an increase in the amplitude of the vibrations of the network, and this has the effect of slowing down the electronic motion. We can restate this fact by saying that the *electron–phonon coupling* is more efficient when the temperature increases.[3] The conductivity of a metal is thus better when the temperature decreases. As for defects, these have the tendency to pin free electrons so that their presence leads to a decrease in the conductivity of the system.

[3] A phonon is a quantum of vibrational energy of the periodic system.

> To summarise, the conductivity of a material essentially depends on four parameters:
>
> - the density of states near the Fermi level
> - the electron-phonon coupling
> - the temperature
> - the defects.

1.2.2 Band structure of real solids

In contrast with the prediction of the free electron model, the band structure for a real periodic solid is not a single band (Fig. 1.5). The allowed energies of a real solid are found within the energy ranges associated with different energy bands separated by forbidden energy gaps (Fig. 1.11).

One of the goals of this book is to adopt a simple approach that allows an understanding of how the nature of the different atoms and the structural details are related to the band structure of a periodic solid. In particular, we will be interested in analysing the orbital nature of the different bands responsible for the properties of the system. For the time being let us assume that these band structures may be obtained and see how they are related to the electrical behaviour of the material.

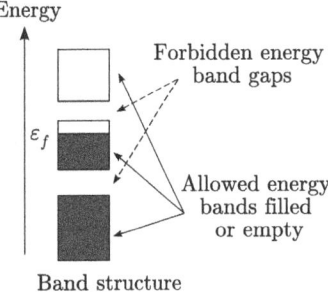

Fig. 1.11
Band structure for a real solid.

1.2.3 Metallic behaviour

When the Fermi level of a system lies inside a band, i.e. in the vicinity of a large number of empty levels, the system is a *metal*. In general, the room temperature resistivity is weak in these cases, of the order of 10^{-6} Ohm.cm.[1] More importantly, the resistivity decreases when the temperature is lowered because the efficiency of the electron–phonon interactions decreases (see Fig. 1.12). This is the main feature characterising the behaviour of a metal.

When considering an energy band of a 3D material built from only one type of orbital, the density of states is usually maximum at the middle of the band.

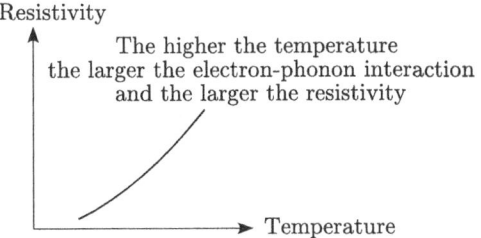

Fig. 1.12
Resistivity vs temperature behaviour for a metallic system.

12   Transport properties of solids

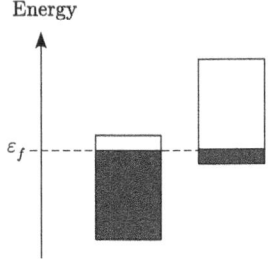

Fig. 1.13
Band structure for a semimetal where the filled states at $T = 0$ K have been represented in black.

[4]Throughout this book we will refer to the Fermi level as the highest occupied level. With such a definition the Fermi level of a semiconductor occurs at the top of a band whereas the thermodynamic definition of the Fermi level places it within the band gap.[1]

Consequently, a system for which the Fermi level occurs near the top or the bottom of a band should not, in principle, be a very good conductor. They are in fact *semimetals* and are characterised (see Fig. 1.13) by the Fermi level lying near the top and bottom of two bands. Their resistivity changes with temperature in the same way as a metal (see Fig. 1.12) but it is generally larger because the density of states around the Fermi level is smaller.

1.2.4   Semiconducting and insulating behaviour

When the Fermi level occurs at the top of a band which does not overlap with another band, [4] the system does not transport current at $T = 0$ K; we are dealing with an *insulator* or a *semiconductor*. If the temperature is different from 0 K, some electrons at the top of the highest filled band (called the *valence band*) are transferred to the bottom of the lowest empty band (called the *conduction band*) by thermal excitation (see Fig. 1.14b). As a consequence, there are filled states at the bottom of the conduction band in the vicinity of empty states which can participate in the transport process. In the same way, the empty states (called *holes*) at the top of the valence band can also participate in the transport process.

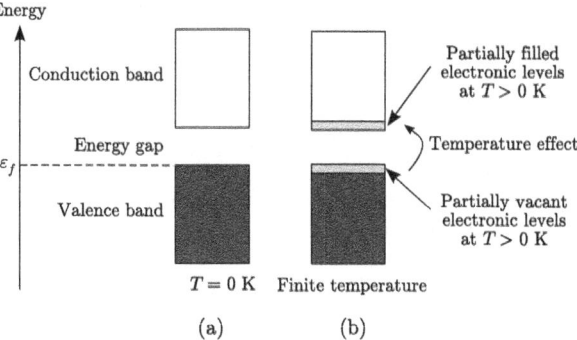

Fig. 1.14
Population of the valence and conduction bands for a semiconducting solid: (a) at $T = 0$ K and (b) at $T \neq 0$ K. The states that are filled are shown in black; those that are partially filled are in grey.

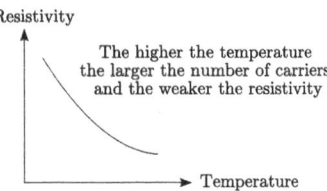

Fig. 1.15
Resistivity vs temperature behaviour for a semiconducting system.

Consequently, the conductivity of a semiconductor originates from the mobility of the electrons (with charge $-e$) of the conduction band and the holes (formally with charge $+e$) of the valence band. However, the number of carriers (with charge $+e$ or $-e$) is weak, and this is more so as the temperature is lowered and as the forbidden energy gap between the valence and conduction bands becomes larger. Thus, in contrast with the situation in a metal, the conductivity of a semiconductor increases with temperature. This is the main characteristic of a semiconductor (Fig. 1.15).

Semiconductors and insulators differ in the magnitude of the band gap between the valence and conduction bands. When this gap is not large, the room temperature conductivity is quite sizable and the system is a semiconductor. In contrast, when the band gap is large, so that the bottom of the conduction band cannot be significantly populated (and the top of the valence band depopulated), the room temperature conductivity will be very low and the system is an insulator. The resistivity of semiconductors is generally between 10^{-2} and 10^9 Ohm.cm, whereas for insulators it is of the order of 10^{14}–10^{22} Ohm.cm. [1]

Table 1.1 Number of carriers per cm^3 at room temperature and main features of the band structure for different bulk materials.

Metal	$> 10^{22}$	Fermi level inside the highest occupied band	Copper
Semimetal	10^{17}–10^{22}	Fermi level occurs where two bands merge or slightly overlap	Graphite
Semiconductor	10^{13}–10^{17}	Fermi level at the top of a band that is not far from the bottom of the next band	Germanium
Insulator	$< 10^{13}$	Fermi level at the top of a band that is far from the bottom of the next band	Diamond

1.2.5 Number of carriers

Reasoning in terms of the number of carriers per volume, materials may be classified as shown in Table 1.1.

Let us emphasise, however, that such a classification is somewhat arbitrary. It is the *variation of conductivity with respect to temperature, i.e. the slope of the ρ vs T curve*, that really determines if the conductivity is activated (semiconductors and insulators) or non-activated (metals and semimetals). For instance it is perfectly possible that the conductivity of a semiconductor at a given temperature is comparable or even higher than that of certain poorly conducting metals.

Exercises

(1.1) Electron trapped in an *infinite potential well*: a possible way to model the behaviour of an electron in a metal is to assume the existence of an infinite potential well outside the system, i.e. outside the interval]− L/2, L/2]. Under such conditions, obtain an expression for the allowed energy states describing the electrons on the metal. Explain in what respect the results obtained may differ from those obtained using the Born–von Karman boundary conditions. Why are the two series of results equivalent for large L values?

References

1. C. Kittel, *Introduction to Solid State Physics*, 7th edition, John Wiley, New York, 1996.
2. H. Ibach, H. Lüth, *Solid State Physics: An Introduction to the Principles of Materials Science*, 2nd edition, Springer Verlag, Berlin Heidelberg, 1995.
3. O. Madelung, *Introduction to Solid State Theory*, 2nd edition, Springer Verlag, Berlin Heidelberg, 1981.
4. N. W. Ashcroft, N. D. Mermin, *Solid State Physics*, Holt, Rinehart and Winston, Philadelphia, 1976.

2 Electronic structure of molecules: use of symmetry

Since our aim is to show that it is possible to analyse the electronic structure of a periodic system following a procedure very similar to that used for molecules, in this chapter we will briefly review some useful notions that are needed to understand the electronic structure of molecules. Of course we will limit ourselves to the notions that are useful in building a bridge between the modern approaches to the electronic structures of, on the one hand, molecules and, on the other, solids. We will begin by recalling some basic aspects of molecular orbital theory. Then we will consider how symmetry may be used to simplify the problem of defining the molecular orbitals by using group theory. The main results will be used in subsequent chapters, taking advantage of the translational symmetry of periodic systems in studying their electronic structure. Once these basic group theory notions have been discussed, we will illustrate their usefulness by considering the example of cyclobutadiene. Finally we will briefly consider some general aspects of the electronic structure of transition metal coordination complexes. Later, these notions will be very useful in discussing the electronic structure of solids containing transition metal atoms.

2.1 Molecular orbital theory

Since electron motion inside molecules is governed by the laws of quantum mechanics, we need to solve the corresponding Schrödinger equation to determine the allowed states. However, this equation cannot be exactly solved for molecules, or more precisely for systems with more than one electron. Approximations are then required to get an acceptable estimate of the solution of the Schrödinger equation. In this chapter we will discuss molecular orbital theory, which provides a very convenient and easily workable description of the electronic structure of molecules. [1]

Let us consider a molecule having N_e electrons, labelled $(e_1, e_2, \ldots, e_{N_e})$, and N_n nuclei. The positions of the electrons e_i and nuclei N_j are described by the $\vec{r}_i$ and $\vec{R}_j$ vectors, respectively.

2.1.1 Born-Oppenheimer approximation

The *Born–Oppenheimer approximation* consists of separating the motion of the electrons from that of the nuclei, based on the fact that electrons move very quickly and adapt instantaneously to the motion of the nuclei, i.e. atomic vibrations. Consequently, for a given geometry of the molecular species characterised by the $\vec{R}_j (j = 1, \ldots, N_n)$ vectors, the electrons' motion is described by an electronic wavefunction $\psi_e(\vec{r}_i, \vec{R}_j)$ $(i = 1, \ldots, N_e\,; \, j = 1, \ldots, N_n)$ in which $\vec{R}_j$ are parameters and $\vec{r}_i$ are variables. The vibrational motion of the nuclei is described by a nuclear wavefunction $\psi_n(\vec{R}_j)$ $(j = 1, \ldots, N_n)$. When the molecule possesses a single equilibrium geometry, the electronic wavefunction $\psi_e(\vec{r}_i; (\vec{R}_j)_{eq})$ $(i = 1, \ldots, N_e\,; \, j = 1, \ldots, N_n)$ is estimated for the nuclei in their equilibrium position, characterised by the vectors $((\vec{R}_j)_{eq}; j = 1, \ldots, N_n)$.

This approximation, although very useful and even indispensable, is not valid when the coupling between the electronic motion and the molecular vibrations is important, for example in transition metal complexes unstable toward a Jahn–Teller distortion. These will be discussed at the end of this chapter.

2.1.2 One-electron approximation

The *one-electron approximation* consists of decoupling the movement of the different electrons assuming that each of them moves in an average potential that represents the average repulsion of all other electrons. This approximation leads to the remarkable result that every electron may be described by an *effective one-electron Hamiltonian* $\hat{h}(e_i)$ that depends only of the position of electron e_i characterised by the vector $\vec{r}_i$. The total electronic wavefunction $\Psi_e(e_1, e_2, \ldots, e_{N_e})$ may be written as an antisymmetrical product of wavefunctions, called a *Slater determinant*. These wavefunctions are molecular orbitals describing the motion of an electron in the molecular species. Such molecular orbitals will be labelled $\phi(\vec{r})$. The quantity $|\phi(\vec{r}_i)|^2$ represents the probability density for electron e_i at a point M characterised by the vector $\vec{r} = \vec{r}_i$. The ground-state electronic configuration is obtained by filling with electrons the lowest energy levels, according to the Pauli principle. Every orbital may be filled with two electrons with opposite spins. Thus, we need to find an appropriate expression for the molecular orbitals.

2.1.3 LCAO approximation

According to the *linear combination of atomic orbitals* approximation (LCAO), a molecular orbital ϕ can be written as a linear combination of N_0 atomic orbitals $\{\chi_j, j = 1, \ldots, N_0\}$ of the different atoms of the molecule:

$$\phi = \sum_{j=1}^{N_0} c_j \chi_j \qquad (2.1)$$

where the coefficients (c_j, $j = 1, \ldots, N_0$) are the constants to be determined. The summation in eqn (2.1) is generally limited to the valence atomic orbitals of the constituent atoms, although occasionally it can also include the first empty atomic orbitals. The N_0 χ_j orbitals can also be fragment orbitals of the molecule.[1]

[1] A fragment is simply a number of atoms that form a chemically convenient building block of the molecule. For instance it is possible to use two CH_2 fragments to build the molecular orbitals of ethylene, C_2H_4.

2.1.4 Secular equations and secular determinant

We are now going to establish the equations that will allow us to determine the coefficients (c_j) and energy (E) associated with the molecular orbital ϕ. If we write the equation verified by this molecular orbital as:

$$\hat{h}\phi = E\phi \tag{2.2}$$

and then express ϕ as in eqn (2.1), we obtain:

$$\sum_{j=1}^{N_0} c_j \left(\hat{h}\chi_j - E\chi_j\right) = 0 \tag{2.3}$$

Let us now multiply the left of this equation by the conjugate of χ_i and then integrate over all spatial coordinates of the wavefunction. We obtain a system of N_0 equations with N_0 unknown coefficients (c_j, $j = 1, \ldots, N_0$) which are the so-called *secular equations*:

$$\sum_{j=1}^{N_0} \left(h_{ij} - ES_{ij}\right) c_j = 0 \quad (i = 1, \ldots, N_0) \tag{2.4}$$

where

$$h_{ij} = \int\int\int \chi_i^* \hat{h} \chi_j d\tau \text{ and } S_{ij} = \int\int\int \chi_i^* \chi_j d\tau \tag{2.5}$$

These expressions may also be written using the Dirac notation as

$$h_{ij} = \langle \chi_i \mid \hat{h} \mid \chi_j \rangle \text{ and } S_{ij} = \langle \chi_i \mid \chi_j \rangle \tag{2.6}$$

These equations are characterised by their *secular determinant*:

$$\text{secular determinant} = \mid \underline{\underline{h}} - E\underline{\underline{S}} \mid \tag{2.7}$$

where $\underline{\underline{h}}$ and $\underline{\underline{S}}$ are the matrix representations of the Hamiltonian and the identity operators in the orbital basis $\{\chi_j, j = 1, \ldots, N_0\}$. The matrix $\underline{\underline{S}}$ is also known as the overlap matrix.

If the secular determinant is non-zero, the solution of the secular equations is nothing other than the trivial solution ($c_j = 0$, $j = 1, \ldots, N_0$). The ϕ wavefunction associated with this solution is physically meaningless since it is nil everywhere. In contrast, if the secular determinant is nil, i.e. the system of equations is bound, nontrivial solutions that are physically meaningful exist. Consequently, only the energies E that lead to a nil secular determinant are associated with possible states of the electrons in the molecule. We are thus left to solve eqn (2.8) to find the allowed energies:

$$\mid \underline{\underline{h}} - E\underline{\underline{S}} \mid = 0 \tag{2.8}$$

Molecular orbital theory

This is an equation of N_0-th order in E the solution of which provides the N_0 values of the allowed energies, E_ℓ ($\ell = 1, \ldots, N_0$). The molecular orbital associated with the energy E_ℓ will be referred to as ϕ_ℓ. The determination of the coefficients associated with molecular orbital ϕ_ℓ involves the solution of the system of secular equations, where the energy E has been replaced by E_ℓ, and normalisation of the ϕ_ℓ function has been imposed.

This procedure, *which must be repeated for every allowed energy* E_ℓ ($\ell = 1, \ldots, N_0$), leads to the determination of all molecular orbitals ϕ_ℓ ($\ell = 1, \ldots, N_0$) as linear combinations of the χ_j ($j = 1, \ldots, N_0$) orbitals.

2.1.5 Basic features of the Hückel and extended Hückel methods

Extended Hückel method

Except as otherwise stated, all numerical results reported in this book have been obtained by using the extended Hückel method. [2] In this approach the interaction term h_{ij} ($i \neq j$) between two Slater orbitals χ_i and χ_j is estimated through the empirical expression:[2]

$$h_{ij} = K S_{ij} \frac{(h_{ii} + h_{jj})}{2} \qquad (2.9)$$

[2] The expression proposed by Slater for the atomic orbitals of polyelectronic atoms is given by $\chi_{n\ell m}(r,\theta,\phi) = N r^{n-1} e^{-\xi r} Y_{\ell,m}(\theta,\phi)$.

in which h_{ii} and h_{jj} are the ionisation energies of a valence electron described by orbitals χ_i and χ_j, respectively. In the Wolfsberg and Helmholtz approximation (eqn (2.9)), [2] K is a constant with value 1.75. In this method, the S_{ij} overlaps are calculated analytically. Despite its empirical character this method leads to a simple and useful description of the electronic structure of molecules. In the following chapters we will adopt this approach to obtain the electronic structure of periodic solids.

Hückel method

An even simpler approach may be used when studying the electronic structure of the π system of conjugated organic molecules, such as butadiene, benzene, etc. This approach was initially proposed by Hückel in the 1930s and is at the origin of the extended Hückel method discussed above. $\sigma - \pi$ separation is assumed and only the π system is considered.

In this very simple approach, only the non-diagonal interaction terms h_{ij} ($i \neq j$) between the atomic orbitals of two adjacent atoms are assumed to be non-zero. They are referred to as β_{ij}. All other interaction terms are neglected. The β_{ij} and S_{ij} terms have opposite signs. In addition the term h_{ii}, denoted α, corresponds to the ionization potential of the χ_i orbital. The α_i energies are always larger in absolute terms than the interaction terms β_{ij}. When the π system under study is exclusively built from carbon atoms and equivalent C–C bonds, the α_i and β_{ij} parameters are simply denoted α and β. Finally, a further simplification involves neglecting the S_{ij} ($i \neq j$) overlaps.

Fig. 2.1
Limiting mesomeric structures for cyclobutadiene. The four hydrogen and carbon atoms are labelled as 1, 2, 3 and 4.

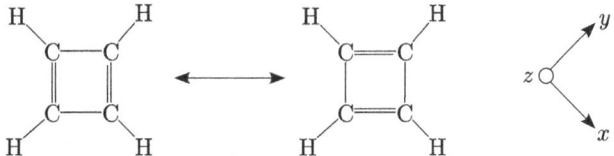

Consequently, when using the Hückel approach to determine the molecular orbitals of a system, eqns (2.4) and (2.8) must be solved assuming that:

$$h_{ii} = \alpha_i$$

$$h_{ij} = \beta_{ij} \text{ if orbitals } \chi_i \text{ and } \chi_j \text{ are adjacent, otherwise } h_{ij} = 0$$

$$S_{ij} = 0 \text{ if } i \neq j \text{ and } S_{ii} = 1 \tag{2.10}$$

The advantage of this method is that it yields simple analytical results for many systems. It also has an invaluable educational interest since the secular equations of any given system may be solved without the need for a computer. Obviously the applicability of the method is considerably more restricted than that of the extended Hückel method. In this book we will frequently use this approach in the study of simple periodic systems.

2.1.6 Symmetry properties of the molecular orbitals

A molecular orbital $\phi(\vec{r}_i)$ describes the behaviour of an electron e_i through the square of its modulus $|\phi(\vec{r}_i)|^2$. This wavefunction must possess symmetry properties compatible with those of the molecular skeleton. [3] Thus, for instance, in planar organic molecules the molecular orbitals of the σ system may be distinguished from those of the π system. Whereas the former are symmetric with respect to the molecular plane, the latter are antisymmetric. Such a distinction considerably simplifies the determination of the molecular orbitals. To illustrate this fact, let us consider a hypothetical regular cyclobutadiene molecule, which may be represented by two Lewis structures as shown in Fig. 2.1.

The molecular orbitals (MO) of regular cyclobutadiene are linear combinations of twenty valence atomic orbitals (AO) of the four carbon and four hydrogen atoms: $\{1s_{H_i}, 2s_{C_i}, 2p_{xC_i}, 2p_{yC_i}, 2p_{zC_i} ; i = 1, \ldots, 4\}$. Among these AOs we may distinguish sixteen AOs that are symmetric with respect to the molecular plane (σ_{xy}), $\{1s_{H_i}, 2s_{C_i}, 2p_{xC_i}, 2p_{yC_i} ; i = 1, \ldots, 4\}$, and four AOs which are antisymmetric with respect to this plane, $\{2p_{zC_i} ; i = 1, \ldots, 4)\}$. Because of this symmetry difference, the system has sixteen symmetric MOs of the σ-type and four antisymmetric MOs of the π-type. These MOs differ also in the nature of the overlap integrals between their constituent AOs.[3]

In the following section we will talk about σ- or π-type orbitals. Since the σ overlaps are more effective than the π overlaps, the bonding molecular orbitals of the σ system are more bonding than those of the π system, whereas the antibonding orbitals of the σ system are more antibonding than those of the π system. The energy diagram for the molecular orbitals of regular

---

[3] The overlap between two atomic orbitals may be characterised by enumerating the number of nodal surfaces contained in the region where the two orbitals overlap. The overlaps denoted σ, π, and δ contain 0, 1, and 2 nodal surfaces. Consequently, the σ-type overlaps are more effective than the π-type overlaps and the latter are more effective than the δ-type overlaps. This is illustrated below, where some examples of σ, π, and δ overlaps associated with s, p, or d AOs are shown.

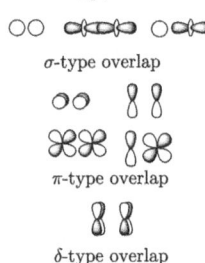

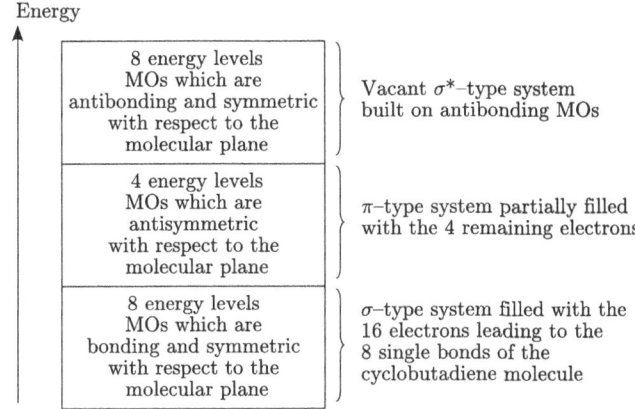

Fig. 2.2
General structure of the molecular orbital diagram of cyclobutadiene.

cyclobutadiene contains three blocks of molecular orbitals (see Fig. 2.2): eight bonding MOs of the σ system, four MOs of the π system, and eight antibonding MOs of the σ system.

Since the system possesses twenty valence electrons, the eight bonding MOs of the σ system are filled and describe the eight C–C and C–H single bonds of cyclobutadiene. The four remaining electrons occupy the lowest energy MOs of the π system according to the Pauli principle. From a chemical point of view, only the highest occupied MOs and the lowest unoccupied MOs are really interesting. This is why the knowledge of the four π MOs of regular cyclobutadiene may provide an explanation for most of the properties of this system.

2.2 A short review of the theory of symmetry point groups

We will now recall and illustrate the main results of group theory, which will be used in the following sections. We will only consider point groups, i.e. groups bearing symmetry operations leaving invariant a given point denoted O.

2.2.1 Different symmetry point groups

The set of symmetry operations leaving a molecule invariant forms a group in the mathematical sense. [1] The different symmetry operations found in molecular systems are outlined in Table 2.1.

The set of symmetry operations generating a group is known as a *generator set*. By this we mean that any symmetry operation of the group may be expressed as the product of symmetry operations of this generator set (see Table 2.2). The number of symmetry elements of the group is called *order of the group* and is referred to as h. In what follows we will also consider two high-symmetry groups with many symmetry operations: the T_d and O_h groups, which leave invariant a tetrahedron and an octahedron, respectively.

Table 2.1 Different point symmetry operations.

Notation	Symmetry operation
E	Identity
i	Inversion of all atoms through the symmetry centre
C_n^m	Order n axis. Rotation by an angle $\frac{2m\pi}{n}$ around an axis.
S_n^m	S_n^1 is the commutative product of a rotation followed by a reflection in a plane $\perp$ to the rotation axis. $S_n^m = (S_n^1)^m$: Improper axis of order n.
σ_d or σ_v	Reflection in a vertical plane containing the Oz axis, which by definition is the main axis, i.e the axis of higher order.
σ_h	Reflection in an horizontal plane $\perp$ to the Oz axis.

Table 2.2 Set of symmetry elements generating the main symmetry point groups.

Group	Generators	Order and characteristics of the group
C_i	$\{i\}$	2 symmetry operations: inversion and identity.
C_s	$\{\sigma\}$	2 symmetry operations: reflection and identity.
C_n	$\{C_n\}$	n operations.
S_n	$\{S_n\}$	$2n$ operations.
C_{nh}	$\{C_n, \sigma_h\}$	$2n$ operations.
C_{nv}	$\{C_n, \sigma_v\}$	$2n$ operations.
D_n	$\{C_n, C_2'\}$ $C_2' \perp C_n$	$2n$ operations. n C_2' axes $\perp$ to the C_n axis.
D_{nh}	$\{C_n, C_2', \sigma_h\}$ $C_2' \perp C_n$	$4n$ operations. n C_2' axes $\perp$ to the C_n axis
D_{nd}	$\{C_n, C_2', \sigma_d\}$ $C_2' \perp C_n$	$4n$ operations. n C_2' axes $\perp$ to the C_n axis. The intersection of a plane σ_d with the horizontal plane is a bisector of the two C_2' axes.

Finally, we also consider two infinite groups:

- the $C_{\infty h}$ symmetry group, which contains all C_ϕ rotations around the Oz axis, as well as all reflections in the planes containing the z-axis (this is the symmetry group of diatomic heteronuclear molecules, for instance)
- the $D_{\infty h}$ symmetry group, which, in addition to all symmetry elements of the $C_{\infty h}$ group, also contains the inversion, the reflection in a plane perpendicular to the Oz axis, an infinite number of C_2' rotations perpendicular to the Oz axis going through O, as well as an infinite number of improper rotations around the Oz axis (this is the symmetry group of diatomic homonuclear molecules, for instance).

2.2.2 Classes

It is possible to classify the different symmetry operations of a group in classes. From the mathematical point of view, two symmetry operations A and B belong to the same equivalence class if there is a symmetry operation X of the group such that B is equal to $X^{-1}AX$. *From a physical point of view, two symmetry operations of the same class are two physically equivalent symmetry operations.*

Let us consider again the regular cyclobutadiene molecule (Fig. 2.3) which is left invariant by the D_{4h} group. The two rotations C_{2a}' and C_{2b}' are related in the following way: $C_{2b}' = \sigma_{d_a}.C_{2a}'.\sigma_{d_a}$, where σ_{d_a} is the reflection in the bisector plane of the two rotation axes C_{2a}' and C_{2b}' (see Fig. 2.3). These two symmetry operations, C_{2a}' and C_{2b}', thus belong to the same class. As can clearly be seen from Fig. 2.1, they are physically equivalent. The two rotations C_{2a}'' and C_{2b}'' belong to the same class as well. However, there is no symmetry operation of the D_{4h} group relating a C_2' and a C_2'' rotation. Thus the four rotations are distributed in two different classes of the D_{4h} group because they are not physically equivalent. Two of them (C_{2a}' and C_{2b}') leave invariant two carbon atoms whereas the other two (C_{2a}'' and C_{2b}'') leave invariant the middle of a C–C bond. Likewise it can be shown that the symmetry operations of the D_{4h} group may be classified as belonging to ten classes: $\{E\}$, $\{C_4, C_4^3\}$, $\{C_2\}$, $\{C_{2a}', C_{2b}'\}$, $\{C_{2a}'', C_{2b}''\}$, $\{i\}$, $\{S_4, S_4^3\}$, $\{\sigma_{xy}\}$, $\{\sigma_{v_a}, \sigma_{v_b}\}$, and $\{\sigma_{d_a}, \sigma_{d_b}\}$. The symmetry planes σ_{d_a} and σ_{d_b} contain the rotation axes C_{2a}'' and C_{2b}'',

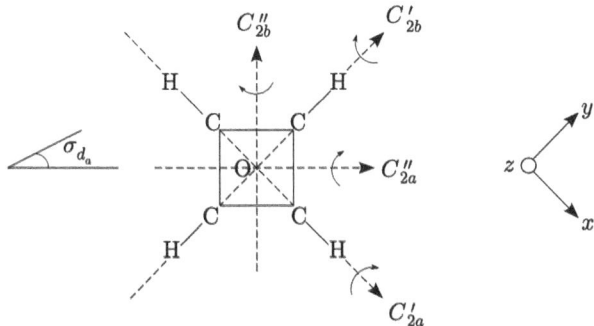

Fig. 2.3
Regular cyclobutadiene. The symmetry planes $\sigma_{v_a}, \sigma_{v_b}, \sigma_{d_a}, \sigma_{d_b}$ contain the Oz axis and the rotation axes $C_{2a}', C_{2b}', C_{2a}'', C_{2b}''$ respectively.

respectively, whereas the symmetry planes σ_{v_a} and σ_{v_b} contain the rotation axes C'_{2a} and C'_{2b}, respectively.

2.2.3 Basis for an irreducible representation

We will now outline the notions of *basis for a representation, representation, and irreducible representation* of a group G. These notions will be illustrated by using the symmetry properties of the $2p_z$ atomic orbitals of the π system of regular cyclobutadiene. The symmetry group of regular cyclobutadiene is D_{4h} which contains sixteen symmetry operations denoted $\{R_j, j = 1, \ldots, 16\}$ ($h = 16$).

Notion of basis for a representation

Let us now consider a set of functions $\underline{f} = \{f_1, f_2, \ldots, f_n\}$, such that the action of any of the symmetry operations of the group G transforms any of the functions, f_i, in a linear combination of the different functions of the $\underline{f}$ set. Such a set is said to be *globally stable* under the action of the symmetry operations of G and constitutes *a basis for a representation of the group G*. From a physical point of view a basis for a representation contains functions equivalent by symmetry. Thus, for instance, the four $2p_z$ orbitals of the carbon atoms of regular cyclobutadiene constitute a basis for a representation of the D_{4h} group as well as of all subgroups of D_{4h}.

Representation of a group

We will now consider the set of matrices $\underline{\underline{M}}_k$ representing the action of the symmetry operations R_k ($k = 1, \ldots, h$) on the basis $\underline{f} = \{f_1, f_2, \ldots, f_n\}$. This set of matrices, $\{\underline{\underline{M}}_1, \underline{\underline{M}}_2, \ldots, \underline{\underline{M}}_h\}$, is called a *representation of the group* and is denoted Γ. In the case of the basis $\{2p_{z_1}, 2p_{z_2}, 2p_{z_3}, 2p_{z_4}\}$ of the D_{4h} group of regular cyclobutadiene (see Figs 2.1, 2.3, and 2.4), one representation of the group consists of sixteen 4×4 matrices representing the action of the different symmetry operations of the D_{4h} group on the set $\{2p_{z_1}, 2p_{z_2}, 2p_{z_3}, 2p_{z_4}\}$. Thus, for instance, the matrix associated with rotation C_4 is that shown in Fig. 2.5.

The number of representations of the group is infinite. For instance, starting with one of the bases, $\underline{f}$, any unitary matrix $\underline{\underline{U}}$ can be used to define another basis, $\underline{f}' = \{f'_1, f'_2, \ldots, f'_n\}$ such that $\underline{f}' = \underline{\underline{U}}\,\underline{f}$. The matrix associated with operation R_k in this new basis, noted $\underline{\underline{M}}'_k$, is equal to $\underline{\underline{U}}\,\underline{\underline{M}}_k\,\underline{\underline{U}}^{-1}$. The set $\{\underline{\underline{M}}'_1, \underline{\underline{M}}'_2, \ldots, \underline{\underline{M}}'_h\}$ is a new representation of the group.

From a physical perspective, these two representations of the group have the same meaning: we thus need to find some common property characterising both of them. This common property is the trace, i.e. the sum of the diagonal elements of the different matrices $\underline{\underline{M}}_k$. We will denote the trace of the $\underline{\underline{M}}_k$ matrix as $\chi(\underline{\underline{M}}_k)$, which in group theory is known as the *character* of the matrix. Using the fact that the products of matrices AB and BA have the same trace, it is easy to show that the matrices $\underline{\underline{M}}_k$ and $\underline{\underline{M}}'_k$ also have the same trace:

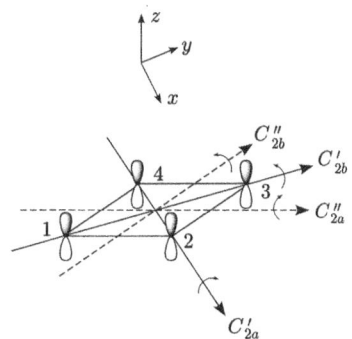

Fig. 2.4
$(2p_z)_i$ orbitals of the four carbon atoms ($i = 1, 2, 3, 4$) of cyclobutadiene.

$$\underline{\underline{M}}_{C_4} = \begin{pmatrix} 0 & 1 & 0 & 0 \\ 0 & 0 & 1 & 0 \\ 0 & 0 & 0 & 1 \\ 1 & 0 & 0 & 0 \end{pmatrix}$$

Fig. 2.5
Matrix representing the action of the C_4 rotation on the basis $\{2p_{z_1}, 2p_{z_2}, 2p_{z_3}, 2p_{z_4}\}$.

$$\underline{\underline{M}}'_k = \underline{\underline{U}}\,\underline{\underline{M}}_k\,\underline{\underline{U}}^{-1}$$

$$\chi\left(\underline{\underline{M}}'_k\right) = \chi\left(\left(\underline{\underline{U}}\,\underline{\underline{M}}_k\right)\left(\underline{\underline{U}}^{-1}\right)\right) = \chi\left(\left(\underline{\underline{U}}^{-1}\right)\left(\underline{\underline{U}}\,\underline{\underline{M}}_k\right)\right) = \chi\left(\underline{\underline{M}}_k\right) \quad (2.11)$$

The set of numbers $\{\chi(\underline{\underline{M}}_1), \chi(\underline{\underline{M}}_2), \ldots, \chi(\underline{\underline{M}}_h)\}$ characterises in a unique way the representation $\{\underline{\underline{M}}_1, \underline{\underline{M}}_2, \ldots, \underline{\underline{M}}_h\}$ as well as any physically equivalent representation, $\{\underline{\underline{M}}'_1, \underline{\underline{M}}'_2, \ldots, \underline{\underline{M}}'_h\}$. Now that we have learnt how to characterise a set of physically equivalent functions by means of their characters, we can try *to find linear combinations of the functions* $\{f_1, f_2, \ldots, f_n\}$ *adapted by themselves to the symmetry of the group.*

Irreducible representation

Let us assume that there is a change of basis allowing the generation of an equivalent basis $\{f'_1, f'_2, \ldots, f'_n\}$ that may be decomposed in several bases Γ_i, every one of which is stable with respect to all the symmetry operations of the group. In that case we may write the representation Γ as a direct sum of representations with lower dimension, Γ_i:

$$\Gamma = \Gamma_1 \oplus \Gamma_2 \oplus \ldots \oplus \Gamma_m \quad (2.12)$$

We can now say that we have *decomposed the reducible representation* Γ into several representations Γ_i. We can then try to decompose these lower-dimension representations, Γ_i. When there is no change of basis allowing the decomposition of the basis Γ_i into several lower-dimension bases, it is said that Γ_i is a basis for an *irreducible representation*. As we will briefly review, group theory makes clear some very useful mathematical properties of irreducible representations.

Properties of irreducible representations

Let us now outline the main properties of irreducible representations of a symmetry group G of order h. In Section 2.3 we will use all these results to study the π-type orbitals of regular cyclobutadiene.

(i) The number of irreducible representations is equal to the number of classes of the group. For instance, the D_{4h} group has ten irreducible representations.
(ii) Matrices associated with symmetry operations of the same class have the same character.
(iii) A reducible representation, Γ, may be decomposed into the different irreducible representations of the group, Γ_i, in the following way:

$$\Gamma = n_1\Gamma_1 \oplus n_2\Gamma_2 \oplus \ldots \oplus n_m\Gamma_m$$

$$\text{with} \quad n_i = \frac{1}{h}\sum_{R_k}\chi_i^*(R_k)\chi(R_k) \quad (2.13)$$

where $\chi_i(R_k)$ and $\chi(R_k)$ ($k = 1, \ldots, h$) are respectively the characters of the representations Γ_i and Γ for the symmetry operations R_k of the group

G. The symbol * in eqn (2.13) means that we must take the complex conjugate of $\chi_i(R_k)$ if the character is complex. *This formula requires knowledge of the characters of the representation Γ. This calculation is generally very simple.*

(iv) Let us assume that the basis $\{f_1, f_2, \ldots, f_n\}$ may be projected onto the Γ_i representation, i.e. $n_i \neq 0$ in eqn (2.13). Then, a linear combination of $\{f_1, f_2, \ldots, f_n\}$ may result from the action of the projection operator $\mathcal{P}_{\Gamma_i}$ on one f_j function: [4]

$$\mathcal{P}_{\Gamma_i}(f_j) = \sum_{R_k} \chi_i^*(R_k) R_k(f_j) \qquad (2.14)$$

[4] Here we use a non-normalised operator. The normalisation of the symmetry-adapted functions will be later carried out in the Hückel approach. Note that this operator may be non-trivial for a degenerate representation. [4]

Application of this formula requires knowledge of how every symmetry operation R_k acts on the function f_j. This calculation is not difficult but may be quite tedious.

(v) If ϕ_i^μ and ϕ_j^ν are two basis elements of two different irreducible representations, Γ_μ and Γ_ν, of the symmetry group of a molecule and $\hat{h}$ is the appropriate one-electron Hamiltonian, the following terms are nil by symmetry:

$$\left\langle \phi_i^\mu \,|\, \hat{h} \,|\, \phi_j^\nu \right\rangle = 0 \text{ and } \left\langle \phi_i^\mu \,|\, \phi_j^\nu \right\rangle = 0 \qquad (2.15)$$

The functions ϕ_i^μ and ϕ_j^ν do not interact by symmetry and possess a nil overlap if they generate different irreducible representations Γ_μ and Γ_ν.

Application to the determination of the molecular orbitals of a molecule

It is essential to bear in mind that any molecular orbital is an element of a basis for an irreducible representation of the symmetry group of the molecule. As a result, it is possible to find the expression of a molecular orbital using group theory through the following procedure:

(i) Determine the symmetry group of the molecule using Table 2.2.
(ii) List the different atomic orbitals that participate in the relevant molecular orbitals of the system.
(iii) Group the different atomic orbitals into sets of symmetry-equivalent atomic orbitals, i.e. into different bases for a representation.
(iv) Decompose such representations (see eqn (2.13)).
(v) Find a basis for every irreducible representation Γ_i (see eqn (2.14)); the different basis elements ϕ_{ij} ($j = 1, \ldots, n_i$) of an irreducible representation Γ_i are linear combinations of atomic orbitals adapted to the symmetry of the molecule.
(vi) The molecular orbitals that are a basis for the irreducible representation Γ_i result from the interaction of the n_i functions ϕ_{ij} ($j = 1, \ldots, n_i$). They may be obtained by solving the secular equations (eqns (2.8) and (2.4)) by considering solely the n_i functions ϕ_{ij} ($j = 1, \ldots, n_i$) of this symmetry. The functions ϕ_{ij} now play the role of the χ_j orbitals in Section 2.1.3. The same procedure must be followed for all other irreducible representations.

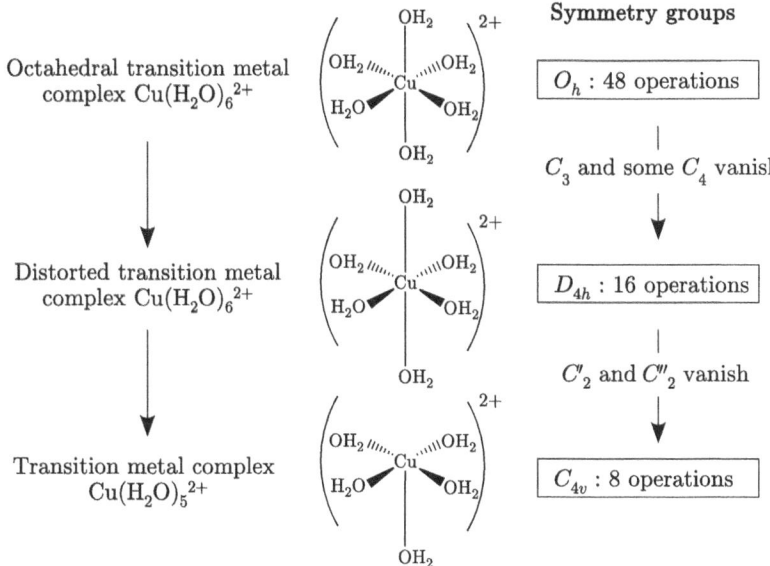

Fig. 2.6
Example of a symmetry lowering.

Symmetry lowering

From the mathematical point of view, a group may be a subgroup of another group of higher order. For instance, group C_{4v} is a subgroup of D_{4h}, which itself is a subgroup of O_h:

$$C_{4v} \subset D_{4h} \subset O_h \tag{2.16}$$

A *symmetry lowering*[5] corresponds to the passage from a group to one of its subgroups. This corresponds to the passage from an object with a given symmetry to another one with a lower symmetry, i.e. one that is invariant through a smaller number of symmetry operations. Thus, an example of the change from O_h to D_{4h} and finally to C_{4v} is provided by the hypothetical distortion of the octahedral compound $Cu(H_2O)_6^{2+}$ through the bond elongation of two ligands (O_h to D_{4h}) and the departure of one of the two apical ligands (D_{4h} to C_{4v}), as shown in Fig. 2.6.

It is important to point out that an irreducible representation for a group G is a representation that might be *reducible* from any subgroup SG. Any basis for a representation of the group G is necessarily stable with respect to the symmetry operations of the subgroup SG.

[5] When discussing symmetry in complexes like these, we will refer to the *local symmetry* around the transition metal atom. In other words, the actual geometry of the H_2O ligands will be ignored and we will consider the geometry of the MO_6 group of atoms.

2.3 Application to the study of the π system of regular cyclobutadiene

In this section we will show how the ideas developed in the previous sections may be used to study the electronic structure of symmetrical molecules by considering the π system of regular cyclobutadiene. Let us point out that the procedure we will follow here will be completely equivalent to the approach that we will follow when determining the orbitals describing the behaviour of an electron in a periodic system (see Section 3.1 of Chapter 3), or when studying related systems like polyacetylene (Chapter 5) or polyacene (Chapter 7).

2.3.1 Decomposition of the $\Gamma(p_z)$ basis

Let us consider the basis for a reducible representation of dimension 4, $\Gamma(p_z) = \{2p_{z_i}, i = 1, \ldots, 4\}$ containing the four atomic orbitals involved in the π system of regular cyclobutadiene. We will reduce this basis within the D_{4h} group. To this end, we need to consider the character table for this group (Table 2.3) and decompose the representation $\Gamma(p_z)$. We thus need to calculate the character of this representation for the different symmetry operations of the group. For example, the C_4 operation does not leave invariant any $2p_{z_i}$ ($i = 1, 2, 3, 4$), as shown in Figs 2.5 and 2.7. The character of $\Gamma(p_z)$ for this operation is 0.

In contrast, the symmetry operation C'_{2a} transforms the $2p_{z_2}$ and the $2p_{z_4}$ to their opposites. At the same time it transforms the two orbitals $2p_{z_1}$ and $2p_{z_3}$ in $-2p_{z_3}$ and $-2p_{z_1}$, respectively. Consequently, the character of $\Gamma(p_z)$ for the C'_{2a} rotation is -2 (Fig. 2.7). Proceeding in this way, we can obtain the characters of the $\Gamma(p_z)$ representation, which are reported in the last line of Table 2.3.

Table 2.3 Character table for the D_{4h} group. The last line contains the characters of the $\{2p_{z_i}, i = 1, \ldots, 4\}$ representation. By convention, the C'_2 axes and the symmetry planes σ_v go through two carbon atoms whereas the C''_2 axes and the symmetry planes σ_d pass in between the atoms.

D_{4h}	E	$2C_4$	C_2	$2C'_2$	$2C''_2$	i	$2S_4$	σ_h	$2\sigma_v$	$2\sigma_d$	
A_{1g}	1	1	1	1	1	1	1	1	1	1	$x^2 + y^2, z^2$
A_{2g}	1	1	1	-1	-1	1	1	1	-1	-1	
B_{1g}	1	-1	1	1	-1	1	-1	1	1	-1	$x^2 - y^2$
B_{2g}	1	-1	1	-1	1	1	-1	1	-1	1	xy
E_g	2	0	-2	0	0	2	0	-2	0	0	(xz, yz)
A_{1u}	1	1	1	1	1	-1	-1	-1	-1	-1	
A_{2u}	1	1	1	-1	-1	-1	-1	-1	1	1	z
B_{1u}	1	-1	1	1	-1	-1	1	-1	-1	1	
B_{2u}	1	-1	1	-1	1	-1	1	-1	1	-1	
E_u	2	0	-2	0	0	-2	0	2	0	0	(x, y)
$\Gamma(p_z)$	4	0	0	-2	0	0	0	-4	2	0	

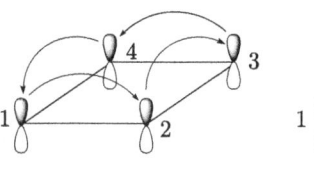
Action of the C_4 rotation

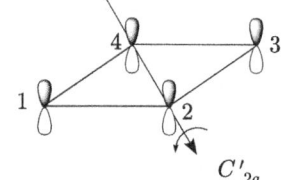
Action of the C'_{2a} rotation

Fig. 2.7
Action of the C_4 and C'_{2a} rotations on the $2p_{z_i}$ ($i = 1, 2, 3, 4$) orbitals.

Now we can decompose the representation $\Gamma(p_z)$ using eqn (2.13):

$$\Gamma(p_z) = n_{A_{1g}} A_{1g} + n_{A_{2g}} A_{2g} + \ldots + n_{E_u} E_u \quad (2.17)$$

$$\text{with} \quad n_i = \frac{1}{h} \sum_{R_k} \chi_i^*(R_k) \chi_\Gamma(R_k)$$

Thus, for instance,

$$n_{A_{1g}} = \frac{1}{16}(1 \cdot (1 \cdot 4) + 2 \cdot (1 \cdot 0) + 1 \cdot (1 \cdot 0) + 2 \cdot (1 \cdot (-2)) + 2 \cdot (1 \cdot 0)$$
$$+ 1 \cdot (1 \cdot 0) + 2 \cdot (1 \cdot 0) + 1 \cdot (1 \cdot (-4)) + 2 \cdot (1 \cdot 2) + 2 \cdot (1 \cdot 0)) = 0$$

Repeating the same calculation for every irreducible representation we obtain the decomposition:

$$\Gamma(p_z) = E_g + A_{2u} + B_{2u} \quad (2.18)$$

2.3.2 Determination of the basis elements for different irreducible representations

What are the linear combinations of $2p_z$ atomic orbitals that form a basis for the irreducible representations E_g, A_{2u}, and B_{2u}? We will answer this question in two different ways: direct and indirect. The direct approach consists of applying the projectors (see eqn (2.14)) for the D_{4h} group. This method can, however, be tedious when the symmetry group possesses a large number of symmetry operations, and an indirect approach may reach equivalent results without the need for long calculations, simply using the character tables.

Direct determination

The linear combinations of $2p_{z_i}$ orbitals of A_{2u} and B_{2u} symmetry may be obtained by application of the projector operators $\mathcal{P}_{A_{2u}}$ and $\mathcal{P}_{B_{2u}}$ (eqn (2.14)) onto, for instance, the $2p_{z_1}$ function. For the E_g representation, the operator $\mathcal{P}_{E_g}$ must be applied onto two functions, for instance the $2p_{z_1}$ and $2p_{z_2}$, to obtain a basis for this doubly degenerate representation. Shown in Table 2.4 are the results of the action of the different symmetry operations of the D_{4h} group on the $2p_{z_1}$ and $2p_{z_2}$ functions.

We can obtain the different (non-normalised) basis elements of the different irreducible representations by means of eqn (2.14):

$$\Psi_{A_{2u}} = N_{A_{2u}} (p_{z_1} + p_{z_2} + p_{z_3} + p_{z_4})$$
$$\Psi_{B_{2u}} = N_{B_{2u}} (p_{z_1} - p_{z_2} + p_{z_3} - p_{z_4})$$
$$(\Psi_{E_g})_1 = N_{E_{g1}} (p_{z_1} - p_{z_3})$$
$$(\Psi_{E_g})_2 = N_{E_{g2}} (p_{z_2} - p_{z_4})$$

Table 2.4 Action of the different symmetry operations of the D_{4h} group on the p_{z_1} and p_{z_2} functions.

D_{4h}	E	C_4	C_4^3	C_2	C'_{2a}	C'_{2b}	C''_{2a}	C''_{2b}
$(p_z)_1$	$(p_z)_1$	$(p_z)_2$	$(p_z)_4$	$(p_z)_3$	$-(p_z)_3$	$-(p_z)_1$	$-(p_z)_4$	$-(p_z)_2$
$(p_z)_2$	$(p_z)_2$	$(p_z)_3$	$(p_z)_1$	$(p_z)_4$	$-(p_z)_2$	$-(p_z)_4$	$-(p_z)_3$	$-(p_z)_1$

D_{4h}	i	S_4	S_4^3	σ_h	σ_{va}	σ_{vb}	σ_{da}	σ_{db}
$(p_z)_1$	$-(p_z)_3$	$-(p_z)_2$	$-(p_z)_4$	$-(p_z)_1$	$(p_z)_3$	$(p_z)_1$	$(p_z)_4$	$(p_z)_2$
$(p_z)_2$	$-(p_z)_4$	$-(p_z)_3$	$-(p_z)_1$	$-(p_z)_2$	$(p_z)_2$	$(p_z)_4$	$(p_z)_3$	$(p_z)_1$

Now we can normalise these functions using, for instance, the Hückel approach:

$$\Psi_{A_{2u}} = \frac{1}{2}\left(p_{z_1} + p_{z_2} + p_{z_3} + p_{z_4}\right) \tag{2.19a}$$

$$\Psi_{B_{2u}} = \frac{1}{2}\left(p_{z_1} - p_{z_2} + p_{z_3} - p_{z_4}\right) \tag{2.19b}$$

$$\left(\Psi_{E_g}\right)_1 = \sqrt{\frac{1}{2}}\left(p_{z_1} - p_{z_3}\right) \tag{2.19c}$$

$$\left(\Psi_{E_g}\right)_2 = \sqrt{\frac{1}{2}}\left(p_{z_2} - p_{z_4}\right) \tag{2.19d}$$

Indirect determination using the information in the right-hand column of the character table

It is sometimes possible to avoid the use of eqn (2.14) by bearing in mind the information found in the right-hand column of the character table. This information assumes that (i) the coordinate system Oxy is centred at the O point, which is invariant after all symmetry operations, and (ii) the z-axis coincides with the main axis of this group. Thus for cyclobutadiene the O point lies at the centre of the cycle and the z-axis is the axis perpendicular to the plane of the molecule. The x- and the y-axes may be taken as going through the carbon atoms labelled 2 and 3 in Fig. 2.4. The information found at the right-hand column of the character tables consists of examples of functions that are bases for different irreducible representations of the group. In particular, the functions $f(x, y, z)$ equal to $x^2 + y^2 + z^2$, x, y, z, xy, xz, yz, $2z^2 - x^2 - y^2$, and $x^2 - y^2$ are always given. These functions possess the same symmetry properties as orbitals ns, np_x, np_y, np_z, nd_{xy}, nd_{xz}, nd_{yz}, nd_{z^2}, and $nd_{x^2-y^2}$, respectively, *centred at the point O , which is left invariant by the different symmetry operations of the group.*

Consequently, the linear combination of the $2p_{z_i}$ ($i = 1, 2, 3, 4$) orbitals basis of the A_{2u} irreducible representation must exhibit the same symmetry as an np_z orbital sitting at the centre of cyclobutadiene. Thus, as shown in Fig. 2.8, it is easy to guess the linear combination of A_{2u} symmetry simply

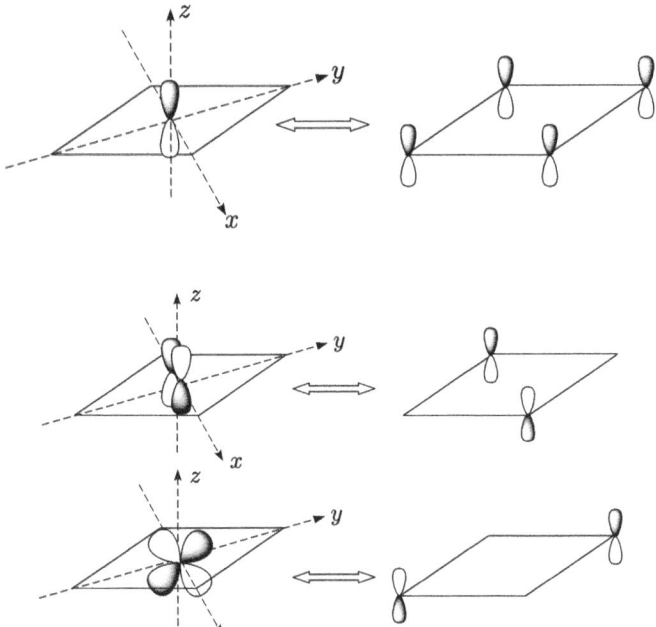

Fig. 2.8
Construction of the linear combination of $2p_z$ orbitals with A_u symmetry of cyclobutadiene using the fact that it has the same symmetry as a fictitious p_z orbital lying at the centre of butadiene. These two functions have the same symmetry properties in D_{4h}.

Fig. 2.9
Construction of the linear combinations of $2p_z$ orbitals of regular cyclobutadiene with E_g symmetry using the fact that they must possess the same symmetry properties as two d_{xz} and d_{yz} orbitals lying at the centre of the cycle.

by looking at the symmetry properties of this fictitious $2p_z$ orbital lying at the centre of the cycle.

It is also possible to find the two linear combinations that are a basis of an irreducible representation of the type E_g by considering the symmetry properties of d_{xz} and d_{yz} orbitals lying at the centre of the cycle (see Fig. 2.9). It is possible to determine the linear combination of B_{2u} symmetry, for which we do not have any information in the character table, by imposing the orthogonality condition with respect to the three functions already determined. Thus, in this favourable example, we can avoid the use of the projectors just by looking at the standard character table. The linear combinations obtained by this procedure may be normalised through the use of the Hückel approach and are given in eqn (2.20).

$$\Psi'_{A_{2u}} = \frac{1}{2}\left(p_{z_1} + p_{z_2} + p_{z_3} + p_{z_4}\right) \quad (2.20a)$$

$$\Psi'_{B_{2u}} = \frac{1}{2}\left(p_{z_1} - p_{z_2} + p_{z_3} - p_{z_4}\right) \quad (2.20b)$$

$$\left(\Psi'_{E_g}\right)_1 = \sqrt{\frac{1}{2}}\left(p_{z_1} - p_{z_3}\right) \quad (2.20c)$$

$$\left(\Psi'_{E_g}\right)_2 = \sqrt{\frac{1}{2}}\left(p_{z_2} - p_{z_4}\right) \quad (2.20d)$$

These linear combinations are identical to those obtained from the projection operator (see eqn (2.19))[6]. As usually happens, this indirect approach leads more easily to linear combinations of atomic orbitals adapted to the symmetry of the D_{4h} group.

[6] The indirect method could have led to a different but equivalent basis for the doubly degenerate representation E_g.

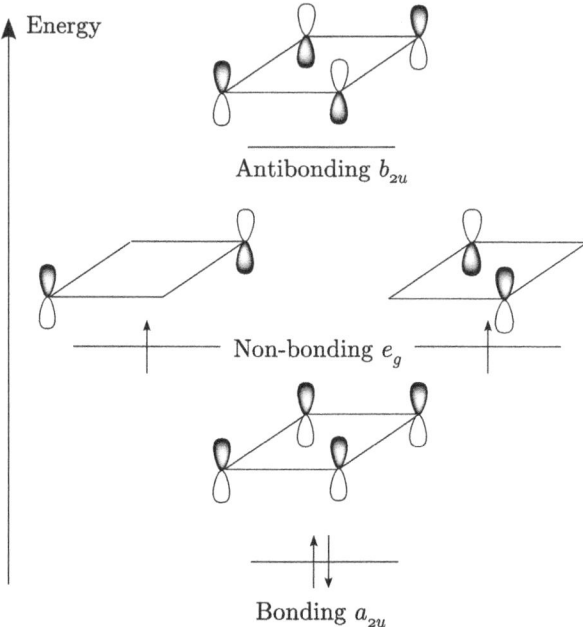

Fig. 2.10
Molecular orbital diagram for the π system of regular cyclobutadiene.

2.3.3 Molecular orbital diagram of the π system of regular cyclobutadiene

The different linear combinations of $2p_z$ atomic orbitals obtained cannot interact among themselves because they are bases of different irreducible representations of the symmetry group of the molecule. These bases may thus be taken as the π molecular orbitals of regular cyclobutadiene. If we now consider the bonding or antibonding character of the interaction between adjacent atomic orbitals in these molecular orbitals, we can propose the energy diagram in Fig. 2.10. Since the π system of the molecule contains four electrons, the ground-state configuration of the system is $(a_{2u})^2(e_{1g})^2$. The delocalised π bonds of regular cyclobutadiene are thus described by three molecular orbitals populated with four electrons, which are delocalised through the molecule. Thus, in this very favourable example, with high symmetry and a small number of orbitals, group theory can provide the detailed expression of the molecular orbitals of the π system in a direct and very simple manner.

2.4 Transition metal complexes

We now turn to the electronic structure of transition metal complexes. Consideration of this problem will help in making clear the approach we will use in later chapters to qualitatively study the electronic structure of relatively complex molecular systems. Some of the results and the general approach will be very useful when we will consider the electronic structure of periodic systems containing transition metal atoms. [4, 5]

2.4.1 Ligands and formal oxidation state

A transition metal complex is made up of a transition metal atom, which behaves as a Lewis acid, and a series of ligands, playing the role of Lewis bases. The ligands that we will consider in this book are the following ions or molecules: H_2O, NH_3, $C_5H_5^-$, CO, CN^-, O^{2-}, Cl^-, I^-, Br^-, ... We will only consider transition metal complexes containing a partially filled $M(nd)$ shell.[7]

Ligands

In a pure ionic picture, a ligand is a chemical species having at least one double-occupied molecular orbital high enough in energy so as to interact with an empty orbital of the transition metal atom.[8] The interaction leads to the formation of a bonding molecular orbital that is mostly localised on the ligand and an antibonding molecular orbital with strong transition metal character (see Fig. 2.11).

The covalent character of the transition metal–ligand interactions increases when the overlap between the associated orbitals increases and the energy difference (ΔE) between these orbitals decreases. All ligands are electron donors through their occupied orbital(s). However, certain ligands also possess an empty orbital, which tends to interact with the transition metal. When this happens the transition metal orbitals receive electrons through interaction with the occupied orbital of the ligand and at the same time lose some of these electrons through interaction with the empty orbital of the ligand, i.e. there is a *back donation*. The transition metal–ligand interaction is stronger when this happens and may be schematised as in Fig. 2.12.

The properties of a ligand may be characterised by describing the donor or acceptor character of the orbitals involved in the interaction with the transition metal, as well as the nature of the associated overlaps (σ, π).

[7] Let us note that there are some examples of transition metal complexes in which the transition metal atom formally plays the role of a Lewis base. However, in the examples studied in this book the transition metal atom always plays the role of a Lewis acid.

[8] This means that only the upper molecular orbitals may be responsible for the ligand properties of a molecule or group. This orbital must be able to interact effectively with the empty level, in general higher in energy, of the Lewis acid.

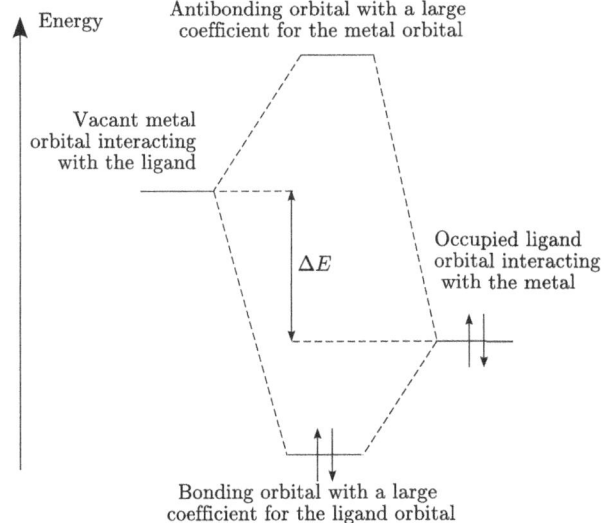

Fig. 2.11
Schematic interaction diagram for an empty transition metal orbital and a filled ligand orbital.

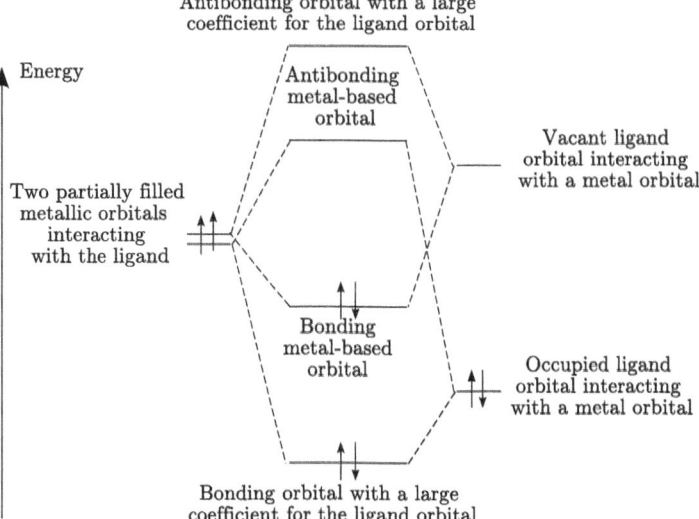

Fig. 2.12
Interaction diagram for two partially filled transition metal orbitals and two ligand orbitals, one of which is empty.

Oxidation state

In chemistry it is usual to characterise an atom in a molecule by means of its *oxidation state*. To determine this state one assumes that the electrons of a bond between two atoms AB *formally* belong to the more electronegative atom, i.e. the atom having the lowest-lying valence orbitals. Thus, in the water molecule, the four electrons associated with the two O–H bonds are attributed to the oxygen atom so that the oxidation state of oxygen is –II whereas that of the hydrogen atoms is +I.

In the same vein, ligand orbitals lower in energy than the metallic orbitals are always assumed to be fully filled when calculating the formal oxidation state of a transition metal. This is why ligands like C_5H_5, CN, Cl, and N, are considered as anionic species ($C_5H_5^-$, CN^-, Cl^-, and N^{3-}) bonded to the metal cation M^{n+}. As an example, the rhenium cation in the $ReNCl_4$ complex is formally Re^{+VII} when bonded to Cl^- and N^{3-}. We will thus consider the rhenium +VII in this complex to have no d electrons (i.e. it is a d^0 transition metal), since the seven electrons it bears in the neutral atom have been formally lost.

It is important to note that the oxidation state is a purely formal concept and that it must not be confused with the actual charge of the transition metal in the complex. The two are only equal if the metal–ligand interactions are completely ionic although we know that the covalent character of this interaction is quite important. However, we will see in the next sections that the concept of oxidation state is invaluable in simplifying our understanding of the electronic structure of complex systems since, in a very easy way, it makes clear the number of *chemically relevant* electrons in a complex molecule.

2.4.2 The ML$_6$ octahedral complex

Case where L is a pure σ donor

We now consider an ML$_6$ octahedral complex (Fig. 2.13a) where L is a pure σ-donor ligand; in other words, a ligand that interacts with the transition metal solely through its occupied orbitals pointing to the transition metal, schematically represented as simple lobes in Fig. 2.13b.

Let us now build the molecular orbital diagram of this complex. Every ligand provides two electrons through the filled orbitals $\{\sigma_i, i = 1, \ldots, 6\}$ while the transition metal, for a given oxidation state, provides m electrons through its $(n+1)s$ and nd orbitals. The character table of the O_h group which leaves the complex invariant is reproduced below (Table 2.5). Consider the two representations Γ_M and Γ_L, which contain the valence orbitals of the transition metal and ligand orbitals:

$$\Gamma_M = \{nd_{xy}, nd_{xz}, nd_{yz}, nd_{x^2-y^2}, nd_{z^2},$$
$$(n+1)s, (n+1)p_x, (n+1)p_y, (n+1)p_z\}$$

and $\quad \Gamma_L = \{\sigma_i\ i = 1, \ldots, 6\}$

The information in the right-hand column of Table 2.5 shows how to decompose the Γ_M representation, which contains atomic orbitals centred on the metal:

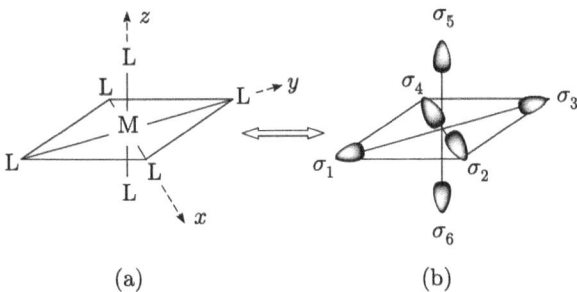

Fig. 2.13
ML$_6$ octahedral complex: (a) structure; (b) schematic representation of the orbitals of the σ-donor ligands.

Table 2.5 Character table of the O_h group and characters of the Γ_L representation (last line).

O_h	E	$8C_3$	$6C_2'$	$6C_4$	$3C_2$	i	$8S_6$	$6\sigma_d$	$6S_4$	$3\sigma_h$	
A_{1g}	1	1	1	1	1	1	1	1	1	1	$x^2 + y^2 + z^2$
A_{2g}	1	1	−1	−1	1	1	1	−1	−1	1	
E_g	2	−1	0	0	2	2	−1	0	0	2	$2z^2 - x^2 - y^2, x^2 - y^2$
T_{1g}	3	0	−1	1	−1	3	0	−1	1	−1	
T_{2g}	3	0	1	−1	−1	3	0	1	−1	−1	(xz, yz, xy)
A_{1u}	1	1	1	1	1	−1	−1	−1	−1	−1	
A_{2u}	1	1	−1	−1	1	−1	−1	1	1	−1	
E_u	2	−1	0	0	2	−2	1	0	0	−2	
T_{1u}	3	0	−1	1	−1	−3	0	1	−1	1	(x, y, z)
T_{2u}	3	0	1	−1	−1	−3	0	−1	1	1	
$\Gamma(L)$	6	0	0	2	2	0	0	2	0	4	

$\{(n+1)s\} \Rightarrow$ basis of A_{1g}

$\{(n+1)p_x, (n+1)p_y, (n+1)p_z\} \Rightarrow$ basis of T_{1u}

$\{nd_{xy}, nd_{xz}, nd_{yz}\} \Rightarrow$ basis of T_{2g}

$\{nd_{x^2-y^2}, nd_{z^2}\} \Rightarrow$ basis of E_g

For representation Γ_L, we can carry out a projection onto the character table of the group O_h using eqn (2.13). We then obtain:

$$\Gamma_L = A_{1g} + T_{1u} + E_g \quad (2.21)$$

We can now estimate a basis of the different representations by taking advantage of the information in the character table.[9] For example, we can find a basis of the A_{1g} representation by considering a linear combination of the σ_i orbitals exhibiting the same symmetry as the $x^2 + y^2 + z^2$ function. We can follow the same approach to find a basis of T_{1u} (same symmetry as the (x, y, z) functions) and E_g (same symmetry as the $(3z^2 - r^2, x^2 - y^2)$ functions). The result of this estimate is shown in Fig. 2.14.

In Table 2.6 we summarise the symmetry properties of the valence orbitals of the transition metal and the ligands, as well as the number, symmetry, and nature of the molecular orbitals resulting from their interaction. Since the overlap between the nd transition metal and ligand orbitals is usually weaker than those involved in the $(n + 1)s$ and $(n + 1)p$ orbitals of the transition metal,

[9] We could also use the appropriate projector operator. However, in this case it would not be the more convenient choice because it would be quite tedious and, in addition, somewhat involved in the case of the degenerate irreducible representations

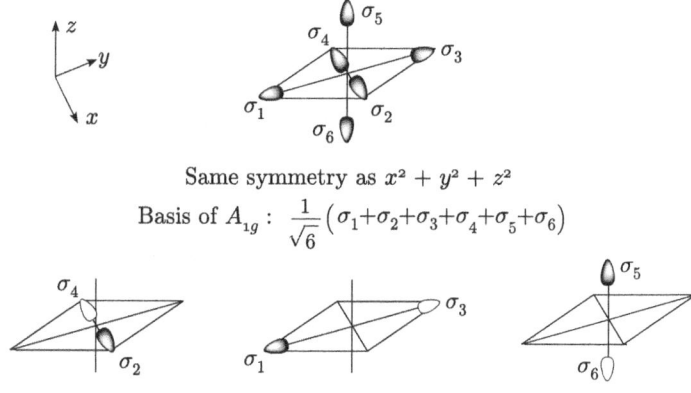

Same symmetry as $x^2 + y^2 + z^2$

Basis of A_{1g}: $\frac{1}{\sqrt{6}}(\sigma_1+\sigma_2+\sigma_3+\sigma_4+\sigma_5+\sigma_6)$

Same symmetry as x Same symmetry as y Same symmetry as z

Basis of T_{1u}: $\frac{1}{\sqrt{2}}(\sigma_2-\sigma_4)$; $\frac{1}{\sqrt{2}}(\sigma_1-\sigma_3)$; $\frac{1}{\sqrt{2}}(\sigma_5-\sigma_6)$

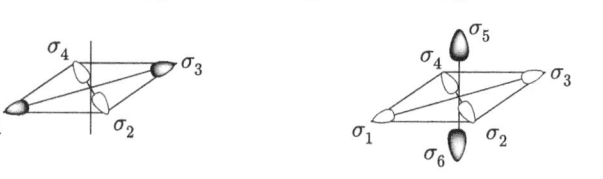

Same symmetry as $x^2 - y^2$ Same symmetry as $2z^2 - (x^2 + y^2)$

Basis of E_g: $\frac{1}{\sqrt{2}}(\sigma_1-\sigma_2+\sigma_3-\sigma_4)$; $\frac{1}{\sqrt{12}}(-\sigma_1-\sigma_2-\sigma_3-\sigma_4+2\sigma_5+2\sigma_6)$

Fig. 2.14
Linear combinations of the σ_i orbitals which form a basis of an irreducible representation of the O_h group. The normalisation of these functions has been carried out according to the Hückel approach.

Transition metal complexes

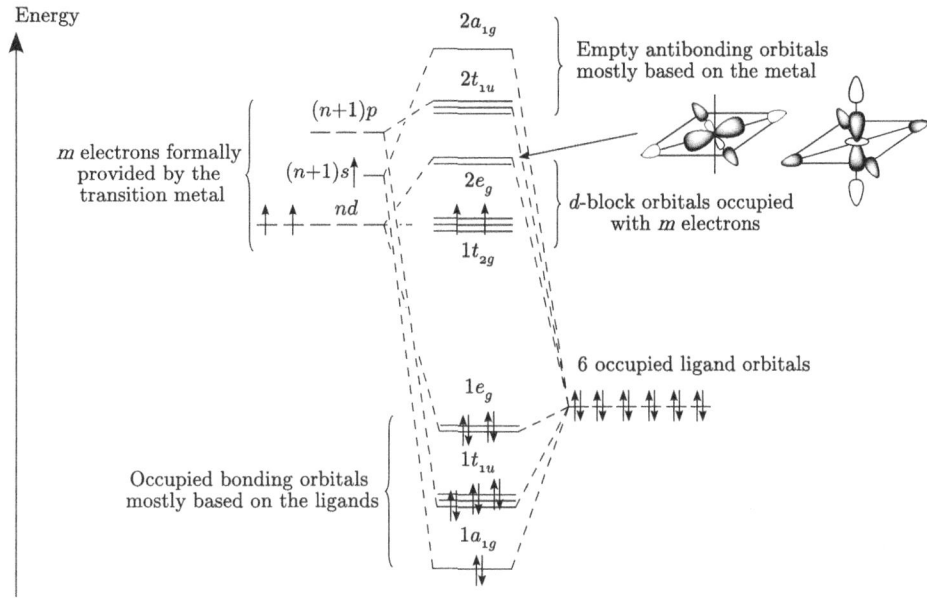

Fig. 2.15
Molecular orbital diagram of the ML$_6$ complex where L is a σ donor ligand. The electron filling corresponds to the case where the metal configuration is $3d^2$ (a Ti^{+2} for instance), i.e. $m = 2$.

the appropriate molecular orbital diagram for the octahedral ML$_6$ transition metal complex is that shown in Fig. 2.15. The ML$_6$ system possesses $m + 12$ electrons, so that the six bonding orbitals are filled and are responsible for the stability of the complex.

As shown in Fig. 2.15, the last m electrons fill the *d-block orbitals*, namely:

- the three $1t_{2g}$ orbitals, nd_{xy}, nd_{xz}, and nd_{yz}, which are pure non-bonding transition metal levels

Table 2.6 Symmetry and nature of the octahedral complex MOs.

Symmetry	Metal AOs	Ligand σ orbitals	Resulting MOs
A_{1g}	$(n+1)s$	1 combination	1 bonding MO 1 antibonding MO
T_{1u}	$(n+1)p_x$ $(n+1)p_y$ $(n+1)p_z$	3 combinations	3 deg. bonding MOs 3 deg. antibonding MOs
E_g	$nd_{x^2-y^2}, nd_{z^2}$	2 combinations	2 deg. bonding MOs 2 deg. antibonding MOs
T_{2g}	$nd_{xy}, nd_{yz}, nd_{xz}$	—	3 deg. non-bonding MOs

- the two antibonding $2e_g$ orbitals, $nd_{x^2-y^2}$, and nd_{z^2}, which are primarily from the transition metal, with a smaller contribution from the ligand orbitals.

This result provides an *a posteriori* justification for the usefulness of characterising the transition metal by its oxidation state through the assumption that the ligands have a closed shell. As a result of this formal partition of the electrons between the transition metal and the ligands, the twelve electrons assigned to the ligands fill the six bonding molecular orbitals, which are mostly ligand in character, whereas the remaining m electrons of the transition metal fill the orbitals with a strong transition metal character. Since in the following we will only be interested in the $1t_{2g}$ and $2e_g$ orbitals, which are responsible for the properties of these octahedral complexes, the number of *chemically active* electrons filling these orbitals is the number of electrons *formally* provided by the metal. From now on we will label Δ_0 the energy difference between the $1t_{2g}$ and $2e_g$ orbitals.

In the rest of the book we will frequently talk about the $2e_g$ orbitals as if they were purely the metal $nd_{x^2-y^2}$ and nd_{z^2} orbitals. However, as mentioned, these orbitals are antibonding molecular orbitals with a strong transition metal character but also with a sizable contribution from the ligands. We will refer to them as $nd_{x^2-y^2}$ and nd_{z^2} orbitals, although they exhibit ligand contributions.

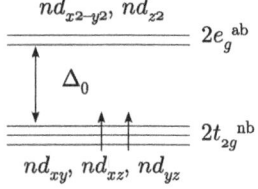

Fig. 2.16
d-block orbitals for the case where one of the ligands is a purely σ donor and a transition metal providing 2 electrons as, for instance, Ti^{+2}.

> In the following we will only consider the five orbitals of the d-block partially filled by the m electrons of the transition metal. These orbitals are responsible for most of the properties of octahedral transition metal complexes (Fig. 2.16).

π effect of the ligands

We will now consider an ML$_6$ complex in which the ligand L may interact with the transition metal through three orbitals: an occupied orbital of σ type pointing towards the transition metal, and two π-type orbitals, either occupied or empty, interacting with the metal through a π overlap. This is the case, for instance, with chloride or carbon monoxide ligands. The Cl$^-$ ligand mainly interacts through its $3p$ orbitals, which are occupied by six electrons: the $3p$ orbital pointing towards the metal plays the role of the σ orbital and the other two play the role of the π orbitals. For CO it is the highest-occupied orbital, essentially non-bonding, which plays the role of the σ orbital. The two empty π^* orbitals also interact with the metal. The shape of these orbitals of CO is displayed in Fig. 2.17.

Whatever the ligand considered, the interaction involving σ-type overlaps dominates and must be considered first. The donor or acceptor π character may be considered as a perturbation and thus can be considered later. Since in general we will be only interested in the d-block orbitals, we will ignore the other occupied or empty orbitals. This is obviously an additional hypothesis, but one that chemists currently employ in their qualitative rationalisations. We will now see how the d-block orbitals (Fig. 2.16) may interact with either donor

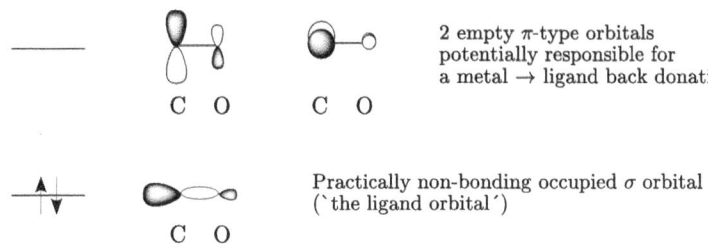

Fig. 2.17
σ and π orbitals of the CO ligand.

or acceptor π ligand orbitals. Fig. 2.18 is a schematic representation of the twelve π ligand orbitals. The basis of representation of these twelve π orbitals may be decomposed in the following way:

$$\Gamma_\pi = T_{1g} + T_{1u} + T_{2g} + T_{2u} \qquad (2.22)$$

Since we are exclusively interested in the orbitals of the d-block, only the interaction between the linear combinations of π orbitals of symmetry t_{2g} can modify the d-block represented in Fig. 2.16. The other combinations, with t_{1g}, t_{1u}, and t_{2u} symmetry, cannot interact with the d-block orbitals. The linear combinations of π-type orbitals of t_{2g} symmetry may be easily obtained since they have the same symmetry as the xy, xz, and yz functions (see Fig. 2.19).

Consequently, every transition metal orbital of $1t_{2g}$ symmetry, i.e. $\{nd_{xy}, nd_{xz}, nd_{yz}\}$, may interact with a linear combination of π orbitals, which is adapted to its symmetry (Fig. 2.19). The orbitals of the d-block are modified, as shown in Figs 2.20a and 2.20b, when the ligand is a π donor (e.g. Cl$^-$) or a π acceptor (e.g. CO), respectively.

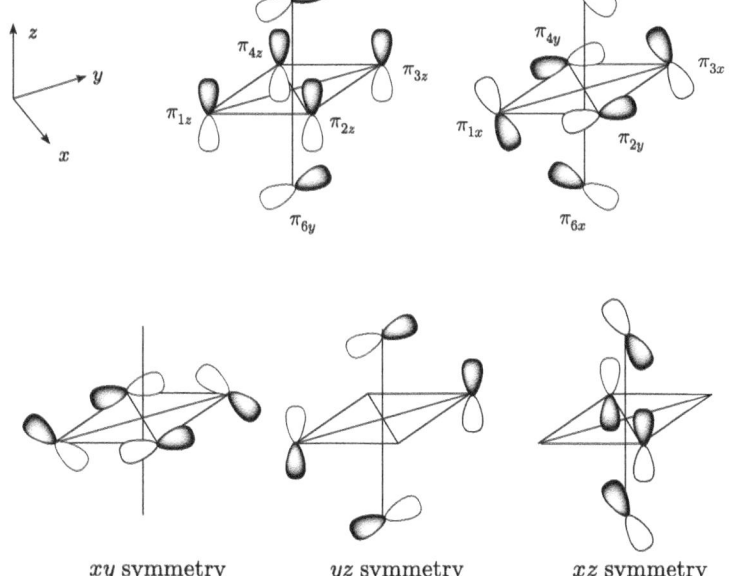

Fig. 2.18
π orbitals of six ligands, either Cl$^-$ or CO, of an octahedral complex with formula ML$_6$. For the CO ligands only the orbital centred on the carbon atom directly interacting with the transition metal is shown (see Fig. 2.17 and Exercise (2.8)).

Fig. 2.19
Combinations of π-type orbitals adapted to the T_{2g} symmetry.

38   Electronic structure of molecules

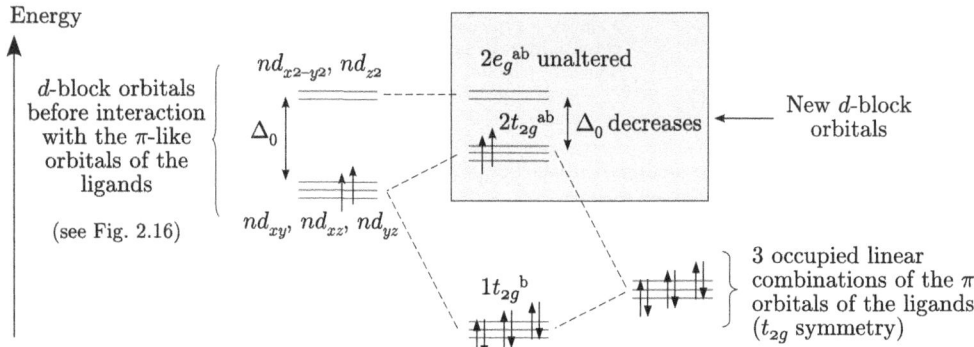

a) π-donor ligand: low-lying and occupied π-type orbitals
Example: Cl⁻

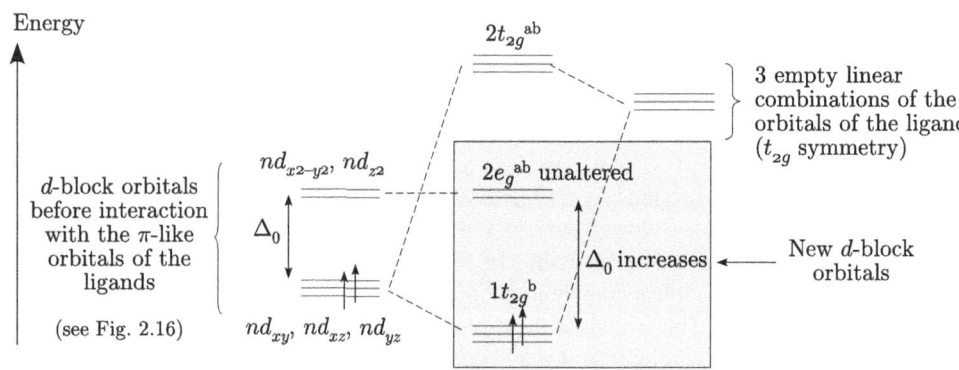

b) π-acceptor ligand: high-lying and vacant π-type orbitals
Example: CO

Fig. 2.20
Metal–ligand interaction: (a) case of a ligand σ and π donor; (b) case of a ligand σ donor and π acceptor. The b and ab labels indicate the bonding and antibonding character of the orbitals. In both (a) and (b) we have chosen to represent the situation for the case where the metal provides two electrons, e.g. Ti^{2+}.

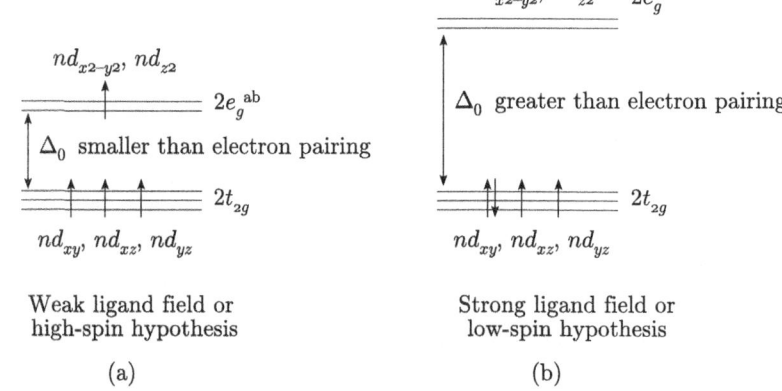

Fig. 2.21
d-block orbitals of a metal complex: strong (a) and weak (b) field situation for a metal with d^4 configuration.

Weak ligand field or high-spin hypothesis
(a)

Strong ligand field or low-spin hypothesis
(b)

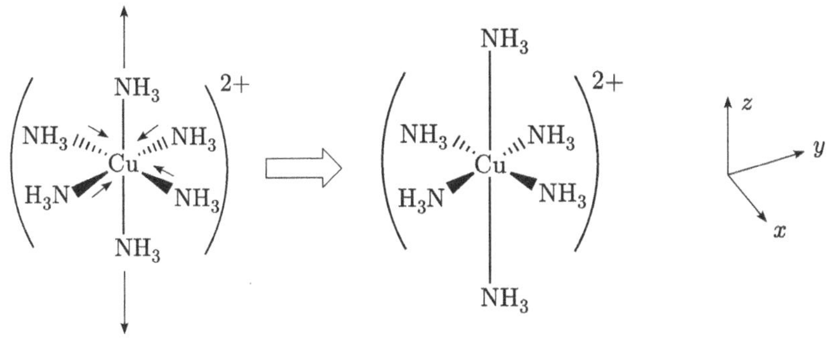

Fig. 2.22
Distortion of an octahedral transition metal complex.

Consider the case of a π-donor ligand (Fig. 2.20a). As a result of the larger strength of the σ interactions, the energy of the antibonding t_{2g} level stays lower than the $2e_g$ orbitals. The π effect of the ligands tends to decrease the energy difference Δ_0 between the $2t_{2g}$ and $2e_g$ levels of the d-block. In contrast, when the ligand is a π acceptor (Fig. 2.20b), the difference in energy between the two molecular orbitals increases because the $1t_{2g}$ is slightly bonding. *Despite this difference, the number of electrons filling the d-block orbitals is the same: it is the number of electrons formally donated by the metal.*

The filling of these levels depends on the nature of the transition metal and ligands. Let us illustrate this fact with the case of a d^4 transition metal. When Δ_0 is smaller than the pairing energy, i.e. the energy needed to put two electrons in the same orbital, the ground state is obtained by filling the molecular orbitals of the d-block with one electron. In this event, the ligand field is *weak* and the transition metal is in the high-spin configuration (Fig. 2.21a). In contrast, when the pairing energy of the two electrons is weaker than Δ_0, the ground state is obtained by filling the lower energy orbital with two electrons. In this event the ligand field is *strong* and the transition metal is in the low-spin configuration (Fig. 2.21b).

2.4.3 Distortions of a complex

Evolution of the d-block orbitals when a distortion occurs

Consider first the octahedral complex $Cu(NH_3)_6^{2+}$, with configuration $(2t_{2g})^6(2e_g)^3$. This complex may be distorted by lengthening the bonds between the metal and the two apical ligands while strengthening those with the four basal ligands. [4, 5] We will assume that the ligand displacements along the z-axis are twice larger than those along the x- and y-axes (see Fig. 2.22).

Since the NH_3 ligand is essentially a σ donor (Fig. 2.16), the $2t_{2g}$ orbitals of the d-block are non-bonding in such way that they are not sensitive to the ligands' displacement. In contrast, the $2e_g$ antibonding orbitals are split in energy as a result of the distortion. The antibonding orbital $3d_{x^2-y^2}$, which exhibits antibonding interactions in the xy plane, will be destabilised when the distortion occurs because the antibonding interactions are increased. The $3d_{z^2}$ orbital has a lobe pointing along the z-axis, with amplitude twice as large as

Fig. 2.23
Comparison of the d-block electron filling for the regular and distorted octahedral complexes.

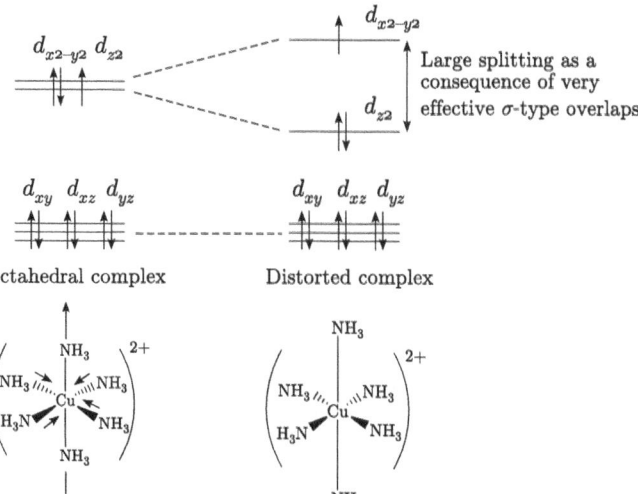

that of the $(x^2 + y^2)$ ring. This, together with the fact that the displacement is assumed to be twice as large along the Oz axis, leads to the stabilisation of the $3d_{z^2}$ orbital.[10] This deformation leads to the splitting of the $2e_g$ pair of orbitals (Fig. 2.23).

From a purely electronic point of view, this distortion stabilises the system because the three $2e_g$ electrons are globally stabilised.

Jahn–Teller effect

The $Cu(NH_3)_6^{2+}$ complex possesses a vibration of e_g symmetry, corresponding exactly to the breathing mode of the molecule in which the transition metal–ligand bond length increases (or decreases) along the Oz axis. Other ligands move towards (or move away from) the metal. This vibration may be represented by the Q vibrational coordinate (see Fig. 2.24).

As a consequence, it is no longer possible to decouple the electronic motion from the vibration represented in Fig. 2.24 because the electronic energy is lowered by such nuclear displacement. Thus we must formally leave the Born–Oppenheimer approximation to satisfactorily rationalise the structural and electronic properties of this complex. In particular, it will be necessary to look at the total energy of the system, i.e. the electronic energy and the kinetic energy of the nuclei. However, after a certain degree of distortion the nuclear repulsion will tend to oppose the displacement of the nuclei so that the total energy of the complex will exhibit a shape not very different from that of Fig. 2.25. There we have assumed that a displacement $+Q$ leads to the same energy consequences as a $-Q$ displacement.

At zero temperature the energy barrier V_0 is large enough to ensure that tunnelling will not be effective and the system in the ground electronic and vibrational states is locked in one of the potential wells. Thus, the system is distorted. In contrast, when the system accesses one of the vibrational

---

[10] In addition to this stabilisation, one must take into account the fact that the $3d_{z^2}$ and $4s$ orbitals can mix in the distorted complex with symmetry D_{4h}, but that such mixing is forbidden in octahedral geometry. This leads to a molecular orbital that is less antibonding and looks like that shown in the scheme below.

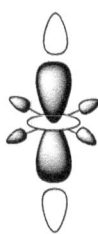

The mixing between the $3d_{z^2}$ and $4s$ orbitals leads to an increase in the amplitude of the lobes pointing towards the far away ligands along the z-axis and to a decrease in the amplitude of the ring in the xy-plane where the nearest ligands are. As a result, this orbital becomes less antibonding and is stabilised by the distortion.

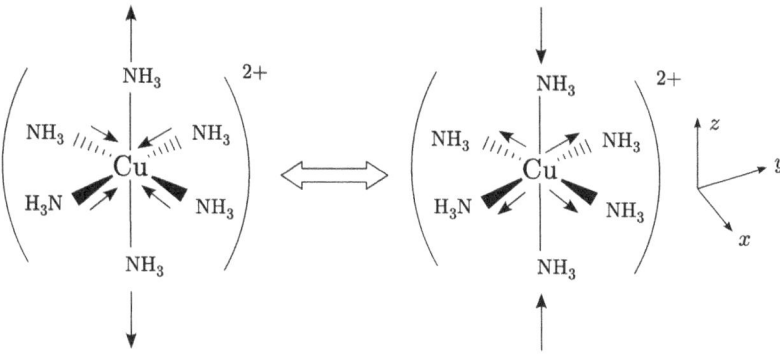

Fig. 2.24
Schematic representation of an E_g type vibration. The displacements along the z-axis have an amplitude twice larger than those along the x- and y-axes.

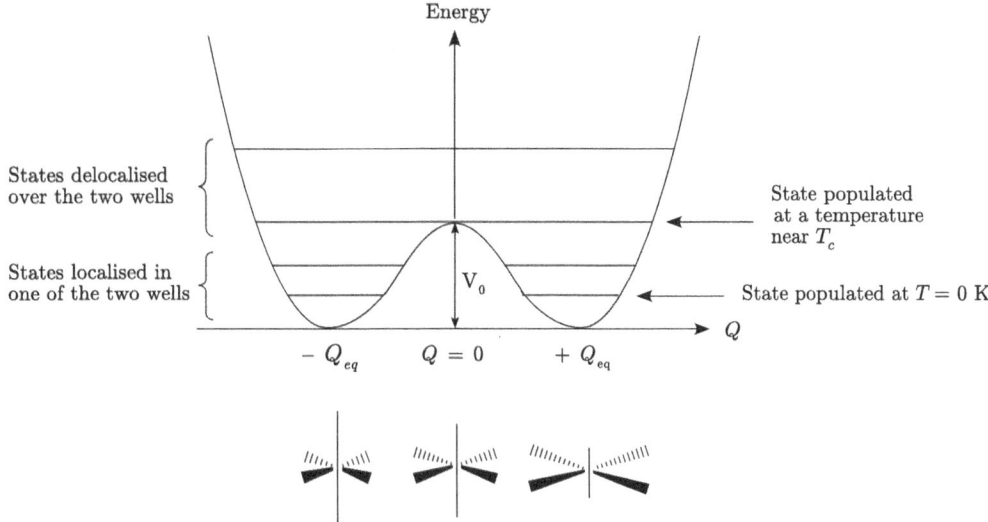

Fig. 2.25
Schematic representation of the total energy (electronic and vibrational) of the Cu(NH$_3$)$_6^{2+}$. The horizontal lines represent the allowed vibrational states of the system.

states near or slightly higher than the potential barrier $+V_0$ through thermal excitation, it is delocalised over the double well. The average geometry, i.e. the octahedral geometry, is then that experimentally observed.

In a more general vein, it can be said that any system where the highest occupied orbital is degenerate and partially filled is susceptible to distortion resulting from the coupling of the electronic motion and one of the vibrations of the system. This interaction allows the splitting of the highest occupied levels and lowers the total energy of the system. This is the so-called *first-order Jahn–Teller effect* and may be schematically illustrated as shown in Fig. 2.26. Such distortion becomes easier to observe as the $+V_0$ barrier increases. At low temperatures the vibrational states that are populated are found below

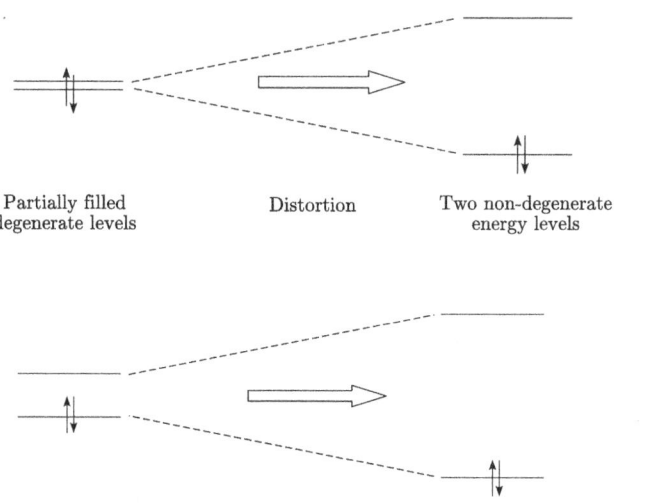

Fig. 2.26
First-order Jahn–Teller effect.

Partially filled degenerate levels   Distortion   Two non-degenerate energy levels

Fig. 2.27
Second-order Jahn–Teller effect.

Electronic levels with different symmetries in the non-distorted compound   Distortion   Electronic levels of equivalent symmetry and larger energy difference

the potential barrier: the system is locked in one of the potential wells and is thus distorted. There is a critical temperature T_c (Fig. 2.25) above which the system is described by a state lying near the potential barrier V_0: the system is then delocalised over the two wells and observed as non-distorted. In our very simple approach we have assumed that the $+Q$ and $-Q$ deformations are equivalent. In fact one of the two is more stable, so that at low temperature only one distorted geometry is observed.

We can talk about a *second-order Jahn–Teller effect*, which is observed when the system is unstable with respect to a distortion that is stabilising the system because such distortion makes possible the interaction between non-degenerate orbitals (see Fig. 2.27). In this case we must consider two levels, one occupied and the other empty, of different symmetry in the undistorted system. If the two orbitals have the same symmetry after the distortion they may interact, the lower (occupied) orbital becoming stabilised while the upper (empty) orbital is destabilised. Such interaction is more stabilising when the energy difference between the occupied and empty orbitals is small. However the stabilisation afforded by a first-order Jahn–Teller effect is generally larger than that resulting from a second-order Jahn–Teller effect.

Exercises

(2.1) Establish an expression for the energy of the π molecular orbitals of ethylene as a function of α and β, according to the Hückel and extended Hückel approaches. Give an expression for the corresponding molecular orbitals. Compare the results obtained by means of the two approaches.

(2.2) Establish an expression for the energy of the π molecular orbitals of a planar allyl group, C_3H_5, as a function of α and β according to the Hückel approach. Give an expression for the corresponding molecular orbitals.

(2.3) Find the 48 symmetry operations leaving invariant an octahedron (for example the $Fe(Cl)_6^{4-}$ complex of Fe(II)).

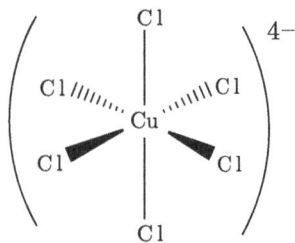

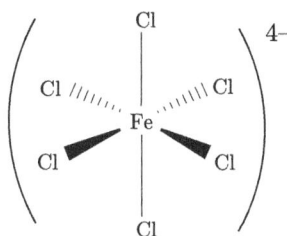

(2.4) What is the appropriate symmetry group for this octahedron when deformed along the vertical axis?

(2.5) Provide physical reasons justifying the existence of different classes in the D_{4h} group.

(2.6) What are the different sets of atomic orbitals of the cyclobutadiene molecule, either from the carbon or hydrogen atoms, constituting bases of a representation of the D_{4h} group?

(2.7) What is the metal oxidation state in the two systems: $Cu(H_2O)_6^{2+}$ and $Pt(CN)_4^{2-}$? How many d electrons formally belong to the metal in these systems?

(2.8) Carry out the decomposition of the basis of representation made of the twelve π orbitals of the ligands represented in Fig. 2.18.

References

1. I. N. Levine, *Quantum Chemistry*, 5th edition, Prentice Hall, London, 2000.
2. R. Hoffmann, *J. Chem. Phys.*, 39, 1397, 1963.
3. Y. Jean and F. Volatron, *Structure Électronique des Molécules*, Ediscience, Paris, 1993; Y. Jean, F. Volatron, and J. K. Burdett, *An Introduction to Molecular Orbitals*, Oxford University Press, New York, 1993.
4. F. A. Cotton, *Chemical Applications of Group Theory*, 3rd edition, John Wiley, New York, 1990.
5. T. A. Albright, J. K. Burdett, M.-H. Whangbo, *Orbital Interactions in Chemistry*, John Wiley, New York, 1985.

Electronic structure of one-dimensional systems: basic notions

The free electron model briefly considered in Chapter 1 reveals that only the electrons near the Fermi level are responsible for the conducting properties of a solid. However, the free electron model is a very useful but grossly simplified approach. In the context of the present work, its major drawback is probably that it does not take into account the real structure of the material. *In what follows we will consider how to build and analyse the band structure of a periodic compound taking into account its crystal structure.*

We will start by considering one-dimensional (1D) compounds whose study simplifies the task of introducing the basic notions. Later on, we will consider more complex two- and three-dimensional (2D and 3D) materials. Throughout this book, except for the last chapter, we will use the linear combination of atomic orbitals (LCAO) approach. In other words, we will describe the electronic behaviour on the basis of a monoelectronic wavefunction written as a linear combination of the valence atomic orbitals of the different atoms of the system. Our approach will be very similar to that used in the first section of Chapter 2 by exploiting as far as possible the symmetry properties (here translational properties) exhibited by periodic 1D compounds. [1, 2, 3, 4] Let us consider a *periodic* 1D system generated by the reference unit cell M_0[1] and a translational vector $\vec{a}$ (see Fig. 3.1). The unit cell M_m is the image of the reference unit cell M_0 after an $m\vec{a}$ translation. The length of the system (L) is assumed to be very large and amounts to $n \cdot a$. We will assume that n is an even integer and we will denote as n' its half (i.e. $n' = \frac{n}{2}$). This is not really a constraint since, from a physical point of view, the only condition that n' must fulfil is to be very large. Using an even value for n renders the reading of this chapter easier without loss of generality. Exactly the same reasoning could have been used with an odd value for n, although at first sight the arguments would seem a little less transparent. The interested readers will see for themselves in Exercise (3.1).

Since n is very large, it is natural to assume that the wavefunction describing the behaviour of the electrons in such a periodic system must reflect the translational symmetry associated with the lattice. However, a formal problem soon appears when trying to account for these symmetry properties. Strictly speaking, a finite system is not left invariant by a translation $\vec{a}$ because the

[1] This reference unit cell may contain several atoms with a given structure, i.e. it is simply a repeat unit.

Bloch and crystal orbitals

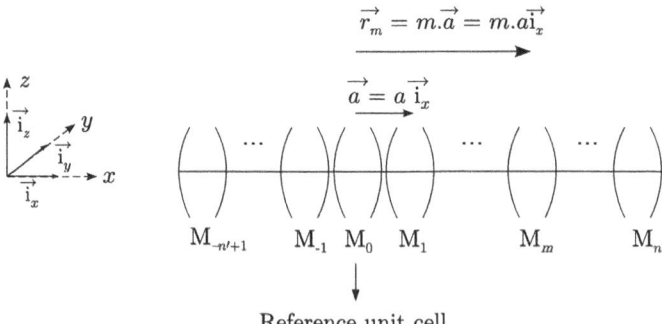

Fig. 3.1
Schematic representation of a periodic 1D system of length L containing n unit cells, n being an even number and $n' = \frac{n}{2}$. Note that $\vec{i}_x, \vec{i}_y, \vec{i}_z$ are the orthonormal basis vectors.

image of unit cell $M_{n'}$ lies outside the system. We have shown in Section 1.1 of Chapter 1 how to tackle this apparent problem by imposing the so-called Born–von Karman boundary conditions of eqn (1.5). These boundary conditions essentially build an equivalence between a linear system (Fig. 3.1) and a cyclic system (Fig. 3.2). The translation associated with a vector $m\vec{a}$ (noted $t_{m\vec{a}}$) for the linear system (Fig. 3.1) is equivalent to the rotation C_n^m for the cyclic system (Fig. 3.2). As a consequence, the translation associated with a vector $n\vec{a}$ ($t_{n\vec{a}}$) may be identified with the rotation C_n^n, i.e. the identity. From the physical point of view, these boundary conditions impose the equivalence of the $-L/2$ and $+L/2$ points. From now on we will indicate the location of cell M_m ($m = 0, \pm 1, \pm 2, \ldots, \pm(n'-1), n'$) by means of an angle $\varphi_m = \frac{2m\pi}{n}$ in the interval $]-\pi, \pi]$.

Since chemists are more familiar with point groups than with translation groups we will generally use the cyclic representation (Fig. 3.2) in order to handle the periodicity of the system. We will use the character table of the C_n symmetry group in order to write the expression for the 'solid orbitals' adapted to the periodic properties of these systems. It is worth noting however, that the symmetry of periodic systems may be higher than that of the C_n group. For instance, the symmetry group for the H_n system generated by a cell containing only one hydrogen atom is the D_{nh} group. However, it is always possible to work with a lower symmetry. In the present case, this allows us to take into account, without loss of generality, the translational symmetry properties common to all periodic 1D systems, whatever the nature of the unit cell involved.

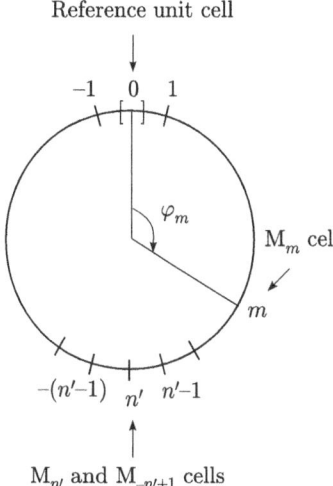

Fig. 3.2
The 1D system in Fig. 3.1 is equivalent to the cyclic system above when the Born–von Karman boundary conditions are imposed.

3.1 Bloch and crystal orbitals

We will now establish an expression for the crystal orbitals, i.e. the monoelectronic wavefunctions describing the behaviour of an electron in a periodic system. We will initially adopt the cyclic representation (Fig. 3.2) of the periodic system. Let us assume that N_0 orbitals ($\phi_1, \phi_2, \ldots, \phi_{N_0}$) per unit cell must be taken into account for the description of the electronic structure of the system. These orbitals may be atomic orbitals or fragment orbitals located in each cell. We will denote the ϕ_j type orbital located in cell M_m as $(\phi_j)_m$. This orbital is the image of the corresponding orbital in the reference cell $(\phi_j)_0$ after a C_n^m rotation. We will build the crystal orbitals using a two-step approach very similar to that used in Chapter 2 to construct the molecular orbitals of a molecule.

- *First, we will take into account the rotational symmetry of the periodic system.* Let us start by taking advantage of the equivalence of the different ϕ_j orbitals: $\{(\phi_j)_{-n'+1}, \ldots, (\phi_j)_0, \ldots, (\phi_j)_{n'}\}$ through rotational symmetry. In this way, we will be able to write the appropriate linear combinations of these orbitals, which are well-adapted to the rotational C_n group. We will refer to these linear combinations, generated by the ϕ_j orbitals and well-adapted to rotational symmetry, as *the Bloch orbitals* BO_j.
- *Second, the mixing between Bloch orbitals of the same symmetry will be considered.* Thus, the next step will require the combination of Bloch orbitals of the same symmetry to obtain the appropriate eigenvectors of the system, i.e. *the crystal orbitals* CO_j.

We may compare this approach to the one that led us to establish the molecular orbital expressions using group theory (Sections 2.2–2.3). *The Bloch orbitals of the periodic system play the same role as the linear combinations of equivalent orbitals that form a basis for an irreducible representation of the molecule symmetry group. The crystal orbitals play the same role as the molecular orbitals.*

Once the wavefunctions of the cyclic system has been established, we will use the equivalence between the cyclic and the linear systems (Figs 3.1 and 3.2) to obtain the expression for the crystal orbitals of the linear system.

3.1.1 Bloch orbitals

Let us now consider the basis for a representation $\Gamma_{\phi_j} = \{(\phi_j)_{-n'+1}, \ldots, (\phi_j)_0, \ldots, (\phi_j)_{n'}\}$ containing the $n\phi_j$ orbitals, which are equivalent through rotational symmetry. The 'symmetry adapted orbitals' of the system can be obtained through the decomposition of the Γ_{ϕ_j} representation using the character table of group C_n containing the C_n^m rotations with angles $\frac{2m\pi}{n}$ ($m \in [-n'+1, n']$).[2] The character table of group C_n, which contains n irreducible representations (IR) Γ_ℓ, labelled by the integer ℓ, with values $0, \pm 1, \pm 2, \ldots, \pm(n'-1), n'$ is shown in Table 3.1. The character of the Γ_ℓ IR with respect to the C_n^m rotation is given by the general formula $\exp(-\frac{2i\pi m \ell}{n})$.

[2] In general it is assumed that the C_n group contains the different C_n^m rotations characterised by the positive integers $m = 1, 2, \ldots, n$. Nevertheless, since the C_n^{m-n} rotation is the same as the C_n^m one, the C_n group also contains the C_n^m rotations with $m = 0, \pm 1, \pm 2, \ldots, \pm(n'-1), n'$ ($n' = \frac{n}{2}$), if n is an even integer. In this chapter we will adopt such a definition.

Table 3.1 Character table for the C_n group (for n even). C_n^m represents a rotation with an angle $\frac{2m\pi}{n}$. The E and C_2 operations correspond to the values $m = 0$ and $m = n' = n/2$, respectively. The last line reproduces the characters of the representation $\Gamma_{\phi_j} = \{(\phi_j)_{-n'+1}, \ldots, (\phi_j)_0, \ldots, (\phi_j)_{n'}\}$ associated with the different rotations.

C_n	E	C_2	C_n	$\ldots$	C_n^m	$\ldots$	$C_n^{-n'+1}$
Γ_ℓ $\ell = 0, \ldots, n'$	1	$(-1)^\ell$	$\exp(\frac{-2i\pi\ell}{n})$	$\ldots$	$\exp(-\frac{2mi\pi\ell}{n})$	$\ldots$	$\exp(\frac{2(n'-1)i\pi\ell}{n})$
Γ_{ϕ_j}	n	0	0	$\ldots$	0	$\ldots$	0

Bloch and crystal orbitals

The decomposition of the representation is thus clear:

$$\Gamma_{\phi_j} = \sum_{\ell=-n'+1}^{n'} \Gamma_\ell \tag{3.1}$$

This decomposition is independent of the nature of the (ϕ_j) orbital. The determination of the linear combination of ϕ_j orbitals that forms a basis for the Γ_ℓ irreducible representation is obtained through the application of the $\mathcal{P}_{\Gamma_\ell}$ projector (eqn (2.14)) on the orbital of the reference cell $(\phi_j)_0$. Since the C_n^m rotation transforms orbital $(\phi_j)_0$ into orbital $(\phi_j)_m$ it is possible to write

$$(\Psi_j)_\ell = \frac{1}{\sqrt{n}} \sum_{m=-n'+1}^{n'} \exp\left(i\frac{2\pi m \ell}{n}\right) (\phi_j)_m \tag{3.2}$$

where $(\Psi_j)_\ell$ is the linear combination of ϕ_j orbitals that forms a basis for the irreducible representation Γ_ℓ. The normalisation coefficient $\frac{1}{\sqrt{n}}$ has been determined in the Hückel approach. Let us note that the sum over m allows consideration of all the cells of the system, the orbital of each cell being preceded by a *phase factor* $\exp(i\frac{2\pi m \ell}{n})$ depending on the position of the cell.

At this point we may adopt the linear representation of the system (Fig. 3.1) by referencing to the position of cell M_m by means of the vector $\vec{r}_m$ equal to $m\vec{a}$ (Fig. 3.1), i.e. its projection r_m on the Ox axis. We will refer to the linear combination of atomic orbitals adapted to the translation symmetry as the *Bloch orbital generated by the ϕ_j orbitals* and note it as $(BO_j)_\ell$. Since the $(\phi_j)_m$ orbital is the image of the $(\phi_j)_0$ as generated by the C_n^m rotation, we may say that the $(BO_j)_\ell$ Bloch orbital is generated by the $(\phi_j)_0$ orbital.

$$(BO_j)_\ell = \frac{1}{\sqrt{n}} \sum_{m=-n'+1}^{n'} \exp\left(i\frac{2\pi \ell}{na} r_m\right) (\phi_j)_m$$

$$(BO_j)_\ell = \frac{1}{\sqrt{n}} \sum_{m=-n'+1}^{n'} \exp\left(i\frac{2\pi \ell}{na} r_m\right) C_n^m (\phi_j)_0 \tag{3.3}$$

where ℓ may adopt the values $0, \pm 1, \pm 2, \ldots, \pm(n'-1), n'$.

The $(BO_j)_\ell$ function is a linear combination of orbitals $(\phi_j)_m$ ($m = 0, \pm 1, \ldots, \pm(n'-1), n'$) modulated by a phase factor. We can write this $\exp(i\frac{2\pi \ell}{na} r_m)$ term as a wave $\exp(i\vec{k}.\vec{r}_m)$ characterised by a wave vector $\vec{k}$ the projection on the Ox axis of which is $\frac{2\pi}{na}\ell$. From now on we will characterise the symmetry properties of the Bloch orbital $(BO_j)_\ell$ by means of the real value k, which describes the periodicity of the phase factors along the chain. Thus we will note this function as $BO_j(\vec{k})$ even if $\vec{k}$ is not a variable but just an index characterising the properties of the Bloch orbital with respect to an integer number of $\vec{a}$ translations:

$$BO_j(\vec{k}) = \frac{1}{\sqrt{n}} \sum_{m=-n'+1}^{n'} \exp(i\vec{k}.\vec{r}_m)(\phi_j)_m \tag{3.4}$$

with $\vec{k} = k\vec{i}_x$ and $k = 0, \pm\dfrac{2\pi}{na}, \pm\dfrac{4\pi}{na}, \cdots, \pm\dfrac{2(n'-1)\pi}{na}, \dfrac{\pi}{a}$

since $k = \dfrac{2\pi}{na}\ell$ and $\ell = 0, \pm1, \pm2, \cdots, n'$

Consequently, k may take n values *regularly spaced* in the interval $]-\tfrac{\pi}{a}, \tfrac{\pi}{a}]$. Since n is very large, k covers all the interval $]-\tfrac{\pi}{a}, \tfrac{\pi}{a}]$. At this point it is useful to introduce the $\vec{a}^*$ vector, defined as:

$$\vec{a}^* = \dfrac{2\pi}{a^2}\vec{a} \quad \Rightarrow \quad \vec{a}^* \cdot \vec{a} = 2\pi \tag{3.5}$$

where a is the norm of a vector $\vec{a}$ which has the dimension of a length. The vector $\vec{a}^*$ of eqn (3.5) has the dimension of the inverse of a length in such a way that it is possible to characterise the wave vector $\vec{k}$ by its real projection k_a on a vector $\vec{a}^*$. Consequently, it is possible to write:

$$\vec{k} = k_a \vec{a}^* = \dfrac{2\pi k_a}{a^2}\vec{a} = \dfrac{2\pi k_a}{a}\vec{i}_x = k\vec{i}_x \quad \rightarrow \quad k_a = \dfrac{a}{2\pi}k$$

with $\quad k_a = 0, \pm\dfrac{1}{n}, \pm\dfrac{2}{n}, \ldots, \pm\dfrac{n'-1}{n}, \dfrac{1}{2}$ (3.6)

We will label the Bloch orbital $\mathrm{BO}_j(\vec{k})$ by means of either the $\vec{k}$ vector (eqn (3.4)) or the real dimensionless number k_a (eqn (3.6)), with possible values $0, \pm\tfrac{1}{n}, \pm\tfrac{2}{n}, \ldots, \pm\tfrac{n'-1}{n}, \tfrac{1}{2}$

$$\mathrm{BO}_j(\vec{k}) = \dfrac{1}{\sqrt{n}} \sum_{m=-n'+1}^{n'} \exp(i 2\pi m k_a)(\phi_j)_m \tag{3.7}$$

since after eqns (3.5) and (3.6), $\vec{k} \cdot \vec{r}_m = (k_a \vec{a}^*) \cdot (m\vec{a}) = 2\pi m k_a$. We will schematically represent the Bloch orbital $\mathrm{BO}_j(\vec{k})$ by writing near the cell M_m the imaginary coefficient preceding orbital $(\phi_j)_m$, i.e. $\exp(i 2\pi m k_a)$ (see Fig. 3.3).

It is very important to distinguish between the $\vec{a}$ and $\vec{a}^*$ vectors. These two vectors generate the so-called direct lattice (i.e. the chain itself) and the reciprocal lattice (i.e. the lattice of the associated $\vec{k}$ vectors), respectively. In addition, vector $\vec{a}$ characterises the periodicity of the real system (see Fig. 3.1) whereas vector $\vec{a}^*$ characterises the periodicity of vector $\vec{k}$.

If we replace $\vec{k} + \vec{a}^*$ for $\vec{k}$ in eqn (3.4) we can establish that:

$$\mathrm{BO}_j(\vec{k}) = \mathrm{BO}_j(\vec{k} + \vec{a}^*) \tag{3.8}$$

In this sense, vector $\vec{a}^*$ characterises the periodicity of the wave vector $\vec{k}$. The interval $]-\tfrac{1}{2}, \tfrac{1}{2}]$, which contains the n allowed values of k_a (see eqn (3.6)), is the *Brillouin zone* of the system.[3]

$$\mathrm{BO}_j(\vec{k}) = \dfrac{1}{\sqrt{n}} \begin{pmatrix} \cdots & -2 & -1 & 0 & 1 & 2 & \cdots \\ & \exp(-i4\pi k_a) & \exp(-i2\pi k_a) & 1 & \exp(i2\pi k_a) & \exp(i4\pi k_a) & \end{pmatrix} \begin{matrix} \leftarrow \text{Cell index} \\ \\ \leftarrow (\phi_j)_m \text{ orbital} \\ \text{coefficient} \end{matrix}$$

---

[3] Note that eqn (3.7) shows that the orbital $\mathrm{BO}(k_a = \tfrac{1}{2})$ is the same as the orbital $\mathrm{BO}(k_a = -\tfrac{1}{2})$. This justifies the Brillouin zone being associated with the $]-\tfrac{1}{2}, \tfrac{1}{2}]$ interval, since points $\tfrac{1}{2}$ and $-\tfrac{1}{2}$ generate the same function.

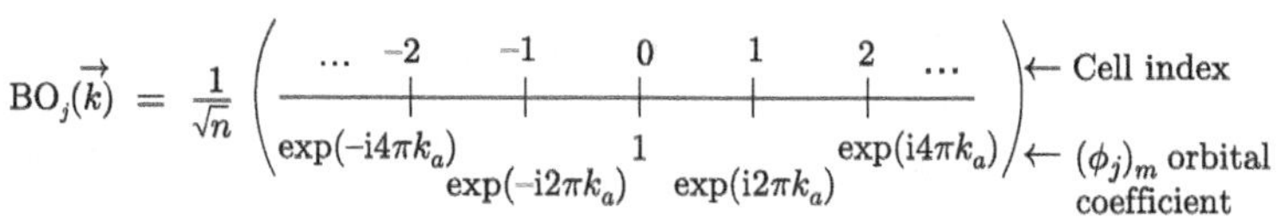

Fig. 3.3
Schematic representation of the Bloch orbital $\mathrm{BO}_j(\vec{k})$.

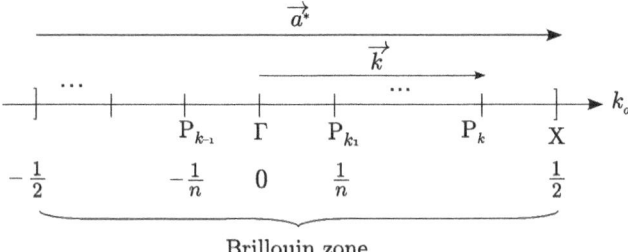

Fig. 3.4

Representation of the Brillouin zone of a 1D system. Above every k point we have indicated the real number k_a that characterises it ($\vec{k} = k_a \vec{a}^*$).

Consequently, if we represent all allowed $\vec{k}$ vectors (Fig. 3.4) using a common origin Γ, it is possible to characterise a given $\vec{k}$ vector by either the location of its extremity P_k or the real number k_a, which is equal to $\frac{m}{n}$. We will often talk about point k even if we do refer to vector $\vec{k}$. We will always refer to the centre of the Brillouin zone, which corresponds to a nil value of k_a, as Γ, whereas the point $k_a = \frac{1}{2}$ will be denoted X when the axis of the 1D system is the axis Ox. Finally, let us note that the $\vec{a}^*$ vector connects the two borders of the Brillouin zone ($k_a = -\frac{1}{2}$ and $k_a = \frac{1}{2}$).

A Bloch orbital $BO_j(\vec{k})$ is a linear combination of ϕ_j orbitals that are equivalent through translational symmetry. The $\vec{k}$ vector, which we will often refer to as the k point, characterises the translational symmetry properties of this Bloch orbital. The n allowed values of the real number k_a are uniformly spaced in the interval $]-\frac{1}{2},\frac{1}{2}]$ (Fig. 3.4) called the Brillouin zone. Two Bloch orbitals $BO_j(\vec{k})$ and $BO_j(\vec{k}')$ associated with different vectors $\vec{k}$ and $\vec{k}'$ possess different symmetry properties. Consequently, the interaction term $< BO_j(\vec{k})|\hat{h}|BO_j(\vec{k}') >$ is nil by symmetry.

3.1.2 Crystal orbitals

The crystal orbitals of the system result from the combination of Bloch orbitals $BO_j(\vec{k})(j = 1, \ldots, N_0)$ of the same translational symmetry, i.e. associated with the same k point. Every crystal orbital $CO(\vec{k})$ is labelled with a $\vec{k}$ vector and is a linear combination of Bloch orbitals $BO_{j'}(\vec{k})$:

$$CO(\vec{k}) = \sum_{j'=1}^{N_0} c_{j'}(\vec{k}) BO_{j'}(\vec{k}) \quad (3.9)$$

The $c_{j'}(\vec{k})$ coefficients as well as the energy $E(\vec{k})$ associated with this function can be obtained by imposing to the $CO(\vec{k})$ function to fulfil the Schrödinger equation (3.10):

$$\hat{h}[CO(\vec{k})] = E(\vec{k}) CO(\vec{k}) \quad (3.10)$$

One-dimensional systems

In other words, by using equation (3.9):

$$\sum_{j'=1}^{N_0} c_{j'}(\vec{k}) \left(\hat{h}\left[BO_{j'}(\vec{k})\right] - E(\vec{k})BO_{j'}(\vec{k}) \right) = 0 \quad (3.11)$$

Let us multiply this equation by the complex conjugate of $BO_j(\vec{k})$ and then integrate the resulting expression over all space. We then obtain the system of N_0 equations in which the N_0 variables are the $c_{j'}(\vec{k})(j' = 1, \ldots, N_0)$ coefficients:

$$\sum_{j'=1}^{N_0} c_{j'}(\vec{k}) \left(h_{jj'}(\vec{k}) - E(\vec{k}) S_{jj'}(\vec{k}) \right) = 0 \quad (j = 1, \ldots, N_0) \quad (3.12)$$

where $h_{jj'}(\vec{k})$ and $S_{jj'}(\vec{k})$ may be written using the Dirac notation (eqns (2.5)–(2.6)) as:

$$h_{jj'}(\vec{k}) = \left\langle BO_j(\vec{k}) | \hat{h} | BO_{j'}(\vec{k}) \right\rangle \text{ and } S_{jj'}(\vec{k}) = \left\langle BO_j(\vec{k}) | BO_{j'}(\vec{k}) \right\rangle \quad (3.13)$$

This system of equations has a solution different from the trivial one ($c_{j'}(\vec{k}) = 0; j' = 1, \ldots, N_0$) if and only if the secular determinant is nil:

$$|\underline{\underline{h}}(\vec{k}) - E(\vec{k})\underline{\underline{S}}(\vec{k})| = 0 \quad (3.14)$$

where $\underline{\underline{h}}(\vec{k})$ and $\underline{\underline{S}}(\vec{k})$ are the matrix representations of the Hamiltonian and overlap operators in the Bloch orbitals $BO_j(\vec{k})(j = 1, \ldots, N_0)$ basis. Solution

Table 3.2 Comparison of the approaches used to obtain the molecular orbitals of a molecule and the crystal orbitals of a periodic system.

Description of the electronic structure of a molecule	Description of the electronic structure of a 1D compound
N_0 atomic or fragment orbitals are used to describe the electronic structure.	N_0 orbitals are used to describe the unit cell. The periodic system contains n cells and thus $n \cdot N_0$ orbitals, denoted $(\phi_j)_m$ ($m = 0, \pm 1, \pm 2, \ldots, \pm(n'-1), n'$) ($j = 1, \ldots, N_0$).
Combination of orbitals equivalent by symmetry allows the construction of N_0 symmetry adapted orbitals, which are a basis for an irreducible representation of the symmetry group of the molecule.	Combination of $(\phi_j)_m$ ($m = 0, \pm 1, \pm 2, \ldots, \pm(n'-1), n'$) orbitals equivalent by translation symmetry allows the construction of n Bloch orbitals $BO_j(\vec{k})$. The $BO_j(\vec{k})$ orbital is a basis for an irreducible representation of the translation group and is characterised by a real number k_a, which may have n different values: $k_a = 0, \pm 1/n, \ldots, \pm(n'-1)/n, \frac{1}{2}$.
Mixing of the symmetry orbitals that form a basis for the same irreducible representation of the symmetry group through solution of the secular equations, leads to the different molecular orbitals that form a basis for this irreducible representation.	Mixing of the Bloch orbitals $BO_j(\vec{k})$ ($j = 1, \ldots, N_0$) associated with the same k point through solution of the secular equations, leads to the different crystal orbitals associated with this k point, $CO_\ell(\vec{k})(\ell = 1, \ldots, N_0)$.

of this N_0-th order equation in $E(\vec{k})$ leads to the determination of the N_0 allowed energies, which we will denote $E_\ell(\vec{k})(\ell = 1, \ldots, N_0)$. When the normalisation condition of the crystal orbitals is also taken into account, the system of equations can be solved for every value of $E_\ell(\vec{k})$, and the coefficients $c_{j'}(\vec{k})(j' = 1, \ldots, N_0)$ of the crystal orbital $CO_\ell(\vec{k})$ associated with this energy are obtained. This procedure must be repeated, in principle, for every point of the Brillouin zone.

Consequently, the crystal orbitals $CO(\vec{k})$ describing the behaviour of electrons in a periodic system may be written as linear combinations of Bloch orbitals, $BO_{j'}(\vec{k})$ (eqn (3.9)). The allowed energies associated with these crystal orbitals may be obtained through the condition that the secular determinant is nil (eqn (3.14)) and their detailed expression requires the solution of the system of eqn (3.12). This approach is completely equivalent to that used to determine the molecular orbitals of a molecule (Section 2.2) as summarised in Table 3.2.

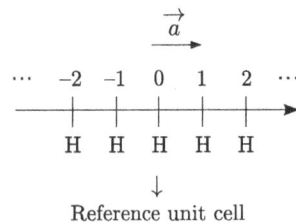

Fig. 3.5
Schematic representation of the H_n chain.

3.2 Electronic structure of the model chain H_n

As a first application we will now consider the electronic structure of the simplest 1D system, *the model chain* H_n (see Fig. 3.5). Every cell of this system may be described by using a single $1s$ orbital centred on the hydrogen atom. Consequently, the crystal orbitals of this system, $CO(\vec{k})$, may be identified with the Bloch orbitals $BO_{1s}(\vec{k})$ generated by the different $1s$ atomic orbitals (eqn (3.7)). The n crystal orbitals of H_n are given by the following formula:

$$CO(\vec{k}) = BO_{1s}(\vec{k}) = \frac{1}{\sqrt{n}} \sum_{m=-n'+1}^{n'} \exp(i2\pi m k_a) 1s_m \qquad (3.15)$$

with $k_a = 0, \pm\frac{1}{n}, \pm\frac{2}{n}, \ldots, \pm\frac{n'-1}{n}, \frac{1}{2}$ and where the $1s_m$ orbital is centred on the hydrogen atom of cell M_m. The different crystal orbitals only differ by their phase factors $\exp(i2\pi m k_a)$. However, for a given m this factor is only real for the Γ ($k_a = 0$) and X ($k_a = \frac{1}{2}$) points. We will start by considering the orbitals for these two points because the physical content of real orbitals is easier to grasp than that of complex ones.

3.2.1 Representation of the CO(Γ) and CO(X) functions

The phase factor at the Γ point is 1 for every value of m. The CO(Γ) orbital may thus be obtained by copying the $1s_0$ orbital of the reference cell on every cell. At the X point the factor $\exp(i2\pi m k_a)$ is $(-1)^m$. Thus the orbital coefficients in two adjacent cells have opposite signs. The CO(X) orbital exhibits alternating positive and negative $1s$ orbitals as schematically shown in Fig. 3.6. The CO(Γ) orbital corresponds to the completely bonding orbital whereas the CO(X) orbital corresponds to the completely antibonding orbital. Since these are the two extreme situations, the other orbitals $CO(\vec{k})$ $\left(k_a \neq 0, \frac{1}{2}\right)$ are necessarily associated with intermediate energies between $E(\Gamma)$ and $E(X)$.

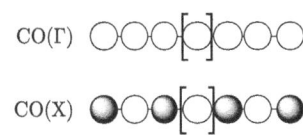

Fig. 3.6
Orbitals of the H_n chain at the Γ and X points.

3.2.2 Energy of the crystal orbitals in the Hückel approach

In this case, the energy $E(\vec{k})$ associated with the crystal orbital CO($\vec{k}$) coincides with $BO_{1s}(\vec{k})$) and is given by:

$$E(\vec{k}) = \frac{\langle BO_{1s}(\vec{k})|\hat{h}|BO_{1s}(\vec{k})\rangle}{\langle BO_{1s}(\vec{k})|BO_{1s}(\vec{k})\rangle} = \frac{h_{1s,1s}(\vec{k})}{S_{1s,1s}(\vec{k})} \quad (3.16)$$

We will use the Hückel approach (see Section 2.1.5 of Chapter 2) to make an estimation of the energy $E(\vec{k})$.

Direct calculation of the energies

Let us start estimating the numerator of eqn (3.16) by developing the Bloch orbital $BO_{1s}(\vec{k})$ at the left of the scalar product using eqn (3.15):

$$h_{1s,1s}(\vec{k}) = \sum_{m=-n'+1}^{n'} \left[\left\langle \frac{\exp(i2\pi m k_a)}{\sqrt{n}} 1s_m \middle| \hat{h} \middle| BO_{1s}(\vec{k}) \right\rangle \right] \quad (3.17)$$

We will denote the term between square brackets A(m). This represents the interaction between the function $\frac{\exp(i2\pi m k_a)}{\sqrt{n}} 1s_m$ located in cell M_m and the Bloch orbital $BO_{1s}(\vec{k})$. We will first show that this term is independent of m. Consider,

$$A(m) = \sum_{m'=-n'+1}^{n'} \frac{\exp(i2\pi(-m+m')k_a)}{n} \langle 1s_m|\hat{h}|1s_{m'}\rangle$$

Using eqn (2.10) and the Hückel approach, this term becomes:

$$A(m) = \sum_{m'=-n'+1}^{n'} \frac{\exp(i2\pi(-m+m')k_a)}{n} (\alpha \delta_{m',m}$$

$$+ \beta(\delta_{m',m+1} + \delta_{m',m-1}))$$

$$= \frac{1}{n}(\alpha + \beta(\exp(i2\pi k_a) + \exp(-i2\pi k_a)))$$

$$A(m) = \frac{1}{n}(\alpha + 2\beta \cos(2\pi k_a)) \quad (3.18)$$

Consequently, the interaction term A(m) is independent of m, i.e. on the cell M_m. This result simply originates from the fact that every cell plays an equivalent role in a periodic system. Let us emphasise that this result is independent of the method used to estimate the energies. As a result, since m takes n different values in the summation of eqn (3.17), the interaction term $h_{1s,1s}(\vec{k})$ is given by n times the A(m) value:

$$h_{1s,1s}(\vec{k}) = n \left(\left\langle \frac{\exp(i2\pi m k_a)}{\sqrt{n}} 1s_m | \hat{h} | BO_{1s}(\vec{k}) \right\rangle \right) = \alpha + 2\beta \cos(2\pi k_a)$$
(3.19)

In addition, the value of the $S_{1s,1s}(\vec{k})$ overlap is 1, since the Bloch orbitals have been defined as linear combinations of normalised orbitals using a Hückel approach. The allowed energies (eqn (3.16)) are given by:

$$E(\vec{k}) = \alpha + 2\beta \cos(2\pi k_a) \quad (3.20)$$

Interaction terms obtained through the Hückel approach

We will now use the results of the previous section *and the Hückel approach* to reach some general results related to the calculation of the overlap, $S_{jj'}(\vec{k})$ and the interaction term $h_{jj'}(\vec{k})$.

It is easy to verify that the basis of Bloch orbitals in the Hückel approach is orthonormalised whatever the nature of the ϕ_j and $\phi_{j'}$

$$S_{jj'}(\vec{k}) = \left\langle BO_j(\vec{k}) | BO_{j'}(\vec{k}) \right\rangle = \delta_{jj'} \quad (3.21)$$

provided that the ϕ_j and $\phi_{j'}$ functions are orthonormal.

In addition, it is always possible to decompose the interaction term $h_{jj'}(\vec{k})$ (see eqn (3.13)) through development of the Bloch orbital $BO_j(\vec{k})$ according to expression (3.7):

$$h_{jj'}(\vec{k}) = \sum_{m=-n'+1}^{n'} \left[\left\langle \frac{\exp(i2\pi m k_a)}{\sqrt{n}} (\phi_j)_m | \hat{h} | BO_{j'}(\vec{k}) \right\rangle \right] \quad (3.22)$$

As a consequence of the equivalence of the different orbitals $(\phi_j)_m$ ($m = 0, \pm 1, \ldots, \pm(n'-1), n'$) ($j = 1, \ldots, N_0$) imposed by the Born–von Karman boundary conditions, the term in the sum of equation (3.22) is independent of m for exactly the same reasons, making A(m) independent of m (see eqn (3.18)). The interaction term $h_{jj'}(\vec{k})$ can be estimated by multiplying by n the interaction term $\left\langle \frac{(\phi_j)_0}{\sqrt{n}} | \hat{h} | BO_{j'}(\vec{k}) \right\rangle$ involving orbital $(\phi_j)_0$ of the reference cell:

$$h_{jj'}(\vec{k}) = n \left\langle \frac{(\phi_j)_0}{\sqrt{n}} | \hat{h} | BO_{j'}(\vec{k}) \right\rangle \quad (3.23)$$

The last term is very easy to evaluate in the Hückel approach because the $(\phi_j)_0$ orbital only interacts with itself and with the orbitals of adjacent atoms. This interaction term may be estimated with the help of a graphical scheme representing the interaction between the function $\frac{1}{\sqrt{n}}(\phi_j)_0$ and the Bloch orbital $BO_{j'}(\vec{k})$.

As an example, we will consider the interaction term $h_{1s,1s}(\vec{k})$ of the H_n system through the use of eqn (3.23). In order to do this we will represent the non-zero interactions between the $\frac{1}{\sqrt{n}} 1s_0$ and the Bloch orbital $BO_{1s}(\vec{k})$ in the Hückel approach (see Fig. 3.7). Three interaction terms are non-zero. They correspond to the 'self-interaction' (α term) of the $1s_0$ orbital and to the interaction

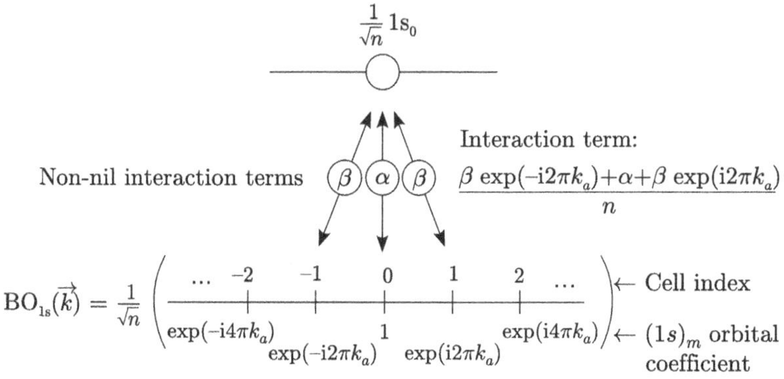

Fig. 3.7
Schematic representation of the orbital interaction between $\frac{1}{\sqrt{n}}1s_0$ and the Bloch orbital $BO_{1s}(\vec{k})$ in the Hückel approach.

(β term) of the $1s_0$ orbital with the adjacent orbitals, $\frac{1}{\sqrt{n}}\exp(-i2\pi k_a)1s_{-1}$ and $\frac{1}{\sqrt{n}}\exp(+i2\pi k_a)1s_1$:

$$\left\langle \frac{1s_0}{\sqrt{n}} \middle| \hat{h} \middle| BO_{1s}(\vec{k}) \right\rangle = \frac{1}{n}\{\alpha + \beta(\exp(i2\pi k_a) + \exp(-i2\pi k_a))\} \quad (3.24)$$

Using eqn (3.23) we can now obtain the $h_{1s,1s}(\vec{k})$ interaction term:

$$h_{1s,1s}(\vec{k}) = n\left\langle \frac{1s_0}{\sqrt{n}} \middle| \hat{h} \middle| BO_{1s}(\vec{k}) \right\rangle = \alpha + 2\beta\cos(2\pi k_a) \quad (3.25)$$

The approach based on the use of eqn (3.23) thus allows a fast and simple derivation of the interaction terms in the Hückel approach.[4] Later on, we will use this method to evaluate the different interaction terms present in the secular determinants in order to avoid tedious calculations.

[4] In the extended Hückel approach, besides the interactions between nearest neighbours, those implicating the second, third...nearest neighbours are also taken into account, up to a given point beyond which they are neglected.

3.2.3 Band structure

Let us now represent the different allowed energies $E(\vec{k})$ (see eqn (3.20)) as a function of k_a (Fig. 3.8). *Let us recall that k_a varies discretely and that only the energies for which k_a is $\frac{\ell}{n}$ ($\ell \in\ ]-n', n']$) are allowed. However, since the number of cells n is very large, the allowed energy values are very close to each other. In practice we have an energy band with a width of 4β*. In addition, it must be noted that all energy levels are doubly degenerate except for $E(\Gamma)$ and $E(X)$. The degeneracy of the $BO(\vec{k})$ and $BO(-\vec{k})$ functions is a general property,[5] which originates from the fact that the k and $-k$ points are physically equivalent. This property provides some justification for the fact that we have defined a Brillouin zone as centred on the Γ point. Equation (3.8) clearly shows that the length of the Brillouin zone making possible the labelling of the different allowed functions must be equal to $\frac{2\pi}{a}$, which is the norm of the reciprocal vector $\vec{a}^*$. However, there is no compelling reason for choosing the interval $]-\frac{\pi}{a}, \frac{\pi}{a}]$ instead of $]0, \frac{2\pi}{a}]$ for instance. We chose the interval $]-\frac{\pi}{a}, \frac{\pi}{a}]$ because it makes transparent (and useful) the energetic equivalence of the states associated with the k and $-k$ points, which are symmetrically placed with respect to Γ.[6] Because of the degeneracy associated with the k and $-k$ points, ($E(\vec{k}) = E(-\vec{k})$), from now on we will only represent half of

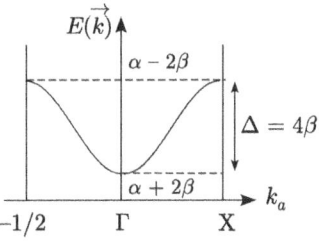

Fig. 3.8
Band structure for the H_n chain. Note that the slope is nil at both the Γ and X points.

[5] In the physics literature, this property is known as 'time reversal'.

[6] The use of a Brillouin zone centred at Γ will be very useful for the study of 2D and 3D systems, where the exploitation of the symmetry may be very helpful.

Model chain H_n

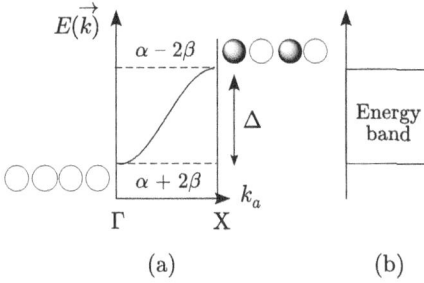

Fig. 3.9
(a) Band structure of the H_n chain.
(b) Schematic representation of the allowed energy levels.

the band structure. For a 1D system we will only represent the energy levels corresponding to k_a values in the interval $[0, \frac{1}{2}]$ (Fig. 3.9). Even if we use such a representation it should not be forgotten that all energy levels $E(\vec{k})$, except those at Γ and X, are doubly degenerate.

3.2.4 Basis for an energy level $E(\pm \vec{k})$

Basis of real functions $\{\sigma(\pm \vec{k}), \delta(\pm \vec{k})\}$

Since the $BO(\vec{k})$ and $BO(-\vec{k})$ functions are associated with the same energy and are complex conjugate, it is possible to build a basis of *real* functions, $\{\sigma(\pm \vec{k}), \delta(\pm \vec{k})\}$, associated with the energy level $E(\pm \vec{k})$

$$\sigma(\pm \vec{k}) = \frac{BO(\vec{k}) + BO(-\vec{k})}{\sqrt{2}} \quad \text{and} \quad \delta(\pm \vec{k}) = \frac{BO(\vec{k}) - BO(-\vec{k})}{i\sqrt{2}} \quad (3.26)$$

The sign $\pm$ before $\vec{k}$ in eqn (3.26) is a reminder that we have combined two Bloch orbitals associated with $\vec{k}$ and $-\vec{k}$ in such a way that the orbitals $\{\sigma(\pm \vec{k}), \delta(\pm \vec{k})\}$ are not individually adapted to the translation symmetry. For the H_n chain these functions are given by eqn (3.27):

$$\sigma(\pm \vec{k}) = \sqrt{\frac{2}{n}} \sum_{m=-n'+1}^{n'} \cos(2\pi m k_a) 1s_m$$

$$\text{and} \quad \delta(\pm \vec{k}) = \sqrt{\frac{2}{n}} \sum_{m=-n'+1}^{n'} \sin(2\pi m k_a) 1s_m \quad (3.27)$$

and are schematically represented in Fig. 3.10.

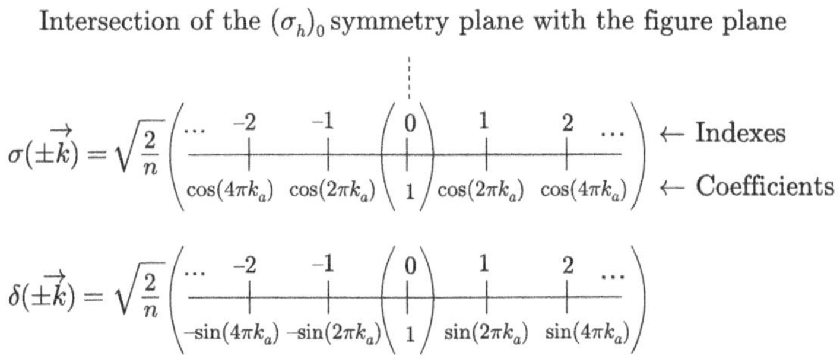

Intersection of the $(\sigma_h)_0$ symmetry plane with the figure plane

$$\sigma(\pm\vec{k}) = \sqrt{\frac{2}{n}} \begin{pmatrix} \cdots & -2 & -1 & 0 & 1 & 2 & \cdots \\ & \cos(4\pi k_a) & \cos(2\pi k_a) & 1 & \cos(2\pi k_a) & \cos(4\pi k_a) & \end{pmatrix} \begin{array}{l} \leftarrow \text{Indexes} \\ \leftarrow \text{Coefficients} \end{array}$$

$$\delta(\pm\vec{k}) = \sqrt{\frac{2}{n}} \begin{pmatrix} \cdots & -2 & -1 & 0 & 1 & 2 & \cdots \\ & -\sin(4\pi k_a) & -\sin(2\pi k_a) & 1 & \sin(2\pi k_a) & \sin(4\pi k_a) & \end{pmatrix}$$

Fig. 3.10
Schematic representation of the basis $\sigma(\pm \vec{k})$ and $\delta(\pm \vec{k})$.

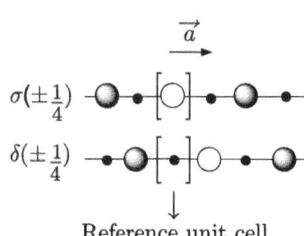

Fig. 3.11
Schematic representation of the real functions $\sigma(\pm 1/4)$ and $\delta(\pm 1/4)$ associated with energy $E(\pm 1/4)$. A dot is used to represent a nil coefficient for a given position.

These two functions have the peculiarity of being symmetric and antisymmetric, respectively, with respect to the symmetry plane $(\sigma_h)_0$, which is perpendicular to the chain axis and cuts the chain at the reference cell atom. Thus, we have built a basis of real functions adapted to the $(\sigma_h)_0$ symmetry plane and associated with the energy $E(\pm \vec{k})$. Let us emphasise that these functions are not individually well adapted to the translational symmetry. Thus, for instance, a basis for the level $E(k_a = \pm \tfrac{1}{4})$ is given by eqn (3.28) and is graphically represented in Fig. 3.11.

$$\sigma(\pm \tfrac{1}{4}) = \sqrt{\frac{2}{n}} \sum_{m=-n'+1}^{n'} \cos(\frac{\pi m}{2}) 1s_m$$

$$\text{and } \delta(\pm \tfrac{1}{4}) = \sqrt{\frac{2}{n}} \sum_{m=-n'+1}^{n'} \sin(\frac{\pi m}{2}) 1s_m \qquad (3.28)$$

in the Hückel approach, these crystal orbitals are non-bonding and are associated with an energy $E(\pm 1/4)$ equal to α, which is the energy of an isolated $1s$ orbital.

Basis of real functions $\{S(\pm \vec{k}), A(\pm \vec{k})\}$

We will now explore a basis for the energy level $E(\pm \vec{k})$, $\{S(\pm \vec{k}), A(\pm \vec{k})\}$, made of real functions that are symmetric and antisymmetric, respectively, with respect to the symmetry plane $(\sigma_h)_{0/1}$. This plane is perpendicular to the chain axis and lies equidistant from the atoms in the cells labelled 0 and 1 (see Fig. 3.12). These two functions are given by eqn (3.29) and are schematically represented in Fig. 3.12.

$$S(\pm \vec{k}) = \frac{1}{\sqrt{2}} \left(\exp(-i\pi k_a) BO(\vec{k}) + \exp(i\pi k_a) BO(-\vec{k}) \right)$$

$$A(\pm \vec{k}) = \frac{1}{i\sqrt{2}} \left(\exp(-i\pi k_a) BO(\vec{k}) - \exp(i\pi k_a) BO(-\vec{k}) \right) \qquad (3.29)$$

As an example, we show in Fig. 3.13 a basis of crystal orbitals, namely $\{S(\pm \tfrac{1}{4}), A(\pm \tfrac{1}{4})\}$, well adapted to the symmetry plane $(\sigma_h)_{0/1}$ obtained by combination of the Bloch orbitals $BO_{1s}(k_a = \tfrac{1}{4})$ and $BO_{1s}(k_a = -\tfrac{1}{4})$.

As will be repeatedly shown throughout this book, the choice of a crystal orbital basis well adapted to the problem at hand can enormously simplify the

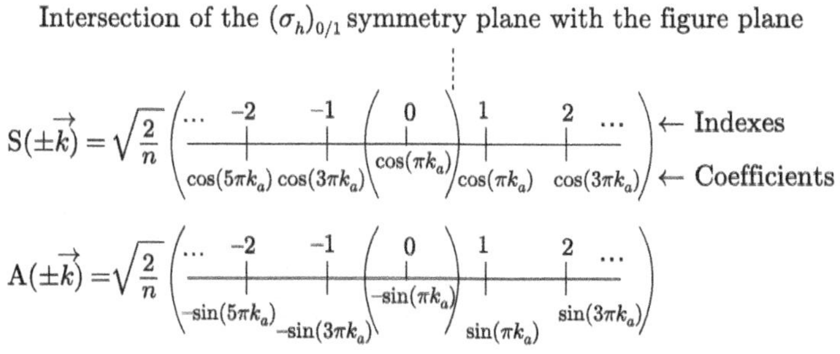

Intersection of the $(\sigma_h)_{0/1}$ symmetry plane with the figure plane

$$S(\pm \vec{k}) = \sqrt{\frac{2}{n}} \left(\begin{array}{ccc|ccc} \cdots & -2 & -1 & 0 & 1 & 2 & \cdots \\ & \cos(5\pi k_a) & \cos(3\pi k_a) & \cos(\pi k_a) & \cos(\pi k_a) & \cos(3\pi k_a) & \end{array} \right) \begin{array}{l} \leftarrow \text{Indexes} \\ \leftarrow \text{Coefficients} \end{array}$$

$$A(\pm \vec{k}) = \sqrt{\frac{2}{n}} \left(\begin{array}{ccc|ccc} \cdots & -2 & -1 & 0 & 1 & 2 & \cdots \\ & -\sin(5\pi k_a) & -\sin(3\pi k_a) & -\sin(\pi k_a) & \sin(\pi k_a) & \sin(3\pi k_a) & \end{array} \right)$$

Fig. 3.12
Schematic representation of the basis $\{S(\pm \vec{k}), A(\pm \vec{k})\}$ made of functions well adapted to the symmetry plane $(\sigma_h)_{0/1}$.

discussion. Since every energy level $E(\vec{k})(k_a \neq 0, \frac{1}{2})$ is doubly degenerate, we can choose $\{BO(\vec{k}), BO(-\vec{k})\}$, $\{\sigma(\pm\vec{k}), \delta(\pm\vec{k})\}$, $\{S(\pm\vec{k}), A(\pm\vec{k})\}$ or any other linear combination that we may find a convenient basis to describe the level.

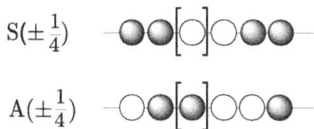

3.2.5 Fermi level of the H_n chain

We will now consider how the n electrons of the H_n chain must be distributed among the n crystal orbitals of the system. The energy band of a 1D system always contains two non-degenerate levels ($k_a = 0$ and $k_a = \frac{1}{2}$) and $(n'-1)$ doubly degenerate levels. *Thus, the band is full when it is filled with $2n$ electrons. This means that every unit cell contributes two electrons to this band.* For the H_n chain, every unit cell can contribute only one electron, so that only half the possible energy levels can be filled at $T = 0$ K. Taking into account that the different k_a allowed values are uniformly distributed in the interval $]-\frac{1}{2},\frac{1}{2}]$, it is clear from the band structure in Fig. 3.8 that all crystal orbitals with k_a values between $-\frac{1}{4}$ and $\frac{1}{4}$ are occupied. This situation may be represented as in Fig. 3.14.

We can now draw the band structure of the H_n chain including the position of the Fermi level (Fig. 3.15). Two different ways of representing the nature of the levels at the Fermi level have been already shown in the previous section (see Figs 3.11 and 3.13).

On the basis of the band structure of Fig. 3.15 we can predict that the H_n chain should be a metal (i.e. a conductor) since there is no gap at the Fermi level.

Fig. 3.13
Schematic representation of the real functions $\{S(\pm\frac{1}{4}), A(\pm\frac{1}{4})\}$ associated with the energy $E(\pm\frac{1}{4})$.

In the framework of the monoelectronic approximation,* when every cell provides m electrons and the different bands do not overlap, $m/2$ bands are filled in the ground state at $T = 0$ K. If some bands overlap, the equivalent of $m/2$ bands are filled (see Exercises (3.7)–(3.9)).

---

*As will be seen in a later chapter, the explicit consideration of the electron–electron repulsions may lead to a different electronic description. Throughout this book we will consider the monoelectronic approximation valid and thus assume the progressive filling of the energy levels so that as many levels as possible are doubly filled. In Chapter 12 we will consider when this approach is expected to break down.

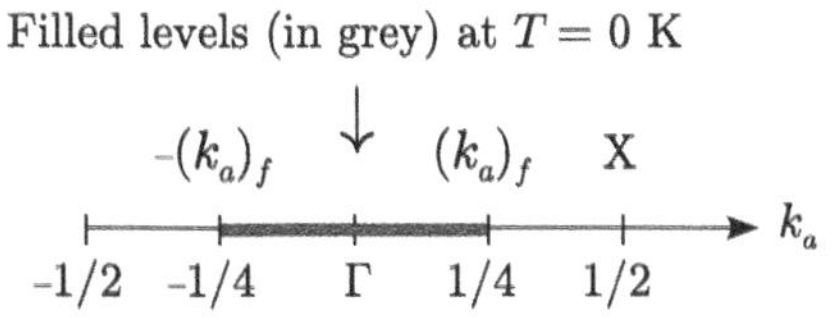

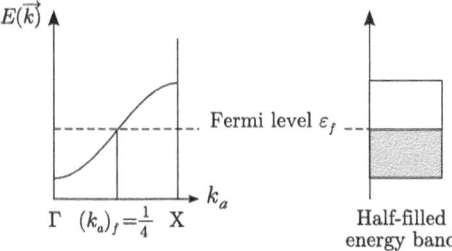

Fig. 3.14
Schematic representation of the occupation of the different levels of the H_n chain. $(k_a)_f$ and $-(k_a)_f$ characterise the crystal orbitals at the Fermi level.

Fig. 3.15
Band structure of the H_n chain. Note that only the states characterised by positive (and zero) values of k_a are shown in the left-hand diagram whereas all states are represented on the right.

3.3 Electronic structure of the dimerised model chain $(H_2)_{n'}$

In order to make some progress we will now consider the H_n system in which an alternation of short and long H–H distances has been introduced (Fig. 3.16). The unit cell of this chain contains two hydrogen atoms arbitrarily noted as ℓ and r (for 'left' and 'right'). Since the chain is dimerised, we will refer to this system as $(H_2)_{n'}$ in order to clearly distinguish it from the regular H_n chain (Fig. 3.5).[7]

Every cell can be electronically described by two $1s$ orbitals centred on the H_l and H_r atoms. These orbitals will be referred to as $1s_\ell$ and $1s_r$. The crystal orbitals of this system result from the combination of the Bloch orbitals $BO_\ell(\vec{k})$ and $BO_r(\vec{k})$ generated by the $1s_\ell$ and $1s_r$ orbitals, respectively. These Bloch orbitals are given by eqn (3.30), where $n'' = n'/2$.

[7] In order to be able to compare the H_n and $(H_2)_{n'}$ systems we will assume that H_n possesses n cells whereas $(H_2)_{n'}$ contains n' ($n' = n/2$) cells of two atoms. In this way the two systems contain exactly the same number of atoms.

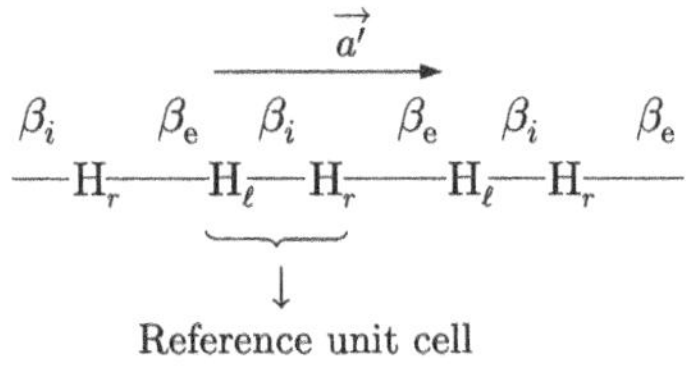

Fig. 3.16
Dimerised system for which $\beta_i < \beta_e < 0$. The hydrogen atoms of every cell have been indexed as ℓ and r.

$$BO_\ell(\vec{k}) = \frac{1}{\sqrt{n'}} \sum_{m=-n''+1}^{n''} \exp(i2\pi m k_{a'})(1s_\ell)_m$$

$$BO_r(\vec{k}) = \frac{1}{\sqrt{n'}} \sum_{m=-n''+1}^{n''} \exp(i2\pi m k_{a'})(1s_r)_m \quad (3.30)$$

if $\vec{k} = k_{a'}\vec{a}'^*$, $\vec{a}'^* = \frac{2\pi}{a'^2}\vec{a}'$, and $k_{a'} = 0, \pm\frac{1}{n'}, \pm\frac{2}{n'}, \ldots, \pm\frac{n''-1}{n'}, \frac{1}{2}$.

We will now work out the band structure of this system following two different approaches:

- firstly, a formal approach by solving the associated secular equations
- secondly, a more qualitative approach leading to the band structure without explicitly solving the secular equations.

3.3.1 Formal determination of the band structure

Energies

Let us impose that the secular determinant of the system is nil, i.e.:

$$\begin{vmatrix} h_{\ell\ell}(\vec{k}) - E(\vec{k})S_{\ell\ell}(\vec{k}) & h_{\ell r}(\vec{k}) - E(\vec{k})S_{\ell r}(\vec{k}) \\ h_{r\ell}(\vec{k}) - E(\vec{k})S_{r\ell}(\vec{k}) & S_{\ell\ell}(\vec{k}) - E(\vec{k})S_{\ell\ell}(\vec{k}) \end{vmatrix} \quad (3.31)$$

where

$$h_{\ell\ell}(\vec{k}) = \left\langle BO_\ell(\vec{k})|\hat{h}|BO_\ell(\vec{k})\right\rangle = \left\langle BO_r(\vec{k})|\hat{h}|BO_r(\vec{k})\right\rangle = h_{rr}(\vec{k})$$

$$h_{\ell r}(\vec{k}) = \left\langle BO_\ell(\vec{k})|\hat{h}|BO_r(\vec{k})\right\rangle = \left(h_{r\ell}(\vec{k})\right)^*$$

$$S_{jj'}(\vec{k}) = \left\langle BO_j(\vec{k})|BO_{j'}(\vec{k})\right\rangle$$

In a dimerised system, it is convenient to distinguish two different types of interactions between adjacent orbitals (Fig. 3.16): β_i, which is associated with the interaction between $1s$ orbitals of adjacent atoms *within a unit cell*, and β_e, which is associated with the interaction between $1s$ orbitals of adjacent atoms *of different unit cells*. Following Fig. 3.16, we will consider that β_i is more negative than β_e because the short H–H bonds have been chosen to be those within the unit cells ($\beta_i < \beta_e < 0$). The $h_{jj'}(\vec{k})$ may be obtained either through direct calculation or by using the diagram associated with the interaction term $< \frac{1}{\sqrt{n'}}(1s_j)_0|\hat{h}|BO_{j'}(\vec{k}) >$ (see Section 3.2.2). Thus, the term $h_{\ell\ell}(\vec{k})$ may be estimated by analysing the interaction between $\frac{1}{\sqrt{n'}}(1s_\ell)_0$ and the Bloch orbital $BO_\ell(\vec{k})$. Let us follow the later approach (see Fig. 3.17).

It follows from Fig. 3.17 that $h_{\ell\ell}(\vec{k}) = n'(\frac{\alpha}{n'}) = \alpha$. This result is not surprising since the Bloch orbital $BO_\ell(\vec{k})$ is non-bonding in the Hückel approach.

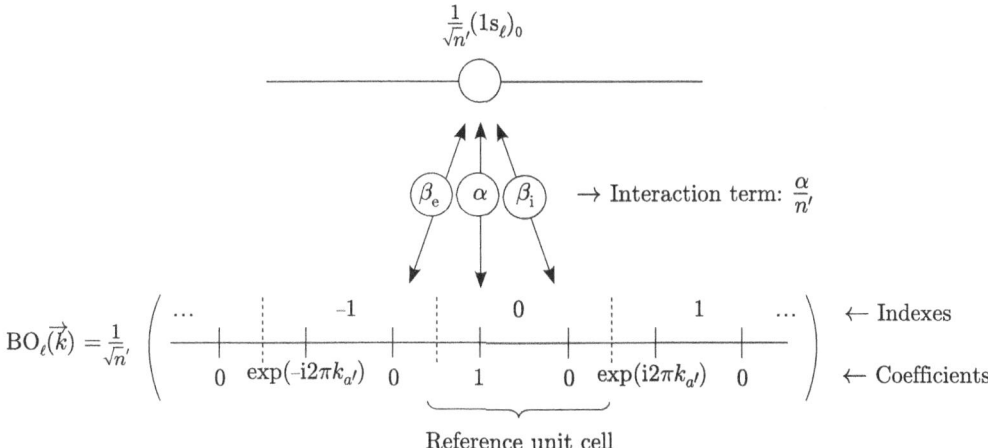

Fig. 3.17
Interaction between $\frac{1}{\sqrt{n'}}(1s_\ell)_0$ and $BO_\ell(\vec{k})$. Only the non-zero interactions according to the Hückel approach are represented in this scheme.

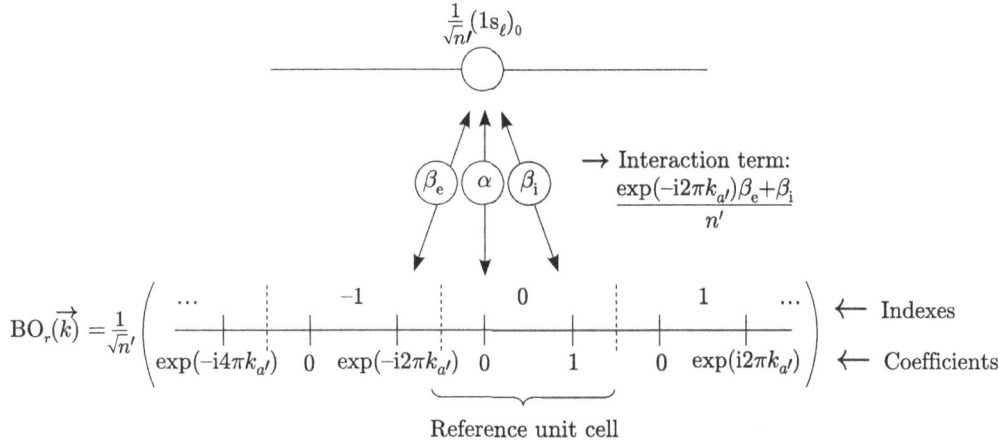

Fig. 3.18
Interaction between $\frac{1}{\sqrt{n'}}(1s_\ell)_0$ and $BO_r(\vec{k})$. Only the non-zero interactions according to the Hückel approach are represented in this scheme.

The interaction term $h_{\ell r}(\vec{k})$ can be obtained as shown in Fig. 3.18. Thus the $h_{jj'}(\vec{k})$ terms can be evaluated as:

$$h_{\ell\ell}(\vec{k}) = h_{rr}(\vec{k}) = \alpha$$

$$h_{\ell r}(\vec{k}) = \beta_i + \beta_e \exp(-i2\pi k_{a'}) = \left(h_{r\ell}(\vec{k})\right)^* \qquad (3.32)$$

Since $S_{jj'}(\vec{k}) = \delta_{jj'}$, development of the secular determinant leads to the equation:

$$\left(h_{\ell\ell}(\vec{k}) - E(\vec{k})\right)^2 - |h_{\ell r}(\vec{k})|^2 = 0$$

$$\left(\alpha - E(\vec{k})\right)^2 - \beta_i^2 - \beta_e^2 - 2\beta_i\beta_e\cos(2\pi k_{a'}) = 0$$

The allowed energies, denoted $E^+(\vec{k})$ and $E^-(\vec{k})$, are given by

$$E^{\pm}(\vec{k}) = \alpha \pm \beta_i\sqrt{1 + \frac{\beta_e^2}{\beta_i^2} + 2\frac{\beta_e}{\beta_i}\cos(2\pi k_{a'})} \quad (\beta_i \text{ is negative}) \qquad (3.33)$$

Finally, the band structure of the $(H_2)_{n'}$ chain may be constructed and is given in Fig. 3.19. The two bands have the same width and do not overlap, except if β_i is exactly the same as β_e. The energy gap ($\Delta = 2(\beta_e - \beta_i)$) becomes larger with the extent of the dimerisation, i.e. when the difference between β_e and β_i increases.

Let us now consider the Fermi level location. Every cell provides two electrons, which thus allows the filling of one band at $T = 0$ K. The Fermi level occurs at the top of the lower band of Fig. 3.19.[8] Consequently, the dimerised chain should not be a metal, in contrast with the situation of the non-dimerised H_n chain (Fig. 3.5). We will come back to this point in Section 3.4.

[8] We emphasise that our use of the Fermi level concept is a purely 'chemical' one, i.e. just to indicate the level occupation at $T = 0$ K.

Crystal orbitals

It is now interesting to determine the crystal orbitals $CO^+(\vec{k})$ and $CO^-(\vec{k})$ associated with the $E^+(\vec{k})$ and $E^-(\vec{k})$ energies, respectively. These crystal orbitals will result from the combination of Bloch orbitals $BO_\ell(\vec{k})$ and $BO_r(\vec{k})$ with coefficients $c_\ell^{\pm}(\vec{k})$ and $c_r^{\pm}(\vec{k})$.

$$CO^{\pm}(\vec{k}) = c_\ell^{\pm}(\vec{k})BO_\ell(\vec{k}) + c_r^{\pm}(\vec{k})BO_r(\vec{k}) \qquad (3.34)$$

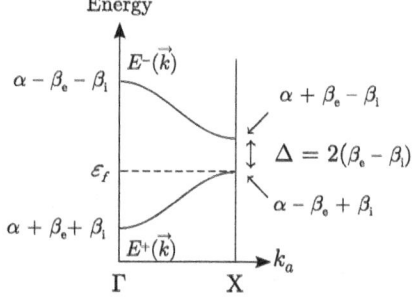

Fig. 3.19
Band structure for the dimerised $(H_2)_{n'}$, i.e. ($\beta_i < \beta_e < 0$).

Determination of these coefficients needs the solution of the corresponding secular equation (3.12). Since this system of equations is associated with a nil determinant, we need only one equation to obtain the eigenvectors. Let us take the first equation:

$$\left(\alpha - E^{\pm}(\vec{k})\right) c_{\ell}^{\pm}(\vec{k}) + (\beta_i + \beta_e \exp(-i2\pi k_{a'})) c_r^{\pm}(\vec{k}) = 0 \qquad (3.35)$$

The coefficients $c_{\ell}^{\pm}(\vec{k})$ and $c_r^{\pm}(\vec{k})$ are related through equation (3.36). If we additionally consider the normalisation condition of the crystal orbitals according to the Hückel approach:

$$\frac{c_{\ell}^{\pm}(\vec{k})}{c_r^{\pm}(\vec{k})} = \frac{\beta_i + \beta_e \exp(-i2\pi k_{a'})}{-\alpha + E^{\pm}(\vec{k})} \quad \text{with} \quad |c_{\ell}^{\pm}(\vec{k})|^2 + |c_r^{\pm}(\vec{k})|^2 = 1 \qquad (3.36)$$

and we replace $E^{\pm}(\vec{k})$ by its expression (eqn (3.33)), the ratio of coefficients:

$$\frac{c_{\ell}^{\pm}(\vec{k})}{c_r^{\pm}(\vec{k})} = \frac{\beta_i + \beta_e \exp(-i2\pi k_{a'})}{\pm \beta_i \sqrt{1 + \frac{\beta_e^2}{\beta_i^2} + 2\frac{\beta_e}{\beta_i} \cos(2\pi k_{a'})}} \qquad (3.37)$$

may be obtained. For the Γ and X points this ratio may be simplified to:

$$\frac{c_{\ell}^{\pm}(\Gamma)}{c_r^{\pm}(\Gamma)} = \pm 1 \quad \text{and} \quad \frac{c_{\ell}^{\pm}(X)}{c_r^{\pm}(X)} = \pm 1 \qquad (3.38)$$

Thus we can finally construct the band structure of the $(H_2)_{n'}$ chain (see Fig. 3.19) with the nature of the orbitals at the centre and border of the Brillouin zone as well as the Fermi level indicated (Fig. 3.20).

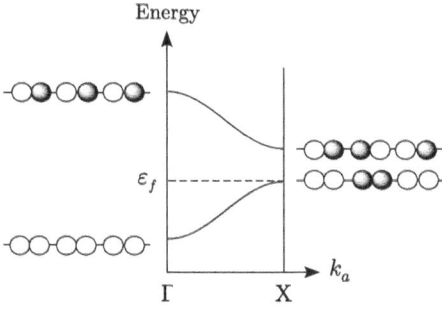

Fig. 3.20
Band structure for the dimerised $(H_2)_{n'}$ chain.

3.3.2 Qualitative determination of the band structure

We will now derive the band structure and crystal orbitals at the Γ and X points for the $(H_2)_{n'}$ chain following a qualitative approach. The basis for this approach is the simple observation that *the interaction between two degenerate orbitals ϕ_1 and ϕ_2 with energy E leads to the formation of a bonding orbital and an antibonding orbital, which are lower and higher in energy than the initial orbitals, respectively.* This is for instance the case when two hydrogen $1s$ orbitals interact, leading to the molecular orbitals of an hydrogen molecule

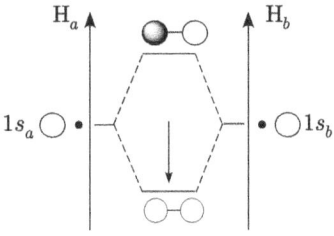

Fig. 3.21
Interaction diagram showing how two $1s$ orbitals lead to the molecular orbitals of the H_2 molecule.

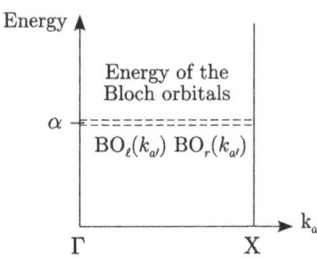

Fig. 3.22
Energy of the Bloch orbitals $BO_\ell(\vec{k})$ and $BO_r(\vec{k})$ as a function of $k_{a'}$.

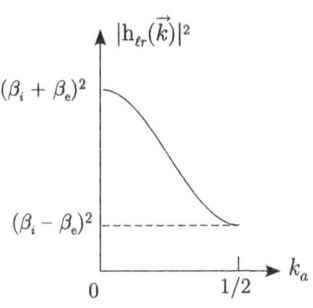

Fig. 3.23
Change of the interaction term $|h_{\ell r}(\vec{k})|^2$ as a function of $k_{a'}$.

Fig. 3.24
(a) Schematic interaction diagram between the Bloch orbitals $BO_\ell(\vec{k})$ and $BO_r(\vec{k})$. (b) Band structure obtained through the interaction of the $BO_\ell(\vec{k})$ and $BO_r(\vec{k})$ Bloch orbitals. The strength of the interaction at different points is indicated by the arrows.

(Fig. 3.21). The stabilisation of the bonding orbital with respect to E is larger as the interaction term h_{ab} ($< 1s_a|\hat{h}|1s_b >$) increases.

We can now use this result in order to obtain the crystal orbitals of the $(H_2)_{n'}$ chain. We will begin by looking at how the energy of the Bloch orbitals $BO_\ell(\vec{k})$ and $BO_r(\vec{k})$ depends on $k_{a'}$. These orbitals are centred on alternate series of hydrogen atoms so that, in the Hückel approach, they are non-bonding. Consequently, their energies are independent of $k_{a'}$ and have the value α (Fig. 3.22).

Now we can proceed to combine the two Bloch orbitals for every k point. In order to do so the value of the interaction term $h_{\ell r}(\vec{k})$ must be known. This term was evaluated in Section 3.3.1 using the simple diagrammatic approach and was shown to be $(\beta_i + \beta_e \exp(-i2\pi k_{a'}))$ (eqn (3.32)). In fact, what we really need is the amplitude of this term, which we may estimate by calculating the square of the modulus, $|h_{\ell r}(\vec{k})|^2$, which amounts to $[\beta_i^2 + \beta_e^2 + 2\beta_i\beta_e \cos(2\pi k_{a'})]$. The variation of this quantity as a function of $k_{a'}$ in the interval $[0, \frac{1}{2}]$ is reported in Fig. 3.23. Figure 3.24a is an interaction diagram of the degenerate Bloch orbitals $BO_\ell(\vec{k})$ and $BO_r(\vec{k})$ leading to the formation of the crystal orbitals $CO^+(\vec{k})$ and $CO^-(\vec{k})$. It schematically shows that the interaction between the two Bloch orbitals decreases along the interval $[0, \frac{1}{2}]$. Thus, the $CO(\vec{k})$ crystal orbitals resulting from the interaction of Bloch orbitals $BO_\ell(\vec{k})$ and $BO_r(\vec{k})$ will be stabilised and destabilised with respect to α, respectively, as $k_{a'}$ decreases. The band structure of the $(H_2)_{n'}$ chain may thus be obtained by 'repelling' the two degenerate curves representing the energy of the Bloch orbitals, the separation increasing as the Γ point is approached (see Fig. 3.24b).

What remains to be done is to determine the crystal orbitals at the Γ and X points. Figure 3.25 is a representation of the interaction of the $BO_\ell(\vec{k})$ and $BO_r(\vec{k})$ Bloch orbitals at Γ and X. Since the two orbitals are degenerate (and thus, equivalent), the resulting crystal orbitals will just be the addition and the difference of the two orbitals, exactly as the molecular orbitals of the H_2 molecule are the addition and the difference of the two atomic orbitals (Fig. 3.21).

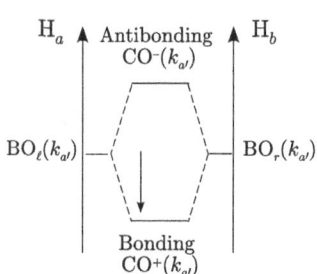

(a)

(b)

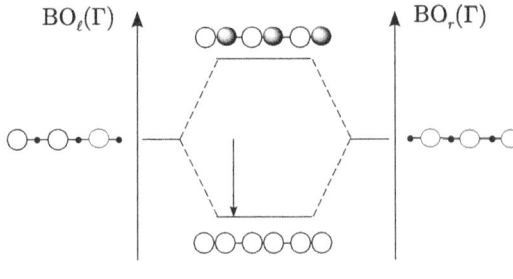

(a) Interaction at Γ

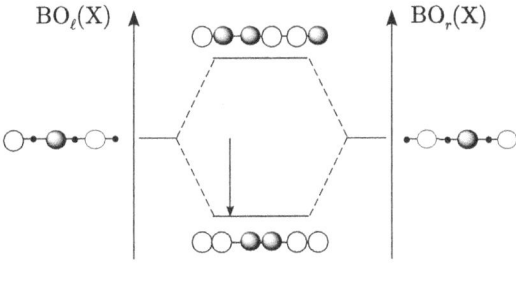

(b) Interaction at X

Fig. 3.25
Schematic representation of how the crystal orbitals at Γ and X may be determined.

Thus, through the use of only the amplitude of the interaction term between two Bloch orbitals, $BO_\ell(\vec{k})$ and $BO_r(\vec{k})$, which can be very simply determined in a diagrammatic way, we have been able to obtain qualitatively the band structure of the $(H_2)_{n'}$ chain without solving the secular equations. Throughout this book we will try, as much as possible, to use a similar qualitative approach in order to study the band structure of different materials.

3.4 Comparison of the regular H_n and dimerised $(H_2)_{n'}$ chains

3.4.1 Comparison of the band structures of the regular H_n chain generated by either a simple or a double unit cell

It is now appropriate to look in more details at the relationship between the band structures of the H_n system in its regular H_n (see Fig. 3.5) or dimerised $(H_2)_{n'}$ (see Fig. 3.16) forms. For this purpose we will start by studying the regular system generated in two different ways: first, using a simple repeat unit made of one hydrogen atom and a translation vector $\vec{a}$ (see representation 1 in Fig. 3.26); second, using a double repeat unit made of two hydrogen atoms and a translation vector $\vec{a}' = 2\vec{a}$. In the latter case we will arbitrarily distinguish the two equivalent hydrogen atoms of the double cell with indices ℓ and r, for 'left' and 'right', respectively (see representation 2 in Fig. 3.26). These two approaches will yield two different band structures describing the same system. We will then show that the two schemes are of course equivalent.

Following our treatment in Section 3.2 for the regular H_n chain (see eqn (3.15)), the crystal orbitals $CO_{(1)}(\vec{k})$ of the regular chain generated by the simple repeat unit are given by:

$$CO_{(1)}(\vec{k}) = \frac{1}{\sqrt{n}} \sum_{m=-n'+1}^{n'} \exp(i2\pi m k_a) 1s_m$$

$$\text{with } k_a \in \left]-\frac{1}{2}, \frac{1}{2}\right] \quad (3.39)$$

The index (1) in crystal orbital $CO_{(1)}(\vec{k})$ is a reminder of the fact that this crystal orbital is obtained in the first representation. A representation of this crystal orbital is given in Fig. 3.27. When the chain is generated by the double repeat unit, we will label the crystal orbitals with the index (2). In this second representation, the crystal orbitals result from the interaction of the Bloch orbitals $BO_{(2)}^{\ell}(\vec{k})$ and $BO_{(2)}^{r}(\vec{k})$ generated by the 1s orbitals centred on either the H_ℓ or H_r atoms (see Fig. 3.26b where $n'' = \frac{n}{4}$). These crystal orbitals will be denoted $CO_{(2)}^{\pm}(\vec{k})$.

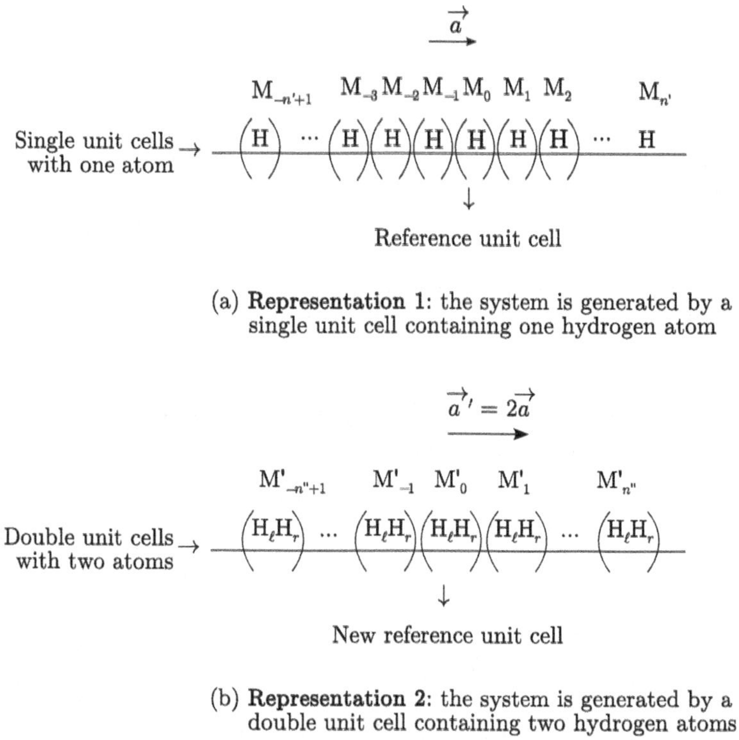

Fig. 3.26
Regular H_n chain generated by a repeat unit of one hydrogen atom (representation 1) or by a double repeat unit of two hydrogen atoms denoted H_ℓ and H_r (representation 2). $n' = n/2$ and $n'' = n'/2$.

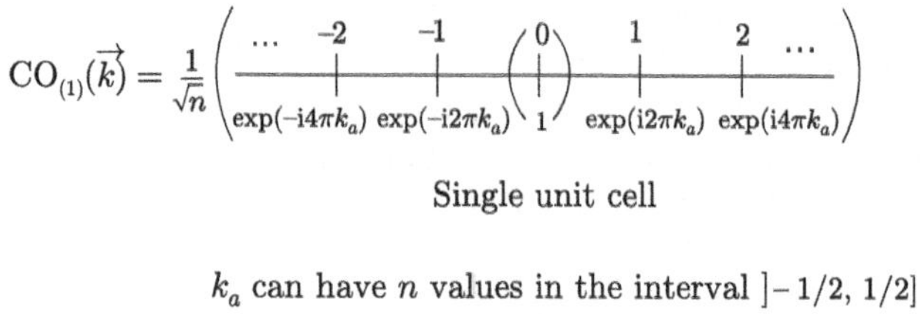

Fig. 3.27
Crystal orbital for the hydrogen chain generated in the first representation.

(a) **Representation 1**: the system is generated by a single unit cell containing one hydrogen atom

(b) **Representation 2**: the system is generated by a double unit cell containing two hydrogen atoms

$$CO_{(1)}(\vec{k}) = \frac{1}{\sqrt{n}} \begin{pmatrix} \cdots & -2 & -1 & 0 & 1 & 2 & \cdots \\ & \exp(-i4\pi k_a) & \exp(-i2\pi k_a) & 1 & \exp(i2\pi k_a) & \exp(i4\pi k_a) & \end{pmatrix}$$

Single unit cell

k_a can have n values in the interval $]-1/2, 1/2]$

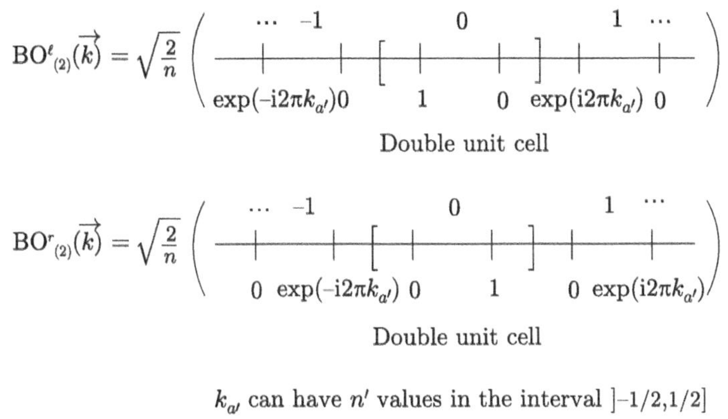

Regular and dimerised chains

$$BO^\ell_{(2)}(\vec{k}) = \sqrt{\frac{2}{n}} \left(\begin{array}{ccccccc} \cdots & -1 & & 0 & & 1 & \cdots \\ \rule{0pt}{2ex}\vphantom{|} & & [& &] & & \\ & \exp(-i2\pi k_{a'})0 & & 1 & 0 & \exp(i2\pi k_{a'}) & 0 \end{array} \right)$$

Double unit cell

$$BO^r_{(2)}(\vec{k}) = \sqrt{\frac{2}{n}} \left(\begin{array}{ccccccc} \cdots & -1 & & 0 & & 1 & \cdots \\ \rule{0pt}{2ex}\vphantom{|} & & [& &] & & \\ & 0\ \exp(-i2\pi k_{a'}) & 0 & & 1 & 0\ \exp(i2\pi k_{a'}) & \end{array} \right)$$

Double unit cell

$k_{a'}$ can have n' values in the interval $]-1/2, 1/2]$

Fig. 3.28
Bloch orbitals for the hydrogen chain obtained in the second representation.

$$BO^\ell_{(2)}(\vec{k}) = \sqrt{\frac{2}{n}} \sum_{m'=-n''+1}^{n''} \exp(i2\pi m' k_{a'})(1s_\ell)_{m'} \quad k_{a'} \in]-\frac{1}{2}, \frac{1}{2}]$$

$$BO^r_{(2)}(\vec{k}) = \sqrt{\frac{2}{n}} \sum_{m'=-n''+1}^{n''} \exp(i2\pi m' k_{a'})(1s_r)_{m'} \quad k_{a'} \in]-\frac{1}{2}, \frac{1}{2}] \quad (3.40)$$

A representation of the two Bloch orbitals is shown in Fig. 3.28.

The crystal orbitals may be obtained using results derived in Section 3.3.1 (i.e. eqns (3.34) and (3.37)) for the case in which $\beta_i = \beta_e = \beta$, since the chain is regular. In this case, the ratio of coefficients is:

$$\frac{c_\ell^\pm(\vec{k})}{c_r^\pm(\vec{k})} = \frac{\beta + \beta \exp(-i2\pi k_{a'})}{\pm \beta \sqrt{2(1 + \cos(2\pi k_{a'}))}} = \pm \exp(-i\pi k_{a'}) \quad (3.41)$$

so that the normalised expression (see eqn (3.36)) for the crystal orbitals as a function of the Bloch orbitals is:

$$CO^\pm_{(2)}(\vec{k}) = \sqrt{\frac{1}{2}} \left(BO_\ell(\vec{k}) \pm \exp(i\pi k_{a'}) BO_r(\vec{k}) \right) \quad (3.42)$$

Since the $\vec{a}'$ vector is $2\vec{a}$, the $\vec{a}'^*$ vector is $\vec{a}^*/2$ (eqns (3.5) and (3.30)). A vector $\vec{k}$ is characterised in the first representation by the real number k_a and in the second representation by the real number $k_{a'}$, which is given by $2k_a$ ($\vec{k} = k_a \vec{a}^* = 2k_a \vec{a}'^* = k_{a'} \vec{a}'^*$).

N.B. *In order to make the comparison of the crystal orbitals obtained in the two representations* (eqns (3.39)–(3.42)) *easier, the crystal orbitals of the second representation, shown in* Fig. 3.29, *are given in terms of the real number k_a instead of $k_{a'}$.*

Since the two representations describe exactly the same system, there must be a clear link between the two series of n crystal orbitals. For every value of k_a in the interval $]-\frac{1}{4}, \frac{1}{4}]$ it is easy to see that the crystal orbitals $CO_{(1)}(k_a)$ (see Fig. 3.27) may be identified as the $CO^+_{(2)}(k_a)$ ones (see Fig. 3.29). Thus the following relation holds:

$$CO_{(1)}(k_a) = CO^+_{(2)}(k_a) \quad \text{with} \quad k_a \in \left]-\frac{1}{4}, \frac{1}{4}\right] \quad (3.43)$$

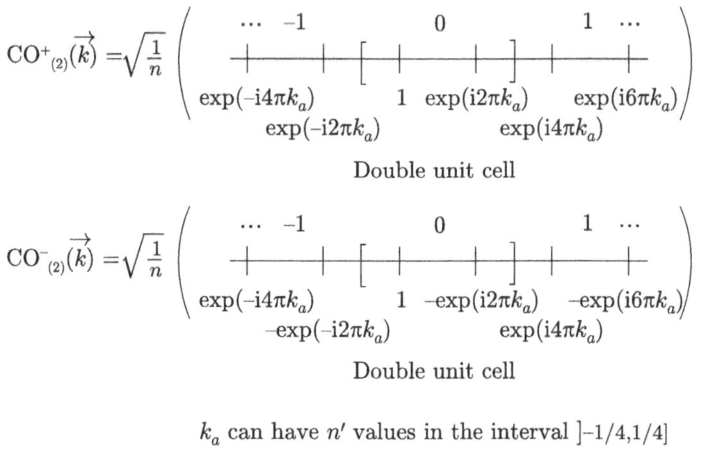

One-dimensional systems

$$CO^+_{(2)}(\vec{k}) = \sqrt{\frac{1}{n}} \begin{pmatrix} \cdots & -1 & & 0 & & 1 & \cdots \\ \exp(-i4\pi k_a) & & 1 & \exp(i2\pi k_a) & & \exp(i6\pi k_a) \\ & \exp(-i2\pi k_a) & & & \exp(i4\pi k_a) & & \end{pmatrix}$$

Double unit cell

$$CO^-_{(2)}(\vec{k}) = \sqrt{\frac{1}{n}} \begin{pmatrix} \cdots & -1 & & 0 & & 1 & \cdots \\ \exp(-i4\pi k_a) & & 1 & -\exp(i2\pi k_a) & & -\exp(i6\pi k_a) \\ & -\exp(-i2\pi k_a) & & & \exp(i4\pi k_a) & & \end{pmatrix}$$

Double unit cell

Fig. 3.29
Crystal orbitals obtained in the second representation in terms of k_a, which is equal to $k_{a'}/2$.

k_a can have n' values in the interval $]-1/4, 1/4]$

In order to find the relationship between the $CO^-_{(2)}(k_a)$ crystal orbitals and the orbitals of the first representation, we will integrate the negative signs of the coefficients (see Fig. 3.29) within the exponentials by using the following equivalences:

$$-\exp(i2\pi k_a) = \exp\left(i2\pi(k_a + \frac{1}{2})\right) = \exp\left(i2\pi(k_a - \frac{1}{2})\right) \quad (3.44)$$

or more generally

$$(-1)^m \exp(i2\pi m k_a) = \exp\left(i2\pi(k_a + \frac{1}{2})m\right) = \exp\left(i2\pi(k_a - \frac{1}{2})m\right) \quad (3.45)$$

Now we can rewrite the $CO^-_{(2)}(\vec{k})$ crystal orbitals in two different ways, as shown in Fig. 3.30. Taking into account that in the first representation k_a varies in the interval $]-\frac{1}{2}, \frac{1}{2}]$, it is possible to identify the $CO^-_{(2)}(\vec{k})$ crystal orbitals with those of the first representation in the following way:

if $k_a \in]0, \frac{1}{4}]$ then $CO^-_{(2)}(k_a) = CO_{(1)}(k'_a)$

with $k'_a = k_a - \frac{1}{2}$ $\left(k'_a \in]-\frac{1}{2}, -\frac{1}{4}]\right)$

if $k_a \in]-\frac{1}{4}, 0]$ then $CO^-_{(2)}(k_a) = CO_{(1)}(k'_a)$

with $k'_a = k_a + \frac{1}{2}$ $\left(k'_a \in]\frac{1}{4}, \frac{1}{2}]\right)$ $\quad (3.46)$

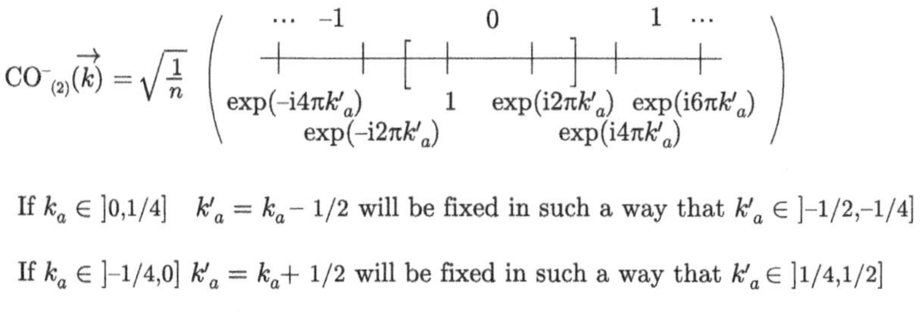

$$CO^-_{(2)}(\vec{k}) = \sqrt{\frac{1}{n}} \begin{pmatrix} \cdots & -1 & & 0 & & 1 & \cdots \\ \exp(-i4\pi k'_a) & & 1 & \exp(i2\pi k'_a) & & \exp(i6\pi k'_a) \\ & \exp(-i2\pi k'_a) & & & \exp(i4\pi k'_a) & & \end{pmatrix}$$

Fig. 3.30
Representation of the $CO^-_{(2)}(\vec{k})$ crystal orbitals.

If $k_a \in]0, 1/4]$ $k'_a = k_a - 1/2$ will be fixed in such a way that $k'_a \in]-1/2, -1/4]$

If $k_a \in]-1/4, 0]$ $k'_a = k_a + 1/2$ will be fixed in such a way that $k'_a \in]1/4, 1/2]$

Regular and dimerised chains

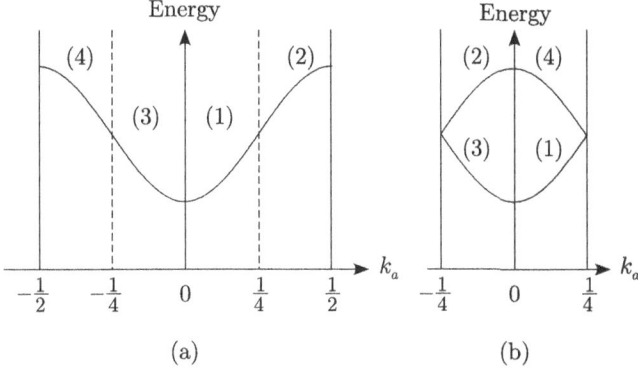

Fig. 3.31
Comparison of the band structures obtained when the regular chain is generated by the simple repeat unit (a) and the double repeat unit (b). The correspondence between the two band structures is made clear by means of the different numbers in parentheses. In the two cases we have represented the crystal orbital energies as a function of the real number k_a in such a way that $\vec{k} = k_a \vec{a}^*$, i.e. by means of a vector of the reciprocal space associated with the H_n system in representation 1 (simple repeat unit).

This is a general result, which does not depend on the computational method used to estimate the energies. It is a consequence of the periodic behaviour of the crystal orbitals. We can now compare the band structures obtained for the two representations (see Fig. 3.31).

According to Fig. 3.31 and eqns (3.43) and (3.46), there is a geometric correspondence between the different lines of these two band structures. For instance, the band denoted (2) in the second band structure ($k_a \in]-\frac{1}{4}, 0]$), may be found in the first band structure as the part of the band denoted (2) in the interval ($k_a \in]\frac{1}{4}, \frac{1}{2}]$).

When the periodicity of the chain is artificially doubled it is said that the band structure must be 'folded'. This is often taken as if the corresponding bands were folded back around the dotted lines centred at the k_a values of $\frac{1}{4}$ and $-\frac{1}{4}$. However, this is not conceptually correct as illustrated in Fig. 3.31. The new periodicity of the system in representation 2 is associated with a new Brillouin zone, which is half that of representation 1. Consequently, the single band of representation 1 must become two different bands in representation 2. The folded upper band originates from a translation by $\pm \vec{a}^*/2$ of the original band (Fig. 3.31a), which in that way is projected into the new, smaller, Brillouin zone (Fig. 3.31b).

The approach followed in this section may at first sight look artificial and useless. Why should we use a double unit cell if we can use a smaller, and thus, simpler one? As we will repeatedly see in this book, it is often very useful to determine the band structure of a system assuming a higher symmetry in order to understand the subtleties leading to the apparently more complex shape of the band structure for a real system with lower symmetry. For the time being, this will lead us to understand the origin of the stabilisation provided by a structural dimerisation for the H_n chain.

3.4.2 Dimerisation in the H_n chain: notion of distortion in a periodic system

We are now ready to compare the band structures of both the regular (Fig. 3.5) and the dimerised (Fig. 3.16) hydrogen chains. In order to make

Fig. 3.32
Comparison of the band structures for the regular (a) and dimerised (b) hydrogen chains assuming that $\frac{\beta_i + \beta_e}{2} = \beta$.

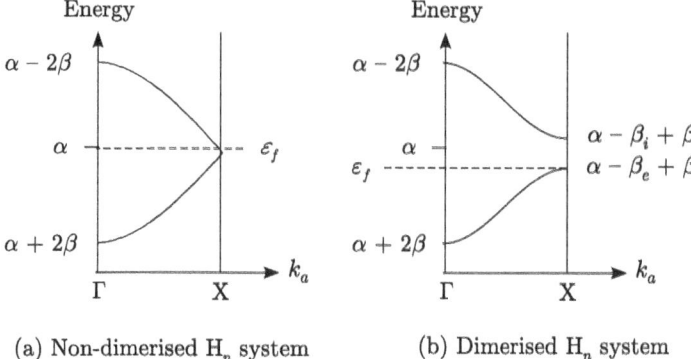

(a) Non-dimerised H_n system
Interaction term β

(b) Dimerised H_n system
Interaction terms β_i and β_e

the comparison easier, it is convenient to generate the two systems using a repeat unit containing two hydrogen atoms and the repeat vector $\vec{a}'$ (i.e. representation 2 in Fig. 3.26). In addition we will assume that the average of the two interaction terms in the dimerised chain (β_i and β_e) coincides with the interaction term in the regular chain (β). The band structures of the two chains are given in Fig. 3.32.

As shown in Fig. 3.32b, the degeneracy seen at the Fermi level (at the X point) of the regular chain no longer occurs for the dimerised system. The dimerisation in the H_n chain leads to the stabilisation of some filled levels and the destabilisation of some empty levels of the regular chain in such a way that the dimerised chain is energetically preferred. This result is the equivalent for a periodic system of the stabilisation provided by a Jahn–Teller distortion in a molecule (see Section 2.4.3). We will show that the regular H_n chain is not stable as a result of the coupling between the electronic motion and the vibrations of the chain. In order to understand this fact, let us consider the nature of the degenerate crystal orbitals at the Fermi level (Fig. 3.33) of the regular chain and the nature of the vibration that could break down the degeneracy.

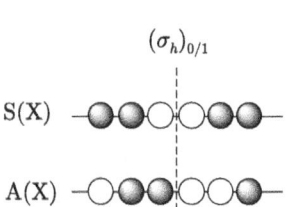

Fig. 3.33
Crystal orbitals at the Fermi level for the regular $(H_2)_{n'}$.

The shape of the crystal orbitals in Fig. 3.33 suggests that a vibration keeping the symmetry plane $(\sigma_h)_{0/1}$ and shortening one every two bonds along the chain will split the orbitals at the Fermi level by lowering the orbital for which the bonding interactions are reinforced and by raising the orbital for which the antibonding interactions are increased. This vibration may be schematically represented as the $(+Q)$ and $(-Q)$ displacements in Fig. 3.34. A $+Q$-type displacement stabilises orbital S(X) and destabilises orbital A(X) while a $-Q$-type displacement leads to the opposite effect. From an electronic perspective the system is unstable with respect to this type of vibrational motion. In contrast, the nuclear repulsions tend to favour the regular geometry

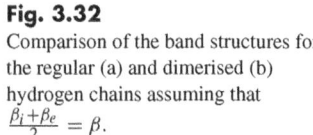

Fig. 3.34
$+Q$ and $-Q$ displacements of the hydrogen atoms corresponding to the vibration splitting the degenerate orbitals at the Fermi level.

($Q = 0$). Finally, there is a compromise between the nuclear repulsions opposing any noticeable displacement of the hydrogens and the electronic structure favouring the dimerisation of the chain. A model curve representing the total energy of the system (electronic plus nuclear repulsion energies) as a function of the deformation coordinate Q is shown in Fig. 3.35.[9]

From this scheme it can be predicted that at $T = 0$ K the system must be distorted and reside in the first vibrational state associated with one of the two potential wells. Thus, the chain is trapped in one of the two dimerised structures and the stability of this structure will increase with the magnitude of V_0. The chain will thus be structurally dimerised and electrically insulating. However, when the temperature increases some excited states will be populated. With these observations in mind we can summarise the situation in two different although equivalent ways:

- First, we can look at the double-well picture and assume that there will be a critical temperature, T_c, for which the thermal motion provides enough energy to the system to populate the states lying very near and slightly higher than the potential energy barrier, V_0. In that case we will get a system which is *delocalised* over the double well. This means that the average structure observed for the system is that of the regular chain. The critical temperature at which the regular structure is observed thus increases with the potential energy barrier, V_0.
- Another way to understand the situation consists in looking at the band structures. At $T = 0$ K only the lower band of the dimerised structure is filled (Fig. 3.36a). As soon as the temperature increases, the bottom levels of the upper band, lying at a higher energy than the bottom levels of the regular system upper band (Fig. 3.36b), start to be filled. In consequence, there is a critical temperature above which it is less energetically costly to adopt the regular structure and populate the bottom part of the upper band[10] (Fig. 3.36c) than to populate the bottom part of the upper band of the dimerised system. This view is summarised in Fig. 3.36.

[9] In Fig. 3.35 it is assumed that the electronic energy dominates over the nuclear repulsions, something which is most likely the case. In principle, however, there can be 1D metallic systems at 0 K if the nuclear repulsion term is strong enough to make any distortion too energetically costly.

[10] Of course we talk about the bottom part of the upper band because we are adopting representation 2 for the regular chain. In the more 'natural' representation 1 of the regular chain, these states are those associated with absolute k_a values of $\frac{1}{4}$ and slightly larger.

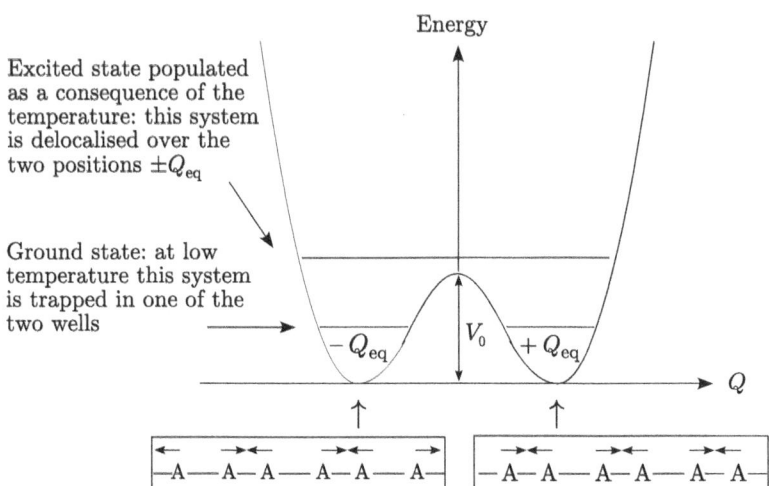

Fig. 3.35
Schematic representation of the total energy of the hydrogen chain as a function of the deformation coordinate Q. The minima are associated with two equivalent equilibrium geometries at low temperature.

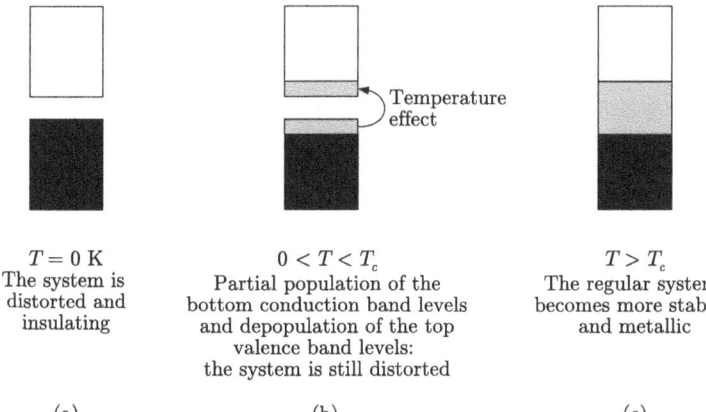

Fig. 3.36
Schematic representation of how an energy gap opens at the Fermi level as a consequence of a distortion. The partially filled states are illustrated in grey.

(a) $T = 0$ K
The system is distorted and insulating

(b) $0 < T < T_c$
Partial population of the bottom conduction band levels and depopulation of the top valence band levels: the system is still distorted

(c) $T > T_c$
The regular system becomes more stable, and metallic

Therefore, there is a temperature below which the system is distorted, i.e. dimerised in the case of the hydrogen chain. It is said that the system undergoes a *first-order Peierls distortion* at the critical temperature T_c. Such a distortion results from the coupling of the electronic structure, which exhibits a partially filled 1D band, and a vibrational motion of the lattice. Such a distortion bears many points in common with the Jahn–Teller distortion in molecules (Section 2.4.3). Peierls-type distortions lead to a change in the size of the repeat unit of periodic systems as well as in their electrical behaviour. In other words, they exhibit a metal-to-insulator phase transition at the critical temperature T_c. Throughout this book we will discuss different examples of periodic systems exhibiting first-order Peierls distortions and we will show that any 1D metallic system tends to exhibit the phenomenon, which leads to a drastic change in their electrical behaviour.

Exercises

(3.1) *Let us assume that the number of cells in a periodic 1D system is odd.

(a) Propose an equivalence between the linear and the cyclic periodic systems for this case.
(b) Prove that if we consider a cycle containing an odd number of cells we can obtain an expression for the Bloch orbitals almost identical to that of eqn (3.2). What is the difference? How important is such a difference knowing that n is very large?

N.B. The solution of the exercise requires the character table of group C_n for n odd ($n' = n/2$) given below.[11]

(3.2) Consider the chain A_n, where every atom A possesses $2p$ valence orbitals. Represent schematically the Bloch orbitals BO_{2p_x} at the Γ and X points generated by the $2p_x$ orbitals. Also represent the Bloch orbitals BO_{2p_y} and BO_{2p_z} at these points. Which are the more bonding and the more antibonding orbitals? Make a schematic energy diagram for these six orbitals.

[11] Character table of group C_n for n odd

C_n (n odd)	E	C_n	...	C_n^m	...	$C_n^{n'-0.5}$
Γ_ℓ	1	$\exp(\frac{-2i\pi\ell}{n})$	...	$\exp(\frac{-2mi\pi\ell}{n})$	....	$\exp(\frac{2(0.5-n')i\pi\ell}{n})$

(3.3) What is the wavelength associated with the $\vec{k}$ wave vectors of the Γ and X points? Correlate this result with the shape of the orbitals.

(3.4) Uniform H_n chain: what is the wavelength of wave vectors associated with the $k_a = \frac{1}{4}$ and $k_a = -\frac{1}{4}$ points? Correlate this result with the shape of the orbitals presented in Figs 3.11 and 3.13.

(3.5) Uniform H_n chain: what is the behaviour of the two functions $\{S(\pm\frac{1}{4}), A(\pm\frac{1}{4})\}$ with respect to the translations associated with vectors $\vec{a}$ and $2\vec{a}$? What are the symmetry operations leaving invariant each of these functions?

(3.6) Bearing in mind the treatment developed in Section 2.3 of Chapter 2, build the molecular orbitals of the cyclic H_4 system. Compare these results with those obtained by replacing n by 4 in eqn (3.15).

(3.7) Consider a partially oxidised H_n chain that can be described as $(H^{0.5+})_n$. In this case, every hydrogen atom contributes half an electron. What is the k_a value corresponding to the Fermi level of this hypothetical system? What is the corresponding value for the $(H^{0.3+})_n$ situation?

(3.8) Consider a two-band system having the band structure given below. Where is the Fermi level when every unit cell contributes 1, 2, 3, or 4 electrons?

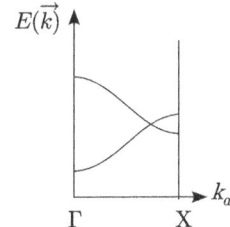

(3.9) *Let us consider in detail the electronic distribution of the H_n system.

(a) Why do we need to assume that the number of cells n may be written as an integer multiple of 4 in order to talk meaningfully about the $k_a = \frac{1}{4}$ point?

(b) Consider that n is a multiple of 4. Determine the precise occupation of the different energy levels of the H_n system and the k point corresponding to the Fermi level.

(c) Consider now the case where n is an even number but not a multiple of 4. Determine the precise occupation of the different energy levels of the H_n system and the k point corresponding to the Fermi level.

(d) Considering that the number of cells is very large, explain why is it possible to say that the Fermi level of H_n is associated with $k_a = \frac{1}{4}$.

References

1. N. W. Ashcroft and N. D. Mermin, *Solid State Physics*, Holt, Rinehart and Winston, Philadelphia, 1976.
2. S. I. Altmann, *Band Theory of Metals*, Pergamon Press, Oxford, 1970.
3. S. I. Altmann, *Band Theory of Solids: An Introduction from the Point of View of Symmetry*, Oxford University Press, Oxford, 1991.
4. A. P. Sutton, *Electronic Structure of Materials*, Oxford University Press, Oxford, 1993.

4 First-order Peierls distortions in periodic 1D systems

In Chapter 3 we saw that the H_n system is unstable with respect to a dimerisation removing the degeneracy at the Fermi level and thus opening an energy gap. In this chapter[1] we will generalise this result showing that any 1D system possessing a partially filled band is susceptible to a periodical structural distortion stabilising the network. [1] This distortion arises from the coupling between the conducting electrons of the undistorted metallic system and one of its lattice vibrations, and is called *electron–phonon coupling*. We will discuss in detail the link between the filling of the band and the nature of the distortion. We will only consider distortions leading to the lifting of the degeneracy at the Fermi level as a result of the interaction between degenerate or quasi-degenerate levels. Such distortions are known as *first-order Peierls distortions*. Other types of distortion will be discussed at the end of Chapter 7.

[1] The discussion in this chapter is not essential for understanding the material in the next three chapters except for the last part of Chapter 7. Thus the reader who is mostly interested in applying the concepts developed in the previous chapter to real materials can proceed to Chapters 5, 6 and 7 and come back to this chapter later on.

4.1 Analysis of the model system $(H^{0.5+})_n$

Let us initially consider the regular model system $(H^{0.5+})_n$ (Fig. 3.1 with M = $H^{0.5+}$), whose band structure is shown in Fig. 4.1. Since every cell provides 0.5 electrons, one-quarter of the band is filled in such a way that the Fermi level is associated with the points $k_a = \pm 1/8$. Figure 4.1 also shows the crystal orbitals $\{S(\pm \frac{1}{8}), A(\pm \frac{1}{8})\}$ of the Fermi level obtained through use of the results in Fig. 3.12. The Brillouin zone of the system, where the $\vec{k}$ vectors associated with filled states at $T = 0$ K are highlighted, is schematically shown in Fig. 4.2.

In this section, among all conceivable first-order Peierls distortions of the system, we will focus on those keeping the $(\sigma_h)_{0/1}$ symmetry plane perpendicular to the propagation axis and being equidistant from the centre of cells M_0 and M_1 (see Fig. 4.1). This is the reason why we have used the crystal orbital basis adapted to such a $(\sigma_h)_{0/1}$ symmetry plane to describe the states at the Fermi level. In Section 4.2 we will show that in fact the results are more general.

Model system $(H^{0.5+})_n$

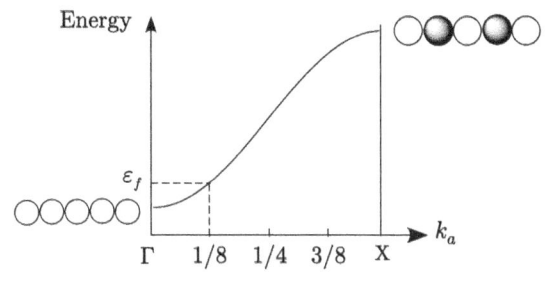

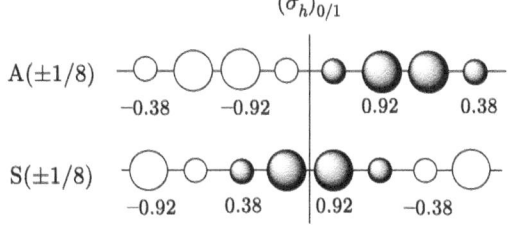

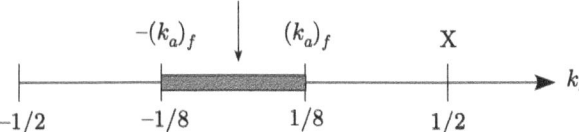

Fig. 4.1
(top) Band structure for the regular system $(H^{0.5+})_n$; (bottom) schematic representation of the crystal orbitals at the Fermi level, which are symmetric, $S(\pm\frac{1}{8})$, and antisymmetric, $A(\pm\frac{1}{8})$, with respect to the $(\sigma_h)_{0/1}$ symmetry plane. The numbers below the different orbitals are the coefficients of the different $1s$ atomic orbitals in the crystal orbital.

Fig. 4.2
Schematic representation of the Brillouin zone of the $(H^{0.5+})_n$ system, where the portion in bold is associated with the filled states.

4.1.1 Effect of a tetramerisation on the Fermi level

What is the nature of the distortion which leads to the lifting of the degeneracy at the Fermi level by making possible the interaction between crystal orbitals? To answer this question let us represent the electron density associated with the crystal orbitals $\{S(\pm\frac{1}{8}), A(\pm\frac{1}{8})\}$, i.e. the quantities $|S(\pm\frac{1}{8})|^2$ and $|A(\pm\frac{1}{8})|^2$ (see Fig. 4.3).

These two out-of-phase curves lead to the appearance of a charge density wave with a periodicity of $4a$. Consequently, a distortion bringing closer together the atoms lying between two consecutive nodal surfaces of the $S(\pm\frac{1}{8})$ function stabilises this crystal orbital and destabilises the $A(\pm\frac{1}{8})$ crystal orbital. Consequently, the tetramerisation represented in Fig. 4.4a lifts the degeneracy at the Fermi level, opening an energy gap as shown schematically in Fig. 4.4b.

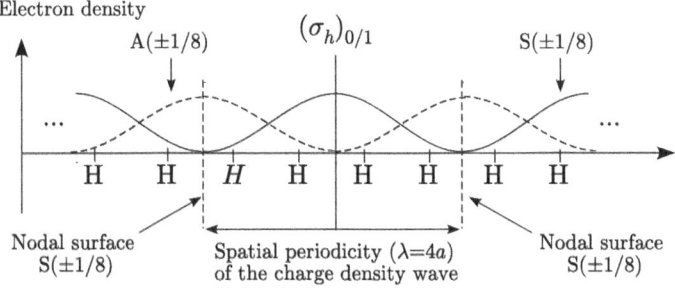

Fig. 4.3
Evolution of the electron density along the chain associated with the $S(\pm\frac{1}{8})$ (continuous line) and $A(\pm\frac{1}{8})$ (broken line) crystal orbitals.

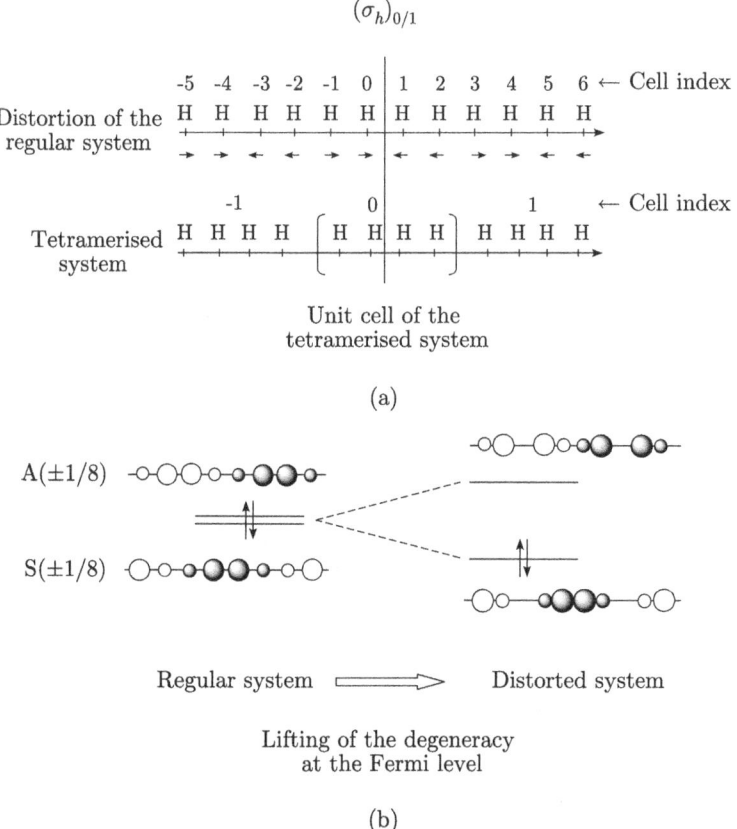

Fig. 4.4
(a) Tetramerisation of the system keeping the $(\sigma_h)_{0/1}$ symmetry plane; atomic displacements are schematically represented by the arrows. Note that the displacements must not necessarily have the same amplitude; (b) degeneracy lifting at the Fermi level as a result of such a tetramerisation.

Up to now, our reasoning has relied on the shape of the crystal orbitals at the Fermi level. However, the stabilisation of a single orbital does not necessarily justify the distortion of the periodic system. Before making further progress we must take into account what happens to the other orbitals of the system.

4.1.2 Effect of a tetramerisation on the states near the Fermi level

Let us start this section by building the band structure of the regular system having a quadruple cell as the repeat unit. To do so let us artificially generate the regular system by means of a cell containing four atoms and locating the origin of the system, O, at the centre of the reference cell (see Fig. 4.5b). We remind ourselves that the symmetry plane perpendicular to the propagation axis of the system and passing through the origin is referred to as $(\sigma_h)_{0/1}$.

The band structure of the regular system generated by a quadruple cell (see Fig. 4.6a) may be obtained from that of the regular system generated by a single unit cell. The only thing we need to do is to fold the initial band appropriately (i.e. the new Brillouin zone is now four times smaller) following the procedure outlined in Section 3.4.1.[2]

Note: To make possible the comparison between the regular and distorted systems, throughout the discussion in this section the $\vec{k}$ vectors will be

[2] Note that a tetramerisation is just a dimerisation of an already dimerised system.

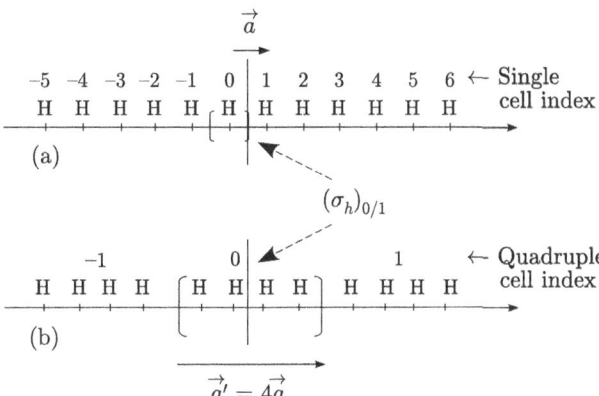

Fig. 4.5
(a) Regular H_n system described with a single unit cell and a repeat vector $\vec{a}$; (b) the $(H_4)_{n''}$, ($n'' = n/4$) system, which is described by a quadruple unit cell and a repeat vector $\vec{a}' = 4\vec{a}$, is equivalent to the regular system in (a).

characterised by the real number k_a, corresponding to the projection of the $\vec{k}$ vector over the $\vec{a}^*$ reciprocal vector ($\vec{k} = k_a \vec{a}^*$) and not over the $\vec{a}'^*$ reciprocal vector corresponding to the $\vec{a}'$ direct vector, which at first sight might seem more natural.

In what follows we must bear in mind the relationship between the band structures of the regular systems generated with the simple and quadruple cells (Fig. 4.6).

Now let us note the effect of a tetramerization (Fig. 4.4) on the crystal orbitals near the Fermi level. For instance, let us consider the crystal orbitals characterised by the same k_1 vector on Fig. 4.6b, i.e. orbitals $CO_1^\circ(\vec{k}_1)$, $CO_2^\circ(\vec{k}_1)$, $CO_3^\circ(\vec{k}_1)$, and $CO_4^\circ(\vec{k}_1)$. The first (point A), is occupied and lies near the Fermi level; the second (point B) is empty and lies near the Fermi level; and the third and fourth (points C and D) are also empty and high-lying in energy. The four points have also been noted in Fig. 4.6a, taking into account the equivalences due to the folding process. It is important to remind ourselves that points A, B, C, and D *in both diagrams* are associated with exactly the same crystal orbitals. The advantage of using the band structure in Fig. 4.6a is that we know the analytical expression of any crystal orbital simply by specifying the k point with which it is associated. In contrast, the use of the second band structure representation will be very useful when trying to compare the band structures of the regular and tetramerised systems.

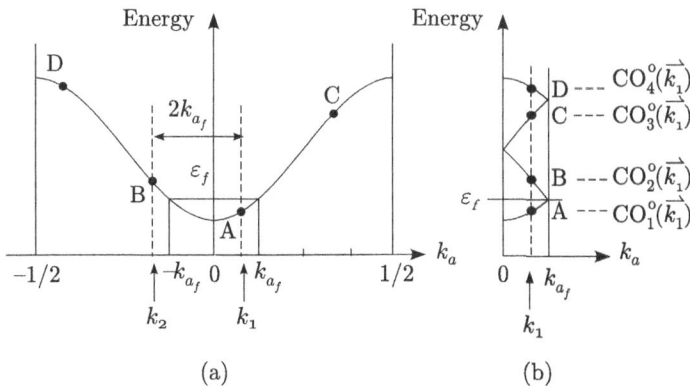

Fig. 4.6
Relationship between the band structures of the regular system generated by a single cell (a) and a quadruple cell (b). In the second diagram only the states associated with positive values of k_a have been shown (i.e. half of the band structure).

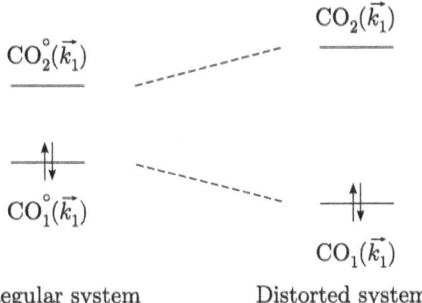

Fig. 4.7
Schematic representation of the evolution of the crystal orbital energies near the Fermi level associated with $\vec{k}_1$ as a result of the tetramerisation.

As soon as the regular system experiences a tetramerisation, so that it is formally generated by a quadruple cell and an $\vec{a}'$ vector equal to $4\vec{a}$, these four orbitals associated with the $\vec{k}_1$ vector may interact, since in the new representation related to $\vec{a}'$ they are associated with the same $\vec{k}$ vector. The filled orbital, $CO_1^\circ(\vec{k}_1)$, mainly interacts with $CO_2^\circ(\vec{k}_1)$, which is the orbital nearest in energy. In what follows we will neglect the weak interaction between $CO_1^\circ(\vec{k}_1)$ and the $CO_3^\circ(\vec{k}_1)$ and $CO_4^\circ(\vec{k}_1)$ orbitals. As a consequence of the distortion, the $CO_1^\circ(\vec{k}_1)$ orbital is stabilised while the $CO_2^\circ(\vec{k}_1)$ orbital is destabilised. This effect will be stronger when the energy difference between the two orbitals is smaller, in other words, when $\vec{k}_1$ approaches the Fermi level. This situation is shown schematically in Fig. 4.7.

Thus, we have qualitatively shown that, when dealing with a first-order Peierls distortion, all filled orbitals near the Fermi level are stabilised by the distortion, as a result of the interaction between quasi degenerate orbitals.

4.1.3 Effect of a tetramerisation on the band structure

Figure 4.8 reproduces the band structure of the regular system $(H_4^{2+})_{n''}$ as well as the crystal orbitals associated with the k_a points 0 and 1/8.

How does the band structure evolve as a result of the tetramerisation? According to the treatment developed in the last two sections, the tetramerisation lies at the origin of the gaps opening at the point with k_a equal to 1/8. In addition, under such a distortion all symmetry planes perpendicular to the direction of propagation that do not go through the centre of the tetramers disappear. Thus, all orbital mixings that are forbidden because of these symmetry

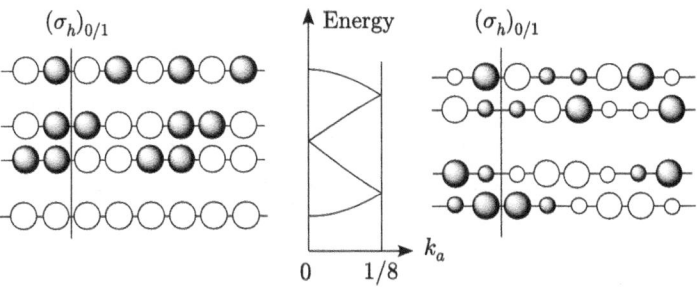

Fig. 4.8
Band structure of the regular $(H_4)_{n''}$ system, where the different crystal orbitals have been obtained according to Fig. 3.12.

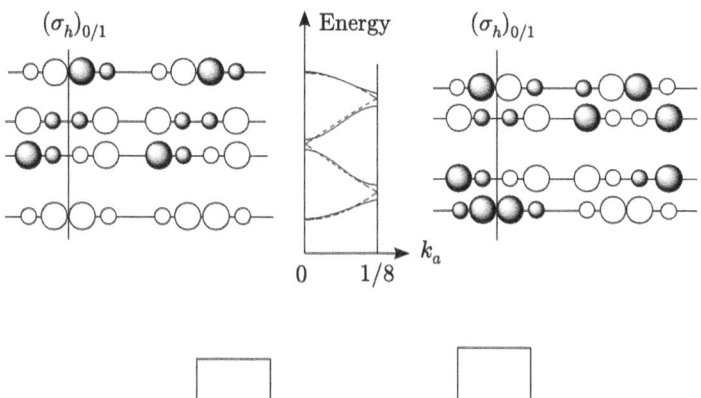

Fig. 4.9
Band structure for the tetramerised system.

Fig. 4.10
As a result of the tetramerisation, which keeps the horizontal symmetry plane, the initial band of the regular system breaks down into four different bands.

planes become allowed in the distorted system. As a result, the shape of the bands is a little bit modified and a band gap separating the second and third bands develops at the Γ point. Figure 4.9 shows the band structure for the tetramerised $(H_4)_{n''}$ system (continuous lines), which is similar to that of the regular system (broken lines).

The tetramerisation leads to the appearance of four different energy bands. Reasoning on the basis of the band structure for the non-distorted system (Fig. 4.1), it can be said that the distortion is associated with the development of gaps at the points $k_a = \pm 1/8$, $\pm 1/4$, and $\pm 3/8$. The relationship between the distorted and non-distorted forms of the system may be schematically shown as in Fig. 4.10.

Since the Fermi level of the regular system $(H_4^{2+})_{n''}$ lies at the k_a points $\pm 1/8$, a tetramerisation leads to a lowering of the electronic energy of the system because the filled orbitals near the Fermi level are stabilised. Careful examination of Fig. 4.10 shows that this result is actually more general, since a band that is three-quarters filled ($k_a = 3/8$) is also stabilised by a tetramerisation. This result is completely general: *1D periodic systems with $1/n$ filled or $1/n$ empty bands exhibit the same type of distortions (n-merisations)*.[3]

4.2 Analysis of first-order Peierls distortion in terms of a charge density wave

The results obtained for the regular $(H^{0.5+})_n$ model system may be generalised for any 1D conductor. We will now show that any tetramerisation of the regular

[3] The next section is a somewhat formal section which can be omitted, except for interested readers. Only the conclusions should be understood.

[4] In fact this section is completely general and could be directly applied to any $(H^{x+})_n$ system and generalised to any 1D system.

$(H^{0.5+})_n$ system,[4] *whatever its nature*, must be accompanied by a modulation of the electronic density of the system. In order to do so, we will analyse the electronic density associated with the $CO_1(\vec{k}_1)$ and $CO_2(\vec{k}_1)$ crystal orbitals (see Fig. 4.7) of the distorted system. [2]

To begin with let us decompose the Hamiltonian of the system, $\hat{H}$, into a zero-order Hamiltonian describing the regular system, $\hat{H}^0$, and a coupling term, $\hat{V}$, describing the coupling between the electronic motion and a vibration allowing the appearance of a tetramerisation. This coupling term allows the interaction between crystal orbitals of the regular system associated with a given $\vec{k}_1$ vector of the band structure in Fig. 4.6. We will only consider the interaction between crystal orbitals $CO_1^\circ(\vec{k}_1)$ and $CO_2^\circ(\vec{k}_1)$ (i.e. points A and B). The mathematical expressions for these orbitals can be easily obtained by exploiting the equivalence between the band structures in Figs 4.6a and 4.6b. In fact, we are looking for the expressions of the crystal orbitals labelled A and B in Figs 4.6a and b. The $CO_1^\circ(\vec{k}_1)$ orbital, denoted A in Fig. 4.6a, is given by eqn (3.4), giving the crystal orbitals of the regular H_n system generated by a single cell:

$$CO_1^\circ(\vec{k}_1) = \sum_{m=-n'+1}^{n'} \exp(i\vec{k}_1 \cdot \vec{r}_m) 1s_m$$

$$= \sum_{m=-n'+1}^{n'} \exp(i2\pi k_{a_1}(m-0.5)) 1s_m \quad (4.1)$$

The last simplification results from the fact that the $\vec{r}_m$ vector characterises the position of the $(1s_m)$ orbital with respect to the origin O located, in the present case (see Fig. 4.5a), in between cells M_0 and M_1 ($\vec{r}_m = (m-0.5)\vec{a}$). Likewise, the $CO_2^\circ(\vec{k}_1)$ orbital (point B) may be expressed as the crystal orbital of the regular H_n system generated by a single cell (eqn (3.4)) associated with point $k_{a_2} = -2k_{a_f} + k_{a_1}$ (see Fig. 4.6a).

$$CO_2^\circ(\vec{k}_1) = \sum_{m=-n'+1}^{n'} \exp(i2\pi(k_{a_1} - 2k_{a_f})(m-0.5)) 1s_m \quad (4.2)$$

The coupling term $\hat{V}$, which is responsible for the tetramerisation, allows the interaction of the two orbitals. After the interaction, the crystal orbitals of the regular form, $CO_1(\vec{k}_1)$ and $CO_2(\vec{k}_1)$, resulting from the interaction of $CO_1^\circ(\vec{k}_1)$ and $CO_2^\circ(\vec{k}_1)$, may be written in the following non-normalised form:

$$\begin{aligned} CO_1(\vec{k}_1) &= CO_1^\circ(\vec{k}_1) + \gamma \exp(-i\phi_\gamma) CO_2^\circ(\vec{k}_1) \\ CO_2(\vec{k}_1) &= \gamma \exp(i\phi_\gamma) CO_1^\circ(\vec{k}_1) - CO_2^\circ(\vec{k}_1) \end{aligned} \quad (4.3)$$

where $\gamma \exp(i\phi_\gamma)$ is a non-nil complex constant depending on k_{a_1}.

Let us consider the electronic density associated with these two crystal orbitals:

$$|CO_1(\vec{k}_1)|^2 = |CO_1^\circ(\vec{k}_1)|^2 + \gamma^2 |CO_2^\circ(\vec{k}_1)|^2$$
$$+ 2\gamma \operatorname{Re}\left(\exp(-i\phi_\gamma) \left(CO_1^\circ(\vec{k}_1)\right)^* CO_2^\circ(\vec{k}_1)\right)$$
$$|CO_2(\vec{k}_1)|^2 = \gamma^2 |CO_1^\circ(\vec{k}_1)|^2 + |CO_2^\circ(\vec{k}_1)|^2$$
$$- 2\gamma \operatorname{Re}\left(\exp(-i\phi_\gamma) \left(CO_1^\circ(\vec{k}_1)\right)^* CO_2^\circ(\vec{k}_1)\right) \quad (4.4)$$

If we develop the term $|CO_1^\circ(\vec{k}_1)|^2$ by means of eqn (4.1):

$$|CO_1^\circ(\vec{k}_1)|^2 = \sum_{m=-n'+1}^{n'} \left(\sum_{m'=-n'+1}^{n'} \exp\left(i2\pi k_{a_1}(m'-m)\right) 1s_m 1s_{m'} \right) \quad (4.5)$$

we can associate with each hydrogen atom the same electronic density, $d_m^1(\vec{k}_1)$, given by:

$$d_m^1(\vec{k}_1) = \sum_{m'=-n'+1}^{n'} \exp\left(i2\pi k_{a_1}(m'-m)\right) 1s_m 1s_{m'} \quad (4.6)$$

Now if we only keep the leading diagonal terms, in other words, if we assume that $(1s)_m (1s)_{m'} = (1s)_m^2 \delta_{m,m'}$, we reach the following expression:

$$d_m^1(\vec{k}_1) = (1s_m)^2 \quad (4.7)$$

A similar result is obtained for $|CO_2^\circ(\vec{k}_1)|^2$. Consequently, the electronic density due to the first two terms in eqn (4.4) is uniformly distributed over the different atoms. Let us now turn our attention to the third term of eqn (4.4):

$$2\gamma \operatorname{Re}\left[\exp(-i\phi_\gamma) \left(CO_1^\circ(\vec{k}_1)\right)^* CO_2^\circ(\vec{k}_1)\right]$$
$$= 2\gamma Re \left[\exp(-i\phi_\gamma) \sum_{m,m'=-n'+1}^{n'} \exp\left(i2\pi k_{a_1}(m'-m)\right. \right.$$
$$\left.\left. -4i\pi k_{a_f}(m'-0.5)\right) 1s_m 1s_{m'} \right] \quad (4.8)$$

If we only keep the leading terms we obtain the following expression:

$$2\gamma \operatorname{Re}\left\{\exp(-i\phi_\gamma)(CO_1^\circ(\vec{k}_1))^* CO_2^\circ(\vec{k}_1)\right\}$$
$$= 2\gamma \operatorname{Re}\left[\sum_{m=-n'+1}^{n'} \exp(-i4\pi k_{a_f}(m-0.5) - i\phi_\gamma)(1s_m)^2 \right]$$
$$= 2\gamma \sum_{m=-n'+1}^{n'} \cos\left(4\pi k_{a_f}(m-0.5) + \phi_\gamma\right) (1s_m)^2 \quad (4.9)$$

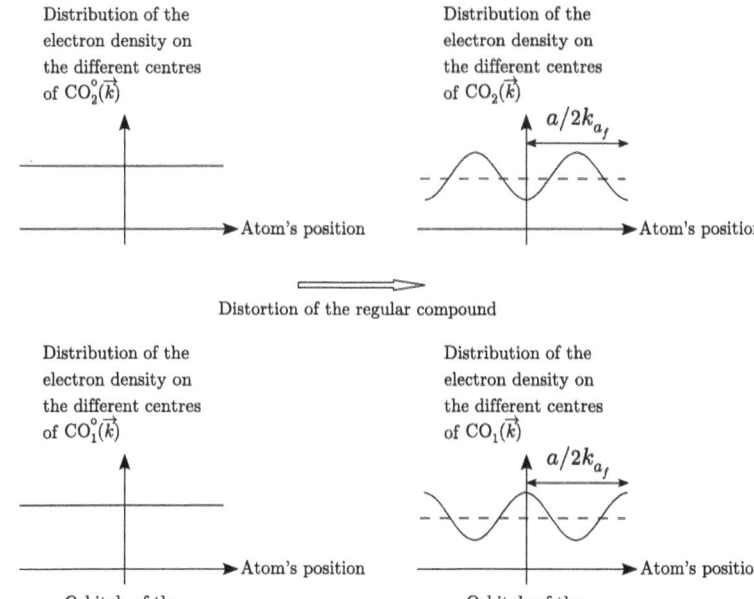

Fig. 4.11
Schematic representation of the electronic distribution among the difference centres of the crystal orbitals of the distorted system. This representation applies to the case where ϕ_γ is nil and γ is positive.

Denoting the contribution of the third term of eqn (4.4) to the electronic density associated with the hydrogen atom of cell M_m as $d_m^3(\vec{k}_1)$, we obtain:

$$d_m^3(\vec{k}_1) = 2\gamma \cos(4\pi k_{a_f}(m - 0.5) + \phi_\gamma)(1s_m)^2 \qquad (4.10)$$

This contribution is not the same for every atom and is at the origin of a modulation of the electron density with periodicity $a/2k_{a_f}$. According to eqn (4.4), a maximum in the electronic density associated with $CO_1(\vec{k}_1)$ corresponds to a minimum in the electronic density associated with $CO_2(\vec{k}_1)$ and vice versa. In Fig. 4.11 we show the electronic density associated with the crystal orbitals of the regular system, $CO_1^\circ(\vec{k}_1)$ and $CO_2^\circ(\vec{k}_1)$, as well as with those of the distorted system, $CO_1(\vec{k}_1)$ and $CO_2(\vec{k}_1)$.

Since only one of these two orbitals is occupied in the distorted system (see Fig. 4.7), the electronic density will exhibit a modulation along the chain (see Fig. 4.12 for a different representation).

We have just shown that, for the $(H^{0.5+})_n$ system, any tetramerisation, whatever its nature, stabilises the crystal orbitals lying at or just below the Fermi level because it allows the interaction between filled orbitals near

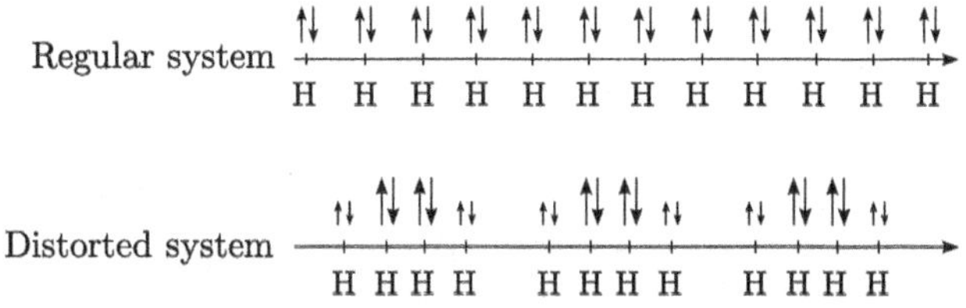

Fig. 4.12
A different representation of the distribution of the electronic density among different centres before and after a tetramerisation.

the Fermi level associated with points k_{a_1}, and also empty orbitals near the Fermi level associated with points $k_{a_2} = k_{a_1} - 2k_{a_f}$ (see Fig. 4.6a). In addition, since only the $CO_1(\vec{k}_1)$ orbitals are filled, the electronic density of the distorted system is modulated, exhibiting a charge density wave of periodicity $a/2k_{a_f}$. Thus, *from the electronic point of view it can be said that the regular system is unstable with respect to a charge density wave of $a/2k_{a_f}$ periodicity*. The kind of tetramerisation that is observed experimentally optimises electronic stabilisation without increasing the energy terms associated with the nucleus-nucleus and/or electron-electron repulsions too much.

4.3 Nesting vector

The results of the previous section can be generalised in the following way. Any 1D metallic conductor possessing one partially filled band tends to accommodate a distortion associated with a charge density wave of $a/2k_{a_f}$ periodicity (see Fig. 4.11). The interaction responsible for the first-order Peierls distortion of the regular metallic system results in an energy gap opening at the Fermi level and thus in the loss of the metallic properties of the distorted system. A metal-to-insulator transition is therefore associated with the structural distortion.

Let us summarise the results by representing the Brillouin zone of the regular system generated by a single cell (Fig. 4.2), with the Fermi level as well as the filling of the levels being schematically shown. We will define as a *nesting vector* the $\vec{q}$ vector connecting a pair of quasi-degenerate orbitals lying near the Fermi level, which interact when the first-order Peierls distortion occurs. As shown in the previous section (see Fig. 4.6a), the orbitals that may interact are those associated with points k_a and $k_a - 2k_{a_f}$ in the band structure of the system generated with a single unit cell. Consequently, the $\vec{q}$ vector is $2\vec{k}_f$ (see Fig. 4.13). It is very important at this point to keep in mind the relationship between the nature of this vector, which in fact is completely determined by the band filling, and the wavelength of the charge density wave responsible for the distortion, i.e. $\lambda = a/q$ since $\lambda = a/2k_{a_f}$.

The notion of nesting vector and the link with the charge density wave responsible for the distortion and the associated change in the transport properties of 1D compounds will be extremely useful later on when we will consider 2D and 3D compounds.

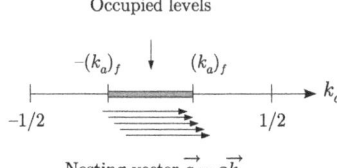

Nesting vector $\vec{q} = 2\vec{k}_f$ linking one empty orbital to an occupied orbital that can interact upon the distortion

Fig. 4.13
Nesting vector connecting those k points that interact upon the distortion.

4.4 Commensurate and incommensurate distortions

4.4.1 Commensurate distortion

As we have shown in the previous section, the electronic energy of a 1D metal may always be lowered by a periodic distortion. This feature results from the

Fig. 4.14
Schematic picture of the charge density wave responsible for the n-merisation. It has been assumed that k_a is $3/40$, i.e. $p = 3$ and $n = 20$. The system is unstable with respect to an 20-merisation, which loosely speaking can be referred to as a 6.666-merisation. The kind of distortion that allows the stabilisation of the orbitals near the Fermi level is schematically represented by means of arrows.

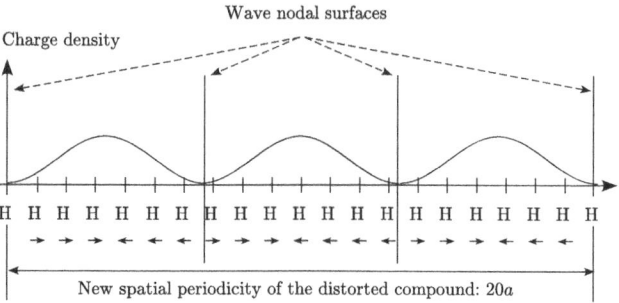

[5] Quite often, the commensurate distortion is considered to be associated with the less general situation where $p = 1$ and $n =$ small integer. In this case, Figs 4.14 and 4.15 of the following discussion would lead to the equivalent of Figs 4.3 and 4.10 for $p = 1$ and $n = 4$.

inherent instability towards a charge density wave of periodicity $a/2k_a$, which leads to the opening of an energy gap at the Fermi level. [3] *If it is possible to write $2k_a$ in the form of a non-reducible fraction p/n, where p and n are integers and $p < n$, the distortion is said to be commensurate with the lattice and the system is unstable with respect to an n-merisation.*[5] Let us consider the particular case where $k_a = 3/40$, i.e. $2k_a = 3/20$ ($p = 3$ and $n = 20$). The charge density wave responsible for the distortion of the system possesses a λ wavelength of $a/2k_a$, i.e. $6.666a$, as schematically represented in Fig. 4.14.

The system has experienced an n-merisation, since two equivalent hydrogen atoms are separated by a distance $n \cdot a$ (in Fig. 4.14, $n = 20$)). The particular distortion suffered by the system is often also referred to as a n/p-merisation because the charge density wave leads to the clustering of atoms between two nodal surfaces of the wave separated by a distance of $(n/p) \cdot a$. Strictly speaking, however, from the structural point of view, the new periodicity of the system is $n \cdot a$. The partially filled energy band of the regular system splits in n different sub-bands, lying on practically the same energy range as the initial energy band (see Fig. 4.15). Consequently, the energy gap separating the different sub-bands will generally decrease when n increases.

Let us now consider the total one-electron energy associated with the regular and distorted (n-merised) systems (Fig. 4.15). To begin with, let us focus on the energetic consequences of the lowest gap opening. This gap opens in between filled states. Thus, some filled states are lowered whereas some others are destabilised. Globally, there is no driving force for the distortion. In fact only the gap opening at the Fermi level provides a driving force for the distortion. Consequently, the energy stabilisation provided by a distortion increases when the gap opening is larger and thus when n decreases. Thus, *the*

Fig. 4.15
Splitting of the partially filled band of the regular system into twenty bands as a result of the 20-merisation undergone by the system.

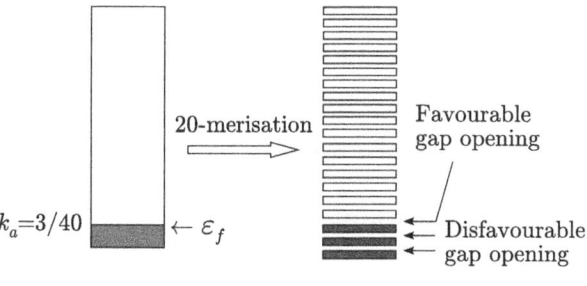

driving force for a first-order Peierls distortion increases when n decreases. Obviously, a distortion will only be observed if the energy increase resulting from the nuclear repulsion concomitant with the development of the distortion is lower than the decrease in energy provided by the electronic energy gain.

4.4.2 Incommensurate distortion

If it is not possible to write $2k_a$ in the form of a non-reducible fraction p/n, where p and n are integers and $p < n$, the distortion is said to be incommensurate with the lattice and will be an n/p-merisation. In that case, if the charge density wave is nil at one hydrogen atom, it will then be non-nil at any other hydrogen atom of the chain. If this is the case, it is not easy to describe pictorially the position of the different atoms of the chain. It is better to keep in mind the idea of a charge density wave tending to approach or separate atoms to stabilise the system as much as possible (i.e. atoms separated by nodal surfaces of the wave will tend to separate whereas atoms between two nodal surfaces will tend to approach).

4.4.3 Comparison

In the case of an incommensurate distortion, every point of the lattice is different with respect to the charge density wave. In contrast, two points of the lattice separated by a length $\ell = n \cdot a$ are equivalent in the case of a commensurate distortion. In a certain way, it can be said that the commensurate distortion is well adapted to the periodicity of the system. As a consequence, *there is always a stronger driving force for commensurate than incommensurate distortions.*

4.5 Conclusions

In this chapter we have shown that any metallic system tends towards a first-order Peierls distortion resulting in the loss of the metallic properties below a certain temperature T_c. From an experimental point of view such a situation may be characterised by the appearance of a change in the slope of the resistivity-versus-temperature curve around T_c (see Fig. 4.16). In addition, by use of diffuse X-ray scattering measurements, it will be possible to observe new Bragg spots (satellites) around T_c that correspond to the new periodicity.

All the distortions considered in this chapter are first-order Peierls distortions involving the interaction between degenerate or quasi-degenerate orbitals lying at or in the vicinity of the Fermi level and which change the conductivity of the material. Throughout this book we will find many examples of such type of distortion. Later, in Chapter 7, we will also examine the case of a system that is unstable with respect to a second-order Peierls distortion involving the interaction between orbitals that are more separated in energy. At the end of Chapter 7 we will summarise the similarities and differences between these two types of distortion.

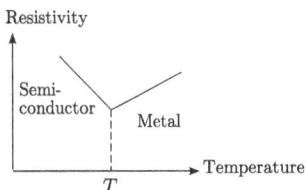

Fig. 4.16
Schematic representation of the resistivity-versus-temperature behaviour of a material undergoing a Peierls distortion at a temperature T_c.

Exercises

(4.1) Model system $(H^{0.5+})_n$: the two wavefunctions $\{S(\pm\frac{1}{8}), A(\pm\frac{1}{8})\}$ are degenerate. This means that there must be a symmetry operation making them equivalent. Discuss what this symmetry operation is.

(4.2) Provide an explanation of the shape of the different bands and the associated crystal orbitals in Fig. 4.9 for a tetramerised system.

References

1. R. E. Peierls, *Quantum Theory of Solids*, Oxford University Press, London, 1955, p.108.
2. M.-H. Whangbo, *J. Chem. Phys.*, 75, 4983, 1981.
3. R. Moret and J.-P. Pouget, in: *Crystal Chemistry and Properties of Materials with Quasi-One-Dimensional Structures*, J. Rouxel, Ed., Reidel, Dordrecht, 1986, p. 87–134.

Application to *trans*-polyacetylene

5

As an application of the ideas developed in Chapter 3 we will now study the electronic structure of the *trans*-polyacetylene chain, $(C_2H_2)_n$. We will assume that the chain is fully planar and has regular geometry in which all C–C distances are identical (Fig. 5.1).

A common and useful way chemists have found to represent the electronic structure of this material is by using either the two mesomeric Lewis structures represented in Figs 5.2a and b or the average structure of Fig. 5.2c. In the latter, all C–C bonds are equivalent, and the dotted lines are used to represent additional C–C bonding with a bond order of 1/2 per C–C pair, delocalised all along the chain.

The latter representation (Fig. 5.2c) is often used to justify the electrical conductivity of this material and, in general, of polymers possessing an extended π system. The basis of this reasoning is that because of the delocalised bonding, the system possesses electrons free to move all along the chain and thus able to act as charge carriers in the conductivity process. We will examine the correctness of this description by building up the band structure of *trans*-polyacetylene step by step. We will begin by considering the electronic structure of ethylene. This study will bring valuable information about the

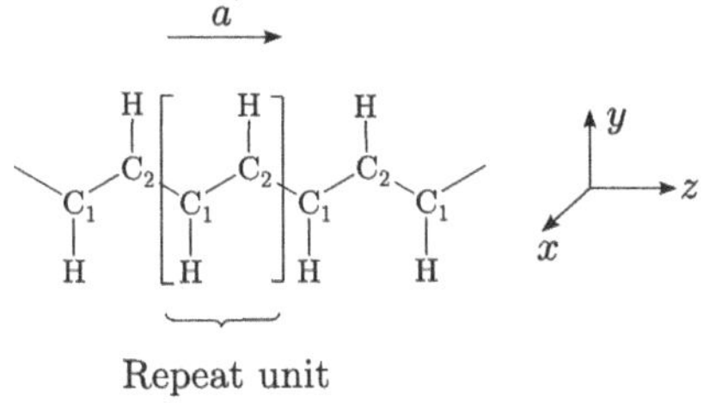

Fig. 5.1
The regular *trans*-polyacetylene chain: the system is generated by a repeat unit containing two carbon and two hydrogen atoms and a repeat vector $\vec{a}$. To distinguish the two carbon atoms of every cell they have been labelled with indices 1 and 2.

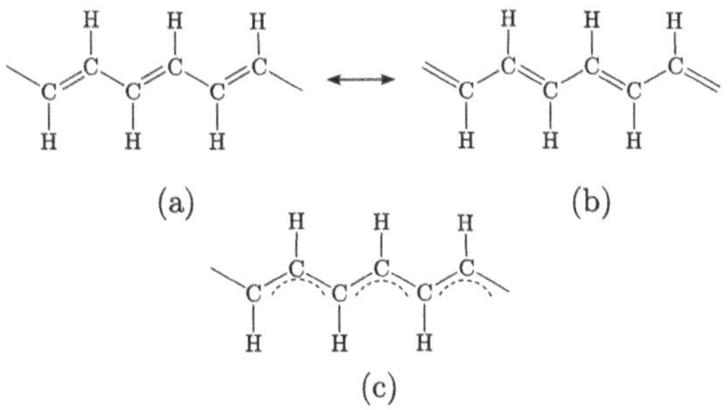

Fig. 5.2
(a)-(b) Mesomeric Lewis structures for *trans*-polyacetylene; (c) average structure of *trans*-polyacetylene.

Fig. 5.3
Coordinate system chosen to describe ethylene and *trans*-polyacetylene (both systems are assumed to be in the xOz plane).

nature of the states that are important to describe the conducting properties of *trans*-polyacetylene, which is a polyacene. We will adopt a common coordinate axis that is convenient to describe the electronic structure of both systems (see Fig. 5.3).

5.1 Electronic structure of ethylene

Figure 5.4 shows the molecular orbital diagram for ethylene. Because of the xOz symmetry plane, every molecular orbital can be classified as symmetric (σ) or antisymmetric (π) with respect to this plane. Thus, ethylene has ten σ and two π molecular orbitals. The σ molecular orbitals are those built from the $2s$, $2p_x$ and $2p_z$ orbitals of the two carbon atoms and the $1s$ orbitals of the four hydrogen atoms. The two π molecular orbitals are those built from the $2p_y$ orbitals of the two carbon atoms. Since the axial σ-type overlap is more effective that the π-type one, the two π molecular orbitals (i.e. bonding, π, and antibonding, π^*) lie in between the five σ bonding and the five σ^* antibonding orbitals (see Fig. 5.4). Since the molecule possesses twelve valence electrons, the six lower lying molecular orbitals must be filled, i.e. five σ bonding and one

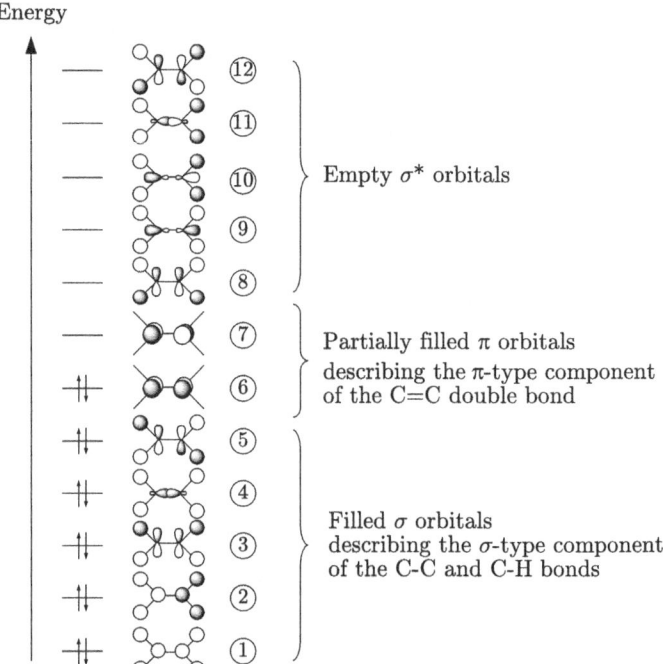

Fig. 5.4
Molecular orbital diagram for ethylene. For a more detailed description the interested reader may consult references [1, 2].

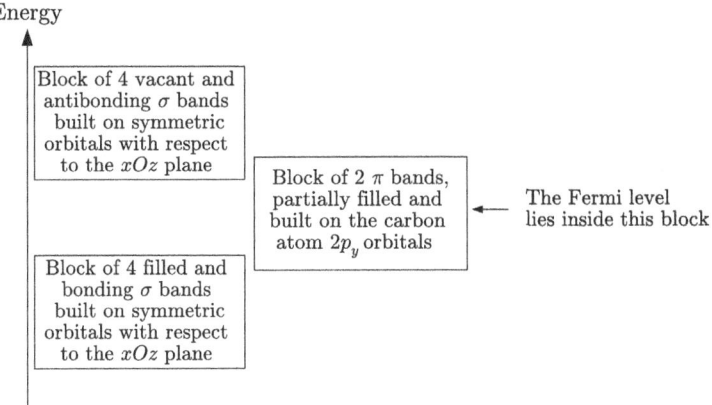

Fig. 5.5
Schematic representation of the band structure for *trans*-polyacetylene.

π bonding orbitals. Thus, the highest occupied molecular orbital of ethylene is a bonding π molecular orbital associated with the carbon–carbon double bond.

5.2 Main aspects of the band structure for *trans*-polyacetylene

We are now ready to extend these ideas to *trans*-polyacetylene, which can be considered as a very large alkene. Let us first consider the general features of its band structure. Since the xOz plane is also a symmetry plane for this system, all crystal orbitals should either be symmetric, i.e. of σ-type, or antisymmetric, i.e. of π-type, with respect to this plane.[1] Since ten atomic orbitals per unit cell (i.e. per C_2H_2 unit) are used to describe the electronic structure of the material, eight of which (the $2s$, $2p_x$ and $2p_z$ of the carbon atoms and the $1s$ orbital of the hydrogens) are symmetric with respect to the xOz plane and two (the $2p_y$ orbital of the carbons) are antisymmetric, the band structure of *trans*-polyacetylene is made of ten valence energy bands, eight of which are of σ-type and two of π-type. By analogy with the electronic structure of alkenes, we can anticipate that there must be three different band blocks. First, a set of four σ-type filled bonding bands. Second, another set of four σ-type empty antibonding bands. Third, two π-type bands, originating from weaker π-type overlaps, which thus lie in between the two σ blocks. In consequence, the band structure of *trans*-polyacetylene may be described, in a very qualitative way, as shown in Fig. 5.5.

Since every unit cell provides ten valence electrons, five energy bands must be filled at $T = 0$ K: the block of four σ bonding bands and the equivalent of one π band. The Fermi level must thus be found within the block of the two π bands. Essentially, the stability of the system mostly originates from the population of the block of σ bands whereas the π block, which contains the higher filled and lower empty levels, controls the conducting properties of the system. For the time being we must consider two possibilities (see Fig. 5.6). First, the two π bands do not overlap or even touch and, consequently, the system is semiconducting (Fig. 5.6a). Second, the two bands overlap and thus the system is metallic or semimetallic (Fig. 5.6b).

[1] See also Chapter 6.

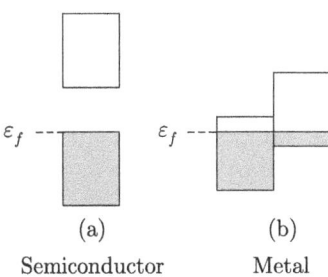

Fig. 5.6
Two possibilities for the π-type band structure of *trans*-polyacetylene: (a) the two bands are well separated; (b) the two bands overlap.

Application to *trans*-polyacetylene

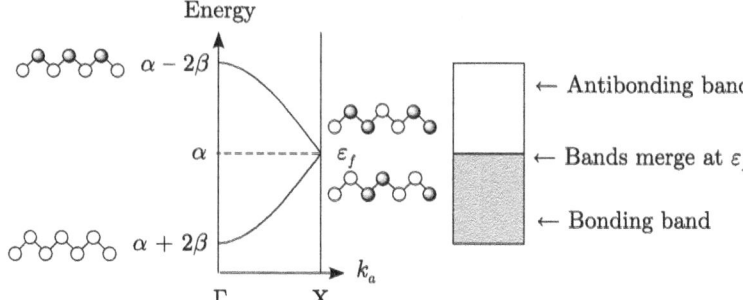

Fig. 5.7
Band structure near the Fermi level for regular *trans*-polyacetylene.

In what follows we will only consider the π bands generated by the carbon $2p_y$ orbitals and we will concentrate on the location of the Fermi level within the π block. All along this chapter only the lobes of the $2p_y$ orbitals pointing upwards are represented so as to simplify the representations.

5.3 Detailed analysis of the band structure of *trans*-polyacetylene

Since by symmetry the $1s$ atomic orbitals of hydrogen and the $2s$, $2p_x$ and $2p_z$ of carbon cannot enter into the crystal orbitals near the Fermi level, our study of *trans*-polyacetylene is equivalent to that of the $(A_2)_n$ system in which atom A is a carbon atom bearing only a $2p_y$ orbital.[2] Thus, our study comes back to the discussion for $(H_2)_{n'}$ (see Sections 3.3 and 3.4), the only difference being the nature of the atomic orbital centred on every atom. Thus using the results of Sections 3.3 and 3.4 for the specific case $\beta_i = \beta_e$ (Fig. 3.32) we obtain the band structure shown in Fig. 5.7 for the π bands of regular *trans*-polyacetylene. [3]

Since the equivalent of a single band must be occupied (i.e. there is one electron per site), the Fermi level occurs at the top of the first band. This means that the Fermi level does occur where the two π bands merge. The two crystal orbitals associated with the Fermi level $CO^{\pm}(\vec{k})$ may be represented as in Fig. 5.8 according to eqn (3.42).

In view of the conclusions reached in Section 3.4.2 and in Chapter 4, it can be concluded that *trans*-polyacetylene is susceptible to a dimerisation (Fig. 3.35), which stabilises the filled levels of the π system near the Fermi level and results in the loss of its metallic properties. However, before predicting that the distortion does occur in practice, it is worth considering if the σ system, which has so far been neglected, precludes the distortion. Let us assume that the distortion has already occurred. This distortion leads to the geometry of one of the two limit structures shown in Fig. 5.2, which have an alternation of single (long) and double (short) bonds.

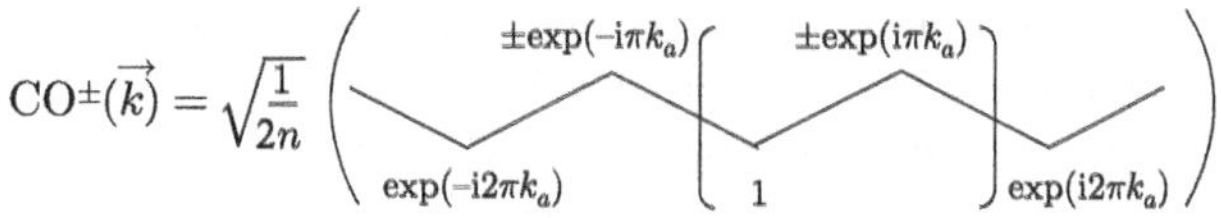

Fig. 5.8
π-type crystal orbitals for the regular *trans*-polyacetylene chain obtained through the Hückel approach and the ideas developed in Chapter 3 (see eqn (3.42)). The + and − signs refer to the lower and upper bands, respectively. Only the coefficients preceding the $2p_y$ AOs are given, n being the number of unit cells in the chain.

[2] To be consistent with the previous chapters, the projection $k_a = 1/2$ will be denoted X, even if the chain axis is now the Oz axis.

$$CO^{\pm}(\vec{k}) = \sqrt{\frac{1}{2n}} \left(\begin{array}{c} \pm\exp(-i\pi k_a) \quad \pm\exp(i\pi k_a) \\ \exp(-i2\pi k_a) \quad 1 \quad \exp(i2\pi k_a) \end{array} \right)$$

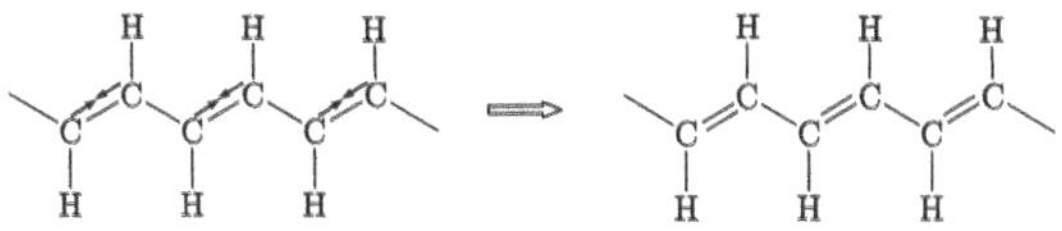

Fig. 5.9
Schematic representation of the regular *trans*-polyacetylene dimerisation process.

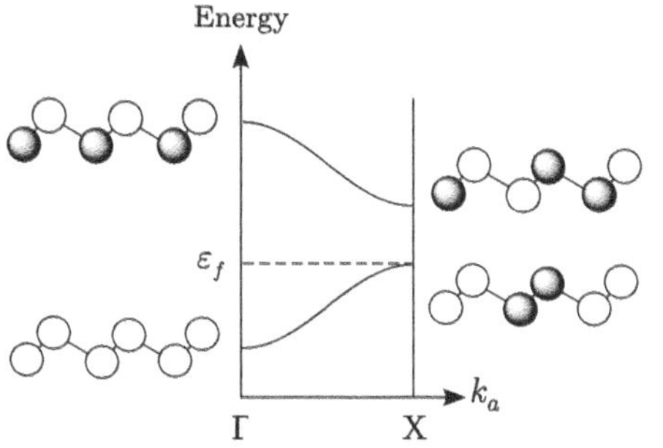

Fig. 5.10
Band structure for distorted *trans*-polyacetylene.

According to the ideas developed in Section 3.4.2, such a deformation leads to the lifting of the degeneracy of the crystal orbitals at the Fermi level (see Figs 3.32 and 3.33), i.e. to the band structure of Fig. 5.10. Analysis of the crystal orbital filled at X shows clearly that the distortion leads to an alternate localisation of single and double bonds.

Using the discussion in Section 3.4.2 we can say that the distortion is characterised by a critical temperature T_c, below which the system is pinned in one of the distorted forms (Fig. 3.35) corresponding to one of the Lewis mesomers represented in Figs 5.2a and b. In the distorted system, the Fermi level lies in a band gap, rendering the system a semiconductor. In contrast, when the system, through thermal excitation, acquires a vibrational energy equal to or larger than the potential barrier V_0 (see Fig. 3.35), the system may be considered as delocalised through the double well. In this latter case the geometry observed corresponds to the average structure represented in Fig. 5.2c. All bond lengths are then equal and the system is metallic (see Fig. 5.7). These results may be summarised as shown in Fig. 5.11.

5.4 Determination of the band structure of *trans*-polyacetylene using the fragment formalism

In this section we will approach the problem of generating the band structure of regular *trans*-polyacetylene in a different way – by using *fragment orbitals* to generate the Bloch orbitals. When trying to understand the electronic structure of a complex molecule it is useful to decompose the system into different fragments for which the electronic structure is easily derived. It is then possible to analyse the interaction between the different fragment orbitals that lead to the orbitals of the molecule. Using this process, a clear understanding of the main electronic factors governing the stability of the molecule is usually reached. For instance, the carbon–carbon bonding of the ethylene molecule may be studied by considering the interaction of two CH_2 fragments. In the

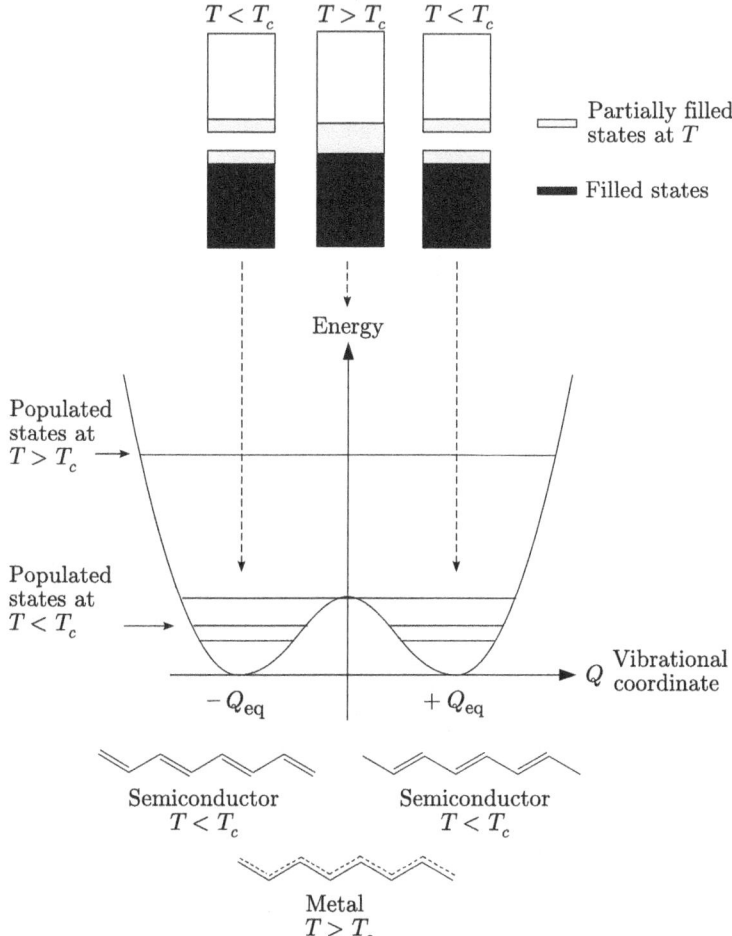

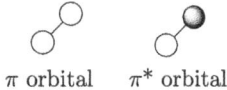
π orbital   π* orbital

Fig. 5.11
A view of the *trans*-polyacetylene distortion based on Lewis structures.

Fig. 5.12
π-type orbitals of the C_2H_2 fragment, with only the upper lobes of the p_y atomic orbitals represented.

same vein we can use the orbitals of a $-CH_2-CH_2-$ fragment to generate the crystal orbitals of *trans*-polyacetylene. To this end, let us consider the π-type normalised molecular orbitals of the $-CH_2-CH_2-$ fragment, i.e. the π and π^* orbitals represented in Fig. 5.12.

We can generate a base of Bloch orbitals $(BO)_\pi(\vec{k})$ and $(BO)_{\pi^*}(\vec{k})$ using the two fragment orbitals π and π^*.

$$BO_\pi(\vec{k}) = \frac{1}{\sqrt{n}} \sum_{m=-n'+1}^{n'} \exp(i2\pi m k_a)(\pi)_m \quad (5.1)$$

$$BO_{\pi^*}(\vec{k}) = \frac{1}{\sqrt{n}} \sum_{m=-n'+1}^{n'} \exp(i2\pi m k_a)(\pi^*)_m \quad (5.2)$$

Since both orbitals are antisymmetric with respect to the xOz plane, the π-type crystal orbitals of the system result from the interaction of these two Bloch orbitals:

$$\mathrm{CO}^{\pm}(\vec{k}) = c_{\pi}^{\pm}(\vec{k})\mathrm{BO}_{\pi}(\vec{k}) + c_{\pi^*}^{\pm}(\vec{k})\mathrm{BO}_{\pi^*}(\vec{k}) \qquad (5.3)$$

We can now estimate the band structure following two different approaches, as we have already done for $(H_2)_{n'}$ (Sections 3.3.1 and 3.3.2):[3]

[3] The reader can thus make a choice.

(i) a mathematical resolution of the secular equations, leading to the analytical expression of the crystal orbitals and their energies, according to the Hückel approach;
(ii) a more qualitative approach based on the analysis of the interaction term between the Bloch orbitals $(BO)_{\pi}(\vec{k})$ and $(BO)_{\pi^*}(\vec{k})$.

5.4.1 Calculation of the band structure by means of the Hückel approach

The $E(\vec{k})$ energies associated with the crystal orbitals may be obtained by imposing that the secular determinant be nil using this new basis:

$$\begin{vmatrix} h_{\pi,\pi}(\vec{k}) - E(\vec{k})S_{\pi,\pi}(\vec{k}) & h_{\pi,\pi^*}(\vec{k}) - E(\vec{k})S_{\pi,\pi^*}(\vec{k}) \\ h_{\pi^*,\pi}(\vec{k}) - E(\vec{k})S_{\pi^*,\pi}(\vec{k}) & h_{\pi^*,\pi^*}(\vec{k}) - E(\vec{k})S_{\pi^*,\pi^*}(\vec{k}) \end{vmatrix} = 0 \qquad (5.4)$$

where

$$h_{\pi,\pi}(\vec{k}) = \left\langle \mathrm{BO}_{\pi}(\vec{k}) | \hat{h} | \mathrm{BO}_{\pi}(\vec{k}) \right\rangle,$$

$$h_{\pi^*,\pi^*}(\vec{k}) = \left\langle \mathrm{BO}_{\pi^*}(\vec{k}) | \hat{h} | \mathrm{BO}_{\pi^*}(\vec{k}) \right\rangle,$$

$$h_{\pi^*,\pi}(\vec{k}) = \left\langle \mathrm{BO}_{\pi^*}(\vec{k}) | \hat{h} | \mathrm{BO}_{\pi}(\vec{k}) \right\rangle = \left(h_{\pi,\pi^*}(\vec{k}) \right)^*$$

$$S_{\pi,\pi}(\vec{k}) = \left\langle \mathrm{BO}_{\pi}(\vec{k}) | \mathrm{BO}_{\pi}(\vec{k}) \right\rangle,$$

$$S_{\pi^*,\pi^*}(\vec{k}) = \left\langle \mathrm{BO}_{\pi^*}(\vec{k}) | \mathrm{BO}_{\pi^*}(\vec{k}) \right\rangle$$

$$S_{\pi^*,\pi}(\vec{k}) = \left\langle \mathrm{BO}_{\pi^*}(\vec{k}) | \mathrm{BO}_{\pi}(\vec{k}) \right\rangle = \left(S_{\pi,\pi^*}(\vec{k}) \right)^*$$

The different terms may be calculated using the procedure developed in Section 3.2.2 and the Hückel approach:

$$h_{\pi,\pi}(\vec{k}) = \alpha + 2\beta \cos^2(\pi k_a) \quad ; \quad h_{\pi^*,\pi^*}(\vec{k}) = \alpha - 2\beta \cos^2(\pi k_a) \qquad (5.5a)$$

$$h_{\pi,\pi^*}(\vec{k}) = i\beta \sin(2\pi k_a) \quad ; \quad S_{\pi,\pi^*}(\vec{k}) = 0 \ ; \ S_{\pi,\pi}(\vec{k}) = S_{\pi^*,\pi^*}(\vec{k}) = 1 \qquad (5.5b)$$

Let us note that $h_{\pi,\pi}(\vec{k})$ and $h_{\pi^*,\pi^*}(\vec{k})$ represent the energies of the Bloch orbitals $(BO)_{\pi}(\vec{k})$ and $(BO)_{\pi^*}(\vec{k})$, respectively, whereas $h_{\pi,\pi^*}(\vec{k})$ describes the interaction between the two orbitals.

The secular equation (eqn (5.4)) may be written in the form:

$$\left(h_{\pi,\pi}(\vec{k}) - E(\vec{k}) \right) \left(h_{\pi^*,\pi^*}(\vec{k}) - E(\vec{k}) \right) - h_{\pi,\pi^*}(\vec{k}) h_{\pi^*,\pi}(\vec{k}) = 0$$

Application to trans-polyacetylene

Developing the different terms by means of eqn (5.5) we obtain the following equation for the allowed energies of the system:

$$E^{\pm}(\vec{k}) = \alpha \pm 2\beta \cos(\pi k_a) \text{ with } k_a \in \left]-\frac{1}{2}, \frac{1}{2}\right] \quad (5.6)$$

Now let us compare the energy of the crystal orbitals (eqn (5.6)) with those of the Bloch orbitals $(BO)_\pi(\vec{k})$ and $(BO)_{\pi*}(\vec{k})$ (eqn (5.5a)). The similarity of expressions (5.6) and (5.5a) is striking: they only differ in the square power of the cosine function. Figure 5.13 is a representation of the variation as a function of k_a of the energy of the $CO^{\pm}(\vec{k})$ crystal orbitals (continuous lines) and the $(BO)_\pi(\vec{k})$ and $(BO)_{\pi*}(\vec{k})$ Bloch orbitals (dotted lines).

There is a strong similarity between the lines representing the energy of the crystal orbitals and those of the Bloch orbitals $(BO)_\pi(\vec{k})$ and $(BO)_{\pi*}(\vec{k})$ as a function of k_a. This means that the band structure for *trans*-polyacetylene may be obtained qualitatively assuming that the Bloch orbitals interact weakly, in other words assuming that the $h_{\pi,\pi*}(\vec{k})$ term can be neglected.

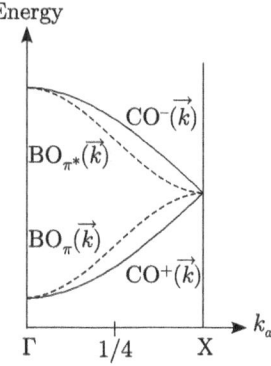

Fig. 5.13
Energy of the Bloch orbitals $\{(BO)_\pi(\vec{k}), (BO)_{\pi*}(\vec{k})\}$ generated from the π and π^* fragment orbitals (dotted lines, see eqn (5.5a)). The energy of the crystal orbitals of the system are also shown (continuous lines, see eqn (5.6)).

5.4.2 Qualitative determination of the band structure

In this section we will try to generate the band structure for regular *trans*-polyacetylene without explicitly solving the secular determinant. We will only consider the shape of the dotted lines in Fig. 5.13, which represent the energy of the Bloch orbitals $(BO)_\pi(\vec{k})$ and $(BO)_{\pi*}(\vec{k})$ as a function of k_a. The interaction between these orbitals is stronger when their energy difference is smaller (i.e. near the X point) and when the interaction term ($h_{\pi,\pi*}(\vec{k})$, see eqn (5.5b)) is large (i.e. near the point associated with k_a equal to 1/4). As a consequence, the mixing between the two Bloch orbitals near Γ is small: the $CO^+(\vec{k})$ and $CO^-(\vec{k})$ crystal orbitals scarcely differ from the $(BO)_\pi(\vec{k})$ and $(BO)_{\pi*}(\vec{k})$ Bloch orbitals and the $c_\pi^+(\vec{k})/c_{\pi*}^+(\vec{k})$ and $c_\pi^-(\vec{k})/c_{\pi*}^-(\vec{k})$ ratios are, in absolute value, considerably larger than one. In contrast, for points not near Γ, an interaction between $(BO)_\pi(\vec{k})$ and $(BO)_{\pi*}(\vec{k})$ develops and leads to a crystal orbital $CO^+(\vec{k})$ more stable than $(BO)_\pi(\vec{k})$ and a crystal orbital $CO^-(\vec{k})$ less stable than $(BO)_{\pi*}(\vec{k})$, as shown in Fig. 5.14.

It may then be concluded that the band structure of regular *trans*-polyacetylene may be generated by 'pushing out' from each other the curves representing the energy dependence of the Bloch orbitals generated by the π and π^* fragments. The 'repulsion' between Bloch orbitals is at a minimum around the center of the Brillouin zone (Γ) and a maximum around the $k_a = 1/4$ point (see Fig. 5.15).

At X the interaction term is nil, so that the Bloch orbitals $BO_\pi(X)$ and $BO_\pi^*(X)$ form a basis for the degenerate level. Near X, although the interaction term is small, the mixing between Bloch orbitals is noticeable because the two Bloch orbitals are quasi-degenerate. This explains the differences between the curves associated with the crystal orbitals and Bloch orbitals around X: whereas the tangent is horizontal for the latter it is non-nil for the former. Again, this means that the Bloch orbital curves 'repel' each other to give the curve associated with the crystal orbitals.

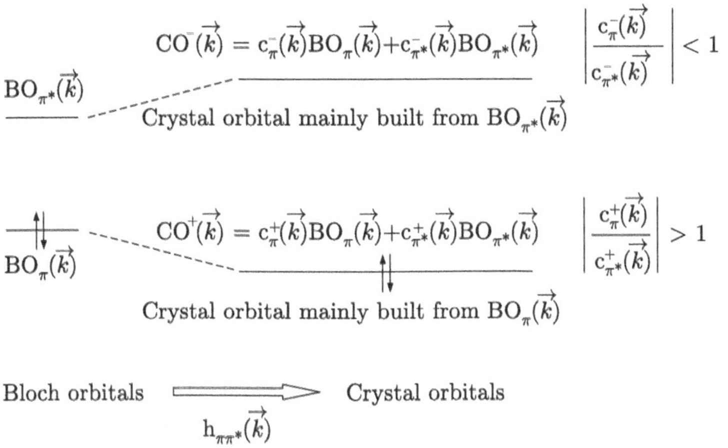

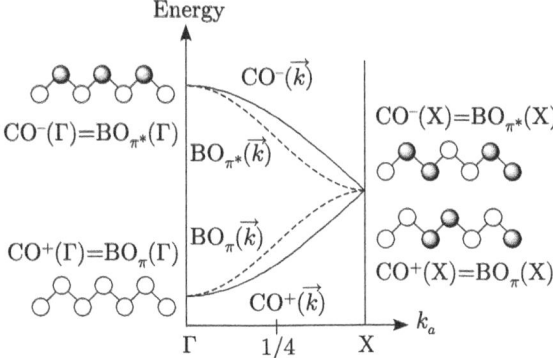

Fig. 5.14
Schematic diagram showing how the $(BO)_\pi(\vec{k})$ and $(BO)_{\pi*}(\vec{k})$ Bloch orbitals interact and lead to the crystal orbitals of the system.

Fig. 5.15
Consequences of the interaction between the Bloch orbitals: the band structure may be obtained by 'repelling' the dotted lines representing the energy of the $(BO)_\pi(\vec{k})$ and $(BO)_{\pi*}(\vec{k})$ Bloch orbitals.

This example illustrates how convenient it can be to develop the crystal orbitals in terms of a basis of Bloch orbitals associated with appropriate fragment orbitals. In favourable cases it allows the determination and qualitative analysis of the band structure of the system. Often the crystal orbitals at Γ and X may be identified with one of the Bloch orbitals generated by appropriate fragment orbitals, thus greatly simplifying the analysis. In the following chapters we will use this idea as much as possible since it gives useful hints on the nature of the band structure of many compounds without having to do any detailed calculations.

5.5 Band gap opening at the Fermi level in *trans*-polyacetylene

The ability to modify the properties of solids may be more rational and effective if it is possible to predict the evolution of the electronic structure when some perturbation occurs. In particular, it is important to be able to understand under what circumstances, of either structural or chemical origin, an energy gap can develop at the Fermi level, since the existence of such a gap completely changes the nature of the conductivity of the system. As we show now, in

Application to *trans*-polyacetylene

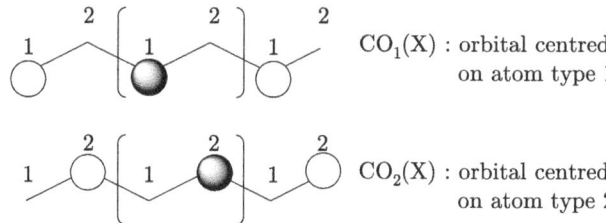

Fig. 5.16
Different crystal orbitals basis appropriate for the description of some features associated with the Fermi level of *trans*-polyacetylene.

order to examine these different circumstances, it is often useful to adapt the way in which we describe the electronic structure at the Fermi level of the initial, unperturbed system, to the kind of perturbation under consideration. For instance, according to the band structure of regular *trans*-polyacetylene (see Fig. 5.7), any perturbation lifting the degeneracy at the Fermi level (i.e. at X) may stabilise the system. Since there are degenerate states at the Fermi level, different crystal orbital bases may be considered. Some of these bases will suggest a quick answer to the question of how to lose the degeneracy. Thus, if the states at the Fermi level are described by means of the two crystal orbitals $CO_\pi(\vec{k}_f)$ and $CO_{\pi^*}(\vec{k}_f)$, as in Fig. 5.7, it is clear that a dimerisation of *trans*-polyacetylene (see Fig. 5.9) will localise the single and double bonds and, as a consequence, will open a gap at the Fermi level (Fig. 5.10).

It is also possible to describe the states at the Fermi level by means of the crystal orbitals $CO_1(X)$ and $CO_2(X)$, which may be expressed as $\frac{1}{\sqrt{2}}(CO_\pi(X) \pm CO_{\pi^*}(X))$, respectively. These are schematically shown in Fig. 5.16. These crystal orbitals concentrate in two different series of atoms of the system.

This representation suggests that if the electronegativity of one of the two series of atoms is changed because of some chemical perturbation, an energy gap will open at the Fermi level because the crystal orbital centred on the more electronegative atom will be stabilised. One way to do this is by using different substituents for the two types of sites in the system, as shown in Fig. 5.17. In the same vein, in Exercise (5.2) we consider a compound derived from *trans*-polyacetylene, where carbon atoms have been replaced by boron and nitrogen atoms.

Fig. 5.17
Scheme of how substituents can be used to open a band gap at the Fermi level of *trans*-polyacetylene.

Exercises

(5.1) Electronic structure of *trans*-polyacetylene: provide a proof for eqns (5.5a) and (5.5b) using the Hückel approach.

(5.2) Study of the $(BNH_2)_n$ chain represented below.

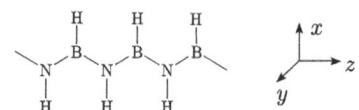

(a) Propose a Lewis structure for this chain.
(b) Why can the states near the Fermi level be appropriately described by using only the $2p_y$ atomic orbitals of the boron and nitrogen atoms?
(c) Determination of the band structure using the Hückel approach: assume that the α_N and α_B constants for nitrogen and boron may be estimated by $\alpha + \beta$ and $\alpha - \beta$, where β is the interaction term between two

adjacent $2p_y$ orbitals and α is the energy of a $2p_y$ orbital of the carbon atom.

  a. Give an expression for the energies allowed.
  b. Determine the crystal orbitals at Γ and X.
  c. Build the band structure.
  d. Where is the Fermi level?
  e. Provide a qualitative discussion of the results.

(d) Qualitative study of the band structure.

  a. Show the energy as a function of k_a for the Bloch orbitals $BO_N(k_a)$ and $BO_B(k_a)$ generated by the $2p_y$ orbitals centred on the nitrogen and boron atoms.
  b. Give an estimate of the interaction term between these two Bloch orbitals.
  c. Use these results to build the band structure of the system.

References

1. Y. Jean and F. Volatron, *Structure Électronique des Molécules*, Ediscience, Paris, 1993, vol. 2, p. 57; The English translation is: Y. Jean, F. Volatron, and J. K. Burdett, *An Introduction to Molecular Orbitals*, Oxford University Press, New York, 1993, p. 200.
2. W. L. Jorgensen and L. Salem, *The Organic Chemist's Book of Orbitals*, Academic Press, New York, 1973.
3. For a discussion of the band structure of *trans*- and *cis*-polyacetylene at the extended Hückel level see: (a) M.-H. Whangbo, R. Hoffmann, and R. B. Woodward, *Proc. Roy. Soc. London*, A366, 23, 1979; (b) M.-H. Whangbo in: *Extended Linear Chain Compounds*, J. S. Miller, Ed., Plenum Press, New York, 1982, vol. 2, ch. 3; (c) J. K. Burdett, *Prog. Sol. State Chem.*, 15, 173, 1984; (d) J. P. Lowe, S. A. Kafafi, and J. P. LaFemina, *J. Phys. Chem.*, 90, 6602, 1986.

Handling the symmetry in 1D compounds

In Chapter 3 we obtained the mathematical expression for the Bloch orbitals of a periodic system by building a linear combination of orbitals adapted to the translational symmetry of the system. We are now going to consider other symmetry properties of 1D compounds. [1, 4] At the end of the chapter, we then outline a general approach to the study of the electronic structure of 1D systems that exploits, as far as possible, the symmetry properties of the system.

6.1 Analysis of the A_n system

Let us consider the periodic system with formula A_n, where A denotes a unit cell of the system, and in which the different atoms are described using a set of orbitals $\{\varphi_1, \varphi_2, \ldots, \varphi_{N_0}\}$. The Bloch orbitals associated with the different orbitals $(\varphi_j)_m$ ($m = 0, \pm 1, \pm 2, \ldots, \pm(n'-1), n'$) will be denoted $(BO)_j(\vec{k})$. We will consider the system in the cyclic and then the linear forms and we will try to establish some simple rules that allow one to predict when two Bloch orbitals $(BO)_j(\vec{k})$ and $(BO)_{j'}(\vec{k}')$ will not interact, i.e. when the interaction term $h_{jj'} = \langle (BO)_j(\vec{k}) \mid \hat{h} \mid (BO)_{j'}(\vec{k}') \rangle$ is nil.

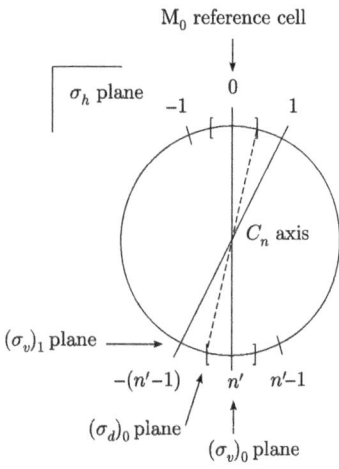

Fig. 6.1
Cyclic system A_n. Possible symmetry planes such as the horizontal σ_h plane and the vertical $((\sigma_v)_0, (\sigma_v)_1)$, and $(\sigma_d)_0$ have been schematically represented. The $(\sigma_v)_0$ symmetry plane has been represented as being at the centre of the reference unit cell but this may not necessarily be so.

6.1.1 Analysis of the cyclic A_n system

Let us begin by considering the system in its cyclic form (Fig. 6.1). Taking into account the symmetry properties of the C_n group it appears that the term $h_{jj'}(\vec{k}, \vec{k}')$ is nil if $\vec{k}$ is different from $\vec{k}'$ because the $\vec{k}$ and $\vec{k}'$ vectors are associated with different irreducible representations of the C_n group. In the following sections we will discuss how to use the possible symmetry operations of the system to predict when the interaction term $\langle (BO)_j(\vec{k}) \mid \hat{h} \mid (BO)_{j'}(\vec{k}) \rangle$ is nil.

Existence of an horizontal σ_h symmetry plane

Here we will explore the consequences of the existence of a σ_h symmetry plane perpendicular to the C_n rotation axis. In this case, the C_{nh} group leaves the system invariant. Under such circumstances we must build a basis of Bloch orbitals, $\{(BO)_j(\vec{k}), j = 1, \ldots, N_0\}$, generated by orbitals, $(\varphi_j)_0$, which are symmetric or antisymmetric with respect to this plane. Since the different orbitals, $(\varphi_j)_m$, have the same symmetry properties with respect to this symmetry plane, the symmetry properties of a Bloch orbital, $(BO)_j(\vec{k})$, are characterised by the k point value and the symmetric or antisymmetric character with respect to σ_h of the orbital in the reference cell, $(\varphi_j)_0$.

> If the cyclic system possesses a horizontal symmetry plane, σ_h, the interaction term $h_{jj'}(\vec{k})$ is nil provided that the $(\varphi_j)_0$ and $(\varphi_{j'})_0$ orbitals do not have the same properties with respect to the σ_h plane.

Existence of vertical σ_v symmetry planes

Now let us assume that the system exhibits a vertical symmetry plane, $(\sigma_v)_0$, containing the rotation axis, C_n, and going through the M_0 and $M_{n'}$ cells. The symmetry group for this system, generated by the C_n (n even) rotation and by this symmetry plane, $(\sigma_v)_0$, is the group C_{nv} (n even). This group contains n rotations C_n^m ($m = 0, \pm 1, \ldots, \pm(n'-1), n'$) and n symmetry operations related to the vertical planes, which may belong to two different classes denoted $\{(\sigma_v)_i \, ; \, i = 0, 1, \ldots, n'-1\}$ and $\{(\sigma_d)_i \, ; \, i = 0, 1, \ldots, n'-1\}$. The planes $(\sigma_v)_i$ and $(\sigma_d)_i$ go through the M_i and $M_{i-n'}$ cells, and the plane $(\sigma_d)_i$ is the bisector of the $(\sigma_v)_i$ and $(\sigma_v)_{i+1}$ planes (Fig. 6.1).

We will use the C_{nv} group to characterise the symmetry properties of the Bloch orbitals. Because of the existence of the $(\sigma_v)_0$ and $(\sigma_d)_0$ symmetry planes passing through the reference cell, we need to build a basis of Bloch orbitals, $\{(BO)_j(\vec{k}) \, ; \, j = 1, \ldots, N_0\}$ generated by the $(\varphi_j)_0$ orbitals, which are symmetric or antisymmetric with respect to either $(\sigma_v)_0$ or $(\sigma_d)_0$. Obviously, a $(\varphi_j)_0$ orbital can only be adapted to one of the two symmetry operations.

We will start by projecting over the character table of group C_{nv} the basis for the Γ_j representation, which is made up of the different φ_j orbitals $\{(\varphi_j)_m, m = 0, \pm 1, \ldots, \pm(n'-1), n'\}$. The different Bloch orbitals $(BO)_j(\vec{k})$ will be labelled by means of the different irreducible representations of the C_{nv} group. As shown in Table 6.1, the character table for the C_{nv} group possesses four non-degenerate irreducible representations (A_1, A_2, B_1, B_2) and $(n'-1)$ doubly degenerate representations, taking into account the fact that the character table of the C_{nv} group may be deduced from that of the C_n group (see Table 3.1). The character of Γ_j is nil for all the rotations C_n^m ($m = \pm 1, \ldots, \pm(n'-1), n'$) because none of the φ_j functions is stable upon their action. Since the different symmetries may be grouped together into two different classes ($n'\sigma_v$ and $n'\sigma_d$), we just need to estimate the character of Γ_j with respect to the $(\sigma_v)_0$ and $(\sigma_d)_0$ symmetries. These characters will be denoted $\chi_{n'\sigma_v}(\Gamma_j)$ and $\chi_{n'\sigma_d}(\Gamma_j)$. Only the $(\varphi_j)_0$ and $(\varphi_j)_{n'}$ may be at the origin of the non-nullity of the character of Γ_j with respect to the $(\sigma_v)_0$ and $(\sigma_d)_0$ symmetries. In addition, the two orbitals $(\varphi_j)_0$ and $(\varphi_j)_{n'}$ necessarily have the same character with respect to these symmetries. Consequently, the decomposition of the Γ_j representation depends on the symmetry

Symmetry in 1D compounds

Table 6.1 The six initial lines reproduce the character table of group C_{nv} (n even). The four last lines are the characters of the Γ_j representation for different possible cases.

C_{nv}	E	C_n C_n^{-1}	C_n^2 C_n^{-2}	...	$C_n^{n'-1}$ $C_n^{1-n'}$	C_2	$n'\sigma_v$	$n'\sigma_d$
A_1		\multicolumn{5}{c	}{Characters of $\Gamma_{i=0}$ of the C_n group ($k_a = 0$)}	1	1
A_2		\multicolumn{5}{c	}{Characters of $\Gamma_{i=0}$ of the C_n group ($k_a = 0$)}	-1	-1
B_1		\multicolumn{5}{c	}{Characters of $\Gamma_{i=n'}$ of the C_n group ($k_a = 1/2$)}	1	-1
B_2		\multicolumn{5}{c	}{Characters of $\Gamma_{i=n'}$ of the C_n group ($k_a = 1/2$)}	-1	1
E_i ($i = 1,\ldots, n'-1$)		\multicolumn{5}{c	}{Characters of $\Gamma_i + \Gamma_{-i}$ of the C_n group ($k_a = \pm i/n$)}	0	0
Γ_j if $(\varphi_j)_0$ is symmetric w.r.t. $(\sigma_v)_0$	n	0	0	...	0	0	2	0
Γ_j if $(\varphi_j)_0$ is antisymmetric w.r.t. $(\sigma_v)_0$	n	0	0	...	0	0	-2	0
Γ_j if $(\varphi_j)_0$ is symmetric w.r.t. $(\sigma_d)_0$	n	0	0	...	0	0	0	2
Γ_j if $(\varphi_j)_0$ is antisymmetric w.r.t $(\sigma_d)_0$	n	0	0	...	0	0	0	-2

properties of $(\varphi_j)_0$ with respect to $(\sigma_v)_0$ and $(\sigma_d)_0$ (i.e. the four last lines of Table 6.1).

Thus we can state that:

- if $(\varphi_j)_0$ is symmetric with respect to $(\sigma_v)_0$

$$\chi_{n'\sigma_v}(\Gamma_j) = 2 \quad \Rightarrow \quad \Gamma_j = A_1 + B_1 + \sum_{i=1}^{n'-1} E_i \quad (6.1)$$

- if $(\varphi_j)_0$ is antisymmetric with respect to $(\sigma_v)_0$

$$\chi_{n'\sigma_v}(\Gamma_j) = -2 \quad \Rightarrow \quad \Gamma_j = A_2 + B_2 + \sum_{i=1}^{n'-1} E_i \quad (6.2)$$

- if $(\varphi_j)_0$ is symmetric with respect to $(\sigma_d)_0$

$$\chi_{n'\sigma_d}(\Gamma_j) = 2 \quad \Rightarrow \quad \Gamma_j = A_1 + B_2 + \sum_{i=1}^{n'-1} E_i \quad (6.3)$$

- if $(\varphi_j)_0$ is antisymmetric with respect to $(\sigma_d)_0$

$$\chi_{n'\sigma_d}(\Gamma_j) = -2 \quad \Rightarrow \quad \Gamma_j = A_2 + B_1 + \sum_{i=1}^{n'-1} E_i \quad (6.4)$$

According to Table 6.1, a Bloch orbital associated with the Γ point is a basis for the representations A_1 or A_2, whereas a Bloch orbital associated with the X point is a basis for the representations B_1 or B_2. A basis for an E_i ($i = 1, \ldots, n' - 1$) representation is made of two degenerate Bloch orbitals, $BO_j(\vec{k})$ and $BO_j(-\vec{k})$, associated with the $\pm k_a$ values equal to $\pm i/n$.

The analysis of the different possible decompositions (eqns (6.1)–(6.4)) shows that at every k-point other than Γ and X, all Bloch orbitals $BO_j(\vec{k})$ ($j = 1, \ldots, N_0$) associated with the same k_a value are a basis for the same irreducible representation $E_i(i = nk_a)$, whatever the nature of the $(\varphi_j)_0$ orbital that generates it. Consequently, the fact that two Bloch orbitals may not interact at points other than Γ and X is not related to these two planes. According to eqns (6.1)–(6.4), at Γ and X the interaction term between two Bloch orbitals $h_{jj'}(\vec{k})$ may be nil because of these vertical symmetry planes. For instance, at Γ, the Bloch orbital $BO_j(\Gamma)$ generated by a $(\varphi_j)_0$ orbital that is symmetric with respect to $(\sigma_v)_0$ (i.e. of A_1 symmetry) cannot interact with a $BO_{j'}(\Gamma)$ Bloch orbital generated by a $(\varphi_{j'})_0$ orbital that is antisymmetric with respect to $(\sigma_v)_0$ (i.e. of symmetry A_2) or with a $BO_{j'}(\Gamma)$ Bloch orbital generated by a $(\varphi_{j'})_0$ orbital that is antisymmetric with respect to $(\sigma_d)_0$) (i.e. of symmetry A_2).

> In conclusion, the existence of vertical symmetry planes, σ_v and σ_d, may be at the origin of the nullity of the interaction terms $h_{jj'}(\vec{k})$ at the Γ and X points. In contrast, for any other point, the σ_v and σ_d symmetry planes do not help in characterising the symmetry properties of the Bloch orbitals.

Using these results

We can now provide an illustration of these results by looking at the A_n system in which A is an atom described by ns and np_z orbitals. Let us consider the interaction term $h_{ns,np_z}(\vec{k})$, between the Bloch orbitals $BO_{ns}(\vec{k})$ and $BO_{np_z}(\vec{k})$ generated by the $(ns)_0$ and $(np_z)_0$ orbitals, respectively, which are centred on the A atom of the reference cell. We denote as np_z the np orbitals the symmetry axis of which is tangential to the circle (see Fig. 6.2a). Such a system possesses a horizontal symmetry plane, as well as vertical symmetry planes $(\sigma_v)_m$ ($m = 0, \ldots, n' - 1$) going through the A_m atoms and $(\sigma_d)_m$ ($m = 0, \ldots, n' - 1$) planes passing in between the A_m and A_{m+1} atoms (see Fig. 6.2b).

The existence of a σ_h symmetry plane cannot lead to the nullity of $h_{ns,np_z}(\vec{k})$ because $(ns)_0$ and $(np_z)_0$ are both symmetric with respect to the σ_h plane. According to the conclusions of the previous section, consideration of the $(\sigma_v)_m$ and $(\sigma_d)_m$ symmetries can only lead to the prediction of the possible nullity of the $h_{ns,np_z}(\vec{k})$ interaction term at the Γ and X points because the $(ns)_0$ and $(np_z)_0$ orbitals are symmetric and antisymmetric, respectively, with respect to $(\sigma_v)_0$ (see Fig. 6.2c). In contrast, at any other point these two Bloch orbitals may interact since these two symmetry operations do not apply (eqns (6.1) and (6.2)). This result may appear surprising because the $(ns)_0$ and $(np_z)_0$ orbitals behave differently with respect to the $(\sigma_v)_0$ plane. To realise that the Bloch orbitals $BO_{ns}(\vec{k})$ and $BO_{np_z}(\vec{k})$ may interact at any k point other than Γ and X we can calculate the interaction term $h_{ns,np_z}(\vec{k})$ using the procedure proposed in Section 3.2.2. Let us recall that:

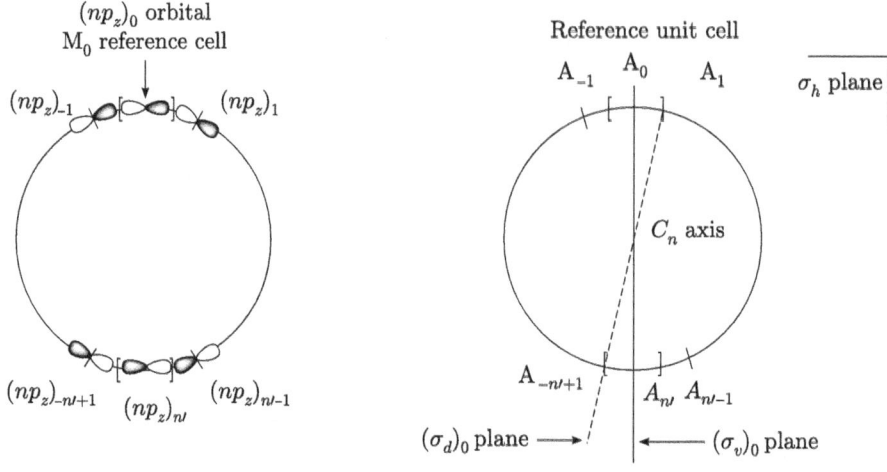

Fig. 6.2
(a) Definition of the np_z orbitals; (b) A_n system generated by an M_0 unit cell centred on the atom A_0; (c) symmetry properties with respect to $(\sigma_d)_0$ of the atomic orbitals of the reference cell.

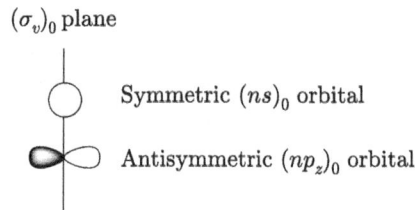

$$h_{ns,np_z}(\vec{k}) = n \left\langle \frac{(ns)_0}{\sqrt{(n)}} \,\bigg|\, \hat{h} \,\bigg|\, BO_{np_z}(\vec{k}) \right\rangle \tag{6.5}$$

To facilitate the discussion, we will use the Hückel approach even if our purpose is completely general and strictly governed by symmetry. We will also use the linear representation of the system to simplify the estimation of the interaction between $\frac{(ns)_0}{\sqrt{n}}$ and the Bloch orbital $BO_{np_z}(\vec{k})$ (see Fig. 6.3).

Because of the occurrence of $\exp(i2\pi k_a)$ and $\exp(-i2\pi k_a)$ coefficients in the expression of the $BO_{np_z}(\vec{k})$ orbital, the interaction term $h_{ns,np_z}(\vec{k})$ (eqn (6.5)) can be estimated to be $2\beta i \sin(2\pi k_a)$. Thus, $h_{ns,np_z}(\vec{k})$ is non-zero if k_a is different from 0 or 1/2. This clearly shows the peculiarity of points Γ and X. For these two points the coefficients associated with orbitals $(np_z)_{+1}$ and $(np_z)_{-1}$ are equal, so that the interaction between the $BO_{ns}(\vec{k})$ and $BO_{np_z}(\vec{k})$ orbitals is effectively nil. For all other points this is not the case: the coefficients associated with the np_z orbitals of $BO_{np_z}(\vec{k})$ at the −1 and +1 positions, $\exp(-i2\pi k_a)$ and $\exp(i2\pi k_a)$, are not equal and thus $h_{ns,np_z}(\vec{k})$ is

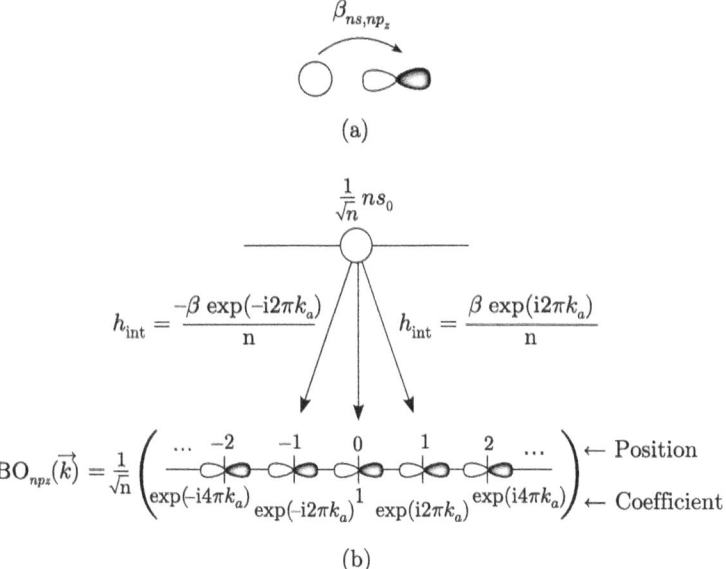

Fig. 6.3
(a) Interaction term β_{ns,np_z} between two adjacent orbitals $(ns)_0$ and $(np_z)_1$; (b) interaction between $\frac{(ns)_0}{\sqrt{n}}$ and the Bloch orbital $BO_{np_z}(\vec{k})$.

not nil. Loosely speaking we can say that the two coefficients $\exp(-i2\pi k_a)$ and $\exp(i2\pi k_a)$ break down the symmetry with respect to the 0 position.

Concluding remarks

We can now generalise the results discussed in the three previous paragraphs. When taking into account the symmetry properties of a cyclic periodic system we must successively consider the points below.

- *The rotations with respect to the main axis perpendicular to the circle. Their consideration allows the construction of the Bloch orbitals. Their associated k point characterises the symmetry properties of these functions with respect to the rotations of the C_n group.*
- *The possible σ_h symmetry leaving invariant every cell. Only the Bloch orbitals possessing the same symmetry properties with respect to this operation may interact.*
- *The possible symmetry operations other than rotations of the C_n group that do not leave invariant every cell*:
  - at Γ and X, only the Bloch orbitals possessing the same symmetry properties with respect to these operations may interact
  - at any other point these symmetry operations may be ignored.

6.1.2 Analysis of the linear A_n system

In the previous section we found it useful to use the cyclic representation of the periodic system because it allowed us to reason on the basis of the properties of point groups. In the following, it will be easier to reason directly using the real geometry of the systems under consideration, i.e. linear geometry. In this section we will show that it is possible to transpose the results of the previous section to linear periodic systems *since they fulfil the Born–von Karman periodic boundary conditions.*

Taking into account the translational properties

Essentially, the Born–von Karman periodic boundary conditions impose the equivalence of a C_n rotation on the cyclic representation and a translation by a vector $\vec{a}$, $t_{\vec{a}}$, on the linear representation, as well as the equivalence of a $t_{n\vec{a}}$ translation and the identity. Consequently, the group of rotations C_n of the cyclic system is equivalent to the group of translations T_n of the linear system. This group contains all translations by vectors that are multiples of the $\vec{a}$ vector, $t_{m\vec{a}}$ ($m = 0, \pm 1, \pm 2, \ldots, (n'-1), n'$). The product tables of the different symmetry operations of the C_n and T_n groups are identical provided that the roles played by the C_n^m rotation and the $t_{m\vec{a}}$ translation are considered identical. These two groups are said to be *isomorphic*. This means that the rotations in the cyclic system play the same role as the translations in the linear system. These two groups possess exactly the same properties, namely the same character table. The Bloch orbitals of the linear system obtained in Chapter 3 (see Fig. 3.3) are consequently bases for the different irreducible representations of the translation group.

Notion of space group

By definition, we will denote as the *space group G* of a periodic system the group containing the symmetry operations leaving the system invariant. Besides the group of translations, T_n, this group also contains the point symmetry operations as well as the combined operations resulting from a product between a point symmetry operation and a translation.

Here we will restrict ourselves to the space symmetry groups G that may be written as the direct product of the translation group T_n and a point group P ($G = T_n \otimes P$). These groups are known as symmorphic. We will consider the existence of non-symmorphic groups in Section 6.4 of this chapter.[1]

Under such conditions the combined symmetry operations resulting from the product between a point symmetry operation and a translation necessarily involve a symmetry operation of the P group and a translation of the T_n group. *In addition, since any point group leaves one point invariant, we choose the P group in such a way that it contains the largest possible number of symmetry operations and leaves invariant at least the point O of the reference cell.*

Space group of the $(Pt(CN)_4)^{2-}$ chain

We will now illustrate the notion of space group by looking at the case of a $(Pt(CN)_4)^{2-}$ chain in an eclipsed configuration, as shown in Fig. 6.4. What space group P leaves invariant this chain as well as a point O of the reference cell? In view of the symmetry of the system it seems natural to place point O either on a platinum atom of the reference cell Pt_0 or at the middle of the Pt_0–Pt_1 segment. Let us place it at the Pt_0 site. It would be

---

[1] Working with space groups which may be written as a direct product $T_n \otimes P$ may involve neglecting some symmetry operations, i.e. those written as the product of a point symmetry operation R and a translation $t_{\vec{b}}$ such that the operations R and $t_{\vec{b}}$ do not leave the system invariant.

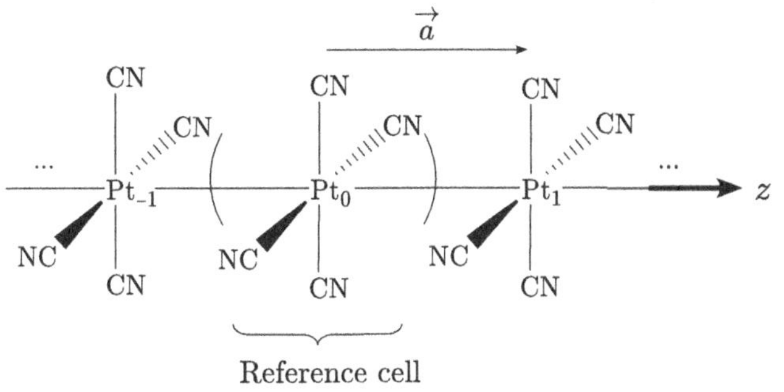

Fig. 6.4
Eclipsed structure of the $(Pt(CN)_4)^{2-}$ chain.

completely equivalent to place it in the middle of the Pt_0–Pt_1 segment (see Exercise (6.2)). The symmetry operations of the D_{4h} group, which leave this O point invariant, also leave the whole chain invariant. We will label this group $(D_{4h})_0$ because it leaves atom Pt_0 invariant. Consequently, the $(D_{4h})_0$ group is the P group that we are looking for.[2] Now let us determine the different symmetry operations contained in the space group G, which is equal to $T_n \otimes (D_{4h})_0$. It is possible to associate the different symmetry operations of $(D_{4h})_0$ in two subgroups E_1 and $(E_2)_0$. E_1 contains the symmetry operations leaving invariant every point of the Oz axis, i.e. $E_1 = \{E, 2C_4, C_2, 2\sigma_v, 2\sigma_d\}$. $(E_2)_0$ contains the symmetry operations that only leave invariant one point of the Oz axis, Pt_0, so that $(E_2)_0 = \{(i)_0, 2(S_4)_0, (\sigma_h)_0, 2(C'_2)_0, 2(C''_2)_0\}$. The index 0 is a reminder of the fact that the symmetry operation leaves the Pt_0 point invariant. The product of a $t_{\vec{a}}$ translation and a symmetry operation of E_1 generates a combined symmetry operation (point symmetry operation–translation). In contrast, the product of a $t_{\vec{a}}$ translation and a symmetry operation of $(E_2)_0$ generates the whole set of point symmetry operations of $(E_2)_{0/1}$, i.e. $\{(i)_{0/1}, 2(S_4)_{0/1}, (\sigma_h)_{0/1}, 2(C'_2)_{0/1}, 2(C''_2)_{0/1}\}$, which leave the middle point of the Pt_0–Pt_1 segment invariant.[3]

Taking into account the different symmetry operations of G

We can now generalise the results discussed in Section 6.1.1 for the cyclic system to 1D compounds.

When taking into account the symmetry properties of a cyclic periodic system we must successively consider the following.

- *The translational symmetry operations* $\{t_{m\vec{a}}, m = 0, \pm 1, \ldots, \pm(n'-1), n'\}$ *of the group* T_n. These allow the construction of the Bloch orbitals, $BO_j(\vec{k})$. Only two Bloch orbitals associated with the same k point may interact.
- *The symmetry operations of P leaving every cell invariant.* These symmetry operations may be rotations around the 1D axis and reflections related to the symmetry planes containing this 1D axis. Only the Bloch orbitals possessing similar properties with respect to such operations may interact.

---

[2] One must be very careful in defining the Oz axis in the cyclic and linear representations. In the cyclic system the Oz axis is the highest order rotation axis, i.e. the C_n rotation axis. In contrast, the Oz axis in the linear system is the axis of the P group which is associated with the highest order rotation axis, i.e. in the present case, the horizontal axis (see Fig. 6.4).

[3] The symmetry operations of $(E_2)_0$ do not commute with the $t_{\vec{a}}$ translation. The product of $t_{\vec{a}}$ and a symmetry operation g of $(E_2)_0$, i.e. $t_{\vec{a}} \cdot g$, generates a symmetry operation of $(E_2)_{0/1}$ leaving the middle point of the Pt_0–Pt_1 segment invariant. However, the product of a symmetry operation of $(E_2)_0$ and $t_{\vec{a}}$, i.e. $g \cdot t_{\vec{a}}$, generates a symmetry operation of $(E_2)_{-1/0}$ leaving invariant the middle point of the Pt_{-1}–Pt_0 segment.

- *The symmetry operations of P which do not leave invariant every cell. These symmetry operations may be the inversion with respect to the O point, the reflection with respect to a plane perpendicular to the 1D axis, and the C_2' rotations with respect to an axis perpendicular to the 1D axis. At Γ and X these symmetry operations may be responsible for the non-interaction between two Bloch orbitals, whereas at any other k point it is possible to neglect them.*

6.1.3 Notion of group of a k point

What are the symmetry operations that must be used to characterise the symmetry properties of a Bloch orbital, $BO(\vec{k})$? Only the symmetry operations of the group P may be responsible for the non-interaction of two Bloch orbitals associated with the same k point. More specifically, in view of the conclusions reached in Section 6.1.2, it is possible to characterise the symmetry of a Bloch orbital by means of the symmetry operations of the P group at the Γ and X points. In contrast, at any other point, only symmetry operations of P leaving every cell invariant may be used to characterise the symmetry of the Bloch orbitals.

By definition, we will use the term *group of the k point* to denote the symmetry group containing the symmetry operations allowing the characterisation of the symmetry properties of the Bloch orbitals at this point. The group of the Γ and X points is the P point group while the group of the other k points is the subgroup of P containing all the symmetry operations of P leaving every cell invariant. Every $BO_j(\vec{k})$ Bloch orbital may be labelled with an irreducible representation of the group of the k point. According to the discussion above, only two Bloch orbitals, $BO_j(\vec{k})$ and $BO_{j'}(\vec{k})$, that are a basis for the same irreducible representation of the group of the k point may interact.

The notion of group of the k point may be illustrated by considering the $(Pt(CN)_4)^{2-}$ chain (see Fig. 6.4). The group of the k points Γ and X is the $(D_{4h})_0$, whereas the group of the other k points is the $(D_{4h})_0$ group from which the inversion $(i)_0$ with respect to Pt_0, the horizontal symmetry plane $(\sigma_h)_0$ perpendicular to the chain axis, the improper rotations $2(S_4)_0$, and the four rotations $(C_2')_0$ and $(C_2'')_0$, whose axes are perpendicular to the 1D axis, have been eliminated. The group of the k points that are different from Γ and X is consequently the C_{4v} group.

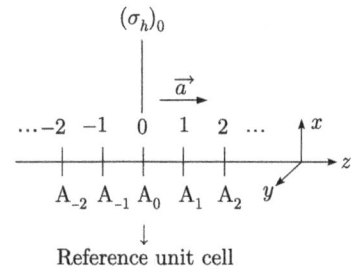

Fig. 6.5
Schematic representation of the periodic A_n system.

6.2 Application to the determination of the band structure for the A_n linear system, where A is an atom

We will now consider a linear periodic system A_n where A is an atom bearing the valence orbitals ns, np_x, np_y, and np_z (see Fig. 6.5). We are going to build the band structure of this system by using, as far as possible, its symmetry

6.2.1 Group of the different k points

This system is stable under the operations of the G space group, which is the direct product of the T_n translation group and the $(D_{\infty h})_0$ group leaving invariant the atom of the reference cell, A_0. Consequently, the appropriate group for Γ and X is the $(D_{\infty h})_0$ group. For any other point, the appropriate group is $C_{\infty v}$, which is a subgroup of $(D_{\infty h})_0$ containing the symmetry operations of $(D_{\infty h})_0$ leaving every cell of the chain invariant.

6.2.2 Symmetry of the different Bloch orbitals

Γ and X points

Looking at the representation of the Bloch orbitals for Γ and X it is easy to determine their symmetry labels according to the $(D_{\infty h})_0$ group (see Fig. 6.6 and Appendix for the character table of the $D_{\infty h}$ symmetry group). The bases $\{BO_{np_x}(\Gamma), BO_{np_y}(\Gamma)\}$ and $\{BO_{np_x}(X), BO_{np_y}(X)\}$ belong to the Π_u symmetry, whereas the Bloch orbitals $BO_{ns}(\Gamma)$ and $BO_{ns}(X)$ are of Σ_g^+ symmetry, and the $BO_{np_z}(\Gamma)$ and $BO_{np_z}(X)$ orbitals belong to the Σ_u^+ symmetry. Consequently, the different orbitals $BO_{np_x}(\vec{k})$, $BO_{np_y}(\vec{k})$, $BO_{np_z}(\vec{k})$, and $BO_{ns}(\vec{k})$ cannot interact at the Γ and X points. Thus the crystal orbitals of the system are nothing more than these Bloch orbitals. The energy ordering of the crystal orbitals at Γ is clearcut:

$$E(BO_{ns}(\Gamma)) < E(BO_{np_x}(\Gamma)) = E(BO_{np_y}(\Gamma)) < E(BO_{np_z}(\Gamma))$$

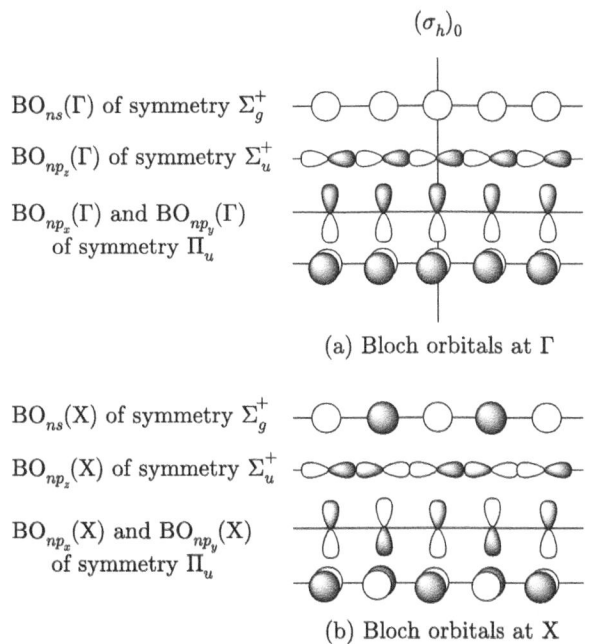

Fig. 6.6
Bloch orbitals of the A_n chain corresponding to the Γ and X points.

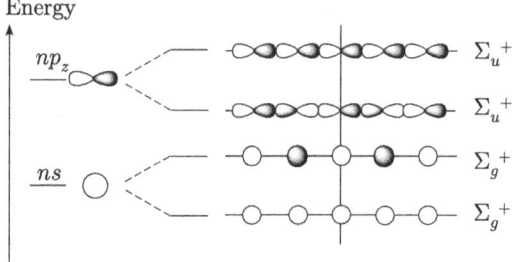

(a) Bloch orbitals based on the ns and np_z orbitals at Γ and X

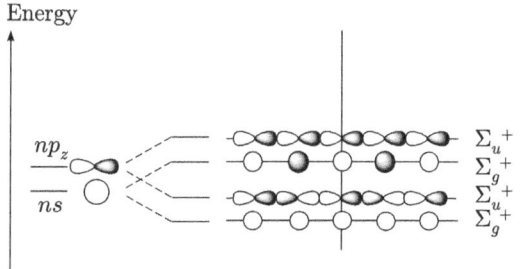

Fig. 6.7
Energies of the BO$_{ns}$ and BO$_{np_z}$ Bloch orbitals for the Γ and X points corresponding to the two different situations considered in the text.

(b) Bloch orbitals based on the ns and np_z orbitals at Γ and X

In contrast, at the X point it is not possible to decide with certainty how the antibonding BO$_{ns}$(X) orbital should be positioned with respect to the bonding BO$_{np_z}$(X) orbital. The energy ordering of these orbitals depends on the energy of the atomic orbitals ns and np. We need to consider the two possibilities shown in Fig. 6.7:

- Case (a): if the energy difference between the ns and np is large and the overlap is moderate, the BO$_{ns}$(X) orbital, despite its antibonding character, lies lower in energy than the BO$_{np_z}$(X) bonding orbital.
- Case (b): if the energy difference between the ns and np is small and/or the overlap is large, the BO$_{ns}$(X) orbital, because of its antibonding character, lies higher in energy than the BO$_{np_z}$(X) bonding orbital.

In both cases, the Bloch orbitals {BO$_{np_x}$(X), BO$_{np_y}$(X)} are associated with a higher energy than the {BO$_{ns}$(X), BO$_{np_z}$(X)} orbitals (except for very special situations in case (b)).

k points other than Γ and X

Since each symmetry operation of the $C_{\infty v}$ group leaves every atom of the chain invariant, we can directly consider the properties of the ns, np_x, np_y, and np_z orbitals in the $C_{\infty v}$ group. Examination of the character table of this group (see Appendix) tells us that the ns and np_z orbitals are bases for the representation Σ^+, whereas the np_x and np_y orbitals are a basis for the representation Π. Consequently, the Bloch orbitals generated by the np_x and np_y orbitals keep their degeneracy at every point and do not interact with the

Table 6.2 Symmetry labels for the Bloch orbitals and crystal orbitals of the A_n chain.

k point	Symmetry label	Bloch orbital	Crystal orbital
Γ and X	Σ_g^+	BO_{ns}	1
Γ and X	Σ_u^+	BO_{np_z}	1
Γ and X	Π_u	$\{BO_{np_x}, BO_{np_y}\}$	2 degenerate orbitals
$ka \neq 0, 1/2$	Σ^+	BO_{ns}, BO_{np_z}	2 orbitals resulting from the interaction of two Bloch orbitals
$ka \neq 0, 1/2$	Π	$\{BO_{np_x}, BO_{np_y}\}$	2 degenerate orbitals

$BO_{ns}(\vec{k})$ and $BO_{np_z}(\vec{k})$ Bloch orbitals. In contrast, the Bloch orbitals $BO_{ns}(\vec{k})$ and $BO_{np_z}(\vec{k})$ can interact at any k point other than Γ and X.

Summary

The main results described in the two previous sections are summarised in Table 6.2.

6.2.3 Bands associated with σ-type overlaps

We will start by considering the two bands generated by the ns and np_z orbitals that are associated with σ-type overlaps.

Energy of the $BO_{ns}(\vec{k})$ and $BO_{np_z}(\vec{k})$ Bloch orbitals

Knowing the shape of the Bloch orbitals at Γ and X (see Figs 6.6 and 6.7), we can represent their energies for any k point (dotted lines in Fig. 6.8). The energy of the $BO_{ns}(\vec{k})$ orbital increases when going from Γ to X whereas the energy of the $BO_{np_z}(\vec{k})$ orbital decreases along the same path. In case (a) (Fig. 6.7) the two dotted lines representing the energy of the Bloch orbitals $BO_{ns}(\vec{k})$ and $BO_{np_z}(\vec{k})$ cannot cross (Fig. 6.8a). However, in case (b) there is a crossing (Fig. 6.8b).

σ-type band structure

Let us now switch to the interaction of the Bloch orbitals $BO_{ns}(\vec{k})$ and $BO_{np_z}(\vec{k})$. This interaction is stronger as the difference in energy between the two Bloch orbitals decreases and the overlap between the orbitals increases. In view of the nil overlap at the points Γ and X, the mixing between the two orbitals will be as much effective as the energies of the two orbitals become similar and as we are farther away from the Γ and X points. Let us begin by considering case (a). The two dotted lines representing the energy of the Bloch orbitals are never very close to each other at points far from X (Fig. 6.8a). Consequently, the mixing between these two Bloch orbitals will be quite modest, i.e. the two dotted lines are only slightly pushed away from each other at any point other than Γ and X. As a result, the lower band will be slightly stabilised and the upper band will be slightly destabilised. The crystal

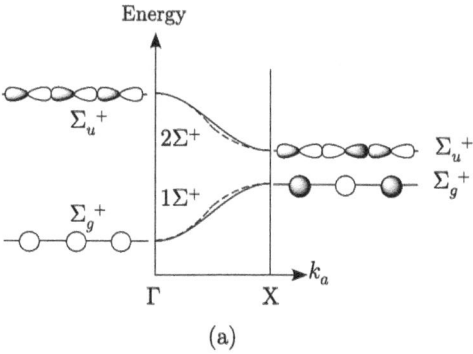

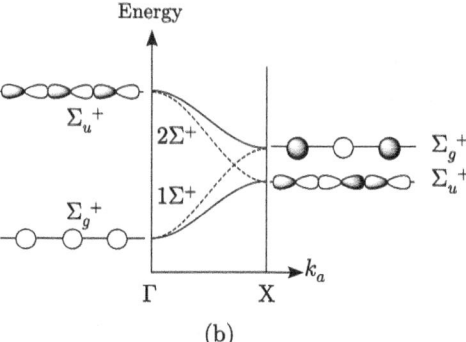

Fig. 6.8
Band structure diagrams associated with the Σ^+-type bands of A_n. The energies of the Bloch orbitals $BO_{ns}(\vec{k})$ and $BO_{np_z}(\vec{k})$ are represented as dotted lines, while the continuous lines represent the energy of the crystal orbitals resulting from the interaction between these two Bloch orbitals. Cases (a) and (b) refer to Figs 6.7a and b, respectively.

orbitals of the lower band are essentially built from orbital $BO_{ns}(\vec{k})$ whereas the crystal orbitals of the upper band are mainly built from orbital $BO_{np_z}(\vec{k})$. In contrast, in the second case (Fig. 6.8b), the Bloch orbitals $BO_{ns}(\vec{k})$ and $BO_{np_z}(\vec{k})$ interact quite strongly near the crossing of the two dotted lines: there is an avoided crossing. The lower band resulting from this interaction is essentially built from orbital $BO_{ns}(\vec{k})$ for k points near Γ, and from orbital $BO_{np_z}(\vec{k})$ for k points near X. For k points near the crossing between the two dotted lines, the crystal orbitals of the two bands result from a strong mixing of the two Bloch orbitals. Thus, in that case, the nature of the predominant Bloch orbital for a given band changes on the way from Γ to X. The main results of the analysis of this section are summarised in Fig. 6.8.

6.2.4 Complete band structure

At this point we need to guess where the degenerate bands involving orbitals np_x and np_y must occur with respect to the other bands. These bands are of Π symmetry for any k point other than Γ and X, and of Π_u symmetry at these two points. Consequently, these bands can cross all other bands, which are of Σ^+ symmetry. On the other hand, in Section 3.2.4 we saw that the Bloch orbitals associated with the points with $k_a = \pm 1/4$ (see Fig. 3.11) are non-bonding and thus their energy coincides with that of the orbital from which they are generated. As a result, the curves representing the energies of the Bloch orbitals generated by the np_x, np_y, and np_z orbitals cross at the points

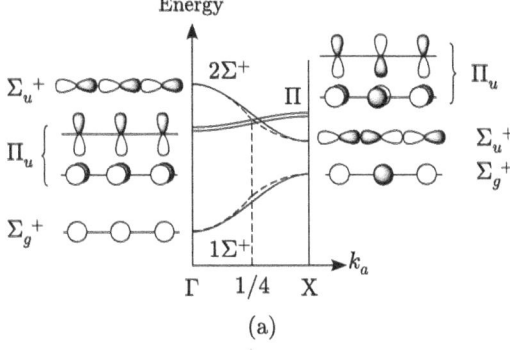

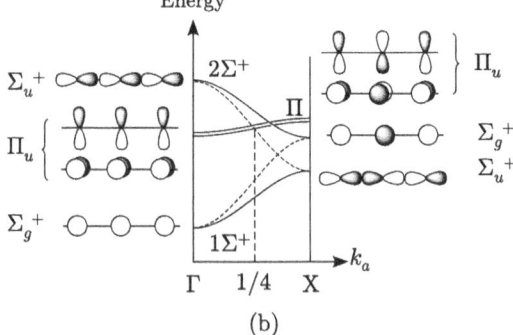

Fig. 6.9
Band structure diagrams for A_n corresponding to the two different situations considered in Fig. 6.7. The dotted lines give the energy dependence of the Bloch orbitals $BO_{ns}(\vec{k})$ and $BO_{np_z}(\vec{k})$.

with $k_a = \pm 1/4$, since the np atomic orbitals are degenerate. Because the lateral overlap between orbitals np_x or np_y is less effective than the axial overlap between the ns and np_z orbitals, the width of the π-type bands is smaller than that of the Σ^+ bands. The present discussion leads us to the two band structures of Fig. 6.9, on which the energies of the Bloch orbitals $BO_{ns}(\vec{k})$ and $BO_{np_z}(\vec{k})$ have also been represented as dotted lines.

This analysis clearly shows that it is possible to determine the band structure of a chain of high symmetry following an approach that is entirely similar to that followed to obtain the electronic structure of a symmetric molecule. The only difference stems from the fact that, for the chain the interactions between the different orbitals must be analysed for every k point. When there is some uncertainty concerning the relative position of two bands, i.e. when simple qualitative reasoning such as that used in our example may not be used, then one must rely on some kind of computation even if it is very simple, for instance the extended Hückel approach. Of course, this is also the case for molecules (see for instance ref. [3].).

6.3 Band structure of the hypothetical $(NaCl)_n$ chain

Following a similar approach, we are now going to consider the electronic structure of the hypothetical regular chain $(NaCl)_n$ in which the chlorine atoms provide five electrons through their $(3p)_{Cl}$ orbitals and the sodium atoms

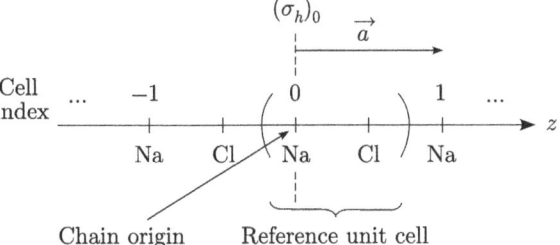

Fig. 6.10
Hypothetical chain with formula NaCl discussed in the text.

provide one electron through the $(3s)_{Na}$ orbital (Fig. 6.10). In this qualitative approach we will neglect the secondary role played by the empty $(3p)_{Na}$ and filled $(3s)_{Cl}$ orbitals. We will qualitatively build the band structure of the system mostly through the use of symmetry-based arguments.

6.3.1 Group of the different k points

This system is stable under application of the symmetry operations of the space group G, which is the direct product of the translation group T_n and the point group $(D_{\infty h})_{Na_0}$ leaving the sodium atom of the reference cell invariant.[4] Consequently, the appropriate group for points Γ and X is $(D_{\infty h})_{Na_0}$.

[4] Here we have placed the origin of the chain at a sodium atom of the reference cell. Of course, it could have been equivalent to place it on the chlorine atom of the reference cell.

Γ and X points

Looking at the Bloch orbitals for these points, schematically represented in Fig. 6.11, their symmetry properties and thus the associated symmetry labels are quite clear. Thus, among all of these orbitals, only $BO_{(3s)_{Na}}(X)$ and $BO_{(3p_z)_{Cl}}(X)$ may interact. In all other cases the Bloch orbitals may be identified with the crystal orbitals of the system.

k points other than Γ and X

For all other k points, the group to be used is $C_{\infty v}$. Looking at the character table for this group it is clear that the $(3s)_{Na}$ and $(3p_z)_{Cl}$ are bases for the same representation, Σ^+, whereas the $(3p_x)_{Cl}$ and $(3p_y)_{Cl}$ orbitals provide a basis for the Π representation. Consequently, the Bloch orbitals generated by the $(3p_x)_{Cl}$ and $(3p_y)_{Cl}$ orbitals are degenerate at any k point and do not interact with $BO_{(3s)_{Na}}(\vec{k})$ and $BO_{(3p_z)_{Cl}}(\vec{k})$. In contrast, the Bloch orbitals $BO_{(3s)_{Na}}(\vec{k})$ and $BO_{(3p_z)_{Cl}}(\vec{k})$ may interact at any k point.

Summary

The symmetry properties of the Bloch orbitals and the crystal orbitals of the $(NaCl)_n$ chain at different k points are summarised in Table 6.3.

6.3.2 Bands associated with σ-type overlaps

We begin this section by looking at the two bands generated by the $(3s)_{Na}$ and $(3p_z)_{Cl}$ orbitals that are involved in σ-type interactions.

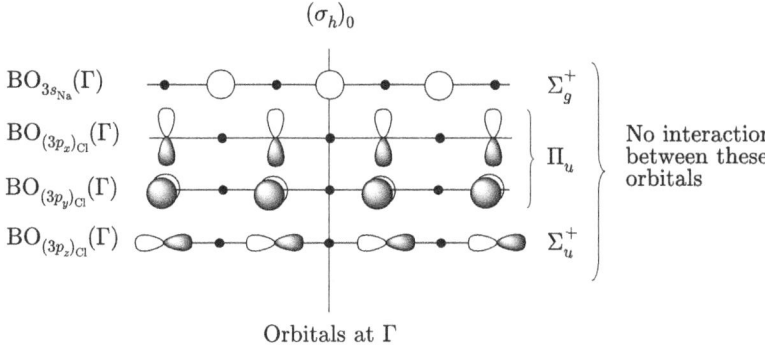

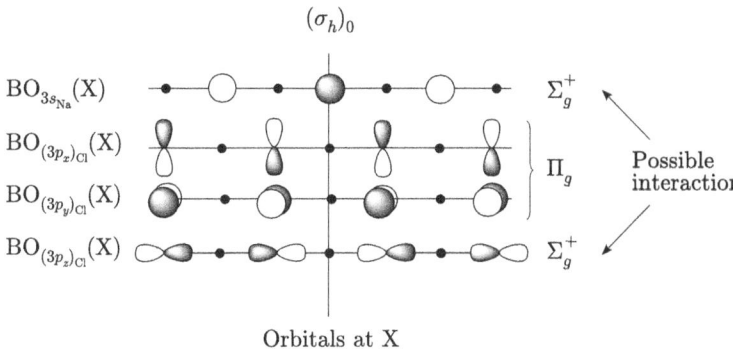

Fig. 6.11
Bloch orbitals for the Γ and X points.

Table 6.3 Symmetry labels for the Bloch orbitals centered on Na (BO_{Na}) and Cl (BO_{Cl}) and for the crystal orbitals (CO) of the $(NaCl)_n$ chain.

k point	Symmetry label	BO_{Na}	BO_{Cl}	CO
Γ	Σ_g^+	$BO_{(3s)_{Na}}$	—	1 non-bonding orbital
Γ	Σ_u^+	—	$BO_{(3p_z)_{Cl}}$	1 non-bonding orbital
Γ	Π_u	—	$BO_{(3p_x)_{Cl}}$ $BO_{(3p_y)_{Cl}}$	2 non-bonding degenerate orbitals
X	Σ_g^+	$BO_{(3s)_{Na}}$	$BO_{(3p_z)_{Cl}}$	2 orbitals resulting from the interaction of two Bloch orbitals
X	Π_g	—	$BO_{(3p_x)_{Cl}}$ $BO_{(3p_y)_{Cl}}$	2 non-bonding degenerate orbitals
$k_a \neq 0, 1/2$	Σ^+	$BO_{(3s)_{Na}}$	$BO_{(3p_z)_{Cl}}$	2 orbitals resulting from the interaction of two Bloch orbitals
$k_a \neq 0, 1/2$	Π	—	$BO_{(3p_x)_{Cl}}$ $BO_{(3p_y)_{Cl}}$	2 non-bonding degenerate orbitals

112 Symmetry in 1D compounds

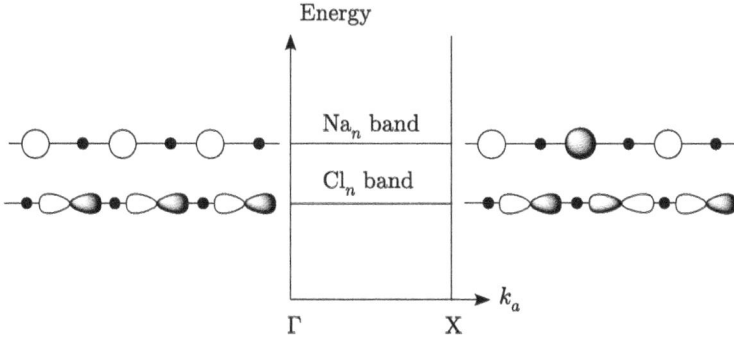

Fig. 6.12
Energies of the $BO_{(3s)_{Na}}(\vec{k})$ and $BO_{(3p_z)_{Cl}}(\vec{k})$ Bloch orbitals.

Energy of the $BO_{(3s)_{Na}}(\vec{k})$ and $BO_{(3p_z)_{Cl}}(\vec{k})$ orbitals

In a Hückel-type approach only those interactions between nearest neighbour atoms are considered. Since the periodic fragments Cl_n and Na_n are not built from nearest neighbour atoms of the $(NaCl)_n$ chain, the energies of the Bloch orbitals $BO_{(3s)_{Na}}(\vec{k})$ and $BO_{(3p_z)_{Cl}}(\vec{k})$ do not change with k and they keep the same value as for the isolated $(3s)_{Na}$ and $(3p_z)_{Cl}$ orbitals, respectively.

Band structure when the $ns - np_z$ interaction is taken into account

At this point we will explore how the Bloch orbitals $BO_{(3s)_{Na}}(\vec{k})$ and $BO_{(3p_z)_{Cl}}(\vec{k})$ can interact. At the Γ point, the two orbitals $BO_{(3s)_{Na}}(\Gamma)$ and $BO_{(3p_z)_{Cl}}(\Gamma)$ cannot interact because they are of different symmetry. In contrast, they can interact at any other point including X. Thus, the two non-interacting bands (dotted lines in Fig. 6.13) repel each other, and this is more effective the farther it is from Γ, the k point under consideration. This is so because the overlap between the Bloch orbitals $BO_{(3s)_{Na}}(\vec{k})$ and $BO_{(3p_z)_{Cl}}(\vec{k})$ increases with k_a in the interval [0, 1/2]. The bonding band is essentially built from orbitals centred on the chlorine atoms whereas the upper antibonding band is essentially made of orbitals centred on the sodium atoms. A schematic band structure summarising the situation is shown Fig. 6.13.

6.3.3 Complete band structure

Now we can combine the results of the previous sections and build the complete band structure for the $(NaCl)_n$ chain. The easiest way to proceed is to remind ourselves that the two degenerate non-bonding π-type bands are

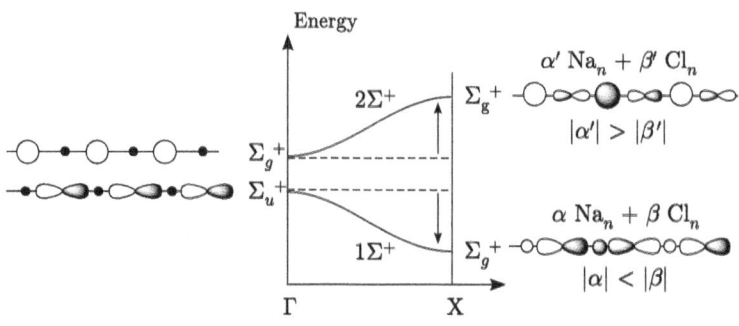

Fig. 6.13
Partial band structure for $(NaCl)_n$ (Σ^+-type bands). The energies of the $BO_{(3s)_{Na}}(\vec{k})$ and $BO_{(3p_z)_{Cl}}(\vec{k})$ are shown as dotted lines. The arrows are used to highlight the effect of the interaction between the two Bloch orbitals.

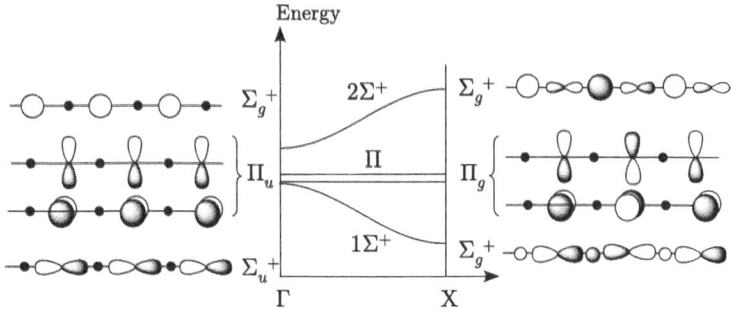

Fig. 6.14
Band structure of $(NaCl)_n$.

flat and have the energy of an isolated $(3p)_{Cl}$ orbital. This leads us to place these two π bands at the top of the σ-bonding lower band. Since we have neglected the contribution of the low-lying $3s$ orbitals of chlorine, there are six electrons to fill these bands. The Fermi level therefore lies at the top of the third band. This completes the information needed to draw the band structure (see Fig. 6.14). It is clear that the system is not metallic because there is a clear band gap at the Fermi level. Now it is interesting to analyse the electronic density of the NaCl chain and compare it with those of isolated Na and Cl atoms. In the chain there are four electrons in the π-type bands localised on the $(3p_x)_{Cl}$ and $(3p_y)_{Cl}$ orbitals, and two electrons on the $1\Sigma^+$ band. This is a filled bonding band, mostly based on the $(3p_z)_{Cl}$ orbitals. Consequently, in the chain there is on average less than one electron in the vicinity of the sodium atom and more than five in the neighbourhood of the chlorine: the chain is thus strongly ionic. The situation is in fact completely analogous to that in the NaCl molecule. An obvious consequence of this situation is that the band gap at the Fermi level for this type of chain will increase with the electronegativity of the anionic partner and with the electropositive character of the cationic partner.

6.4 Consequences of the existence of a glide plane

In Chapter 5 we discussed the band structure of *trans*-polyacetylene on the basis of a Hückel approach. This analysis enabled us to show that this chain is susceptible to a Peierls-type distortion. Here we will try to recover the band structure solely on the basis of symmetry considerations, i.e. following the method developed in Section 6.1.[5] In so doing we will learn that the formal account of the symmetry of the system is not as simple as one might have expected. This is because the space group is not simply written as a direct product of the translation group T_n and a point group P, i.e. the space group is not symmorphic.

6.4.1 Using point group symmetry properties in *trans*-polyacetylene

For the time being let us follow the approach of Section 6.1.1 as applied in Sections 6.2 and 6.3: we first determine a space group G that is symmorphic

[5] The discussion presented in this section is not really needed to understand most of the material in this book, except for Exercise (8.4) of Chapter 8. Consequently, this section can be skipped in a first reading. In contrast, it is important to carefully consider Section 6.5.

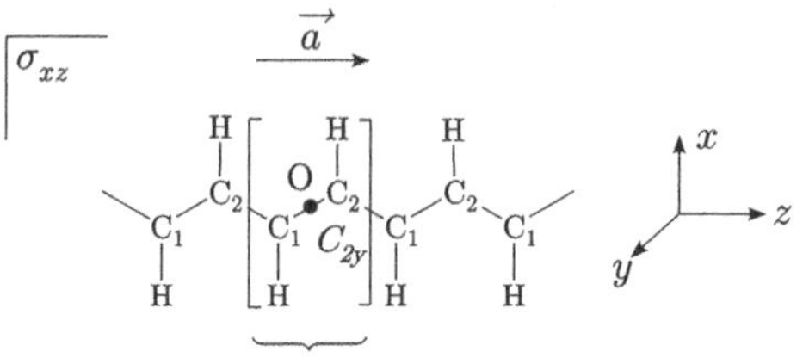

Fig. 6.15
Schematic representation of *trans*-polyacetylene. The origin of the chain lies in the middle of the C–C bond of the reference cell.

and can be written as the direct product $T_n \otimes P$, and which leaves the regular *trans*-polyacetylene chain invariant.

Symmorphic space group

We must choose a symmetry group P leaving invariant both the chain and one of the points of the reference cell, denoted O. The point O and the group P are chosen so that the symmetry group P contains the maximum number of symmetry operations. For *trans*-polyacetylene there are several (P,O) combinations fulfilling this condition. In this section we will use one of them and study another in Exercise (6.5). In that way we will verify that the choice of the (P,O) pairing does not matter so long as it is properly used. As shown in Fig. 6.15, we have chosen to place point O at the centre of the reference cell, i.e. in the middle of the C–C bond. The group P that leaves the chain and the O point invariant is the C_{2h} group, which contains the identity (E), the rotation around the Oy axis (C_{2y}), the inversion with respect to O (i), and the symmetry with respect to the plane of the polymer (σ_{xz}) (see Fig. 6.15).[6] The appropriate group for points Γ and X is C_{2h}, $\{E, C_{2y}, i, \sigma_{xz}\}$, while for the other k points it is C_s, $\{E, \sigma_{xz}\}$. The crystal orbitals associated with the π system are all antisymmetric with respect to σ_{xz} in such a way that all crystal orbitals for k points other than Γ and X are of A'' symmetry in the C_s group. No band crossing between π-type bands is allowed in the interval $]0, 1/2[$. In contrast, the crystal orbitals at the Γ and X points may be of A_u or B_g symmetry because the two irreducible representations have a -1 character with respect to σ_{xz}. We will try to determine the crystal orbitals for these two points by constructing linear combinations of Bloch orbitals that are bases for the A_u or B_g representations.

Determination of the crystal orbitals at Γ and X

Let us now represent the Bloch orbitals $(BO)_\pi(\vec{k})$ and $(BO)_{\pi*}(\vec{k})$ built from the π and π^* fragment orbitals, respectively, at the Γ and X points (see Fig. 6.16) and analyse their symmetry properties according to the C_{2h} group. According to the symmetry labels corresponding to the C_{2h} group the Bloch orbitals cannot mix. Consequently at Γ and X we may identify the $(BO)_\pi(\Gamma)$, $(BO)_{\pi*}(\Gamma)$, $(BO)_\pi(X)$, and $(BO)_{\pi*}(X)$ Bloch orbitals with the crystal orbitals. Thus, a simple symmetry-based reasoning leads to the crystal orbitals of the chain for the Γ and X points. In addition, it is clear that the $(BO)_\pi(\Gamma)$ orbital is bonding whereas the $(BO)_{\pi*}(\Gamma)$ orbital is antibonding. The orbitals at X are degenerate and non-bonding.

---

[6] The main axis of this group is the C_2 rotation axis. We will denote it Oy and not Oz as is more usual. This does not produce a problem when using the character table of the C_{2h} group: the C_{2y} rotation plays the role of the C_2 rotation while the σ_{xz} symmetry plays the role of the σ_h horizontal plane. For this particular example we label the axis along the chain direction Oz.

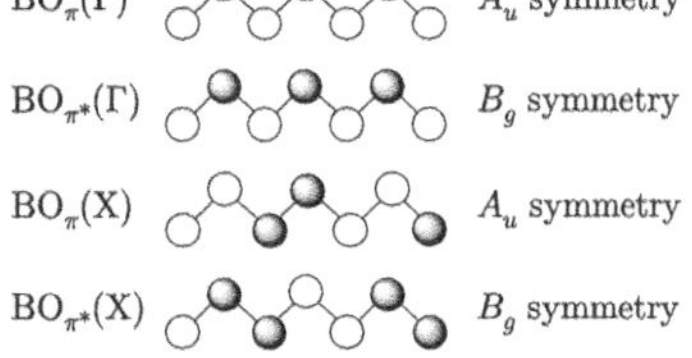

$BO_\pi(\Gamma)$ A_u symmetry

$BO_{\pi*}(\Gamma)$ B_g symmetry

$BO_\pi(X)$ A_u symmetry

$BO_{\pi*}(X)$ B_g symmetry

Fig. 6.16
$BO_\pi(\vec{k})$ and $BO_{\pi*}(\vec{k})$ Bloch orbitals at Γ and X. Only the upper lobes of the p_y orbitals are shown.

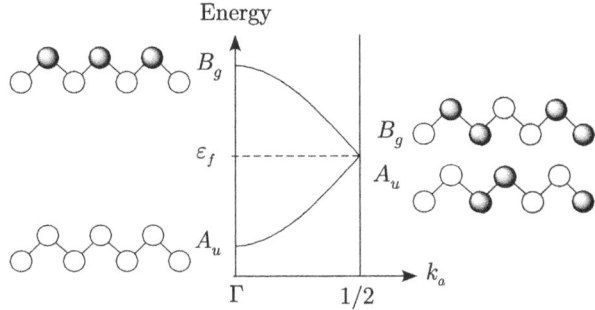

Fig. 6.17
Qualitative band structure obtained exclusively on the basis of symmetry arguments. Only the upper lobes of the p_y orbitals are shown.

Shape of the band structure

Knowing that the bands cannot cross for k points in the interval $[0, 1/2]$, we must connect the two non-degenerate crystal orbitals at Γ with the two degenerate crystal orbitals at X, avoiding any crossing. Thus the qualitative π-type band structure of *trans*-polyacetylene is as shown in Fig. 6.17. Let us remind ourselves that this band structure has been obtained solely on the basis of symmetry-based reasoning and following the approach of Section 6.1.1.

6.4.2 Complete space group (non-symmorphic) of *trans*-polyacetylene

The two crystal orbitals at X are degenerate even if, using the C_{2h} group, they are found to have different symmetry (Fig. 6.17). This type of degeneracy is rarely accidental, but is usually the manifestation of some additional symmetry of the system. Such a degeneracy is not predicted by the type of analysis we have performed using the C_{2h} group. However, this is not the result of an inappropriate choice of the P group but originates in the existence of symmetry operations that cannot be written as the product of a point symmetry operation leaving the system invariant and a translation with vector $m\vec{a}$.

Existence of a glide plane

The chain of *trans*-polyacetylene is stable under the action of a reflection on the σ_{yz} plane followed by a translation by half the repeat vector of the chain, $t_{\vec{a}/2}$ (see Fig. 6.18). Such a combined symmetry operation is known as a *glide plane*, and it will be referred to as g_σ ($g_\sigma = t_{\vec{a}/2}.\sigma_{yz} = \sigma_{yz}.t_{\vec{a}/2}$). Under such an operation, every type 1 carbon atom is transferred to one adjacent type 2 carbon atom and vice-versa (see Fig. 6.18). It is worth noting that *the chain is not stable under the individual actions of the σ_{yz} and $t_{\vec{a}/2}$ operations. Only the product, g_σ, is a symmetry operation of the system.*

Complete space group of *trans*-polyacetylene

The full space group of *trans*-polyacetylene cannot be written as a direct product of a purely translational group T_n and a point group P because of the existence of the g_σ symmetry operation. The full space group is said to be *non-symmorphic*. This space group may be written in the form $T_n \otimes C_{2h} \otimes \{E, g_\sigma\}$. What are the symmetry operations contained in such a space group? To identify

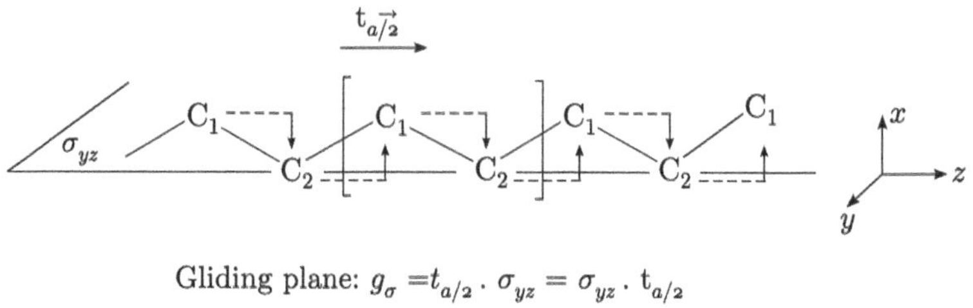

Fig. 6.18
Schematic diagram illustrating the effect of a glide plane g_σ on the carbon chain of *trans*-polyacetylene lying on the xOz plane.

Gliding plane: $g_\sigma = t_{a/2} \cdot \sigma_{yz} = \sigma_{yz} \cdot t_{a/2}$

Table 6.4 Product of the symmetry operations of the C_{2h} group (leaving invariant the C–C bond of the reference unit cell) and the g_σ operation. The symmetry operations of the $(C_{2x})_i$ ($i = 1, 2$) group are the rotations around an axis parallel to Ox and going through the carbon atom C_i of the reference cell. The $g_{C_{2z}}$ symmetry operation is the commutative product of the C_{2z} rotation and a half-translation $t_{\vec{a}/2}$ ($g_{C_{2z}} = C_{2z} \cdot t_{\vec{a}/2}$).

Symmetry operations (R) of the C_{2h} group	Result of the product $R.g_\sigma$	Result of the product $g_\sigma.R$
E	g_σ	g_σ
i	$(C_{2x})_1$	$(C_{2x})_2$
C_{2y}	$(\sigma_{xy})_1$	$(\sigma_{xy})_2$
σ_{xz}	$g_{C_{2z}}$	$g_{C_{2z}}$

the different symmetry operations generated by g_σ, we build Table 6.4, which is the product table for the direct product $C_{2h} \otimes \{g_\sigma\}$.

Thus, the product of the symmetry operation g_σ and the symmetry operations i and C_{2y} leaving invariant the point O, generate the symmetry operations $(C_{2x})_i$ and $(\sigma_{xy})_i$ leaving invariant the carbon atom C_i of the reference cell ($i = 1, 2$). In contrast, the product of symmetry σ_{xz} and the symmetry operation g_σ may be identified with the symmetry operation noted $g_{C_{2z}}$, equal to the commutative product ($t_{\vec{a}/2} \cdot C_{2z}$). The latter symmetry operation is a *screw axis*.

The complete space group of *trans*-polyacetylene thus contains:

- the translations of the T_n group
- the point symmetry operations leaving invariant either the middle of a C–C bond or a carbon atom
- the product of the different translational and point symmetry operations
- the two sliding symmetry operations g_σ and $g_{C_{2z}}$.

Let us note that it is perfectly possible to generate the full *trans*-polyacetylene chain by using a C–H unit and the full space group of the system including the gliding symmetry operations g_{2z} and g_σ.

Characterisation of the symmetry properties of the crystal orbitals by means of the complete space group

As a consequence of the factorisation of the full space group of *trans*-polyacetylene in the form $T_n \otimes C_{2h} \otimes \{E, g_\sigma\}$ the symmetry properties of the different crystal orbitals at Γ and X may be described by a C_{2h} symmetry label and the behaviour with respect to the non-point symmetry operation g_σ. For k points other than Γ and X, the crystal orbitals may be characterised by a C_s symmetry label and the properties with respect to g_σ.

6.4.3 Crystal orbitals of *trans*-polyacetylene by means of the non-symmorphic space group $G = T_n \otimes C_{2h} \otimes \{E, g_\sigma\}$

Determination of the linear combinations of orbitals adapted to the glide plane

Let us now build linear combinations $\Psi_i(\vec{k})$ of $2p_y$ orbitals that are well adapted to the glide plane g_σ and the translation through a vector $\vec{a}$, i.e. $\Psi_i(\vec{k})$ functions that are eigenvectors of g_σ and $t_{\vec{a}}$. Since the $\Psi_i(\vec{k})$ function is characterised by the k vector, this means that $\Psi_i(\vec{k})$ is a combination of Bloch orbitals associated with the same k vector. Thus, $\Psi_i(\vec{k})$ is an eigenstate of $t_{\vec{a}}$ associated with the eigenvalue $\exp(-i2\pi k_a)$ (see Exercise (6.1)). In the following analysis we will denote as $\lambda_i(\vec{k})$ the eigenvalue (unknown) of $\Psi_i(\vec{k})$ with respect to g_σ. The $\Psi_i(\vec{k})$ functions will be used to describe the crystal orbitals of *trans*-polyacetylene.

$$g_\sigma \Psi_i(\vec{k}) = \lambda_i(\vec{k}) \Psi_i(\vec{k}) \quad (6.6a)$$

$$t_{\vec{a}} \Psi_i(\vec{k}) = \exp(-i2\pi k_a) \Psi_i(\vec{k}) \quad (6.6b)$$

Since the square of g_σ may be identified with $t_{\vec{a}}$, we can write

$$(g_\sigma)^2(\Psi_i(\vec{k})) = \lambda_i(\vec{k})^2 \Psi_i(\vec{k}) = t_{\vec{a}}(\Psi_i(\vec{k})) = \exp(-i2\pi k_a) \Psi_i(\vec{k})$$

Consequently, the eigenvalues $\lambda_i(\vec{k})$ can only adopt the values $\exp(-i\pi k_a)$ or $-\exp(-i\pi k_a)$. We will denote as $\Psi_+(\vec{k})$ and $\Psi_-(\vec{k})$ the linear combinations associated with the eigenvalues $\exp(-i\pi k_a)$ and $-\exp(-i\pi k_a)$, respectively. Since the glide plane transfers a $2p_y$ orbital of a carbon atom to a $2p_y$ orbital centred on an adjacent atom on its right, a non-normalised expression of the $\Psi_+(\vec{k})$ and $\Psi_-(\vec{k})$ functions that are solutions of eqn (6.6) may be easily obtained (see Fig. 6.19).

Determination of the crystal orbitals at k points other than Γ and X

For any k point other than Γ and X the crystal orbitals of the π system of *trans*-polyacetylene must be adapted to the properties of the C_s group and the glide plane g_σ. Because of the antisymmetric character of the p_y orbitals of the

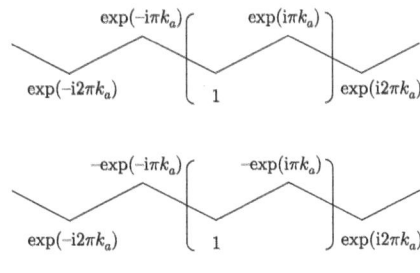

Linear combination $\Psi_+(k)$ of $2p_y$ orbitals, associated with the eigenvalue $\exp(-i\pi k_a)$ with respect to g_σ

Linear combination $\Psi_-(k)$ of $2p_y$ orbitals, associated with the eigenvalue $-\exp(-i\pi k_a)$ with respect to g_σ

Fig. 6.19
$\Psi_+(\vec{k})$ and $\Psi_-(\vec{k})$ functions adapted to the translation symmetry as well as to the glide plane. Only the coefficients of the $2p_y$ atomic orbitals are shown.

carbon atoms, any of the π-type crystal orbitals is of A'' symmetry in group C_s. Consequently, only the glide plane may characterise the symmetry of a crystal orbital. Since, in the previous section, we obtained two linear combinations of different symmetry, $\Psi_+(\vec{k})$ and $\Psi_-(\vec{k})$, these two combinations may be identified with the crystal orbitals of *trans*-polyacetylene so long as the normalisation constant is not taken into account. Indeed, it can easily be verified that, neglecting the normalisation constant, the $\Psi_+(\vec{k})$ and $\Psi_-(\vec{k})$ functions coincide exactly with the crystal orbitals resulting from the resolution of the secular determinant discussed in Chapter 5 (see Fig. 5.8). Thus, without need for any calculation, just taking into account the symmetry properties of *trans*-polyacetylene, it has been possible to obtain the crystal orbitals for any k point other than Γ and X.

Determination of the crystal orbitals at Γ

At the Γ point the crystal orbitals CO(Γ) of the π system of *trans*-polyacetylene must be adapted to the properties of the C_{2h} group and the glide plane g_σ. Let us schematically represent the $\Psi_+(\Gamma)$ and $\Psi_-(\Gamma)$ linear combinations adapted to the glide plane g_σ and analyse their symmetry within the C_{2h} group (see Fig. 6.20).

It is clear that the two linear combinations have different symmetry properties with respect to both the symmetry operations of the C_{2h} group and the glide plane g_σ. They can therefore be identified with the π-type crystal orbitals of *trans*-polyacetylene at Γ.

Determination of the crystal orbitals at X

At the X point the crystal orbitals CO(X) of the π system of *trans*-polyacetylene must also be adapted to the properties of the C_{2h} group and the glide plane g_σ. Let us schematically represent the $\Psi_+(X)$ and $\Psi_-(X)$ linear combinations adapted to the glide plane g_σ and analyse their symmetry with respect to the symmetry operations of the C_{2h} group (see Fig. 6.21). These two functions are not bases for an irreducible representation of the C_{2h} group. For instance, the effect of a C_{2y} rotation over these functions is the following:

$$C_{2y}(\Psi_+(X)) = i\Psi_-(X) \quad \text{and} \quad C_{2y}(\Psi_-(X)) = -i\Psi_+(X)$$

The two functions $\Psi_+(X)$ and $\Psi_-(X)$ interchange under the effect of a rotation (we disregard the $\pm i$ constant in the functions) and they are not adapted to the C_{2h} symmetry. It is, however, easy to verify that the combinations:

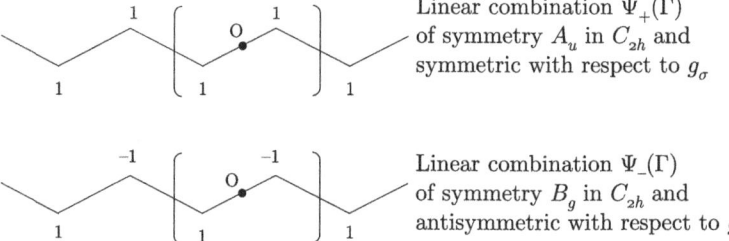

Fig. 6.20 Schematic representation of the $\Psi_+(\Gamma)$ and $\Psi_-(\Gamma)$ functions. Only the coefficients of the $2p_y$ atomic orbitals are shown. The origin chosen, O, is marked.

Linear combination $\Psi_+(\Gamma)$ of symmetry A_u in C_{2h} and symmetric with respect to g_σ

Linear combination $\Psi_-(\Gamma)$ of symmetry B_g in C_{2h} and antisymmetric with respect to g_σ

Linear combination $\Psi_+(X)$ symmetry adapted to g_σ but not to C_{2h}, and keeping O invariant.

Linear combination $\Psi_-(X)$ symmetry adapted to g_σ but not to C_{2h}, and keeping O invariant.

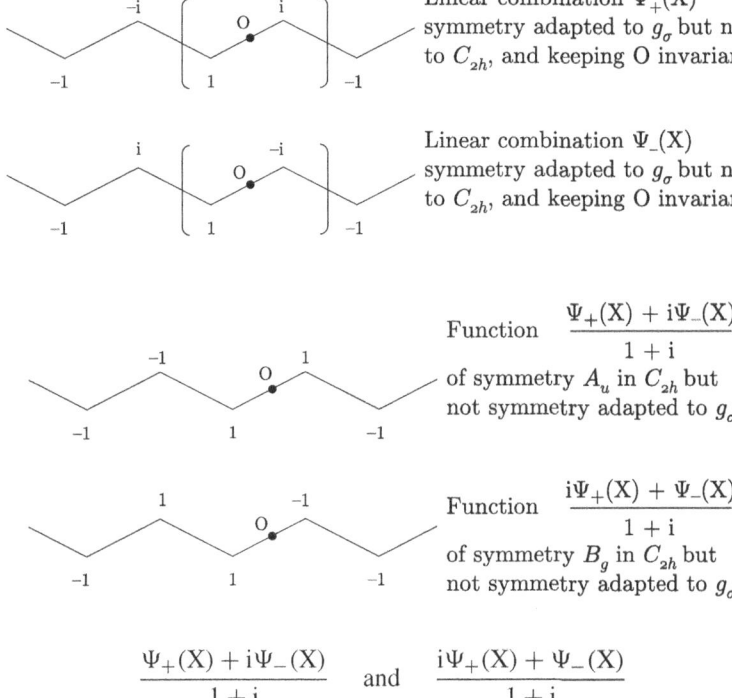

Fig. 6.21
Symmetry properties of the $\Psi_+(X)$ and $\Psi_-(X)$ functions. O is the chosen origin.

Function $\dfrac{\Psi_+(X) + i\Psi_-(X)}{1+i}$ of symmetry A_u in C_{2h} but not symmetry adapted to g_σ

Function $\dfrac{i\Psi_+(X) + \Psi_-(X)}{1+i}$ of symmetry B_g in C_{2h} but not symmetry adapted to g_σ

Fig. 6.22
Schematic representation of the $\frac{\Psi_+(X)+i\Psi_-(X)}{1+i}$ and $\frac{i\Psi_+(X)+\Psi_-(X)}{1+i}$ functions. The O point is left invariant by the C_{2h} group.

$$\frac{\Psi_+(X) + i\Psi_-(X)}{1+i} \quad \text{and} \quad \frac{i\Psi_+(X) + \Psi_-(X)}{1+i}$$

are bases for the irreducible representations A_u and B_g of the C_{2h} group (see Fig. 6.22), respectively, but that they are not eigenfunctions of g_σ. Consequently, it is impossible to find a basis of orbitals that is stable with respect to the glide symmetry g_σ while being a basis for an irreducible representation of the C_{2h} group. *Such a situation is typical of a doubly degenerate representation.* Only when taking into account the complete space group, including the symmetry operation g_σ, may the degeneracy at the X point be understood. In Section 6.4.1 we worked with a subgroup of the complete space group: this is the reason why we obtained two degenerate states with different symmetry (Fig. 6.17). Usually there is nothing wrong in working with a subgroup of the complete space group so long as it is understood that this is what is being done.

6.4.4 Concluding remarks

The analysis developed in this chapter shows that the complete space group of *trans*-polyacetylene must be written in the form $T_n \otimes C_{2h} \otimes \{E, g_\sigma\}$, i.e. it cannot be given in the form of a direct product $T_n \otimes P$ where P is a point group. Consequently, it is a *non-symmorphic* group. This analysis has also revealed the existence in periodic systems of glide planes or screw axes and shown how they can be taken into account by looking at the example of the π orbitals of *trans*-polyacethylene. For this system we obtained an analytical expression of the crystal orbitals for any k point. In addition we showed that the two crystal orbitals must be degenerate at the X point. Thus, for any k point

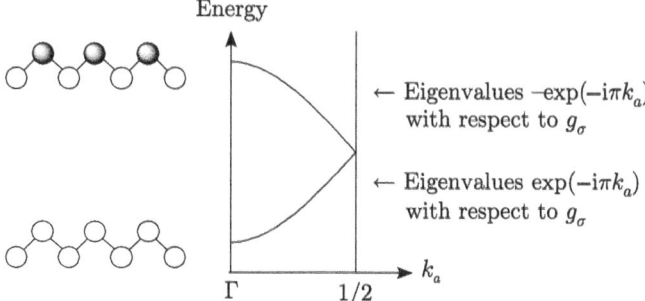

Fig. 6.23
Symmetry properties of the crystal orbitals with respect to the glide plane g_σ.

other than X, the crystal orbitals may be labelled according to their symmetry behaviour with respect to the g_σ glide plane (see Fig. 6.23).

From a broader perspective, to formally take into account symmetry operations such as the screw axes or glide planes requires mathematical techniques that are clearly outside the scope of this book. The interested reader will find these developments in many different books, among which references [4] and [5] are recommended. Here we have merely tried to show how this kind of symmetry operation influences the electronic structure, by considering the π system of *trans*-polyacethylene. In addition, we have shown that neglecting these symmetry operations, i.e. working in a group that is just a subgroup of the complete one, is always possible, although this approach may lead to some ambiguities.

6.5 Work plan for the study of a 1D system

At this point we outline a general approach that makes possible a qualitative determination of the band structure of a periodic 1D system in a simple way, something that should help in understanding the results of precise computer-generated band structures. The different steps of this approach are:

- Define the reference cell of the periodic system as well as the repeat translational vector $\vec{a}$ leading to the generation of the chain.
- Analyse the symmetry of the chain and characterise the P group, i.e. the group possessing the largest number of point symmetry operations leaving both the 1D system and one point of the reference cell invariant. If the group is non-symmorphic it is useful to define it as the direct product of the T_n group, the P group, and one symmetry operation such as a screw axis or glide plane. If the group is symmorphic then it is useful to determine the symmetry group associated with any k point.
- Determine a set of orbitals $\{(\varphi_i)_0, \ i = 1, \ldots, N_0\}$ adapted to the symmetry of the system and the number of electrons that every cell brings to the system.
- Build the Bloch orbitals generated by these orbitals and associate each one with a symmetry label. Predict the number and symmetry of the crystal orbitals of the system.
- Build the band structure of the system considering those of different symmetries separately. Analyse the nature of the orbitals at Γ and X. Determine

the qualitative shape of the band structure. If the orbital interactions are complex, it may be worth looking at the system numerically.
- Determine the Fermi level at $T = 0$ K and predict the conductivity behaviour.
- Discuss what happens in the eventuality of a distortion of the system.

Exercises

(6.1) Verify that the Bloch orbitals are stable with respect to the symmetry operations of the T_n group. What is their character with respect to a translation? What is the situation for a crystal orbital?

(6.2) Show that the $(D_{4h})_{0/1}$ group, which leaves invariant the midpoint of the Pt_0–Pt_1 segment of the chain in Fig. 6.4, also leaves invariant the full chain. Show that the G space group generated by $(D_{4h})_{0/1}$ and T_n is the same as that generated by $(D_{4h})_0$ and T_n.

(6.3) Let us consider the sodium chloride type chain:

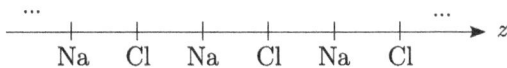

Determine the more complete P point group leaving invariant both a point of the reference cell and the full chain. Is there only one group fulfilling this requirement?

(6.4) Sodium chloride chain: how will the energies of the $BO_{(3s)_{Na}}(\vec{k})$ and $BO_{(3p_z)_{Cl}}(\vec{k})$ Bloch orbitals in Fig. 6.12 change if the extended Hückel approach was used instead of the Hückel one?

(6.5) *Trans*-polyacetylene: show that it could had been possible to choose another O point and another P group also possessing four symmetry operations other than those discussed in Section 6.4.1.

(6.6) Why is not necessary to take into account the screw axis $g_{C_{2z}}$ when discussing the symmetry of the crystal orbitals of *trans*-polyacetylene?

References

1. F. A. Cotton, *Chemical Applications of Group Theory*, 3rd edition, John Wiley, New York, 1990.
2. S. I. Altmann, *Band Theory of Solids: An Introduction from the Point of View of Symmetry*, Oxford University Press, Oxford, 1991.
3. Y. Jean and F. Volatron, *Structure Électronique des Molécules*, Ediscience International, Paris, 1993; Y. Jean, F. Volatron, and J. Burdett *An Introduction to Molecular Orbitals*, Oxford University Press, New York, 1993.
4. M. Tinkham, *Group Theory and Quantum Mechanics*, McGraw-Hill, New York, 1964.
5. H. Jones, *Theory of Brillouin Zones and Electronic States in Crystals*, 2nd edition, North Holland, Amsterdam, 1975.

7 Application to polyacene

In this chapter, we will describe and analyse the electronic structure of polyacene $(C_4H_2)_n$, with the planar structure shown in Fig. 7.1, using the ideas developed in Chapters 3 and 6.

This system may be associated with several Lewis structures with single and double C–C bonds, which differ in the location of the double bonds. We will start our analysis by considering the regular structure of polyacene in which the C–C bonds are formally equivalent. We will represent the delocalised π bonds by inscribing a circle within the hexagons (see Fig. 7.2).

Fig. 7.1
The polyacene system. The compound is generated by a unit cell containing four carbon and two hydrogen atoms and a translation vector denoted $\vec{a}$.

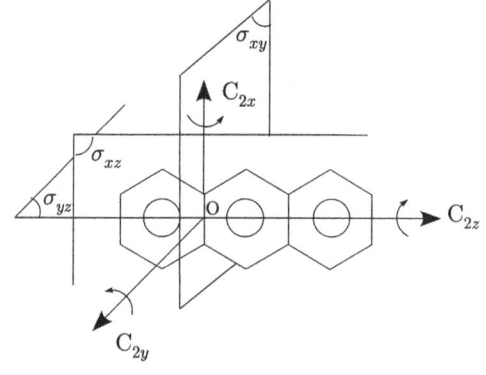

Fig. 7.2
Structure of regular polyacene, assumed to be in the xOz plane. The main symmetry operations of the group P leaving the chain invariant, as well as the origin O, lying at the centre of a vertical C–C bond, are indicated.

Band structure

We will start by considering the band structure of regular polyacene and then we will analyse what kind of distortions can provide additional stability. For reasons similar to *trans*-polyacetylene, we can assume that the crystal orbitals around the Fermi level are those built from the π-type orbitals of the system. Consequently, throughout this chapter we will only consider the $2p_y$ atomic orbitals of the carbon atoms.

7.1 Band structure near the Fermi level

In this section we will qualitatively build up the π-type band structure of regular polyacene following the step-by-step approach outlined in Section 6.5.

7.1.1 Unit cell definition

The unit cell and repeat vector ($\vec{a}$) needed to generate the regular polyacene chain are shown in Fig. 7.1. Every carbon atom contributes one $2p_y$ atomic orbital and one electron to the π system of the chain. Since polyacene is generated by a C_4H_2 unit cell, a natural starting point to build the π-type bands is the set of π-type Bloch orbitals of the C_4H_2 fragment, in a geometry not very different from that of *cis*-butadiene (C_4H_6).

7.1.2 Symmetry analysis of the chain

The space group of polyacene may be written as the direct product of T_n, the group of translations, and the D_{2h} group leaving invariant the O point in the middle of the C–C bond shared by two adjacent six centred rings (see Fig. 7.2). The symmetry operations associated with the latter group are the identity, the inversion with respect to the O point, the C_{2x}, C_{2y}, and C_{2z} rotations around the Ox, Oy, and Oz axes, and the reflections with respect to the xOy, xOz, and yOz planes (σ_{xy}, σ_{xz}, and σ_{yz}). The character tables for the D_{2h} and C_{2v} groups are given in the Appendix.

The appropriate group for points Γ and X is D_{2h}, whereas for the other k points it is C_{2v}, which is the subgroup leaving every cell invariant. The symmetry operations associated with this group are the identity, the C_{2z} rotation around the Oz axis, and the reflections with respect to the xOz and yOz planes (σ_{xz} and σ_{yz}). Since all π-type crystal orbitals $CO(\vec{k})$ ($k_a \neq 0, 1/2$) are antisymmetric with respect to the xOz plane, the crystal orbitals of polyacene are either of B_2 or A_2 symmetry depending on their symmetric or antisymmetric nature, respectively, with respect to the yOz plane (see Fig. 7.2).

7.1.3 Appropriate fragment orbitals

The symmetry analysis of polyacene suggests that an appropriate set of fragment orbitals to build the π band structure of this chain are the four π orbitals of the C_4H_2 fragment. These orbitals are either symmetric or antisymmetric with respect to the σ_{yz} operation, which is the appropriate symmetry operation for all k points except Γ and X, for which the symmetry is higher, but

Polyacene

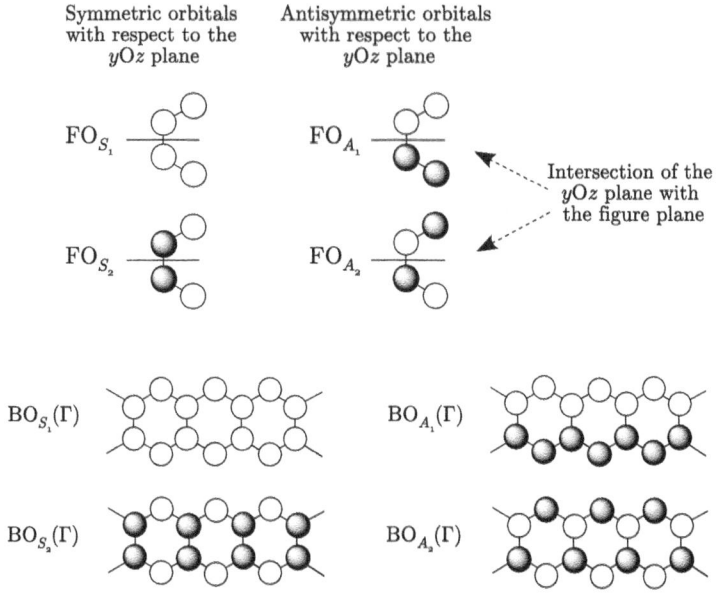

Fig. 7.3
Fragment orbitals chosen to build the Bloch orbitals for regular polyacene.

Fig. 7.4
The $BO_{S_1}(\Gamma)$, $BO_{S_2}(\Gamma)$, $BO_{A_1}(\Gamma)$, and $BO_{A_2}(\Gamma)$ Bloch orbitals of regular polyacene. These orbitals are characterised by the lower indexes S_1, S_2, A_1, and A_2 depending on the nature of the fragment orbitals from which they derive (FO_{S_1}, FO_{S_2}, FO_{A_1}, and FO_{A_2}). The origin (O) has been chosen to be at the centre of a C–C bond (see Fig. 7.2).

the symmetry group also contains this operation. These orbitals are shown schematically in Fig. 7.3.

Note: As usual, throughout this chapter only the lobes of the $2p_y$ orbitals pointing outside the plane of the paper are represented.

7.1.4 Crystal orbitals at the Γ and X points

Since the crystal orbitals exhibit a higher symmetry at the Γ and X points, we will start by looking at the shape of the crystal orbitals for these two points. First, let us represent the Bloch orbitals associated with the Γ point generated by the chosen fragment orbitals. Let us begin by analysing these orbitals from the viewpoint of the symmetry properties of the point group appropriate for Γ, i.e. the D_{2h} group (see Fig. 7.4).

Both the Bloch orbitals $BO_{S_1}(\Gamma)$ and $BO_{S_2}(\Gamma)$ are a basis for the same B_{2u} representation of the D_{2h} group. They may therefore interact and lead to two crystal orbitals, $CO_{1b_{2u}}(\Gamma)$ and $CO_{2b_{2u}}(\Gamma)$. Since the energies of the Bloch orbitals $BO_{S_1}(\Gamma)$ and $BO_{S_2}(\Gamma)$ lie relatively far from each other, their interaction is not very strong, so that the bonding $CO_{1b_{2u}}(\Gamma)$ crystal orbital reminds very much the $BO_{S_1}(\Gamma)$ orbital, and the antibonding $CO_{2b_{2u}}(\Gamma)$ crystal orbital reminds very much the $BO_{S_2}(\Gamma)$ orbital. The same reasoning applies to the orbitals that are antisymmetric with respect to the yOz plane. Useful as it is, this qualitative reasoning does not allow us to determine the exact expression of the crystal orbitals at Γ, but only their shape.

Let us now perform the same analysis for the X point. First, we will consider the two $BO_{S_1}(X)$ and $BO_{S_2}(X)$ orbitals, which are symmetric with respect to the yOz plane (See Fig. 7.5).

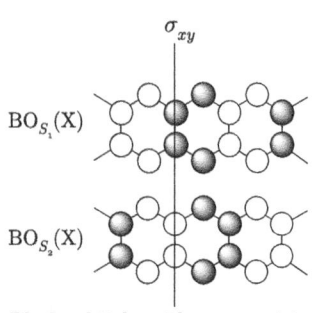

Bloch orbitals neither symmetric nor antisymmetric with respect to σ_{xy}

Fig. 7.5
$BO_{S_1}(X)$ and $BO_{S_2}(X)$ of regular polyacene. The origin (O) has been chosen to be at the centre of a C–C bond (see Fig. 7.2).

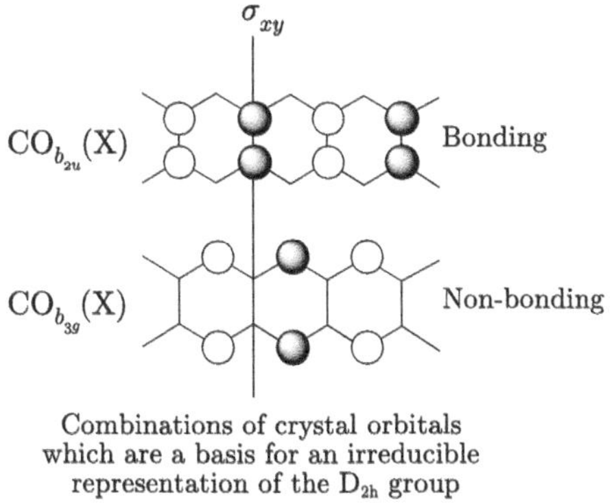

Band structure

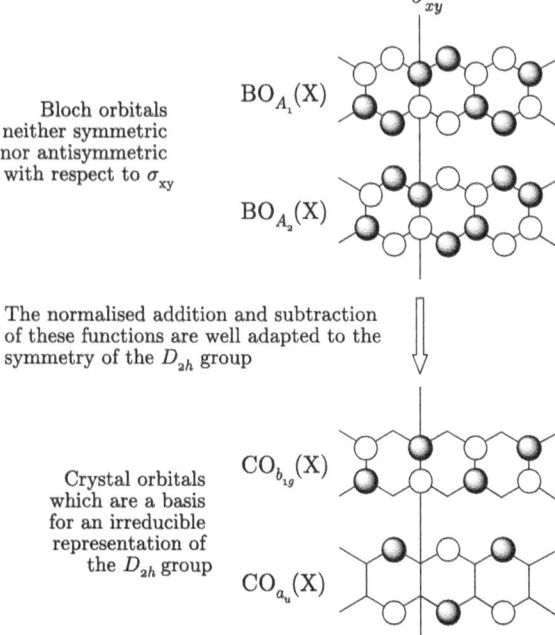

Combinations of crystal orbitals which are a basis for an irreducible representation of the D_{2h} group

Fig. 7.6
Crystal orbitals for the X point, which are symmetric with respect to the horizontal plane obtained as the addition and subtraction of the $BO_{S_1}(X)$ and $BO_{S_2}(X)$ Bloch orbitals. The origin (O) has been chosen to be at the centre of a C–C bond (see Fig. 7.2).

The Bloch orbitals $BO_{S_1}(X)$ and $BO_{S_2}(X)$ do not provide a basis for a representation of the D_{2h} group because they are neither symmetric nor antisymmetric with respect to the xOy plane. We must combine them to obtain functions well adapted to the D_{2h} symmetry, i.e. functions that are a basis for an irreducible representation of this group. In this case we must simply add and subtract the two Bloch orbitals; this leads to functions that are a basis for the B_{2u} and B_{3g} irreducible representations of the D_{2h} group (Fig. 7.6). These two orbitals cannot mutually interact and they are already crystal orbitals that are symmetric with respect to the horizontal symmetry plane: we will denote them $CO_{b_{2u}}(X)$ and $CO_{b_{3g}}(X)$.

The same approach can be used to obtain the two Bloch orbitals that are a basis for the irreducible representations B_{1g} and A_u of the D_{2h} group from the two Bloch orbitals antisymmetric with respect to the horizontal plane. The process is summarised in Fig. 7.7. For symmetry reasons the functions

Fig. 7.7
Generation of the crystal orbitals that are antisymmetric with respect to the horizontal plane for the X point. The origin (O) has been chosen to be at the centre of a C–C bond (see Fig. 7.2).

represented at the bottom of Fig. 7.7 cannot interact, so that they are already crystal orbitals of regular polyacene and will be denoted as $CO_{b_{1g}}(X)$ and $CO_{a_u}(X)$.

Thus using only well-known orbital interaction concepts and simple symmetry arguments we have been able to determine the crystal orbitals of regular polyacene at Γ and X without carrying out any detailed calculations. We are now ready to write a qualitative π band structure for regular polyacene.

7.1.5 π-type band structure of polyacene

How can we draw the π band structure of regular polyacene on the basis of the results of Section 7.1.4? For every k point other than Γ and X, the crystal orbitals may be labelled according to the different irreducible representations of the C_{2v} group. Since all π-type orbitals are antisymmetric with respect to σ_{xz}, the crystal orbitals may be separated according to their symmetry with respect to σ_{yz}. Those that are symmetric will be labelled B_2, while those that are antisymmetric will be labelled A_2. The crystal orbitals of B_2 symmetry result from the mixing of the $BO_{S_1}(\vec{k})$ and $BO_{S_2}(\vec{k})$ Bloch orbitals, which are symmetric with respect to σ_{yz}. Those of A_2 symmetry result from the interaction of the $BO_{A_1}(\vec{k})$ and $BO_{A_2}(\vec{k})$ Bloch orbitals, which are antisymmetric with respect to this plane. Thus two symmetric and two antisymmetric crystal orbitals with respect to σ_{yz} will be obtained. Since two bands of the same symmetry cannot cross, in building the qualitative band structure we must connect the two crystal orbitals at Γ that are symmetric with respect to σ_{yz}, i.e. $CO_{1b_{2u}}(\Gamma)$ and $CO_{2b_{2u}}(\Gamma)$, with those at X that are also symmetric, i.e. $CO_{b_{2u}}(X)$ and $CO_{b_{3g}}(X)$. This process allows us to obtain the two continuous lines in Fig. 7.8.

Now we must repeat the same process for the crystal orbitals that are antisymmetric with respect to the horizontal plane. We obtain the two bands represented with dotted lines in Fig. 7.8. The relative energy of the four crystal orbitals at Γ, as well as that of the two crystal orbitals $CO_{b_{1g}}(X)$ and $CO_{b_{2u}}(X)$, is clearcut. However, to determine a full qualitative band structure we must first have an idea about the energy of the two crystal orbitals $CO_{a_u}(X)$ and $CO_{b_{3g}}(X)$. Taking into account only the nearest-neighbour interactions (according to the Hückel approach) these two orbitals should be degenerate.

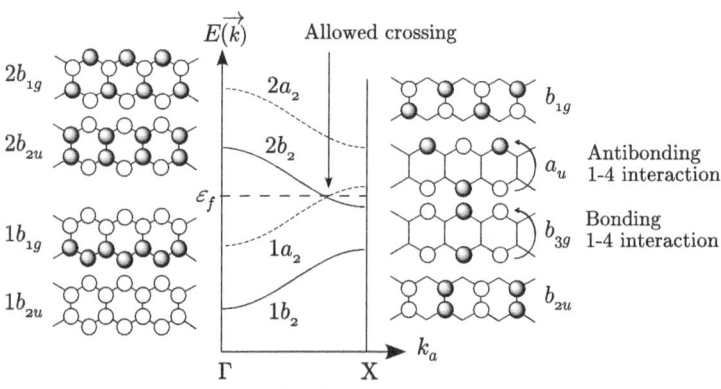

Fig. 7.8
Qualitative band structure for regular polyacene. The symmetry labels in the crystal orbitals at Γ and X are those corresponding to the D_{2h} group. The symmetry labels in the band structure diagram are those appropriate for each band according to the C_{2v} group.

However, if the weak 1–4 type interactions are also considered, then the $CO_{b_{3g}}(X)$ orbital is slightly more stable than the $CO_{a_u}(X)$ (see Fig. 7.8).

Since every cell contributes four electrons to the π system of regular polyacene, the equivalent of two bands must be fully filled. Consequently, the Fermi level lies at the crossing of the $1a_2$ and $2b_2$ bands (see Fig. 7.8), which occurs very near the X point. The band structure of regular polyacene can thus be schematically represented as shown in Fig. 7.9. This is typical of a semimetallic system.[1]

In view of this band structure, it must be considered likely that a distortion will open a band gap by stabilising band $1a_2$ and destabilising band $2b_2$. This is schematically shown in Fig. 7.10.

In principle, the band structure represented in Fig. 7.10 could be obtained in the following cases:

- The b_{2u} and b_{3g} orbitals at X interact, as well as the a_u and b_{1g} ones, as a result of a distortion that removes the xOy symmetry plane but keeps the yOz. Under such circumstances the $1b_2$ and $1a_2$ bands would be stabilised whereas the $2b_2$ and $2a_2$ bands would be destabilised.
- The crossing between the $1a_2$ and $2b_2$ bands (see Fig. 7.8) becomes forbidden. This means that the distortion removes one of the symmetry operations

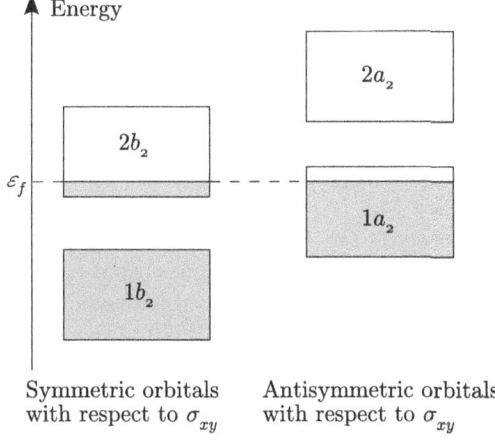

Fig. 7.9
Schematic diagram showing the occupation of the different π bands of regular polyacene at $T = 0$ K. The filled levels are shown in grey.

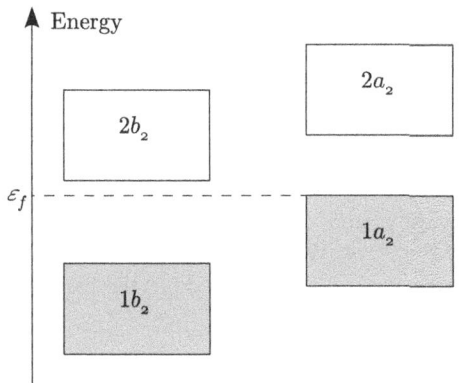

Fig. 7.10
Slightly modified schematic band structure for polyacene as a result of a hypothetical distortion stabilising the $1a_2$ band and destabilising the $2b_2$ band.

of the C_{2v} group, as, for instance, the reflection with respect to the yOz horizontal plane.

In the next section we will consider these possible distortions in detail, since they would have the important consequence of rendering the polyacene chain non-metallic.

7.2 Distortions in polyacene

7.2.1 Disappearance of the σ_{xy} symmetry plane

We consider now the distortion which leads to the appearance of localised double bonds (i.e. short bonds) as shown in Fig. 7.11.

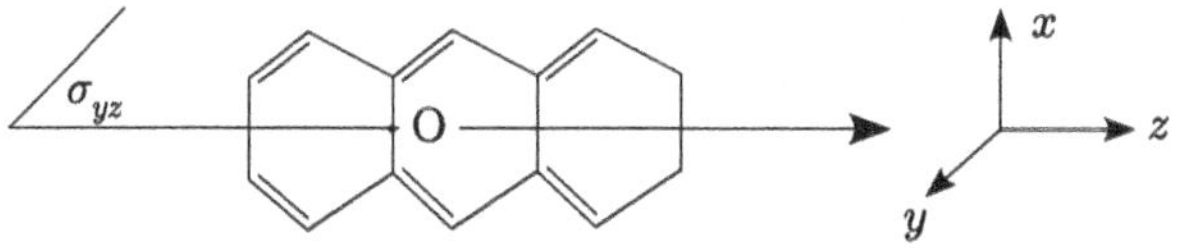

Fig. 7.11
Structure of a distorted polyacene exhibiting localised double bonds and keeping the σ_{yz} symmetry plane.

Such a distortion removes the σ_{xy} symmetry plane, the i inversion, and the C_{2x} and C_{2y} rotations, but leaves the O point invariant. The symmetry group appropriate for all k points is now C_{2v}. The orbitals of the regular structure, which were symmetric with respect to σ_{yz}, $CO_{b_{2u}}(X)$, and $CO_{b_{3g}}(X)$ (see Fig. 7.8), are now of B_2 symmetry and thus, having lower symmetry, they can interact. As a result, there is one crystal orbital that is stabilised with respect to $CO_{b_{2u}}(X)$ and another crystal orbital that is destabilised with respect to $CO_{b_{3g}}(X)$ (see Fig. 7.12a). The orbitals of regular polyacene that were antisymmetric with respect to σ_{yz} are of the same symmetry, A_2, in the new structure and thus they can also interact. As a consequence, there is one crystal orbital that is stabilised with respect to $CO_{a_u}(X)$ and another orbital that is destabilised with respect to $CO_{b_{1g}}(X)$ (see Fig. 7.12b).

If the distortion is strong enough, band $1a_2$ lies below band $2b_2$ all along the band structure. The new band structure, which is shown in Fig. 7.13, is very similar to that shown in Fig. 7.10. Such a distortion should thus stabilise the system. In contrast with the *trans*-polyacetylene distortion, in the present case the orbitals interacting during the distortion are initially non-degenerate (see Fig. 7.12): we are dealing with a *second-order* Peierls distortion.

This scheme shows that the orbital mixings, made possible in the occupied bands because of the symmetry lowering associated with the distortion, strengthen the bonding character of the C–C bonds, which become shorter. In *trans*-polyacetylene, the localisation of the double bonds tends to stabilise the π system and leads to a loss of the metallic properties.

7.2.2 Disappearance of the σ_{yz} symmetry plane

Now let us consider the kind of distortion that leads to a localisation of double bonds as shown in Fig. 7.14. We leave it to the reader, as a guided exercise, to consider how this type of distortion modifies the band structure of regular polyacene.

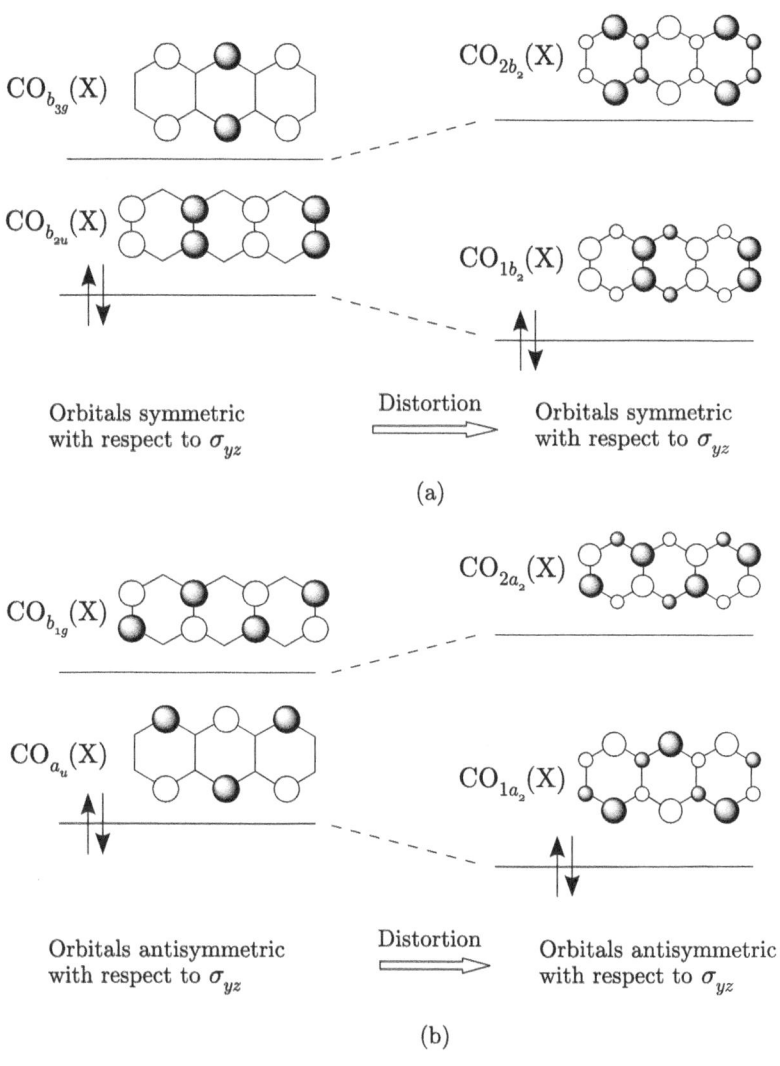

Fig. 7.12
Schematic representation of the interaction of the crystal orbitals of regular polyacene at X when it distorts as in Fig. 7.11.

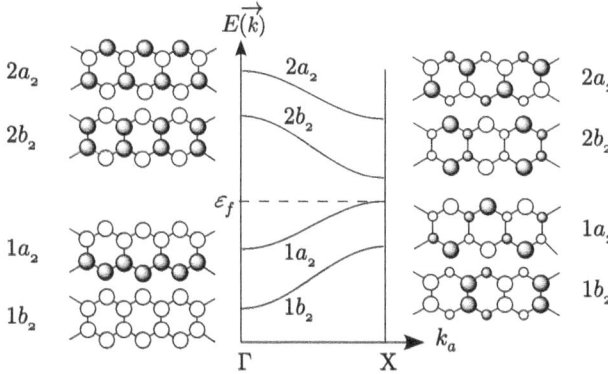

Fig. 7.13
Band structure for polyacene when it is distorted as in Fig. 7.11.

Polyacene

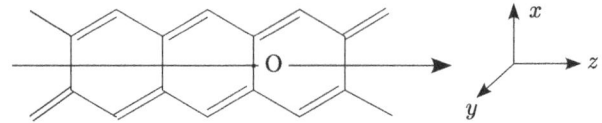

Fig. 7.14
A different distorted polyacene, exhibiting localised double bonds but keeping the C_{2y} symmetry axis.

7.3 General remarks concerning Peierls distortions

We now compare the results obtained in this chapter with those concerning *trans*-polyacetylene (Chapter 5) and the H_n model chain (Chapter 3). This will allow us to summarise the main differences between first-order and second-order Peierls distortions.

7.3.1 First-order Peierls distortions

First-order Peierls distortions make possible very efficient orbital mixings between degenerate or quasi-degenerate crystal orbitals that lie at or near the Fermi level. Such distortions lead to changes in the size of the unit cell (i.e. they affect the translational properties of the system), which result in the opening of a band gap at the Fermi level, thus changing the transport properties of the system, as we saw in Chapter 4.

For instance, the H_n system studied in detail in Chapter 3 is an example of a 1D system exhibiting a first-order Peierls distortion. The distortion results in the opening of a band gap at the Fermi level and a unit cell twice larger. The consequences of such a dimerisation can be summarised as shown in Fig. 7.15. Starting with the band structure of the regular system, we can say that the dimerisation makes possible the interaction between levels that are near the Fermi level, i.e. that are associated with k points that lie near $k_f = \pm 1/4$, because the band is half-filled (Fig. 7.15b). Since every pair of k points that are separated by $2k_f$, i.e. half of the Brillouin zone, correspond to one filled and one empty levels, their interaction necessarily leads to the opening of a band gap around the Fermi level. Since the unit cell generating the distorted chain is twice as large as the undistorted one, we must represent the band structure of the distorted system using the new Brillouin zone appropriate for this system, i.e.] − 1/4, 1/4] (Fig. 7.15c).

Another example of a system exhibiting a first-order Peierls distortion is *trans*-polyacetylene (cf. Chapter 5). Here, a dimerisation allows the stabilisa-

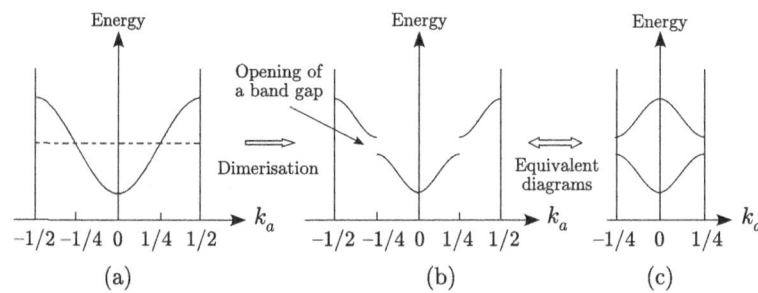

Fig. 7.15
Band structure for the H_n chain: (a) in its regular structure; (b) and (c) in its dimerised structure. The band structure in (c) is the 'folded' version of the band structure in (b). In the rest of the book we will always use representation (c).

tion of the occupied orbitals around the Fermi level and the destabilisation of the empty ones, resulting in a change from metallic to semiconducting behaviour. The dimerisation changes the translational properties of the system since the screw axis and glide planes disappear as a consequence of the distortion.

7.3.2 Second-order Peierls distortions

Second-order Peierls distortions are associated with interactions between non-degenerate crystal orbitals, which provide the driving force for the stabilisation afforded by such a distortion. They are associated with changes in the point group symmetry but do not modify the translational properties of the system.

The possible distortions for polyacene that are considered in this chapter provide examples of this type of process. For simplicity we will assume that bands $1a_2$ and $2b_2$ are degenerate at X, i.e. we will neglect the weak 1–4 interactions. This does not introduce any serious flaws in our discussion. The first distortion we have considered (Fig. 7.11) allows the mixing at X of the $CO_{b_{2u}}(X)$ and $CO_{b_{3g}}(X)$ crystal orbitals of the regular structure, as well as of the $CO_{a_u}(X)$ and $CO_{b_{1g}}(X)$ ones. More generally, such a distortion allows the interaction of the levels of the $1a_2$ and $2a_2$ bands as well as those of the $1b_2$ and $2b_2$ all along the Brillouin zone, but especially around X. These interactions are at the origin of the appearance of a band gap and thus of the lowering of the electronic energy of the system.

The second distortion considered (Fig. 7.14) allows the mixing of the $CO_{b_{2u}}(X)$ and $CO_{a_u}(X)$ crystal orbitals of the regular structure as well as the $CO_{b_{3g}}(X)$ and $CO_{b_{1g}}(X)$ ones. More generally, such a distortion avoids any band crossing inside the Brillouin zone. These interactions also lead to the opening of a band gap at the Fermi level.

For the two distortions considered, the two crystal orbitals of regular polyacene lying at the Fermi level, which are essentially non-bonding ($CO_{b_{3g}}(X)$ and $CO_{a_u}(X)$), become slightly bonding and slightly antibonding, respectively, after the distortion (Fig. 7.12). The two distortions considered for regular polyacene allow the interaction of non-degenerate crystal orbitals: this is the fingerprint of a second-order Peierls distortion that does not change the translational properties of the system.

In general, such second-order Peierls distortions are associated with lower energy gains than those associated with first-order Peierls distortions, simply because they result from weaker orbital interactions as a consequence of the larger energy difference between the interacting orbitals.

Exercises

(7.1) Outline different Lewis structures for polyacene.

(7.2) Draw a schematic representation of the π-type molecular orbitals of *cis*-butadiene and give their occupation in the fundamental state. What are the symmetry properties of these orbitals?

(7.3) Is there a direct product of groups $\{P, O\}$, other than that discussed in Section 7.1.2, that is appropriate to generate the space group of regular polyacene?

(7.4) Draw an energy diagram with the energies of the different C_4H_2 fragment orbitals.

(7.5) What is the difference between the fragment orbitals of Exercise (7.4) and the π molecular orbitals of *cis*-butadiene?

(7.6) Indicate a different set of fragment orbitals that could have been chosen to build the band structure of polyacene.

(7.7) Establish an interaction diagram for the Bloch orbitals of polyacene at Γ, showing how they lead to the different crystal orbitals. Draw an energy diagram of the different crystal orbitals.

(7.8) Draw the energies of the different crystal orbitals at X in the diagram you produced in Exercise (7.7).

(7.9) Redraw the qualitative band structure of Fig. 7.8 by using the P group obtained in Exercise (7.3). What are the main conclusions of the exercise?

(7.10) Determination of the band structure of polyacene distorted as in Fig. 7.14. We will follow a similar procedure as in Section 7.2.1, i.e. we will assume that the electronic structure of regular polyacene is already known and treat the distortion as a perturbation leading to the mixing of π orbitals, which in the regular structure could not interact.

(a) What symmetry operations disappear when going from the regular (Fig. 7.1) to the distorted (Fig. 7.14) system?

(b) Is the band crossing shown in Fig. 7.8 possible in the distorted system?

(c) How will the crystal orbitals for regular polyacene, which are shown in Fig. 7.8, be modified as a result of the distortion?

(d) Show where the new Fermi level lies. Show that the nature of the new occupied crystal orbitals at Γ and X demonstrate that there is a strengthening of the C–C bonds, which become shorter in the new structure.

References

1. For qualitative discussions of the band structure of polyacene see for instance: (a) M.-H. Whangbo, R. Hoffmann, and R. B. Woodward, *Proc. Roy. Soc. London*, A366, 23, 1979; (b) J. K. Burdett, *Prog. Sol. State Chem.*, 15, 173, 1984; (c) M. Kertesz, Y. S. Lee, and J. P. Stewart, *Int. J. Quant. Chem.*, XXXV, 305, 1989.

Electronic structure of selected inorganic chains

8

We will now analyse the electronic structure of selected inorganic chains and the relationship between this structure and the conduction properties. We have chosen to study two examples in detail and to treat three additional systems as exercises. It is essential that after considering the two detailed analyses of 1D examples in this chapter the reader also attempts the exercises. This is the best prelude to consideration of the more complex 2D and 3D materials discussed in the following chapters. Here we will follow the approach outlined at the end of Chapter 6 (see Section 6.5). First, we will establish the band structure of the system under consideration, then we will consider the possible structural distortions that may stabilise the system and finally we will consider how the resulting band structure may give a hint on the transport properties of the material.

8.1 KCP

Let us start by considering the electronic structure of $K_2[Pt(CN)_4]Br_{0.3} \cdot 3H_2O$, which is usually referred to as KCP. It is the best known of the inorganic metallic chains built on the basis of square-planar tetracyanoplatinate units, with general formula $C_x[Pt(CN)_4]A_y \cdot zH_2O$, where C is a cation and A is an anion. [1] KCP is a typical example of a real compound (i.e. a 3D material). It is made of structurally independent chains with formula $(Pt(CN)_4^{1.7-})_n$, in between which the K^+ cations, Br^- anions and water molecules reside. The system may be considered 1D because the interchain interactions are very weak. These materials also provide a nice illustration of the notion of doping. Doping of a material is the process by which the number of electrons is modified, by adding either a donor or an acceptor. For instance, $K_2[Pt(CN)_4] \cdot 3H_2O$ is a non-doped compound whereas KCP is doped because of the addition of 0.3 bromines per platinum atom. These bromine atoms are found in isolation between the chains of KCP. Since the bromine atoms are more electronegative than the platinum chains, they are found as Br^- anions. Thus doping of the platinum chains results in a removal of electrons, i.e. an oxidation. Consequently, every $Pt(CN)_4$ fragment of KCP bears a charge of −1.7 instead of −2.0 in the non-doped material. The materials of this family, $C_x[Pt(CN)_4]A_y \cdot zH_2O$,

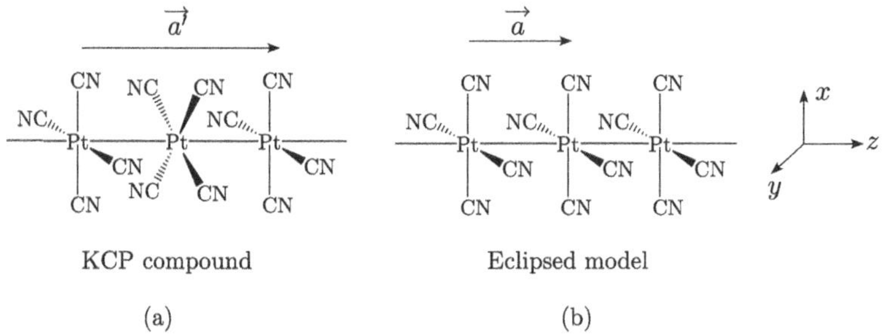

Fig. 8.1
Structure of the Pt(CN)₄ chain: (a) staggered structure observed in KCP; (b) eclipsed structure of the model chain.

KCP compound (a)

Eclipsed model (b)

[1] The Pt–Pt bond length is 3.47 Å in the non-doped compound but 2.87 Å in KCP.

may be doped with anions (A = Br⁻, Cl⁻, F⁻, FHF⁻, etc), cations (C = K⁺, Cs⁺, NH_4^+, etc.) or both simultaneously. The conductivity properties of these materials depend on the degree of doping. Thus the non-doped compound, $K_2[Pt(CN)_4]\cdot 3H_2O$, is an insulator whereas KCP is metallic at room temperature and becomes semiconducting at lower temperatures, as a result of the development of a Peierls distortion.

Even if doping does not induce major modifications in the structure of these materials, there is a considerable shortening of the Pt–Pt bond [1] as well as a change in the relative conformation of two adjacent Pt(CN)₄ units (eclipsed in the non-doped compound but staggered in KCP). From a structural viewpoint, KCP is built from $(Pt(CN)_4^{1.7-})_n$ chains generated by a repeat unit consisting of two Pt(CN)₄ square-planar units in a staggered configuration and a repeat vector $\vec{a}'$ (see Fig. 8.1a). Before considering the staggered chain we will study the eclipsed chain shown in Fig. 8.1b. The latter is generated by a vector $\vec{a}$ ($\vec{a} = \vec{a}'/2$) and a repeat unit of just one $(Pt(CN)_4^{1.7-})$ unit. Then we will examine the influence of the CN⁻ ligands' orientation on the conductivity of the system. In principle, since the Pt–Pt distances are of the order of 3 Å, the ligand orbitals do not overlap. Hence, the conductivity process must essentially involve only the platinum orbitals. Consequently, the ligand orientation must play a secondary role in the conductivity. Thus, a study of the band structure of the structurally simpler eclipsed model should be a convenient starting point to build the band structure for KCP.

8.1.1 Band structure of the eclipsed chain $[Pt(CN)_4]^{(2-\delta)-}$

Symmetry

As shown in Section 6.1.2, the space group of this chain may be written as a direct product $T_n \otimes D_{4h}$, the D_{4h} group leaving the platinum atom of the reference cell invariant. Consequently, the appropriate group for the Γ and X points is D_{4h}, while it is C_{4v} for the other k points.

Choice of fragment orbitals generating the Bloch orbitals

Let us start our analysis by generating the Bloch orbitals of the system from the $Pt(CN)_4^{(2-\delta)-}$ fragment orbitals adapted to the C_{4v} symmetry, which must hold for any k point. For instance, we can build a basis of Bloch orbitals from the orbitals of the metal complex $Pt(CN)_4^{2-}$, which is of D_{4h} symmetry and thus well adapted to the C_{4v} symmetry. The mainly metal-based orbitals of

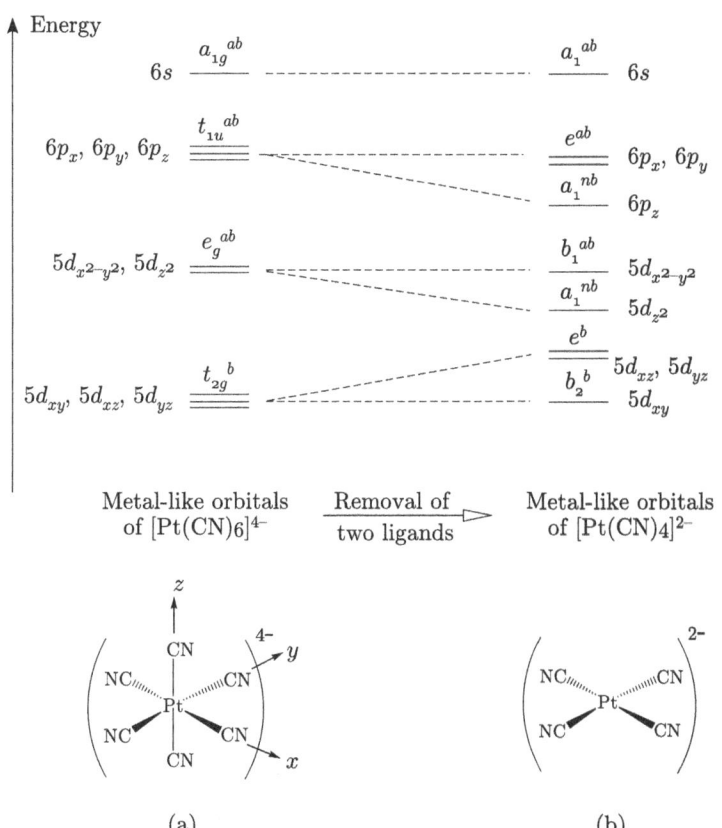

Fig. 8.2
Molecular orbitals with strong metal component (a) of the octahedral $Pt(CN)_6^{4-}$; (b) of the square-planar $Pt(CN)_4^{2-}$ compound. For clarity, the ligand contributions have not been shown. The symmetry labels corresponding to the C_{4v} group, which is the appropriate symmetry group to label the Bloch orbitals generated from these fragment orbitals, are also given. The labels b, nb, and ab denote the bonding, non-bonding, or antibonding character of the associated orbitals.

this complex may be deduced from those for the $Pt(CN)_6^{4-}$ octahedral complex shown at the left side of Fig. 8.2.

In the octahedral compound, the orbitals with strong metal character implicate either the $(5d_{x^2-y^2}, 5d_{z^2})$ orbitals of e_g symmetry, the $(5d_{xz}, 5d_{yz}, 5d_{xy})$ orbitals of t_{2g} symmetry, the $(6p_x, 6p_y, 6p_z)$ orbitals of t_{1u} symmetry, or the $6s$ orbital of a_1 symmetry. If we now remove the two ligands in the Oz axis, the square-planar complex is generated. The departure of the two ligands will cause the stabilisation of the molecular orbitals of the octahedral complex that are antibonding along the Oz direction, and a destabilisation of the molecular orbitals that are bonding along this direction. The molecular orbitals of the square-planar complex with symmetry labels corresponding to the C_{4v} group are shown at the right side of Fig. 8.2. It is clear that the departure of two axial ligands is associated with a stabilisation of the orbital $5d_{z^2}$, which becomes almost non-bonding in the square-planar complex because of the loss of metal–ligand antibonding character along the axial direction and the mixing of the $6s$ orbital which in this geometry is allowed by symmetry. Also worthy of a mention is a strong stabilisation of the molecular orbital denoted $6p_z$, which in the square-planar complex does not exhibit any σ metal–ligand antibonding interaction. Finally let us note that in $Pt(CN)_4^{2-}$ all bonding or non-bonding orbitals of the d-block are filled whereas the strongly antibonding

$5d_{x^2-y^2}$ orbital is empty.[2] In our analysis of the band structure of the chain we will consider all orbitals of the $5d$ block and the almost non-bonding orbital $6p_z$ involving the metal. *Why do we need to retain all these orbitals of the square-planar fragment?* To begin with we do not wish to skip any step of the reasoning in this case, the first example in which we consider a real chain containing transition metal atoms. In addition, all fragment orbitals that may play a role in the electronic structure must of course be retained. Since we wish to comment on the conductivity of the system, it is mandatory to retain all orbitals of the $Pt(CN)_4^{(2-\delta)-}$ unit that may be involved in strong interactions along the chain direction, i.e. the Oz axis. This is why in addition to the d-block orbitals we have retained the empty orbital mostly based on the metal $6p_z$ orbital. Even if this orbital lies relatively high in energy for the isolated fragment, the bottom part of the energy band based on this orbital may occur at energies of the same order, or even lower, than the top of some bands based on the $5d$ block orbitals.

Analysis of the Bloch orbitals for the Γ and X points

We will start by looking at the Bloch orbitals for the Γ and X points generated by the fragment orbitals retained. We will characterise the symmetry of these orbitals with respect to the D_{4h} group that leaves the platinum atom of the reference cell invariant, i.e. the symmetry group of the Γ and X points (see Fig. 8.3). Since all these Bloch orbitals possess different symmetry properties,

[2] In this complex, the platinum oxidation state is +II, i.e. its electronic configuration is d^8.

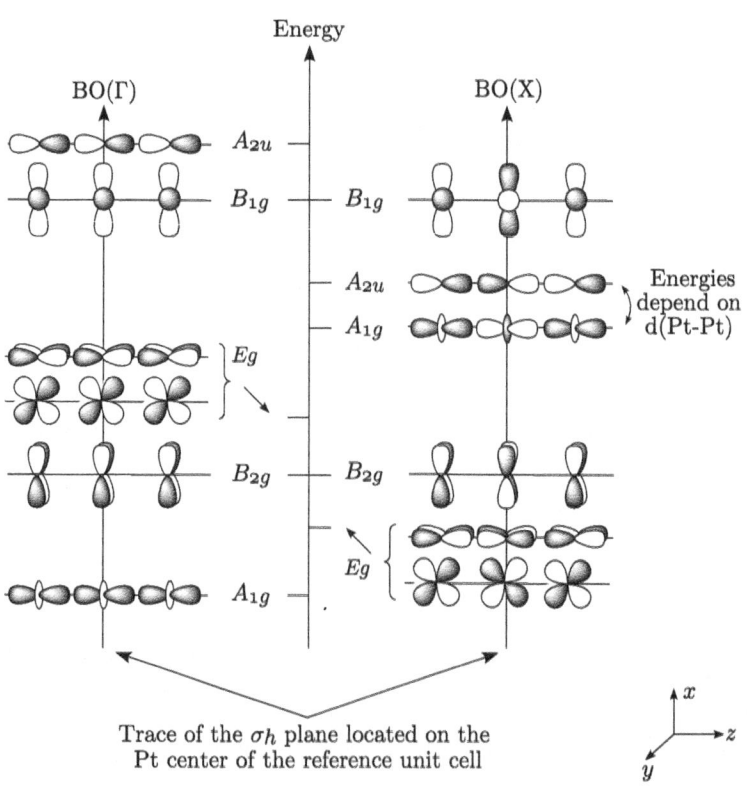

Fig. 8.3
Bloch orbitals at the Γ and X points. The symmetry labels according to the D_{4h} group are given.

they may be identified with the crystal orbitals of the system. At X the relative energies of the $BO_{d_{z^2}}(X)$ and $BO_{p_z}(X)$ depend on the Pt–Pt distance, which may vary as a function of the degree of doping. *In the following schemes we will assume that the $BO_{d_{z^2}}(X)$ orbital is lower in energy than the $BO_{p_z}(X)$ orbital* (Fig. 8.3). This is the case when the Pt–Pt distance is not very short. The opposite case will be discussed in Section 8.1.2.

Band structure of the eclipsed chain

Since the symmetry group for k points other than Γ and X is the C_{4v} group leaving every platinum atom invariant, the symmetry label for the different Bloch orbitals is the same as the symmetry label of the fragment orbital from which it originates (Fig. 8.2). Consequently, only the bands generated from the $6p_z$ and $5d_{z^2}$ fragment orbitals of a_1 symmetry may interact for any k point other than Γ and X.[3] All other orbital mixings are precluded by symmetry, so that the Bloch orbitals of symmetry other than a_1 may be identified with the crystal orbitals of the system.

How can the different bands be positioned on an energy scale? Let us consider first the a_1-type bands resulting from the interaction of the $BO_{d_{z^2}}(\vec{k})$ and $BO_{p_z}(\vec{k})$ Bloch orbitals. The energies of the $BO_{d_{z^2}}(\vec{k})$ and $BO_{p_z}(\vec{k})$ Bloch orbitals as a function of k_a are shown as dotted lines in Fig. 8.4. The energy of the $BO_{d_{z^2}}(\vec{k})$ Bloch orbital increases when k_a goes from 0 to 1/2, while the energy of the $BO_{p_z}(\vec{k})$ Bloch orbital decreases along the same interval. Since these orbitals interact at any point other than Γ and X, the two dotted lines repel each other from Γ to X giving rise to the continuous lines, which represent the energy of the crystal orbitals resulting from the mixing of the two Bloch orbitals.

Since the crystal orbitals of b_1, b_2, and e symmetries coincide with the Bloch orbitals, the shape of the curves representing their energies may be obtained by connecting the energies of the corresponding crystal orbitals at Γ and X through a cosine-type line (see eqn (3.20)). Thus, knowing the energies of the crystal orbitals at Γ and X (Fig. 8.3) as well as the shape and relative energy of the a_1-type bands (Fig. 8.4), we may arrive at the band structure of Fig. 8.5 for the eclipsed $[Pt(CN)_4]_n$ chain. According to this diagram the $1a_1$ and $2a_1$ bands are considerably wider than the $1e$ bands which, in turn, are wider than

[3] In this qualitative treatment we are not taking into account the $6s$ orbitals which are also of a_1 symmetry and thus can mix with the $6p_z$ and $5d_{z^2}$ orbitals. However, the mixing is not very important because the $6s$ orbitals are high in energy. A similar argument applies to the $6p_x$ and $6p_y$ orbitals which are of the same symmetry as the $5d_{xz}$ and $5d_{yz}$.

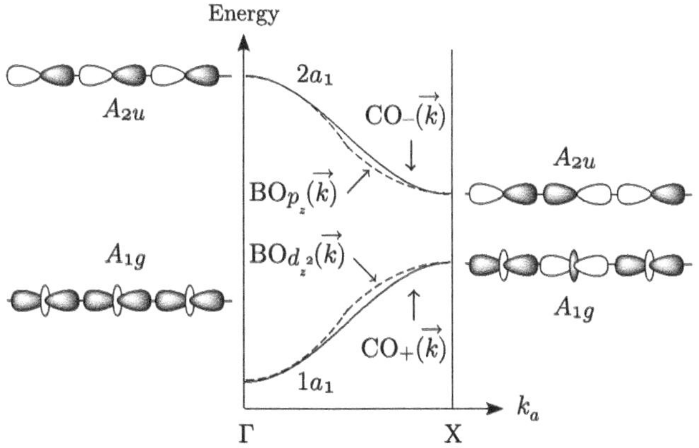

Fig. 8.4
Energy bands of a_1-type symmetry: the dotted lines are the Bloch orbital energies while the continuous lines are the energies of the crystal orbitals resulting from their interaction.

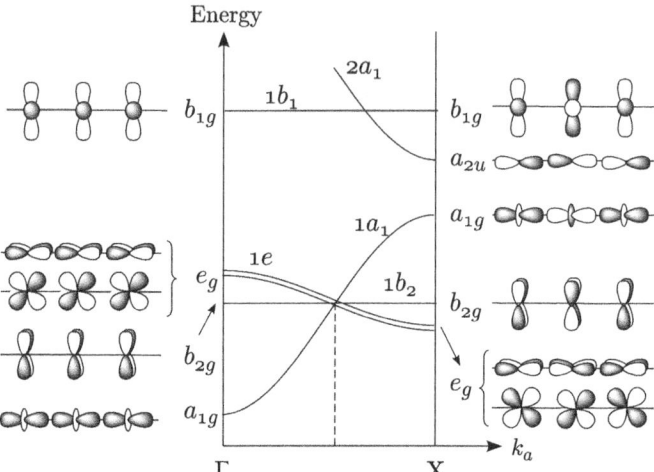

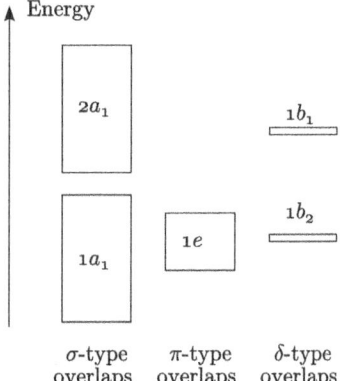

Fig. 8.5
Schematic band structure for the eclipsed [Pt(CN)$_4$]$_n$ chain.

Fig. 8.6
General band structure for the eclipsed [Pt(CN)$_4$]$_n$ chain.

the $1b_1$ and $1b_2$ bands. These differences originate from the different nature of the overlaps associated with each type of band: very strong (σ) for the $1a_1$ and $2a_1$ bands, moderate (π) for the $1e$ bands, and finally very weak (δ) for the $1b_1$ and $1b_2$ bands. In general, a flat band suggests very weak interactions along the chain. However, even if flat, this band contains as many states as any other band, for instance the strongly dispersive ones. The band structure of the system can also be represented in a simplified way as shown in Fig. 8.6.

This qualitative approach does not allow us to be completely sure about the relative positioning of all bands. It is important to determine which features of the band structure can be derived from the simple qualitative approach and which aspects can only be firmly established by computation. It follows from the analysis in this section that the band structure must necessarily exhibit the following features:

(i) non-existence of a crossing between the $1a_1$ and $2a_1$ bands
(ii) relative band widths in agreement with the overlap involved.

Among the uncertainties we have at this point, the positioning of the bottom part of the $2a_1$ band (essentially $6p_z$ in character) with respect to the very flat $1b_1$ band (essentially $d_{x^2-y^2}$ in character) is not clear from purely qualitative reasoning. This feature depends on the Pt–Pt bond length, which will vary as a function of the degree of doping or as a function of pressure. The reader can verify that the properties of these chains do not depend on the relative position of the $2a_1$ and $1b_1$ bands. If this were not the case, a computation using the actual geometry of the chain would be necessary in order to establish the band structure of the chain for certain.

Electronic structure of the eclipsed chain

At this point we must look at the filling of the different bands for the non-doped K$_2$Pt(CN)$_4$·3H$_2$O compound at $T = 0$ K. Every unit with formula Pt(CN)$_4^{2-}$ contributed eight electrons, i.e. its electronic configuration is formally $(1a_1)^2(1b_2)^2(1e)^4$. This being the case, the four lower bands of Fig. 8.5 must be filled and thus the Fermi level occurs at the top of the $1a_1$ band, which is essentially the d_{z^2} band. Consequently, the non-doped compound must be non-metallic since there is a band gap at the Fermi level.

In the doped compound with formula $K_2Pt(CN)_4Br_{0.3}\cdot 3H_2O$ the addition of bromine results in the oxidation of the compound, and thus the partial emptying of the highest-occupied band. Now every cell contributes only 7.7 electrons, since 0.3 electrons have, on average, been transferred to the bromine atoms, i.e. the formal electronic configuration of every square-planar unit is $(1a_1)^{1.7}(1b_2)^2(1e)^4$. The $1a_1$ band is thus only 85% filled, and the Fermi level occurs at $k_{a_f} = \pm 0.425 = \pm 0.5 \mp 0.075$. Since the $1a_1$ band is 15% empty and given that a 15% empty band behaves like a 15% filled band as far as a potential Peierls distortion is concerned (see Chapter 4), we can conclude that this system must be unstable with respect to a 6.666-merisation (see Fig. 4.14). Consequently, it is expected that when the temperature is lowered this compound exhibits a Peierls distortion and loses its metallic properties and exhibits a modulation of its structure.

In view of the previous analysis, the hypothetical eclipsed compound, doped as in KCP, must be metallic at high temperature but should undergo a 6.666-merisation at lower temperature, which makes the system semiconducting.

8.1.2 Band structure of KCP (staggered chain)

How do the electronic properties of real KCP (Fig. 8.1a) differ from those of the eclipsed model (Fig. 8.1b)?

Band structure of the staggered chain generated by a double cell

To facilitate a comparison of the properties of the two compounds represented in Figs 8.1a and 8.1b we must generate the band structure of an eclipsed model compound with a double cell, $(Pt(CN)_4)_2$, and a repeat vector $\vec{a}'$ equal to $2\vec{a}$. The band structure of the staggered compound may then easily be obtained from the band structure of the eclipsed model compound with the double cell. In fact, this band structure can simply be obtained by folding (see Fig. 3.31) the band structure of Fig. 8.5.

The band structure of Fig. 8.7 contains ten energy bands and the bottom part of the mainly p_z band. In the model eclipsed $K_2Pt(CN)_4Br_{0.3}\cdot 3H_2O$ compound every $(Pt(CN)_4^{1.7-})_2$ unit provides 15.4 electrons. Thus, seven bands must be full while the upper band, which is mostly d_{z^2} in character, is 70% filled. Consequently, the Fermi level for the model eclipsed compound is characterised by $k_{a'_f} = 0.15$. The $5d_{z^2}$ and $6p_z$ bands and their filling at $T = 0$ K for the model eclipsed $K_2Pt(CN)_4Br_{0.3}\cdot 3H_2O$ compound are shown in Fig. 8.8.

Band structure of KCP

We should now compare the electronic structure of the model above – the eclipsed $K_2Pt(CN)_4Br_{0.3}\cdot 3H_2O$ compound – with that of the real, staggered KCP compound. Analysis of the eclipsed system shows that the Fermi level cuts a band mostly involving the $5d_{z^2}$ and $6p_z$ orbitals of the platinum. Since the ligands do not directly interact, the only interactions occurring along the

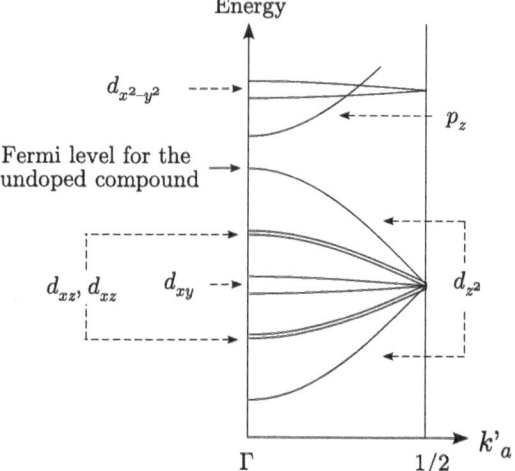

Fig. 8.7
Band structure of the eclipsed $[Pt(CN)_4]_n$ chain generated with a double unit cell. Note that the horizontal axis refers to $k_{a'}$, i.e. the projection of the k vector on the reciprocal vector $\vec{a}'^*$, instead of k_a ($\vec{a}'^* = \vec{a}^*/2 \Rightarrow k_{a'} = 2k_a$).

chain are those associated with the platinum orbitals. Consequently, the two bands of a_1 symmetry (Fig. 8.8) are independent of the orientation of the ligands. In addition, the other bands associated with the metallic orbitals that do not possess this cylindrical symmetry will be only weakly modified and their width will stay practically unaltered when moving from the eclipsed to the staggered compound. Consequently, the upper, partially filled band for the staggered compound will be in essence a d_{z^2} band. As a result the reasoning used above applies as well and the Fermi level must be associated with $k_{a'_f} = 0.15$ (see Fig. 8.8).

Using the analysis in Section 4.4.1 we can predict that KCP must be unstable towards a 3.333-merisation of the chain represented in Fig. 8.1a since $k_{a'_f}$ is $\pm 3/20$ ($2k_{a'_f} = \pm 3/10$). Such a distortion clusters an average of 6.666 platinum atoms along the chain since every repeat unit contains two platinum atoms. This deformation is at the origin of the metal-to-insulator

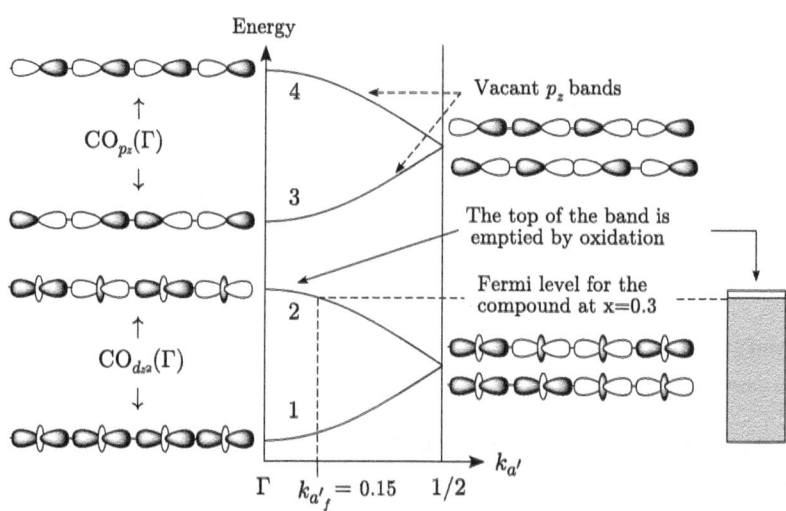

Fig. 8.8
d_{z^2} and p_z bands of the eclipsed chain generated by a cell containing two $Pt(CN)_4$ fragments. The Fermi level shown corresponds to an eclipsed chain with formula $K_2Pt(CN)_4Br_{0.3}\cdot 3H_2O$.

transition observed experimentally below 150 K. The distortion has been demonstrated experimentally by an X-ray study, which clearly revealed the new periodicity. [2]

Role of doping and pressure

The analysis of the conducting properties of KCP provides an excellent example of why doping may be an efficient way to control the conductivity of certain materials. Essentially, doping a material is a way to modify the Fermi level by either oxidising or reducing the material, something which in the present case may be brought about through the use of an halogen or an alkaline metal, respectively. It is an experimental observation that the Pt–Pt distances change according to the degree of doping: for instance the Pt–Pt distance decreases as the degree of doping with the bromine anion increases. This observation may be very simply explained by noting that the oxidation of the chain empties the top part of the d_{z^2} band, which is made up of Pt–Pt antibonding levels.

In principle, any increase of the external pressure on a crystal of KCP should lead to a decrease of the Pt–Pt distances as well as to an increase in the interactions along the chain: the bonding orbitals should be stabilised and the antibonding orbitals should be destabilised. Thus, the pressure effect should be similar to that of doping with an oxidising agent since it tends to shorten the Pt–Pt distances. However, *pressure does not modify the Fermi level*. In particular, an increase of pressure or doping results in a stabilisation of the bonding $BO_{p_z}(\Gamma)$ crystal orbital at the bottom of the p_z band and a destabilisation of the antibonding $BO_{d_{z^2}}(\Gamma)$ crystal orbital at the top of the d_{z^2} band (see Fig. 8.8). When doping or pressure are low, the $BO_{d_{z^2}}(\Gamma)$ orbital is more stable, while at high pressure or doping, the $BO_{p_z}(\Gamma)$ should be more stable. Now the question is: can bands 2 and 3 of Fig. 8.8 cross at high pressure or doping? The space group of KCP may be written as a direct product $T_n \otimes D_{4h} \otimes \{E, g_{C_{8z}}\}$, where $g_{C_{8z}}$ refers to the screw axis around the chain direction ($g_{C_{8z}} = t_{\vec{a}/2} \cdot C_{8z} = C_{8z} \cdot t_{\vec{a}/2}$). For any k point other than Γ and X' the crystal orbitals must be adapted to the symmetry of the C_{4v} group and the $g_{C_{8z}}$ operation. Since the crystal orbitals of bands 2 and 3 of Fig. 8.8 possess the same symmetry properties with respect to these symmetry operations, the crossing of these bands is forbidden (see Exercise (8.4)). In contrast, at Γ the $BO_{d_{z^2}}(\Gamma)$ and $BO_{p_z}(\Gamma)$ orbitals do not have the same symmetry properties with respect to the D_{4h} group (in particular, with respect the σ_h symmetry plane). Consequently, an accidental degeneracy between the $BO_{d_{z^2}}(\Gamma)$ and $BO_{p_z}(\Gamma)$ orbitals is allowed.

The evolution of the a_1-type bands when pressure or doping increases is shown schematically in Fig. 8.9. This figure clearly demonstrates that an increase in pressure or doping leads initially to a decrease in the gap between the two bands (i.e. moving from case (a) to case (b)) and then, to an increase, when $BO_{p_z}(\Gamma)$ becomes more stable than $BO_{d_{z^2}}(\Gamma)$ (i.e. moving from case (b) to case (c)). A priori, there is a pressure and doping level at which the two a_1-type bands touch at Γ. *Except for this particular point, the two bands cannot touch – the non-doped compound cannot be metallic whatever the pressure.* The compound can only become a semiconductor with a small band gap or a

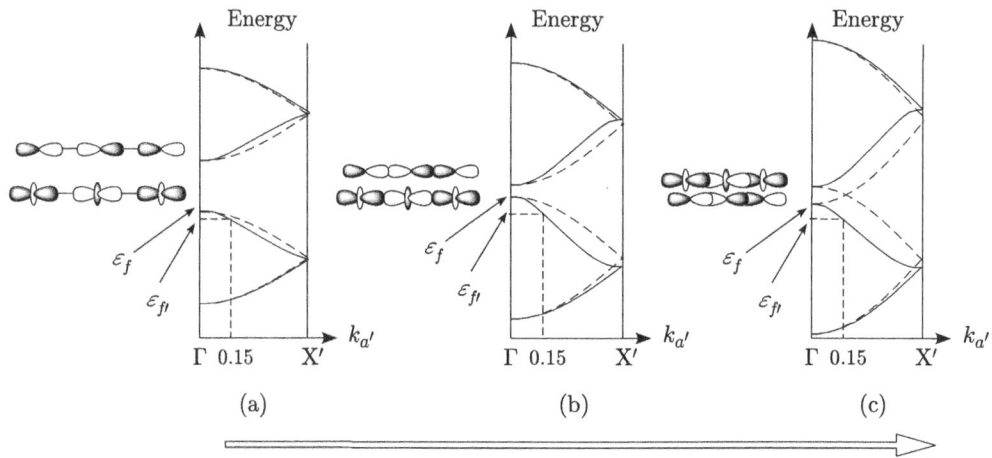

The increase of pressure (or doping) induces a shortening of the Pt-Pt distances and then the increase of the interactions along the 1D chain axis

Fig. 8.9
Influence of pressure or doping on the band structure of the staggered [Pt(CN)$_4$]$_n$ chain. The dotted lines are the Bloch orbital energies whereas the continuous lines are the energies of the crystal orbitals. ϵ_f and ϵ'_f are the Fermi levels for the non-doped and doped chains discussed in the text, respectively.

[4] For very high doping levels, the width of the bands should in principle decrease as shown in Fig. 8.9c. Tight-binding calculations [3] show that in the case of KCP the dispersion of bands d_{z^2} and p_z is almost optimised and that the situation is intermediate between those of cases (b) and (c) of Fig. 8.9.

semimetal when the pressure responsible for the accidental degeneracy of the BO$_{d_{z^2}}(\vec{k})$ and BO$_{p_z}(\vec{k})$ orbitals is approached.

Figure 8.9 provides an illustration of the doping effect on the chain by means of an oxidant. Doping not only partially empties the upper filled band but also decreases the Pt–Pt distances and thus makes the bands wider.[4] The fact that the partially filled band is both partially filled and wide is at the origin of the metallic-type conductivity of KCP.

As will be shown in Chapter 12, when a band is narrow, i.e. when the intercell interactions are weak, the system may be semiconducting or even insulating even if the band is partially filled. Understanding this situation, which at first sight may appear surprising, requires consideration of the role of electronic repulsions, and thus go beyond the monoelectronic approximation used throughout this chapter. For the present case it might be expected that if doping is such that the Pt–Pt distances are long, the system may be semiconducting and that pressure application may render the compound metallic.

8.1.3 Conclusions

The compound with formula K$_2$[Pt(CN)$_4$]Br$_{0.3}$· 3H$_2$O, usually known as KCP, which may be considered as a doped version of K$_2$[Pt(CN)$_4$]· 3H$_2$O, is a metal at room temperature as a consequence of the doping, which partially empties the upper filled and dispersive band. KCP becomes semiconducting at low temperatures as a result of a distortion that can be described as a 3.33-merisation, i.e. the clustering of 6.66 platinum atoms on average. The occurrence of a Peierls distortion associated with a metal-to-semiconductor transition is a common feature of all doped compounds of this family, the only

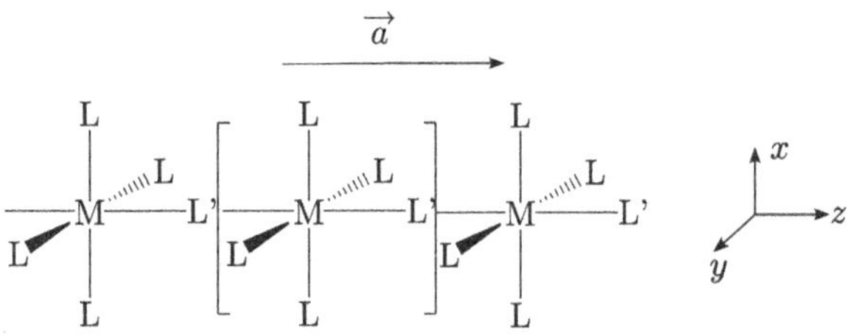

Fig. 8.10
Regular structure of the $(ML_4L')_n$ chain.

difference being the nature of the distortion, which is related to the doping degree and thus the filling of the upper band.

8.2 $(ML_4L')_n$ chains

Let us now consider the electronic structure of chains with formula $(ML_4L')_n$, where L refers to a ligand which is a σ (and possibly π) donor, and L' is a monoatomic anion with an np^6 configuration (L' = N^{3-}, Cl^-, I^-). In particular we will be interested in the $(ReCl_4N)_n$ and $(Pt(NH_3)_4Cl^{2+})_n$ chains whose geometries derive from the regular structure shown in Fig. 8.10. This structure is generated by a repeat unit of formula ML_4L' which will be decomposed into an ML_4 fragment and an L' ligand to facilitate the analysis. The two compounds that we will study in this section, $(ReCl_4N)_n$ and $(Pt(NH_3)_4Cl^{2+})_n$, differ in the number of electrons provided by the metal: the metal is formally d^0 in the first case but d^7 in the second.

We will start by working at the qualitative band structure of a regular chain with generic formula $(ML_4L')_n$, and then we will use it to interpret the structural and electronic properties of the $(ReCl_4N)_n$ and $(Pt(NH_3)_4Cl^{2+})_n$ chains.

8.2.1 Symmetry

The symmetry group leaving the chain with formula $(ML_4L')_n$ as well as a point O of the reference cell invariant is the D_{4h} group. We have chosen to place the point O on the metallic atom of the reference cell, although it would be completely equivalent to place it on the L' atom of this cell. The space group of this chain may be written as a direct product of the type $T_n \otimes D_{4h}$. Consequently, the appropriate group for the Γ and X points is D_{4h}, which leaves the point O invariant. The appropriate group for the other k points is C_{4v}.

8.2.2 Choice of the fragment orbitals to generate the Bloch orbitals

The Bloch orbitals of this system must be generated by using orbitals adapted to the C_{4v} symmetry, which holds for any k point. Consequently, we will generate the Bloch orbitals by means of the orbitals of the ML_4 fragment and the np orbitals of the L' ligand. Since we are interested in the crystal orbitals near the Fermi level, we will only consider the d-block metal orbitals of the ML_4 fragment when the ligand is a group that is a σ (and possibly π) donor.

The familiar orbitals for this fragment, where the ligand is a σ donor and a π acceptor, are shown in Fig. 8.2. Consequently, the energy diagram for the d-block orbitals of the ML$_4$ fragment used in our analysis is, in fact, identical to that shown in Fig. 8.2, except for the fact that the d_{xy}, d_{xz}, and d_{yz} orbitals are non-bonding if the L ligand is a purely σ donor, whereas the d_{xy} orbital is slightly higher in energy than the d_{xz} and d_{yz} orbitals if the L ligand is a π donor (see Exercise (8.6)).

8.2.3 Analysis of the Bloch orbitals at the Γ and X points

The symmetry labels for the Bloch orbitals generated from the ML$_4$ fragment orbitals (BO$_{d_{z^2}}$, BO$_{d_{x^2-y^2}}$, BO$_{d_{xy}}$, BO$_{d_{xz}}$, and BO$_{d_{yz}}$) in the D_{4h} group, which leave the metal atom of the reference cell invariant, are shown in Fig. 8.2. Let us note that the Bloch orbital energies in (ML$_4$L')$_n$ compounds differ from those represented in Fig. 8.3 because in the chains under consideration the metal atoms are not adjacent. *All Bloch orbitals are non-bonding in a purely Hückel approach: their energies are those of the same orbitals in the isolated fragment.*

The Bloch orbitals generated from the np_x, np_y, and np_z atomic orbitals of the L' ligand (BO$_{p_x}(\vec{k})$, BO$_{p_y}(\vec{k})$, and BO$_{p_z}(\vec{k})$) are shown in Fig. 8.11. In order to build the qualitative band structure of the system we need to find the symmetry labels of these orbitals in the D_{4h} group. To facilitate the task, the location of the symmetry plane σ_h has been indicated in Fig. 8.11.

8.2.4 Symmetry of the Bloch orbitals

Since the appropriate group for k points other than Γ and X is the C_{4h} group leaving every metal atom and every L' ligand invariant, the symmetry label of a Bloch orbital can be identified with the label of the fragment orbital from which it originates. A summary of the symmetry labels for the different Bloch orbitals is given in Table 8.1.

A consequence of the symmetry information in this table is that no mixing between Bloch orbitals is allowed for Γ. In contrast, for any other k point, the

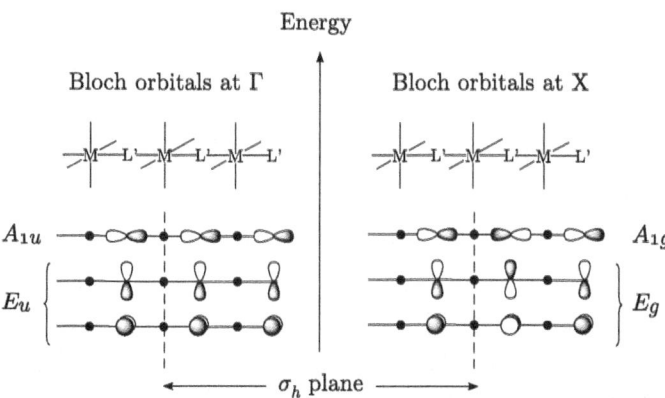

Fig. 8.11
Bloch orbitals generated by the np orbitals of the L' ligands. The symmetry labels according to the D_{4h} group are given.

$(ML_4L')_n$ chains

Table 8.1 Symmetry of the Bloch orbitals (BO) generated by the different ML_4 and L' fragment orbitals (FO) labelled in the D_{4h} or C_{4v} symmetries.

BOs generated from the given FOs	Γ point in D_{4h}	X point in D_{4h}	Other k points in C_{4v}
$d_{x^2-y^2}$ (metal)	B_{1g}	B_{1g}	B_1
d_{z^2} (metal)	A_{1g}	A_{1g}	A_1
d_{xy} (metal)	B_{2g}	B_{2g}	B_2
d_{xz}, d_{yz} (metal)	E_g	E_g	E
p_x, p_y (ligand L')	E_u	E_g	E
p_z (ligand L')	A_{1u}	A_{1g}	A_1

$BO_{d_{z^2}}(\vec{k})$ and $BO_{p_z}(\vec{k})$ Bloch orbitals as well as the pairs of orbitals ($BO_{d_{xz}}(\vec{k})$, $BO_{d_{yz}}(\vec{k})$) and ($BO_{p_x}(\vec{k})$, $BO_{p_y}(\vec{k})$) can interact.

8.2.5 Band structure

Energy of the Bloch orbitals

Let us now show schematically the energy of the different Bloch orbitals as a function of k_a (see Fig. 8.12).

Bands of a_1-type symmetry

Because of their symmetry (Table 8.1), the Bloch orbitals $BO_{d_{z^2}}(\vec{k})$ and $BO_{p_z}(\vec{k})$ can interact for every k point other than Γ. This interaction is more effective as we approach the X point, where the overlap between the Bloch orbitals is maximum. As a result there is, at X, a bonding crystal orbital mainly

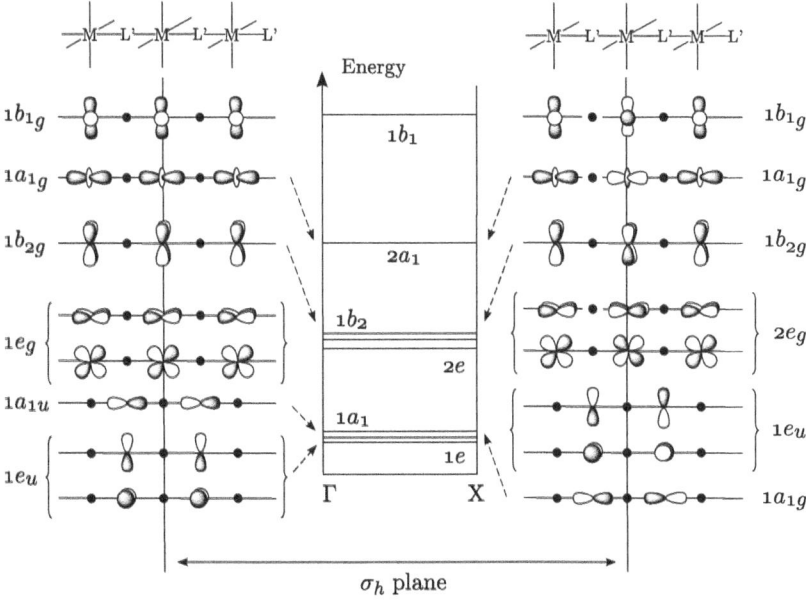

Fig. 8.12
Energies of the Bloch orbitals for the ML_4L' chain. The $1b_2$ and $2e$ bands are degenerate if the ligand L is a pure σ donor, whereas the $2e$ band is slightly lower in energy if ligand L is both a σ and π donor.

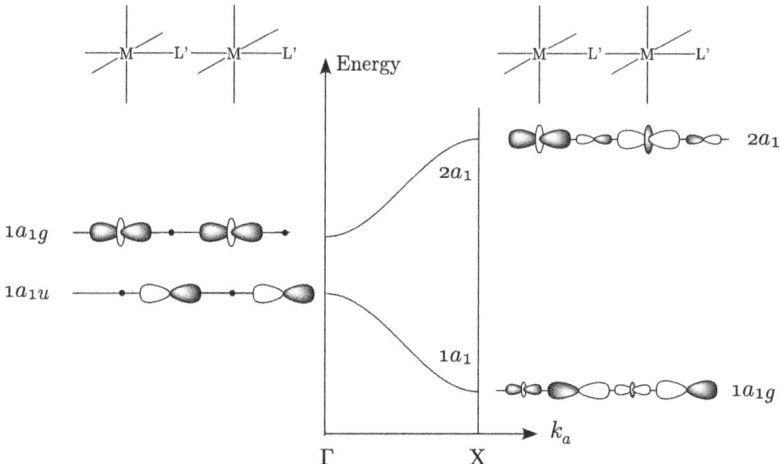

Fig. 8.13
Bands of a_1 symmetry of the ML$_4$L′ chain.

centred on the p_z orbitals of the L′ ligands, and an antibonding crystal orbital, essentially centred on the d_{z^2} orbitals of the metal atoms (Fig. 8.13).

Bands of *e*-type symmetry

The pairs of Bloch orbitals $(\text{BO}_{d_{xz}}(\vec{k}), \text{BO}_{d_{yz}}(\vec{k}))$ and $(\text{BO}_{p_x}(\vec{k}), \text{BO}_{p_y}(\vec{k}))$ are of the same symmetry (Table 8.1) and consequently they can interact for any k point other than Γ. The mixing is stronger as the X point is approached because at that point the overlaps are optimised. Thus at the X point two pairs of crystal orbitals result, one of which is mostly centred on the (p_x, p_y) orbitals of the L′ ligands and the other on the (d_{xz}, d_{yz}) orbitals of the metal atoms (Fig. 8.14).

Band structure of the system

Since the different Bloch orbitals with symmetries other than a_1 and e coincide with the crystal orbitals, the band structure of Fig. 8.15, which takes into account the results of Fig. 8.13 and 8.14, may be arrived at.

This diagram provides the general shape of the band structure of chains with formula ML$_4$L′. Of course the details of the computed band structures for real systems of this type may exhibit many differences in detail, but the

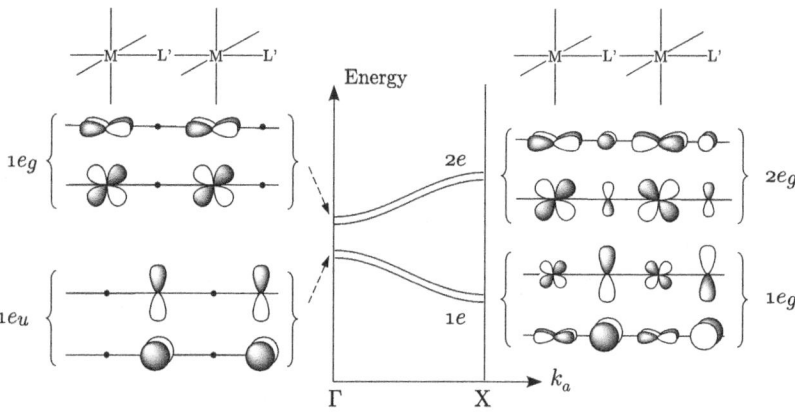

Fig. 8.14
Bands of *e* symmetry of the ML$_4$L′ chain.

general shape should be correct. It must be noted that many aspects of the band structure depend on the nature of the two ligands (L and L'), the metal (M), and the metal–ligand bond lengths. However, what can be ascertained is that any of the calculated band structures for this type of chain must possess:

- three bands ($1a_1$ and $1e$) essentially centred on the L' ligands going down in energy from Γ to X
- three bands essentially centred on the d_{xy} (band $1b_2$, essentially flat), d_{xz}, and d_{yz} orbitals (degenerate pair of bands rising in energy from Γ to X)
- a d_{z^2} band (antibonding $2a_1$ band going up from Γ to X)
- a flat $d_{x^2-y^2}$ ($1b_1$) band.

Since the σ-type overlaps are more effective than the π-type ones, the bands of a_1 symmetry are wider than the bands of e symmetry.

As simple as this diagram is, it will be very helpful in providing a rationale for the structural and conductivity properties of real compounds of this type. Band structures calculated using the extended Hückel approach for $(MoNCl_4^-)_n$ and $(PtL_4X^{2+})_n$ (X = Cl, Br, and L = NH_3) can be found in the references. [4, 5]

8.2.6 Study of the $(ReCl_4N)_n$ chain

The rhenium atom in this chain should be assigned a +VII oxidation state (i.e. it is a d^0 metal atom) so that only the six electrons formally provided by the N^{3-} ligand fill the bands represented in Fig. 8.15. Consequently, a convenient first step in describing the electronic structure of this chain would be to say

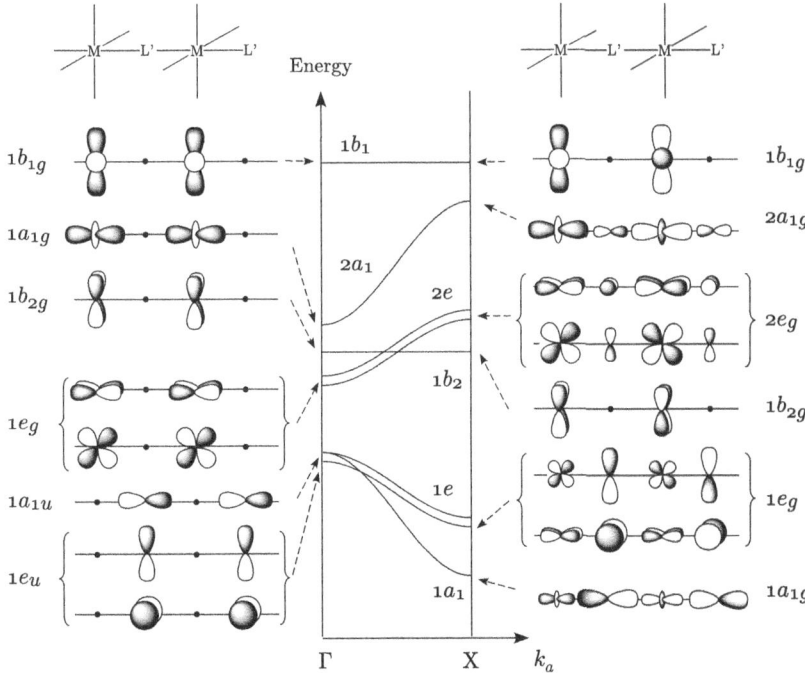

Fig. 8.15
Band structure for the ML$_4$L' chain. The relative position of the $2e$ and $1b_2$ bands depends on the nature of the ligands. If the L ligand is a purely σ donor, these bands are degenerate at Γ. However, if the ligand is a σ and π donor then the $1b_2$ band is slightly higher in energy than band $2e$.

that it is a $(1e)^4(1a_1)^2$ example of the band structure shown in Fig. 8.15. The Fermi level is located on top of the three bands mainly centred on the N^{3-} ligands. This compound should thus be insulating. Since the Cl^- ligand is a σ and π donor, band $1b_2$ at Γ lies at a slightly higher energy level than band $2e$. Consequently, the band gap between the valence bands ($1a_1$ and $1e$) and the conduction bands ($2e$) is essentially given by the difference in energy between a $2p$ orbital of the nitrogen atom and the (d_{xz}, d_{yz}) orbitals of the ReCl$_4$ fragment.

The first question we may ask is if there is some distortion that can stabilise this system. The $1a_1$ and $1e$ bands may be stabilised by a distortion, making possible the interaction between Bloch orbitals at Γ. In the regular chain such interaction is not allowed because of the existence of the σ_h symmetry plane in the D_{4h} group. A distortion leading to the disappearance of this symmetry plane will lower the symmetry at Γ from D_{4h} to C_{4v} and, consequently, will make possible the interaction between the metal-based and ligand-based a_1 and e bands. As a result, bands $1a_1$ and $1e$ near Γ will be stabilised and bands $2a_1$ and $2e$ near Γ will be destabilised. An example of this kind of distortion is shown in Fig. 8.16, where short and long Re–N bonds alternate along the chain axis. Thus, the symmetry group of the new chain no longer contains the σ_h plane and the group leaving both the distorted chain and a point O of the reference cell invariant is C_{4v}. The space group of this chain may be written as a direct product $T_n \otimes C_{4v}$. The appropriate group for all k points is C_{4v} so that the $BO_{d_{z^2}}(\Gamma)$ and $BO_{p_z}(\Gamma)$ Bloch orbitals can always interact, leading to a filled bonding crystal orbital essentially based on the L' ligand orbitals, and an empty antibonding crystal orbital mostly centered on the metal atoms. For the same reason, the pairs of orbitals ($BO_{d_{xz}}(\Gamma)$, $BO_{d_{yz}}(\Gamma)$) and ($BO_{p_x}(\Gamma)$, $BO_{p_y}(\Gamma)$) interact to lead to one pair of filled bonding orbitals and one pair of empty antibonding orbitals. The filled bonding crystal orbitals resulting from such an interaction at Γ are shown in Fig. 8.17. Such a distortion allows the interaction of all crystal orbitals of the $1a_1$ and $1e$ bands near the Γ point with the corresponding crystal orbitals of the $2a_1$ and $2e$ bands. The distortion shown in Fig. 8.16 thus increases the band gap between the valence and conduction bands. The analysis of the crystal orbitals in Fig. 8.17 clearly shows that such a distortion is accompanied by a localisation of the electrons on the short Re–N bonds. Experimentally, the compound is observed to exhibit a distortion of this kind, [6] although more pronounced, since the ML$_4$ fragments are no longer locally planar but pyramidal (see Fig. 8.18). The distortion from local square-planar to pyramidal ReCl$_4$ allows the hybridization of the $6p$ and

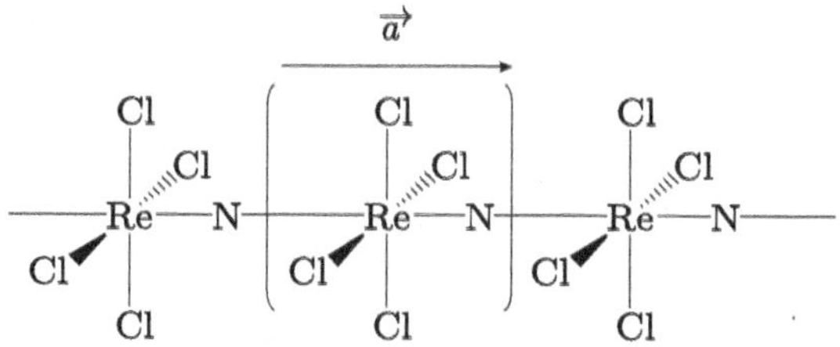

Fig. 8.16
Distortion of the (ReCl$_4$N)$_n$ chain leading to the disappearance of the symmetry plane perpendicular to the chain axis.

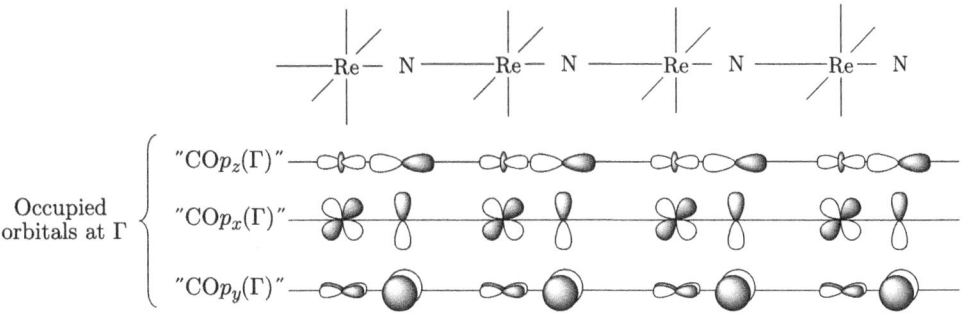

Fig. 8.17
Occupied orbitals at Γ after development of the distortion shown in Fig. 8.16.

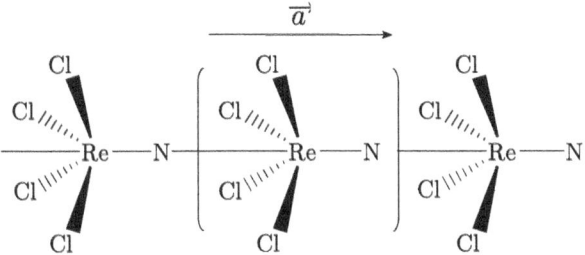

Fig. 8.18
The observed pyramidalised structure.

$5d$ metal orbitals leading to the new fragment orbitals shown in Fig. 8.19. [8] The pyramidalisation modifies somewhat the position of the empty metal-based bands of this system and strengthens the bonding character of the orbitals at the top of the $1a_1$ and $1e$ bands. Thus, for instance, after the distortion the filled bonding orbitals at Γ, essentially centred on the ligands, have the shape shown in Fig. 8.20. Consequently, the pyramidalisation provides an additional stabilisation to the system. All the filled levels near the Fermi level are stabilised when such a distortion occurs. This chain is an excellent example of a real compound stabilised by a second-order Peierls distortion that allows the interaction between crystal orbitals of empty and filled bands, i.e. non-degenerate orbitals, increasing the band gap at the Fermi level.

The consequences of the distortion may be summarised as shown in Fig. 8.21.

8.2.7 Electronic structure of the $(Pt(NH_2Et)_4Cl^{2+})_n$ chain

In this chain the platinum metal atoms have a formal +III oxidation state. Since every $[Pt(NH_2Et)_4]^{3+}$ fragment provides seven electrons and every chloride anion possess six $3p$ electrons, every unit cell contributes 13 valence electrons. Consequently, a convenient first step in describing the electronic structure of this chain consists in considering it as a $(1a_1)^2(1e)^4(1b_2)^2(2e)^4(2a_1)^1$ example of the band structure in Fig. 8.15. The Fermi level lies in the middle of the d_{z^2} band. Thus, in principle, a dimerisation of the system should occur.

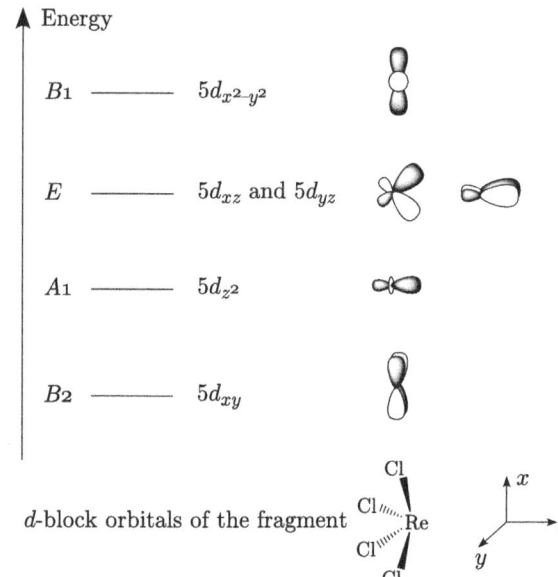

Fig. 8.19
Relevant orbitals of the pyramidalised $ReCl_4$ fragment. Only the metal contributions are shown.

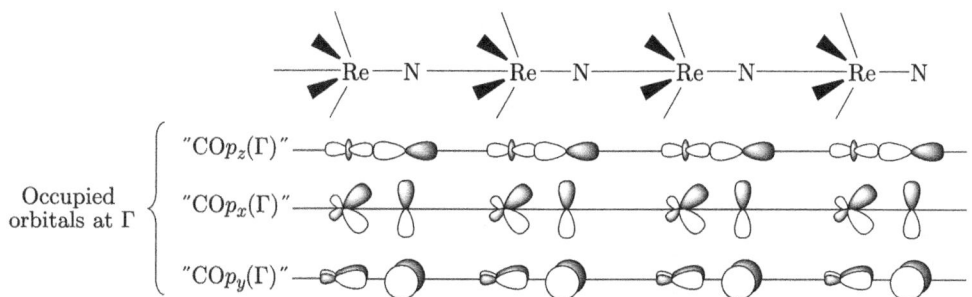

Fig. 8.20
Occupied orbitals at Γ for the observed structure.

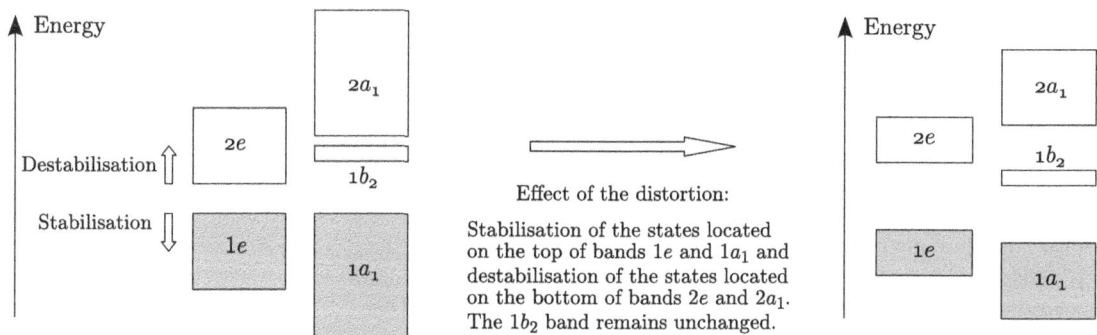

Effect of the distortion:

Stabilisation of the states located on the top of bands $1e$ and $1a_1$ and destabilisation of the states located on the bottom of bands $2e$ and $2a_1$. The $1b_2$ band remains unchanged.

Fig. 8.21
Evolution of the band structure for the $(ReCl_4N)_n$ chain as a result of the distortion shown in Fig. 8.18.

$(ML_4L')_n$ chains

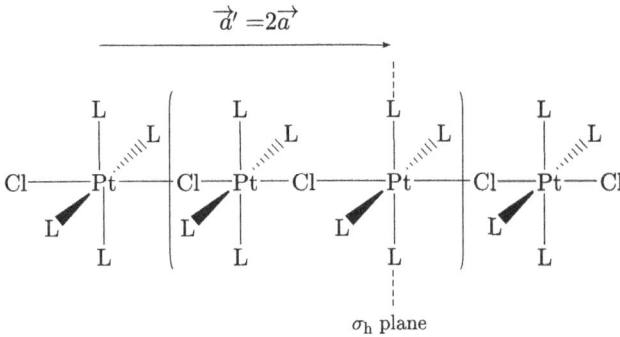

Fig. 8.22
Distorted structure observed for the $(PtL_4Cl^{2+})_n$ (L = NH$_2$Et) chain.

Indeed, this chain does exhibit a dimerised structure (Fig. 8.22). In such a distorted structure, one of every two platinum atoms is coordinated to six ligands (four NH$_2$ET and two Cl$^-$) while the other platinum atoms interact significantly with only four EtNH$_2$ ligands. To understand the nature of the observed dimerisation, [7] let us look in some detail at the d_{z^2} and p_z bands of a chain generated by a double unit cell, i.e. with formula $(Pt(NH_2Et)_4Cl)_2^{4+}$. As shown in Fig. 8.23 the Fermi level should lie at the middle of the folded upper band, which is mostly d_{z^2} in character. How can we obtain a basis for the crystal orbitals at the Fermi level that is well adapted to the symmetry of the distorted compound, i.e. with respect to the σ_h plane? It may be interesting to take advantage of the equivalence of the diagrams of Figs 8.23 and 8.13, obtained by generating the same chain with a double cell (Fig. 8.23 and representation 2 in Section 3.4.1) or a single cell (Fig. 8.13 and representation 1 in Section 3.4.1). As we have already established in Chapter 3, the orbitals associated with point $k_{a'} = 1/2$ in the second representation (double cell) correspond to the orbitals associated with the points $k_a = \pm 1/4$ in the first representation (single cell). The crystal orbitals at the Fermi level of Fig. 8.23, which are thus associated with the point with $k_{a'}$ equal to 1/2, are linear combinations of the Bloch orbitals $(BO_{d_{z^2}}(k_a = 1/4), BO_{d_{z^2}}(k_a = -1/4))$ and $(BO_{p_z}(k_a = 1/4), BO_{p_z}(k_a = -1/4))$ obtained by means of the first representation.

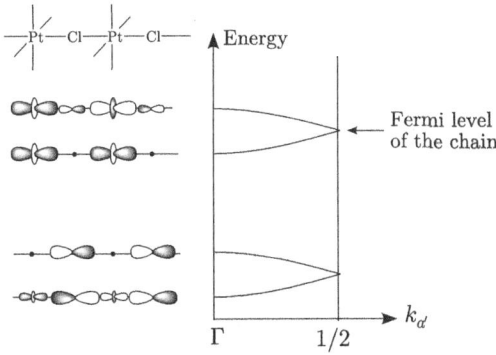

Fig. 8.23
d_{z^2} and p_z bands of the $(PtL_4Cl^{2+})_n$ (L = NH$_2$Et) chain generated by a double unit cell. The horizontal axis refers to $k_{a'}$, i.e. the projection of the k vector on the reciprocal vector $\vec{a}'^*$ ($\vec{a}'^* = \vec{a}^*/2$).

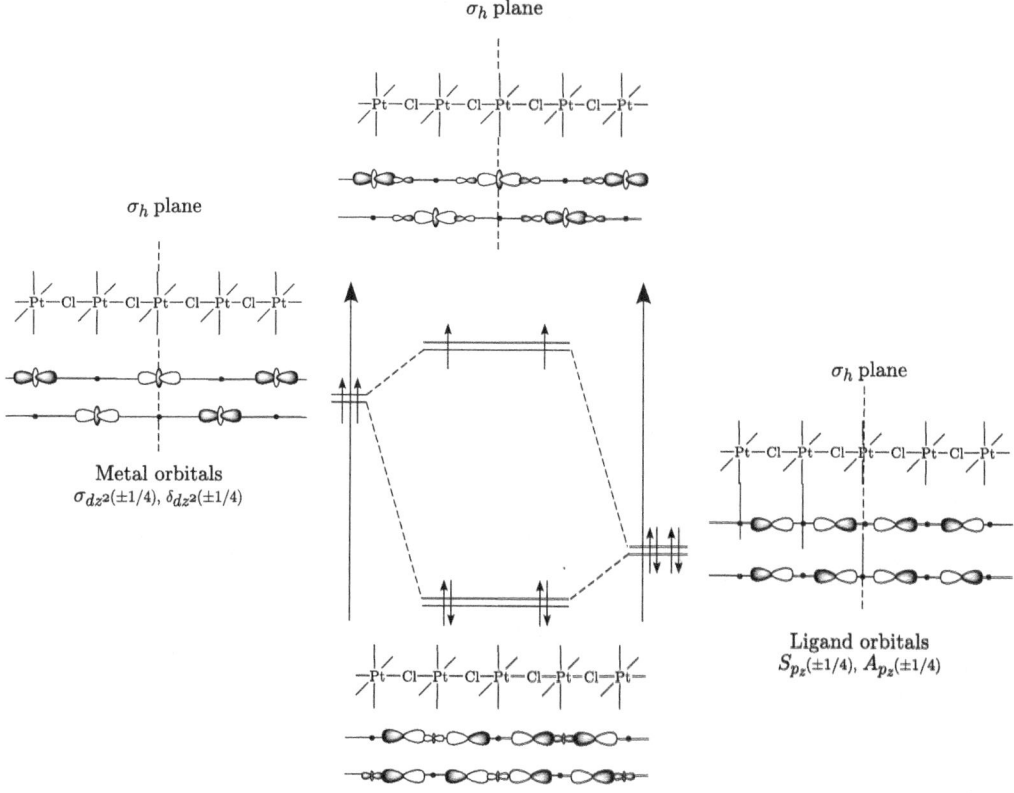

Fig. 8.24
Bloch orbital interaction diagram showing the nature of the crystal orbitals near the Fermi level.

In addition, it is completely equivalent to consider the interaction between orbitals $\{\sigma_{d_{z^2}}(k_a = \pm 1/4), \delta_{d_{z^2}}(k_a = \pm 1/4)\}$ (Fig. 3.10) and $\{S_{p_z}(k_a = \pm 1/4), A_{p_z}(k_a = \pm 1/4)\}$ (Fig. 3.12), which are pairs of orbitals equivalent to the Bloch orbitals and adapted to the symmetry plane σ_h, i.e. adapted to the distortion that we want to consider. Figure 8.24 is a representation of the interaction between the pairs of orbitals $\{\sigma_{d_{z^2}}(k_a = \pm 1/4), \delta_{d_{z^2}}(k_a = \pm 1/4)\}$ and $\{S_{p_z}(k_a = \pm 1/4), A_{p_z}(k_a = \pm 1/4)\}$ leading to a filled bonding orbital pair, mostly centred on the ligands, and the antibonding orbitals at the Fermi level mostly centred on the metals.

Taking into account the shape of the crystal orbitals at the Fermi level (Fig. 8.24), it is easy to see that the distortion represented in Fig. 8.22 causes the disappearance of the degeneracy and this results in the stabilisation of the system (see Fig. 8.25).

After such a distortion the system is no longer metallic. If the localisation of the electrons in the upper filled band is examined and particularly the crystal orbitals at the Fermi level, it is clear that the distortion stabilising the system tends to localise the d_{z^2} electrons on those platinum atoms coordinated with four ligands $(Pt(NH_2Et)_4)^{2+}$, while the d_{z^2} orbitals of the six-coordinated platinum atoms, $(Pt(NH_2Et)_4Cl_2)^{2+}$, are empty. We thus formally obtain an

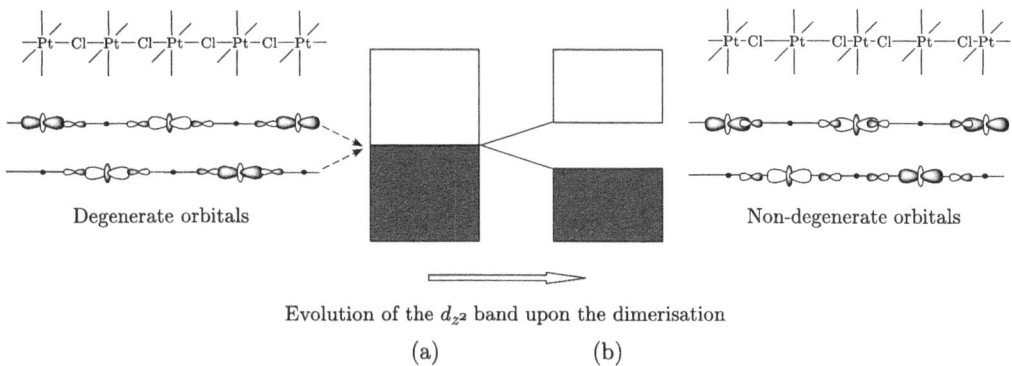

Fig. 8.25
Illustration of how the degeneracy at the Fermi level disappears as a result of the distortion shown in Fig. 8.22.

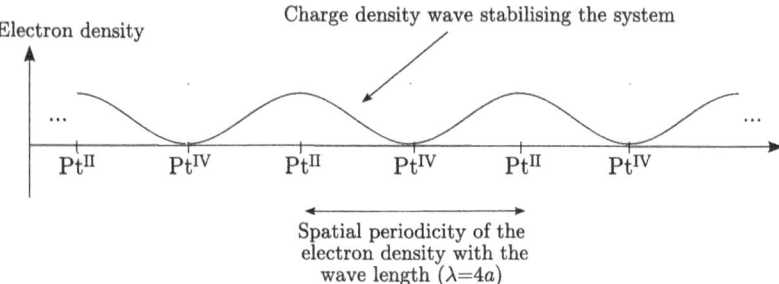

Fig. 8.26
Schematic representation of the charge density wave stabilising the system, highlighting the mixed valence nature of the chain after the distortion.

alternation of platinum atoms with oxidation degrees of +II and +IV: the distorted compound is a mixed-valence compound. The distortion leading to the change in the conductivity regime is thus linked with the localisation of the d_{z^2} electrons on alternate platinum atoms.

This system clearly illustrates the notion of first-order Peierls distortion modifying the translational symmetry of the system and changing the conductivity regime. The structural deformation leads to the appearance of a charge density for which the maxima and minima are located on the platinum atoms as illustrated in Fig. 8.26. It is worth noting that the distortion has already occurred at room temperature so that the chains with formula $(Pt(NH_2Et)_4X^{2+})_n$ (X = Br, Cl) are semiconductors. This suggests that the energy gain for these chains as a result of the Peierls distortion is larger than that associated with the 3.333-merisation of KCP. This observation seems reasonable because a dimerisation is usually more favorable than any other n-merisation ($n > 2$).

8.3 Suggested studies

We will now suggest the study of other examples in the exercises. The questions are written in a manner that is detailed enough to make the analysis easy.

Exercises

(8.1) Using the Hückel approach show that the interaction term $< BO_{6p_z}(\vec{k}) \mid \hat{h} \mid BO_{5d_{z^2}}(\vec{k}) >$ between the two Bloch orbitals generated from the $6p_z$ and $5d_{z^2}$ fragment orbitals in Fig. 8.4 increases as the k point gets farther from the Γ and X points.

(8.2) An X-ray analysis reveals the existence of three different Pt–Pt bond lengths in the compound $Rb_{1.67}[Pt(C_2O_4)] \cdot \frac{3}{2}H_2O$, with values 2.72, 2.83 and 3.02 Å. Provide an explanation for this experimental observation.

(8.3) Consider the hypothetical $K_2Pt(CN)_4Li_{0.2}$ compound. Does this compound conduct electricity? Is the material susceptible to distortion? How do the Pt–Pt distances differ from those in the non-doped compound?

(8.4) Consider the staggered KCP structure (Fig. 8.1a).

(a) Verify that the complete space group of KCP may be written as the direct product $T_n \otimes D_{4h} \otimes \{E, g_{C_{8z}}\}$, where $g_{C_{8z}}$ refers to the screw axis along the chain direction.

(b) Determine the symmetry of the Bloch orbitals $BO_{d_{z^2}}(\Gamma)$ and $BO_{p_z}(\Gamma)$ with respect to the D_{4h} group and the $g_{C_{8z}}$ symmetry operation. Can these two orbitals interact? Can they be accidentally degenerate?

(c) Determine the symmetry of the Bloch orbitals $BO_{d_{z^2}}(X)$ and $BO_{p_z}(X)$ with respect to the D_{4h} group and the $g_{C_{8z}}$ symmetry operation. Can these two orbitals interact?

(d) Taking into account the analysis in Section 3.4.1, establish the expression for the Bloch orbitals $BO_{d_{z^2}}(\vec{k})$ and $BO_{p_z}(\vec{k})$ for values of k_a other than 0 and 1/2. Deduce the symmetry characteristics of these orbitals with respect to the group C_{4v} and the $g_{C_{8z}}$ symmetry operation. Can these two orbitals interact?

(e) Determine the shape of the d_{z^2} and p_z bands. Can these two bands possibly cross?

(8.5) Determine the oxidation state of Re and Pt in the $(ReCl_4N)_n$ and $(Pt(NH_3)_4Cl^{2+})_n$ compounds. How many valence electrons are associated with the repeat unit of these materials?

(8.6) Consider the ML_4L' chain

(a) Place the d-block orbitals of the ML_4 fragment when the ligand L is a purely σ donor (for example NH_3) or σ and π donor (for instance Cl^-).

(b) Suggest an alternative set of fragment orbitals that could have been used to generate the Bloch orbitals of this chain.

(c) Why can the role played by the $6p_z$ orbital of the metal be neglected in a first approximation?

(d) Why can the role played by the ns orbital of the L' ligand (L' = Cl^- and N^{3-}) be neglected in a first approximation?

(e) Up to now the origin of the chain has been located at the M atom of the reference cell. Is any other location possible for the origin? If the answer is positive, show that this alternative choice would have led to different symmetry labels for the Bloch orbitals but would have not changed the type of mixing allowed, i.e. it would have led to the same crystal orbitals.

(f) Describe the main sources of uncertainty about the location of the different energy bands.

(8.7) Consider the $ReCl_4N$ chain. Provide a justification for the orbitals shown in Fig. 8.19.

(8.8) Consider the dimerised $Pt(NH_2Et)_4Cl^{2+}$ chain.

(a) What molecular argument can show that the Pt^{II} atoms interact with four ligands (i.e. $Pt(NH_2Et)_4$) while the Pt^{IV} atoms interact with six ligands (i.e. $Pt(NH_2Et)_4Cl_2$)?

(b) Could some other dimerisation be proposed? How could it be shown that the observed dimerisation is the most likely?

(8.9) Consider a hypothetical $(C_4H_2)_n$ system.

(a) Propose different Lewis structures for this compound.
(b) Which atomic orbitals must be taken into account to provide a rationale of the electronic properties of this system?
(c) What is the unit cell of the system?
(d) Work out the symmetry of the system: what is the appropriate space group for the different k points?
(e) Which fragment orbitals are most convenient to use to generate the Bloch orbitals of the chain?
(f) Produce a qualitative band structure for this chain indicating the main uncertainties concerning the relative positioning of the bands.
(g) What kind of conductor should this system be?

(8.10) Consider the hypothetical $(B_2N_2H_2)_n$ system.

Now consider also the following series of compounds: $(B_2N_2H_2)_n$, $(B_2N_2H_2Li)_n$, and $(B_2N_2H_2Br)_n$. The Li and Br atoms are intercalated between the $(B_2N_2H_2)_n$ chains and it is assumed that their only role is as electron donors and acceptors, respectively. Build the band structure for these systems. What kind of conductivity should be expected for these compounds? *Note: Since in a purely qualitative approach it is not often possible to be sure of the relative positioning of all the bands, different qualitative band structures can be proposed if necessary.*

(8.11) Consider the hypothetical regular chain $(VO_5^{6-})_n$, shown in the following figure:

(a) What is the unit cell of the system?
(b) Work out the symmetry of the system: what is the appropriate space group for different k points?
(c) Which fragment orbitals are most convenient to use to generate the Bloch orbitals of the chain?
(d) Produce a qualitative band structure for this chain, indicating the main uncertainties concerning the relative positioning of the bands.
(e) What kind of conductor should the $(VO_5^{6-})_n$ chain be? Could this system undergo some kind of distortion?

(8.12) Consider the $(NbX_4)_n$ (X = Cl, Br, I) chains

Let us try to understand why $(NbX_4)_n$ chains are semiconducting and exhibit a dimerised structure. The non-distorted, ideal structure of the chain is schematically shown below, with every niobium atom having an ideal octahedral environment.

Regular $(NbX_4)_n$ (X=Cl,Br,I)

In fact, $(NbX_4)_n$ (X = Cl, Br, I) chains are distorted: the solid exhibits a dimerisation, as a consequence of which there are two different alternating Nb–Nb distances as shown below:

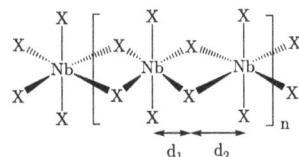

Distorted $(NbX_4)_n$ (X=Cl,Br,I)

(a) Regarding the regular chain:
  (i) Knowing that the X ligand is a halogen, what is the oxidation state of the niobium atoms in this chain?
  (ii) The niobium environment in the chain is octahedral. How can it be shown that an appropriate treatment of the electronic structure of this chain takes into account only the d_{xz}, d_{yz}, and $d_{x^2-y^2}$ orbitals of the metal?

(b) Consider the building up of the band structure of the ideal system. In the following we will only consider the interaction between the p atomic orbitals of the bridging halogenide ligands and the d_{xz}, d_{yz}, and $d_{x^2-y^2}$ orbitals of the metal atoms.
  (i) Justify the choice of orbitals used to build the band structure.
  (ii) Work out the symmetry of the system: what is the appropriate space group for the different k points?
  (iii) Construct the crystal orbitals at the Γ and X points and build a qualitative band structure for the chain.

(c) Interpretation of the band structure. The real d-block band structure of the ideal $NbCl_4$ chain obtained using the extended Hückel approach [5] is shown in the following figure.

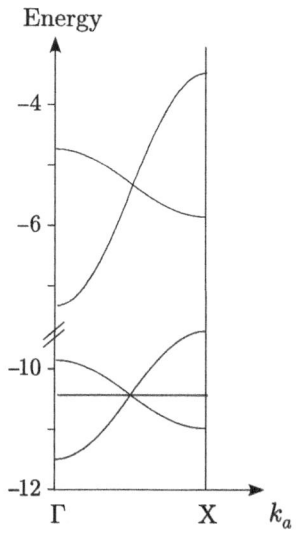

(i) How can it be shown that all bands in that figure cross for $k_a = 1/4$?
(ii) How many electrons should these bands contain? Where would the Fermi level occur? Would the NbCl$_4$ chain be insulating or metallic with this geometry?
(iii) How would the band structure change if a dimerisation occurs?
(iv) Which band is most strongly affected by the distortion? Which band is least affected by the distortion?
(v) How can it be shown that NbI$_4$ is semiconducting at room temperature but becomes metallic at high pressure?

References

1. J. R. Ferraro and J. M. Williams, *Introduction to Synthetic Electrical Conductors*, Academic Press, Orlando, 1987, Chapter 4.
2. R. Comès, M. Lambert, and H. R. Zeller, *Phys. Rev. B*, 88, 2036, 1973.
3. M.-H. Whangbo, R. Hoffmann, *J. Am. Chem. Soc.*, 100, 6093, 1978.
4. R. A. Wheeler, M.-H. Whangbo, T. R. Hughbanks, R. Hoffmann, J. K. Burdett, and T. A. Albright, *J. Am. Chem. Soc.* 108, 2222, 1986.
5. M.-H. Whangbo and M. J. Foshee, *Inorg. Chem.*, 20, 113, 1981.
6. W. Liese, K. Dehnicke, I. Walker, and J. Strähle, *Z. Naturforsch. B*, 34, 693, 1979.
7. H. J. Keller, *Extended Linear Chain Compounds*, J. S. Miller, editor, Plenum Press, New York, 1982, Vol. 1, Chapter 8.
8. T. A. Albright, J. K. Burdett, and M.-H. Whangbo, *Orbital Interactions in Chemistry*, John Wiley, New York, 1985.

Electronic structure of 2D and 3D systems

9

In Chapters 3–8 we considered in detail how to build the electronic band structure of 1D systems from the viewpoint of their orbital interactions. However, most systems of interest for a chemist or physicist are 2D or 3D. The reason why we have devoted so much attention to 1D systems is that they provide the simplest way to introduce the basic methodology for studying periodic materials. Extension to more than one dimension is a quite obvious exercise, which we will briefly discuss in this chapter. It is true that working with a 3D periodic system leads automatically to consideration of a 3D reciprocal space (and thus, a 3D Brillouin zone), which makes the problem a bit more complex. Sometimes it will be necessary to use tools that allow us to reduce the large amount of information we need to handle. This will lead to the introduction of the density of states concept, which will be discussed in the next chapter. Exploiting the symmetry of the system is another technique that can considerably simplify the task of analysing 3D systems and which we must consider in slightly more depth. In this chapter we briefly consider the extension of the main notions developed in Chapter 3 to 3D systems from a very practical point of view. [1, 2, 3] We hope that the reader will soon realise that little new knowledge is necessary to enter into the realm of the analysis of 3D solids.

9.1 Basic concepts

9.1.1 Direct and reciprocal lattices

An infinite periodic 3D system is built by repeating a basic unit, the reference unit cell (M_0), along three different directions of space, a, b, and c. Thus, to characterise a given image of this reference cell we need to specify how many times the cell has been translated along the different directions. In other words, any lattice point of the *direct lattice* of this general 3D system, i.e. the lattice defined by all these replicas, may be represented by a vector $\vec{r}$

$$\vec{r} = m\,\vec{a} + p\,\vec{b} + q\,\vec{c} \quad (9.1)$$

where $\vec{a}$, $\vec{b}$, and $\vec{c}$ are the translational vectors along the three different directions, i.e. the *direct lattice vectors*, and m, p, and q are the number of translations along the given direction to generate the lattice point. The *primitive*

2D and 3D systems

cell is defined as the smallest part of the lattice that would generate the whole structure when repeated. Since the position of any lattice point is characterised by a vector with three components, the $\vec{k}$ vector that we must define to build the Bloch orbitals of the system will also have three different components. In general we may write the $\vec{k}$ vector as:

$$\vec{k} = k_a\,\vec{a}^* + k_b\,\vec{b}^* + k_c\,\vec{c}^* \tag{9.2}$$

where $\vec{a}^*$, $\vec{b}^*$, and $\vec{c}^*$ are the three *reciprocal lattice vectors* and k_a, k_b, and k_c are the three components of the vector associated with the given *reciprocal lattice* point. In general, for a given *direct lattice*, its *reciprocal lattice* is defined as a set of vectors satisfying the relationship:

$$\exp(\vec{k}\cdot\vec{r}) = 1 \tag{9.3}$$

which is fulfilled when the product $\vec{r}\cdot\vec{k}$ is an integral multiple of 2π. Then, the three reciprocal lattice vectors are defined through the relationships:

$$\vec{a}^* = 2\pi\,\frac{\vec{b}\times\vec{c}}{\vec{a}\cdot(\vec{b}\times\vec{c})} = \frac{2\pi}{V}\,\vec{b}\times\vec{c}$$

$$\vec{b}^* = 2\pi\,\frac{\vec{c}\times\vec{a}}{\vec{a}\cdot(\vec{b}\times\vec{c})} = \frac{2\pi}{V}\,\vec{c}\times\vec{a}$$

$$\vec{c}^* = 2\pi\,\frac{\vec{a}\times\vec{b}}{\vec{a}\cdot(\vec{b}\times\vec{c})} = \frac{2\pi}{V}\,\vec{a}\times\vec{b} \tag{9.4}$$

so that they fulfill:

$$\vec{a}\cdot\vec{a}^* = \vec{b}\cdot\vec{b}^* = \vec{c}\cdot\vec{c}^* = 2\pi \tag{9.5}$$

$$\vec{a}^*\cdot\vec{b} = \vec{b}^*\cdot\vec{c} = \vec{c}^*\cdot\vec{a} =$$

$$\vec{a}\cdot\vec{b}^* = \vec{b}\cdot\vec{c}^* = \vec{c}\cdot\vec{a}^* = 0 \tag{9.6}$$

As a simple illustration of this process let us construct the reciprocal lattice of the direct orthorhombic lattice of Fig. 9.1a defined by three lattice vectors $\vec{a}$, $\vec{b}$, and $\vec{c}$:

$$\vec{a} = a\,\vec{u}$$
$$\vec{b} = b\,\vec{v}$$
$$\vec{c} = c\,\vec{w} \tag{9.7}$$

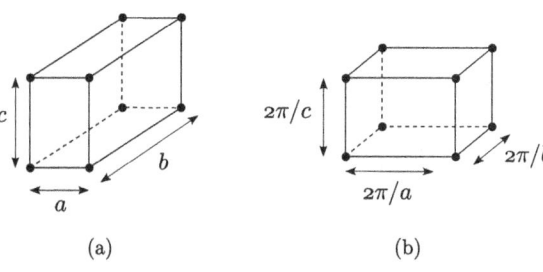

Fig. 9.1
Schematic illustration of the direct (a) and reciprocal (b) lattices for an orthorhombic system.

where $\vec{u}$, $\vec{v}$, and $\vec{w}$ are the unit vectors along the a, b, and c directions. Then the three reciprocal lattice vectors $\vec{a}^*$, $\vec{b}^*$, and $\vec{c}^*$ are simply:

$$\vec{a}^* = \frac{2\pi}{a} \vec{u}$$

$$\vec{b}^* = \frac{2\pi}{b} \vec{v}$$

$$\vec{c}^* = \frac{2\pi}{c} \vec{w} \qquad (9.8)$$

and the reciprocal lattice may be represented as in Fig. 9.1b. Consequently, the reciprocal lattice of an orthorhombic lattice is also an orthorhombic lattice, although the relative values of the lattice vectors are different.

The latter definition is completely general and can also be applied when the direct lattice vectors are not an orthogonal set. If we have a direct lattice generated by $\vec{a}$, $\vec{b}$, and $\vec{c}$ vectors that are not orthogonal to each other, then the reciprocal lattice vectors are those given in eqn (9.4). The lengths and directions of the vectors $\vec{b} \times \vec{c}$, $\vec{c} \times \vec{a}$, and $\vec{a} \times \vec{b}$ are easy to obtain. For instance, the vector $\vec{a} \times \vec{b}$ is a vector with length equal to the area of the parallelogram defined by the vectors $\vec{a}$ and $\vec{b}$ and direction perpendicular to both $\vec{a}$ and $\vec{b}$ (see Fig. 9.2). The positive direction of the vector may be obtained through the right-hand screw convention. For a 2D case we can use exactly the same scheme once a unit vector $\vec{c} = \vec{w}$ is added along the direction of $\vec{a} \times \vec{b}$.

9.1.2 Bloch and crystal orbitals

In general, the translation group can be written as a simple product group of translations along the principal directions of the direct lattice. In other words, when building the Bloch orbital associated with a given orbital of the reference cell for a 3D lattice, the exponential phase factor can be written as the product of three exponential phase factors:

$$\exp(i2\pi m k_a) \cdot \exp(i2\pi p k_b) \cdot \exp(i2\pi q k_c) \qquad (9.9)$$

so that, using exactly the same reasoning as in Section 3.1, the Bloch orbital can be expressed as:

$$\mathrm{BO}_j(\vec{k}) = \frac{1}{\sqrt{N}} \sum_{m=-n'+1}^{n'} \sum_{p=-n'+1}^{n'} \sum_{q=-n'+1}^{n'} \exp(i2\pi m k_a)$$

$$\cdot \exp(i2\pi p k_b) \cdot \exp(i2\pi q k_c) (\phi_j)_{m,p,q} \qquad (9.10)$$

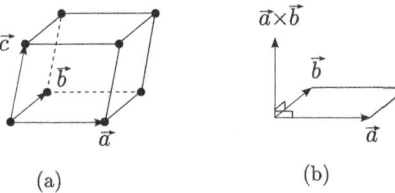

Fig. 9.2
Graphical illustration of the construction of the reciprocal vectors for a general non-orthogonal set of direct lattice vectors.

where n is the number of cells in each direction, $n' = \frac{n}{2}$, N is the total number of cells, and $(\phi_j)_{m,p,q}$ is the ϕ_j orbital in the cell defined by the m, p, and q indexes. A more compact expression is:

$$BO_j(\vec{k}) = \frac{1}{\sqrt{N}} \sum_{m=-n'+1}^{n'} \sum_{p=-n'+1}^{n'} \sum_{q=-n'+1}^{n'} \exp(i\vec{k} \cdot \vec{r})(\phi_j)_{m,p,q} \quad (9.11)$$

The different k_a, k_b, and k_c values allow the description of all possible phase relationships between the ϕ_j orbitals along the $\vec{a}$, $\vec{b}$, and $\vec{c}$ directions. Because of its periodicity, the allowed values of k_a, k_b, and k_c are those in the]-1/2, 1/2] interval, which constitutes the Brillouin zone of the system. We may schematically represent the Bloch orbital $BO_j(\vec{k})$ as in Fig. 9.3, where every dot represents a cell of the 3D system and where for each cell we give the (m, p, q) indices characterising the cell and the corresponding phase factor.

The crystal orbitals of a system with N_0 atomic orbitals in the reference cell result from the combination of Bloch orbitals $BO_j(\vec{k})$ ($j = 1, \ldots, N_0$) of the same translational symmetry, i.e. associated with the same k point. Every crystal orbital $CO(\vec{k})$ is labelled with a $\vec{k}$ vector (with three components) and is a linear combination of Bloch orbitals $BO_{j'}(\vec{k})$:

$$CO(\vec{k}) = \sum_{j'=1}^{N_0} c_{j'}(\vec{k}) BO_{j'}(\vec{k}) \quad (9.12)$$

As was shown in Section 3.1.2, the $c_{j'}(\vec{k})$ ($j' = 1, \ldots, N_0$) coefficients for the different crystal orbitals $CO_\ell(\vec{k})$ ($\ell = 1, \ldots, N_0$), as well as the associated energies $E_\ell(\vec{k})$ ($\ell = 1, \ldots, N_0$), can be obtained by resolution of the secular equations:

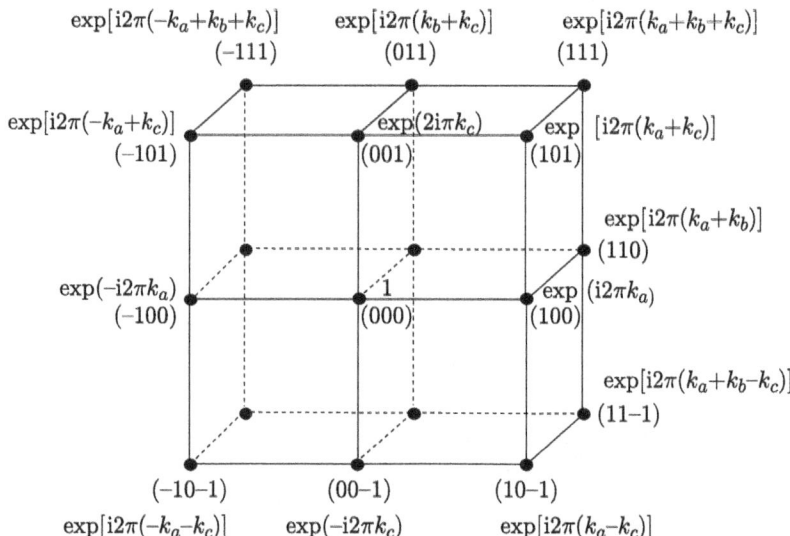

Fig. 9.3
Schematic representation of the Bloch orbital $BO_j(\vec{k})$ for a 3D system.

$$\sum_{j'=1}^{N_0} c_{j'} \left(\hat{h}[BO_{j'}(\vec{k})] - E(\vec{k})BO_{j'}(\vec{k}) \right) = 0 \qquad (9.13)$$

where $h_{jj'}(\vec{k})$ and $S_{jj'}(\vec{k})$ are given by:

$$h_{jj'}(\vec{k}) = \langle BO_j(\vec{k}) \mid \hat{h} \mid BO_{j'}(\vec{k}) \rangle$$
$$S_{jj'}(\vec{k}) = \langle BO_j(\vec{k}) \mid BO_{j'}(\vec{k}) \rangle \qquad (9.14)$$

Thus we need to solve the N_0-th order equation in $E(\vec{k})$ associated with the secular determinant:

$$\mid \underline{\underline{h}}(\vec{k}) - E(\vec{k})\underline{\underline{S}}(\vec{k}) \mid = 0 \qquad (9.15)$$

where $\underline{\underline{h}}(\vec{k})$ and $\underline{\underline{S}}(\vec{k})$ are the matrix representations of the Hamiltonian and overlap operators in the Bloch orbitals $\{BO_j(\vec{k}) \, (j = 1, \ldots, N_0)\}$ basis.

This enables us to determine the N_0 allowed energies $E_\ell(\vec{k}) (\ell = 1, \ldots, N_0)$ and sets of coefficients $c_{j'}(\vec{k}) \, (j' = 1, \ldots, N_0)$ of the associated crystal orbitals $\{CO_\ell(\vec{k}) (\ell = 1, \ldots, N_0)\}$ for the particular value of $\vec{k}$. The procedure must be repeated, in principle, for all combinations of the allowed k_a, k_b, and k_c values, i.e. for all k points of the Brillouin zone. Except for the increase in complexity inherent in the use of a 3D Brillouin zone, the process of determining the electronic structure of a 3D periodic solid does not bring any new difficulties beyond those already discussed for 1D materials.

9.1.3 Brillouin zone

Because of the periodicity properties of the $\vec{k}$ vector, the electronic structure of the system can be completely described by considering the Bloch orbitals associated with the set of allowed k_a, k_b, and k_c values:

$$-\frac{1}{2} < k_a \leq \frac{1}{2}$$
$$-\frac{1}{2} < k_b \leq \frac{1}{2}$$
$$-\frac{1}{2} < k_c \leq \frac{1}{2} \qquad (9.16)$$

However, note that because of this periodicity, it would be equally correct to take a different set of $\vec{k}$ vectors, for example those defined by:

$$0 < k_a \leq 1$$
$$0 < k_b \leq 1$$
$$0 < k_c \leq 1 \qquad (9.17)$$

In other words, there are many different and equally valid ways to define the basic repeat unit of the reciprocal space of the system.

For many applications it is convenient to use as repeat unit a primitive cell that displays the full symmetry of the lattice. This is the so-called Wigner–Seitz

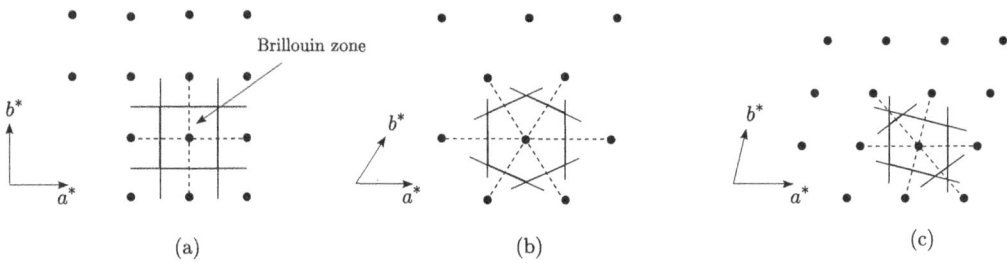

Fig. 9.4
Illustration of the construction of the Brillouin zone for three different 2D reciprocal lattices: (a) square, (b) hexagonal, and (c) oblique lattices.

[1] Note that the Wigner–Seitz cell can be constructed for both the direct and reciprocal lattices.

cell.[1] Conventionally, the Wigner–Seitz cell of the reciprocal space is known as the *first Brillouin zone*. Since higher-order Brillouin zones are not important in the context of this book, we will simply refer to the *Brillouin zone*. The way in which it is constructed is illustrated in Fig. 9.4 for the case of square, hexagonal, and oblique 2D lattices. Simply speaking, it is made by drawing perpendicular planes at the midpoint of the lines joining a reference lattice point with all other lattice points. The smallest enclosed area (or volume for 3D cases) is the Brillouin zone. In practice, a small number of close lattice points are needed. For instance, using the second nearest-neighbours for the square reciprocal lattice of Fig. 9.4a does not reduce the area of the Brillouin zone obtained by using the nearest neighbours. For 3D lattices the procedure is exactly the same. For instance, the Brillouin zone of the orthorhombic reciprocal lattice is a rectangular prism centred around one point of reciprocal space (see Fig. 9.5).

9.1.4 Symmetry and the Brillouin zone

As discussed in Section 3.2.3, time reversal (i.e. $E(\vec{k}) = E(-\vec{k})$) allows us to consider just half of the Brillouin zone in describing the electronic structure of any 1D periodic system. The other half can be obtained from the first by application of this type of symmetry. The fraction of the Brillouin zone

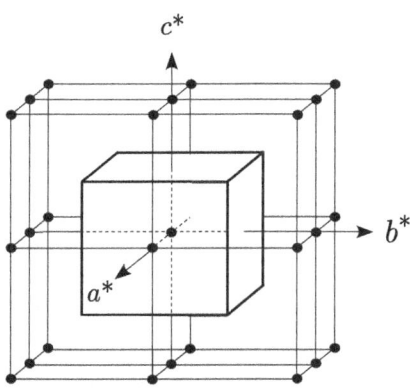

Fig. 9.5
Brillouin zone for the orthorhombic reciprocal lattice of Fig. 9.1b.

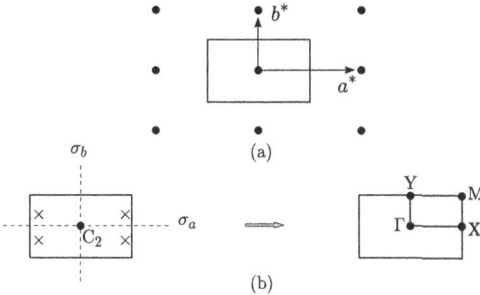

Fig. 9.6
(a) Brillouin zone for a rectangular lattice. (b) Determination of the irreducible part by looking at the set of k points equivalent by symmetry to a general k point.

that may generate the whole Brillouin zone by application of the different symmetry operations of the system is known as the *irreducible Brillouin zone*. Time reversal is a general property associated with the imposed boundary conditions and this kind of symmetry therefore also applies to any 2D or 3D system. However, the portion of the Brillouin zone that must be considered in actual computations is usually considerably smaller once the symmetry properties of the system have been fully exploited. [4, 5] This can save a lot of computational effort in studying real materials and we must therefore briefly consider this point here.

Let us consider for instance the case of a rectangular lattice with a and b lattice vectors. In this example the reciprocal space is also a rectangular lattice, with lattice vectors a^* and b^*, and the Brillouin zone is a rectangle centred on one of the lattice points (see Fig. 9.6). The rectangular lattice possesses a symmetry plane (σ_h) that is the plane of the lattice, C_2 axes perpendicular to this plane, symmetry planes perpendicular to the a (σ_a) and b directions (σ_b), and inversion centres (i). In other words, the operations of the D_{2h} symmetry group. What is the effect of these symmetry operations on a Bloch orbital associated with a general point (k_a, k_b) of the Brillouin zone?

The Bloch orbital associated with this k point is:

$$\frac{1}{\sqrt{N}} \sum_{m=-n'+1}^{n'} \sum_{p=-n'+1}^{n'} \exp(i2\pi m k_a) \exp(i2\pi p k_b)(\phi)_{m,p} \quad (9.18)$$

It can be graphically represented as in Fig. 9.7a. Here, on top of each lattice point, the appropriate phase factor ($\exp(i2\pi m k_a) \cdot \exp(i2\pi p k_b)$) is shown, where m and p are the two indices characterising any direct lattice point. Let us first consider the C_2 axis. Under the effect of this symmetry element the (k_a, k_b) point of the Brillouin zone moves to the ($-k_a$, $-k_b$) point. The Bloch orbital associated with the ($-k_a$, $-k_b$) point is:

$$\frac{1}{\sqrt{N}} \sum_{m=-n'+1}^{n'} \sum_{p=-n'+1}^{n'} \exp(-i2\pi m k_a) \exp(-i2\pi p k_b)(\phi)_{m,p} \quad (9.19)$$

and can be represented as shown in Fig. 9.7b. The Bloch orbitals associated with the two k points are different but completely equivalent. The BO($-k_a$, $-k_b$) can be obtained from the BO(k_a, k_b) by application of the C_2

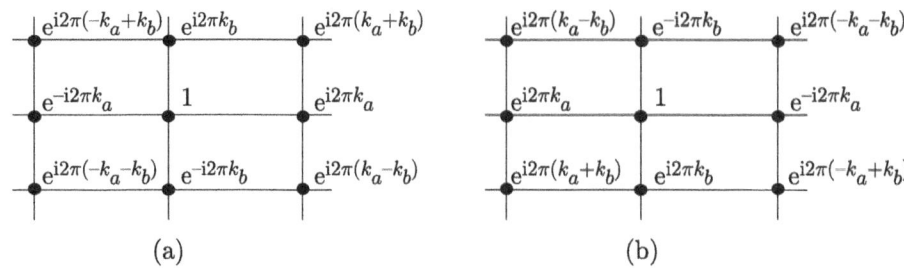

Fig. 9.7
Schematic representation of the Bloch orbitals BO(k_a, k_b) (a) and BO($-k_a$, $-k_b$) (b) of the rectangular lattice.

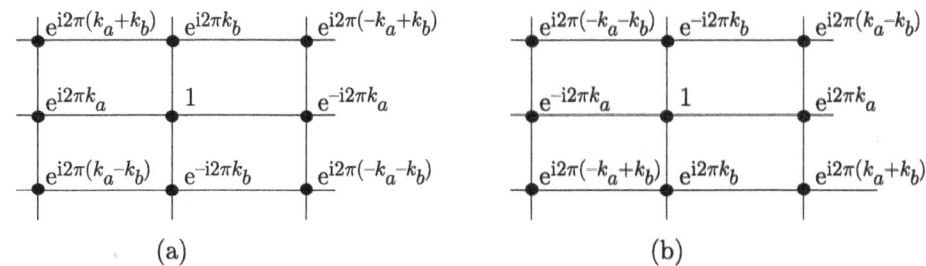

Fig. 9.8
Schematic representation of the Bloch orbitals BO($-k_a$, k_b) (a) and BO(k_a, $-k_b$) (b) of the rectangular lattice.

symmetry operation. In a similar way, the Bloch orbitals corresponding to the $(-k_a, k_b)$ and $(k_a, -k_b)$ points of the Brillouin zone (see Fig. 9.8) can be obtained from the initial one (Fig. 9.7a) by application of the σ_b and σ_a symmetry operations, respectively.

In general, when a set of k points of the Brillouin zone can be generated from each other by application of the symmetry operations of the lattice, the Bloch orbitals for all these k points can be obtained from just one of them by using the appropriate symmetry operations. Thus, the Bloch orbitals for all these k points are equivalent and, of course, degenerate in energy.

Consequently, to determine the irreducible part of the Brillouin zone for a given system (i.e. the minimum number of k points for which the crystal orbitals are needed to generate the crystal orbitals associated with the different k points of the full Brillouin zone), one just needs to consider how a general k point of the Brillouin zone is transformed under application of all the symmetry operations of the system. In addition, since the time-reversal operation is, by construction, a general property of the Bloch orbitals, the inversion symmetry should always be used even if it does not occur in the system under consideration. As illustrated in Fig. 9.6b, in the case of the rectangular lattice, application of all the symmetry operations to a general (k_a, k_b) point leads to a set of four equivalent k points. Consequently, the area of the irreducible Brillouin zone is one-quarter the area of the full Brillouin zone.

Several points that lie on faces, edges, or vertexes of the Brillouin zone are usually given special labels and the centre is always referred to as Γ. These points are usually high-symmetry points. For instance, in the case of the rectangular lattice the X, Y, and M labels designate the (k_a, k_b) points (1/2, 0), (0, 1/2), (1/2, 1/2), respectively. It must be noted that since the repeat unit of most periodic systems contains more than one atom, the symmetry properties of the actual system can be lower than the symmetry properties of the lattice (see Exercise (9.4)). Systems with many symmetry elements have irreducible Brillouin zones that are a small portion of the Brillouin zone. For instance, the volume of the irreducible Brillouin zones of simple cubic or hexagonal systems with the full symmetry properties of the lattice are $\frac{1}{48}$ and $\frac{1}{24}$ of the volume of the Brillouin zone.

Crystal orbitals may be classified according to their behaviour with respect to the symmetry operations of the lattice. This may be very useful in generating the band structure (i.e. determination of avoided or allowed band crossings, etc.), studying optical transitions, etc. As discussed in Chapter 6, the crystal orbitals for most of the points of the Brillouin zone do not exhibit the full symmetry properties of the lattice. Thus, it is important to know which symmetry elements are appropriate for classifying the crystal orbitals at different points of the Brillouin zone.

A crystal orbital is an eigenfunction of a symmetry operator if, under application of this operator, the same crystal orbital multiplied by a constant is generated. As noted above, when two k points of the Brillouin zone are related by a symmetry operation of the lattice, the crystal orbitals of the two k points may be generated from each other by application of this symmetry operation. Thus, because of the presence of the k vector in the phase factors, a crystal orbital can only be an eigenfunction of symmetry operators leaving the k point unaltered or transforming it into a translationally equivalent one (the crystal orbitals corresponding to k points separated by one reciprocal lattice vector formally differ by a constant $\exp(i2\pi) = 1$).

For instance, consider the s band of a rectangular lattice with one atom per cell. The general expression for a crystal orbital of this system is given in Fig. 9.7a. In the case of the Γ point, where k_a and k_b are 0, the crystal orbital is simply a collection of s orbitals with the same sign at every lattice point (Fig. 9.9a). This crystal orbital is, for example, symmetric with respect to the C_2 axis going through one of the lattice points. For the Y point, where k_a is 0 and k_b is 1/2, the crystal orbital is a collection of s orbitals with the same sign along the a direction but alternating positive and negative signs along the b direction (Fig. 9.9b). This crystal orbital is also symmetric with respect to the same C_2 axis. In contrast, it is clear from Fig. 9.7a that a general crystal

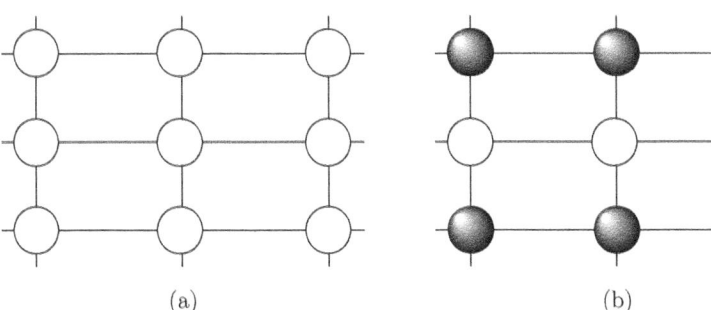

Fig. 9.9
Schematic representation of the Bloch orbitals BO(Γ) (a) and BO(Y) of the rectangular lattice.

2D and 3D systems

orbital with k_a and/or k_b values other than 0 or 1/2 is neither symmetric nor antisymmetric with respect to this axis. This is not surprising since the Γ point is left unaltered by the C_2 rotation and the Y point evolves under the action of the C_2 axis into a point ($k_a = 0$, $k_b = -1/2$), which is equivalent to Y under a translation, b^*. In contrast, a general k point of the Brillouin zone that does not coincide with the Γ, X, Y, or M points is not left unaltered nor transformed into a translationally equivalent point by the C_2 rotation operation. The reader may use the general crystal orbital in Fig. 9.7a to verify that the crystal orbitals for the Γ, X, Y, and M points are symmetric or antisymmetric with respect to the symmetry operations of the D_{2h} group, those for a general point along the Γ → X, Γ → Y, X → M, and Y → M lines can be classified according to the symmetry operations of the C_{2v} subgroup and those for all other points of the Brillouin zone according to the symmetry operations of the C_s subgroup.

9.2 Analysis of the electronic structure of 2D model systems

9.2.1 The square lattice $^2_\infty[H_n]$ system

As a simple application of these concepts let us consider the electronic structure of a model $^2_\infty[H_n]$ square lattice, i.e. the 2D extension of the analysis in Section 3.2 of the $(H)_n$ model chain.

Brillouin zone

The Brillouin zone of the square lattice was determined in Fig. 9.4a. Before analysing the band structure of the system we must consider the irreducible part. The square lattice is a very symmetric lattice in which all the symmetry operations of the D_{4h} symmetry group exist: C_4 axes perpendicular to the plane of the lattice, a symmetry plane (σ_h), which is the plane of the lattice, symmetry planes perpendicular to the a and b directions (σ_v), symmetry planes perpendicular to the $(a + b)$ and $(a - b)$ diagonal directions (σ_d), inversion centres (i), etc. When we look at how these symmetry operations transform a general point of the Brillouin zone (see Fig. 9.10) it is found that the irreducible part is just one-eighth of the Brillouin zone.

Representation of the CO(Γ), CO(X), and CO(M) functions

Since there is just a $1s$ orbital per repeat unit, the crystal orbitals of the system may be identified with the Bloch orbitals $BO_{1s}(\vec{k})$ generated by the different $1s$ atomic orbitals. The crystal orbitals of the $^2_\infty[H_n]$ system are thus given by:

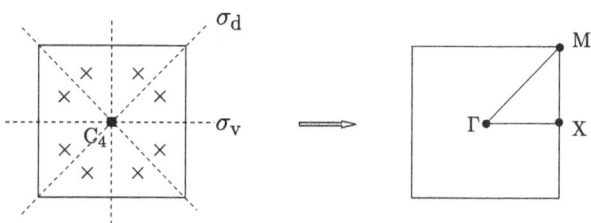

Fig. 9.10 Determination of the irreducible Brillouin zone for the 2D square lattice.

$$\text{CO}(\vec{k}) = \text{BO}_{1s}(\vec{k})$$

$$= \frac{1}{\sqrt{N}} \sum_{m=-n'+1}^{n'} \sum_{p=-n'+1}^{n'} \exp(i2\pi m k_a) \exp(i2\pi p k_b)(1s)_{m,p} \quad (9.20)$$

The different crystal orbitals only differ by their phase factors $\exp(i2\pi m k_a) \cdot \exp(i2\pi p k_b)$. For the Γ, X, and M points the crystal orbitals are real. At the Γ point the phase factor is simply 1 for every value of m and p. Thus the CO(Γ) orbital can be obtained by repeating the $(1s)_{0,0}$ orbital of the reference cell on every cell (see Fig. 9.11). Consequently, this is the most bonding combination of the system. At the X point the phase factor is $\exp(i2\pi m(1/2)) \cdot \exp(i2\pi p(0))$, i.e. $(-1)^m \cdot (1)^p$. Consequently, the orbital coefficients in two adjacent cells along the a direction have opposite signs. However, they have the same sign along the perpendicular direction. In other words, as shown in Fig. 9.11, this crystal orbital is completely antibonding along the a direction but completely bonding along the perpendicular direction. Finally, at M the phase factor is $\exp(i2\pi m(1/2)) \cdot \exp(i2\pi p(1/2))$, i.e., $(-1)^m \cdot (-1)^p$. The crystal orbital exhibits an alternation of positive and negative signs along both the a and b directions. As shown in Fig. 9.11 this is the most antibonding crystal orbital.

Energy of the crystal orbitals in the Hückel approach

The energy $E(\vec{k})$ associated with a given CO($\vec{k}$), which in this case coincides with $\text{BO}_{1s}(\vec{k})$, is given by:

$$E(\vec{k}) = \frac{\langle \text{BO}_{1s}(\vec{k}) | \hat{h} | \text{BO}_{1s}(\vec{k}) \rangle}{\langle \text{BO}_{1s}(\vec{k}) | \text{BO}_{1s}(\vec{k}) \rangle} = \frac{h_{1s,1s}}{S_{1s,1s}} \quad (9.21)$$

The evaluation of the energy as a function of the $\vec{k}$ vector is very easy when using a Hückel approach. For this we need to remind ourselves of the main notions developed in Section 3.2.2. For instance, because of the equivalence of all sites, in evaluating the Hamiltonian and overlap matrix elements it is enough to consider the interaction of the $1s$ orbital in a given cell and those of all other cells. There will be N identical terms of this type contributing to both the numerator and denominator of eqn (9.21) but, since the crystal orbitals are normalised, this N factor will be cancelled. So long as we use the Hückel approach, and since the $1s$ orbitals are normalised, the denominator is

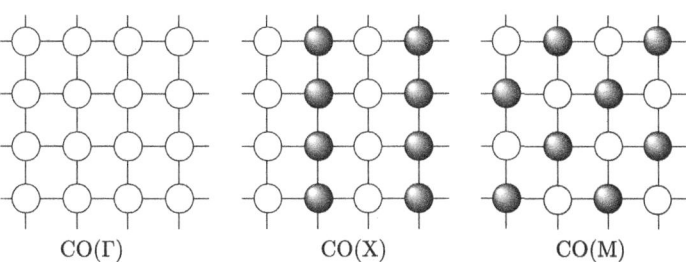

Fig. 9.11
Schematic representation of the crystal orbitals CO(Γ), CO(X), and CO(M) of the $_\infty^2[\text{H}_n]$ system.

always 1 because the overlaps between 1s orbitals of other cells are neglected. Concerning the numerator, only five terms are retained: a) the 'self-interaction' of the 1s orbital with itself, b) the two terms associated with the interaction of the 1s orbital of the reference cell with those of the nearest neighbours in the a direction, and c) the two terms associated with the interaction of the 1s orbital of the reference cell with those of the nearest neighbours in the b direction. The first term is simply α, the second and third terms are $\beta(\exp(i2\pi k_a) + \exp(-i2\pi k_a)) = 2\beta \cos(2\pi k_a)$, and the last two terms are $\beta(\exp(i2\pi k_b) + \exp(-i2\pi k_b)) = 2\beta \cos(2\pi k_b)$. Consequently the allowed energies for this system are given by:

$$E(\vec{k}) = \alpha + 2\beta \left(\cos(2\pi k_a) + \cos(2\pi k_b) \right) \quad (9.22)$$

Using this formula, the energy of the Γ, X, and M points is $\alpha + 4\beta$, α, and $\alpha - 4\beta$, which is in agreement with the qualitative discussion in the previous subsection. Thus the total energy width of the band is $8|\beta|$, twice as large as in the case of a chain. This is so because now every 1s orbital is engaged in interactions with the four nearest neighbours, instead of two in the chain. Using eqn (9.22), the $E(\vec{k})$ vs $\vec{k}$ may be calculated for any k point and thus the band structure of Fig. 9.12 can be generated. For 2D and 3D systems only certain lines along the irreducible Brillouin zone, usually high-symmetry lines, are plotted in the band structure.

Symmetry properties of the crystal orbitals

Now let us look at the symmetry exhibited by the different Bloch orbitals of the square lattice with respect to the point symmetry operations. As considered in some detail in Chapter 6, because of the existence of the phase factors necessary to describe the translational symmetry, some symmetry elements of the point group of the system are not valid symmetry elements for points of the Brillouin zone other than the Γ point.

According to what was discussed in Section 9.2.1, the appropriate symmetry group for the Γ point is D_{4h}, and it is easy to check that the Bloch orbital for this point, shown in Fig. 9.10, is symmetric with respect to all of the symmetry elements of this group.

The situation is different for the X point. Here the C_4 axis moves the X point to the Y point. Although the wavefunctions for these two points must be equivalent because there is a symmetry element that relates them, the Y point cannot be transferred back to the X point using the translational symmetry.

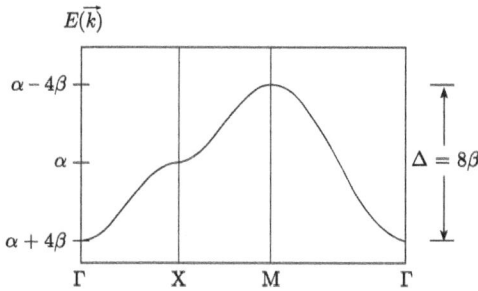

Fig. 9.12
Band structure of the $^2_\infty[H_n]$ square lattice system.

Thus the C_4 axis is not a symmetry element valid for the X point. However the lower-order C_2 axis perpendicular to the plane of the lattice is an appropriate symmetry operation since it transforms the X point into the (–1/2, 0) point, which is equivalent to the first because the two points may be superposed by a translation along a^*. In the same way, it is easy to see that the plane of the lattice and the planes perpendicular to the lattice containing the a^* and b^* axes are appropriate symmetry planes since they transform the X point into itself or a translationally equivalent point. In contrast, the planes and symmetry axes along the diagonal directions are not acceptable symmetry elements for this point. Thus the appropriate symmetry group for the X point is D_{2h}. Coming back to the Bloch orbital for the X point drawn in Fig. 9.11, it is easy to see that this orbital is always symmetric or antisymmetric with respect to all symmetry elements of the D_{2h} group.

The M point is kept unaltered or is transferred to a translationally equivalent point by all the symmetry elements of the group and D_{4h} is therefore the appropriate symmetry group for this point. This can be easily confirmed by checking that the Bloch orbital for the M point shown in Fig. 9.11 is always symmetric or antisymmetric with respect to all symmetry elements of the D_{4h} group.

The fact that D_{4h} is the appropriate symmetry group for both Γ and M points does not ensure that it will also be so for intermediate points along the Γ → M line of the Brillouin zone. An arbitrary point along this line is only left unchanged or transported to a translationally equivalent point under the action of the plane of the lattice, the plane perpendicular to the lattice containing the $(a + b)$ diagonal direction, and the C_2 axis along this diagonal direction. Thus the appropriate symmetry group is C_{2v}. Since k_a is equal to k_b, along this line, the general Bloch orbital can be schematically represented as in Fig. 9.13a. It is clear that this orbital is symmetric or antisymmetric with respect to all symmetry operations of this group. In the same way, it can be seen that the appropriate symmetry group along the Γ → X and X → M lines is also C_{2v} (see Fig. 9.13b and c). A general point of the Brillouin zone out of these high symmetry lines can only be classified according to the plane of the lattice. This has no relevance for a lattice with only s orbitals, but may have consequences when the lattice also has p orbitals.

9.2.2 The square lattice $^2_\infty [A_n]$ system

Let us now consider the case of a square lattice of atoms A, where A is an atom bearing one s and three p orbitals. If we initially assume that there is no s–p interaction we must deal only with the three bands generated by the three p orbitals. We will assume that the a and b directions of the lattice coincide with the x and y directions, respectively. Thus, one of the three p orbitals, p_z, is antisymmetric with respect to the plane of the lattice, while both p_x and p_y are symmetric with respect to this plane. This means that the p_z band behaves completely independently of the p_x and p_y bands. In other words, the formula describing the energy of the different p_z Bloch orbitals is:

$$E(\vec{k}) = \alpha_p + 2\beta_\pi (\cos(2\pi k_a) + \cos(2\pi k_b)) \qquad (9.23)$$

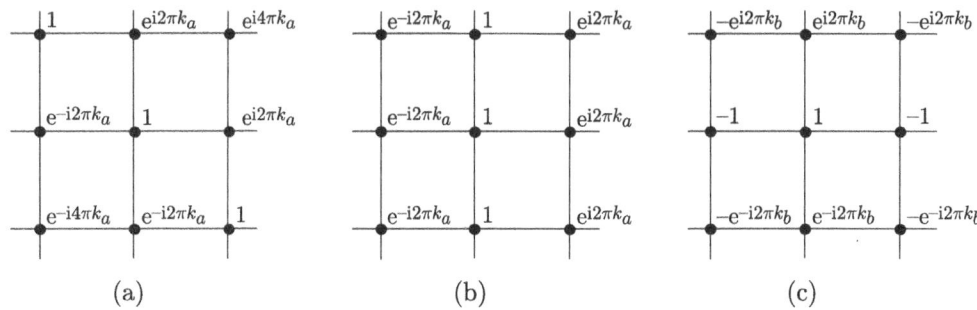

Fig. 9.13
Schematic representation of the crystal orbitals for a general point along the $\Gamma \to M$ (a) and $\Gamma \to X$ (b), and $X \to M$ (c) lines of the Brillouin zone.

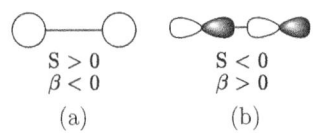

Fig. 9.14
Illustration of the different signs of the overlap and transfer integrals for two s orbitals (a) or two p orbitals making σ interactions.

Fig. 9.15
$E(\vec{k})$ vs $\vec{k}$ dependence of the p_x and p_y bands of the $^2_\infty[A_n]$ system assuming no s–p interaction and taking into account only σ-type nearest-neighbour interactions.

where β_π is the interaction term (also called the transfer integral) associated with the relatively weak π-type interactions of the p_z orbitals and α_p is the energy of the p orbital. The crystal orbitals $CO_{p_z}(\Gamma)$, $CO_{p_z}(X)$, and $CO_{p_z}(M)$ are those shown in Fig. 9.11, where every circle represents a top-down view of a p_z orbital.

We are left with two p orbitals per site, p_x and p_y. Every one of these two orbitals makes strong σ-type interactions along one of the main directions of the lattice but weaker π-type interactions along the perpendicular direction. If we start by assuming that the π-type interactions are nil (in which case the band associated with the p_z orbital is completely flat), the $E(\vec{k})$ vs $\vec{k}$ dependence of the bands is simply given by:

$$E_{p_x}(\vec{k}) = \alpha_p + 2\beta_\sigma (\cos(2\pi k_a))$$
$$E_{p_y}(\vec{k}) = \alpha_p + 2\beta_\sigma (\cos(2\pi k_b)) \quad (9.24)$$

where β_σ is the transfer integral associated with the σ-type interactions of the p orbitals. Note that the second term in eqns (9.22) and (9.24) contribute with a different sign to the total energy, although the equations are written equivalently. This simply results from the different signs of the σ-type overlaps between two s or two p orbitals (see Fig. 9.14). The $E(\vec{k})$ vs $\vec{k}$ dependence of the bands is shown in Fig. 9.15 and the shape of the Bloch orbitals, which in this case coincide with the crystal orbitals at Γ, X, and M, using the phase factors discussed above for the s band, are shown in Fig. 9.16a, b, and c, respectively.

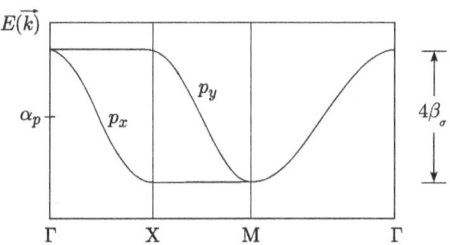

Note that the crystal orbitals of Fig. 9.16 are well adapted to the analysis of their symmetry properties along the $\Gamma \to X$ and $X \to M$ lines but not along the $\Gamma \to M$ line. Since for every atom A of the lattice, the p_x and p_y orbitals are completely equivalent, two linear combinations $\frac{1}{\sqrt{2}}(p_x + p_y) = p_{x+y}$ and $\frac{1}{\sqrt{2}}(p_x - p_y) = p_{x-y}$ can be equally well used. The crystal orbitals generated by repeating these two orbitals with the phase factors appropriate for the $\Gamma \to M$ line provide a pair of crystal orbitals that are well adapted to the symmetry (see for instance Fig. 9.17 for the new crystal orbitals at Γ and M). Note also that real functions can always be generated by making linear combinations of the crystal orbitals for k points related by time reversal. The argument developed in Section 3.2.4 for 1D systems is completely general, as the reader can easily confirm by using the crystal orbitals shown in Fig. 9.7. These real expressions can be of value in the analysis of computational results (for instance in calculating and understanding analytical tools such as charges or overlap populations, which will be considered in the next chapter).

The band structure of Fig. 9.15 is easily understood by looking at the shape of the crystal orbitals in Fig. 9.16 and counting the number of σ interactions per site. This band structure is typical of a system that is made of the superposition of two 1D systems although it is structurally 2D (i.e. the p_x orbitals, which interact along a but behave independently along the perpendicular direction, and the p_y orbitals, which behave in the opposite way).

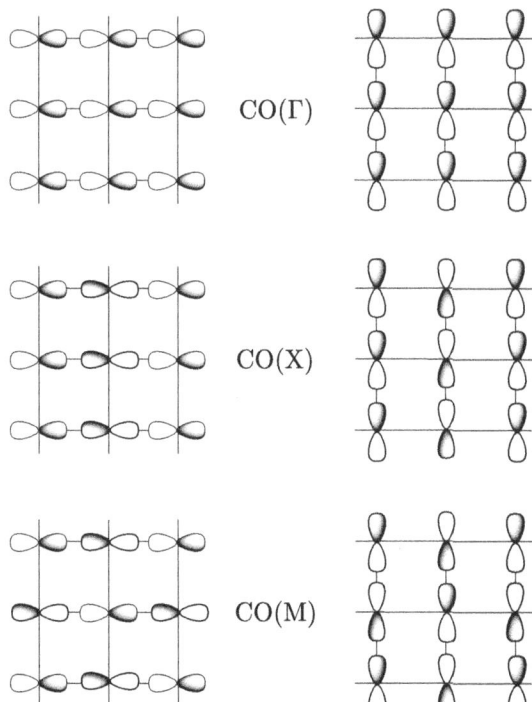

Fig. 9.16
Schematic representation of the crystal orbitals CO(Γ), CO(X), and CO(M) associated with the p_x and p_y orbitals of the $^2_\infty[A_n]$ system.

Now we can switch on the weaker π-type interactions. Denoting the transfer integral associated with the π interactions between two nearest-neighbour p orbitals as β_π, it is easy to write the equations for the $E(\vec{k})$ vs $\vec{k}$ dependence of the bands:

$$E_{p_x}(\vec{k}) = \alpha_p + 2\beta_\sigma(\cos(2\pi k_a)) + 2\beta_\pi(\cos(2\pi k_b))$$
$$E_{p_y}(\vec{k}) = \alpha_p + 2\beta_\pi(\cos(2\pi k_a)) + 2\beta_\sigma(\cos(2\pi k_b))$$
$$E_{p_z}(\vec{k}) = \alpha_p + 2\beta_\pi(\cos(2\pi k_a)) + 2\beta_\pi(\cos(2\pi k_b)) \quad (9.25)$$

Consequently, the $E(\vec{k})$ vs $\vec{k}$ dependence of the p_x and p_y bands as well as those of the p_z bands are those shown in Fig. 9.18a and b, respectively.

Under the effect of the π interactions the p_x and p_y bands are no longer purely 1D. Since each of the two orbitals makes two σ interactions with nearest neighbours *and* two π interactions, and the two types of transfer integrals are of opposite sign, the total dispersion of each individual band is retained but the total dispersion of the two bands $(4|\beta_\sigma| + 4|\beta_\pi|)$ increases by $4|\beta_\pi|$ as a result of the introduction of the π interactions. The total dispersion of the p_z bands is $8|\beta_\pi|$ because these orbitals make four nearest-neighbour π interactions.

Introduction of the s–p interaction is conceptually easy. The p_z band is not affected at all, for symmetry reasons. For the s, p_x, and p_y bands, a 3×3 determinant must be solved and four different transfer integrals (β_s, β_σ, β_π, and β_{s-p}) must be considered in building the different elements of the secular determinant. However, solution of the secular determinant leads to cumbersome expressions for the $E(\vec{k})$ vs $\vec{k}$ dependence, and they will not therefore be outlined here since they do not provide additional insights with respect to the discussion concerning the s–p interaction in a chain (see Section 6.2). Purely on symmetry grounds it is clear that the s–p interaction does not affect the energy of the Γ, X, and M points, as is also the case for the high-symmetry points, Γ and Z, in the chain. However at lower-symmetry points there can be s–p mixing. For instance, let us consider the $\Gamma \rightarrow$ M line. The bands along this line may be classified according to the C_{2v} group, since a general k point of this line is left unaltered by the symmetry plane of the lattice, the C_2 axis in this plane along the $(a+b)$ diagonal direction, and a plane perpendicular to the lattice and containing the $(a+b)$ diagonal direction. As shown in Fig. 9.17, the Bloch orbitals built from the p_{x+y} and p_{x-y} orbitals are of different symmetry with respect to the C_2 axis and the vertical symmetry plane. This is not only true for Γ and M but for any point along the $\Gamma \rightarrow$ M line (see Fig. 9.13a). The Bloch orbitals along $\Gamma \rightarrow$ M based on the s orbital have the same symmetry as those based on the p_{x+y} orbital, except at the Γ and M points. Thus the p_{x-y} band will be left unaltered when the s–p interaction is switched on, but the p_{x+y} band will be slightly destabilised. Consequently, the two bands of Fig. 9.18a will no longer have the same energy along the $\Gamma \rightarrow$ M line. In analysing the consequences of s–p mixing in actual calculations, one should consider the limiting situations developed in Section 6.2.

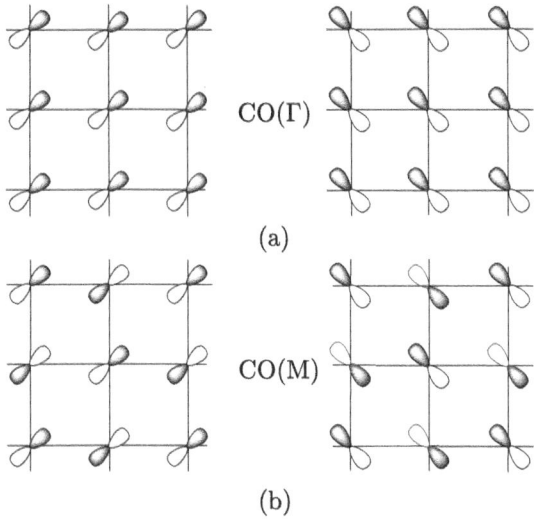

Fig. 9.17
Schematic representation of the crystal orbitals CO(Γ) (a) and CO(M) (b) associated with the p_{x+y} and p_{x-y} orbitals of the $^2_\infty[A_n]$ system.

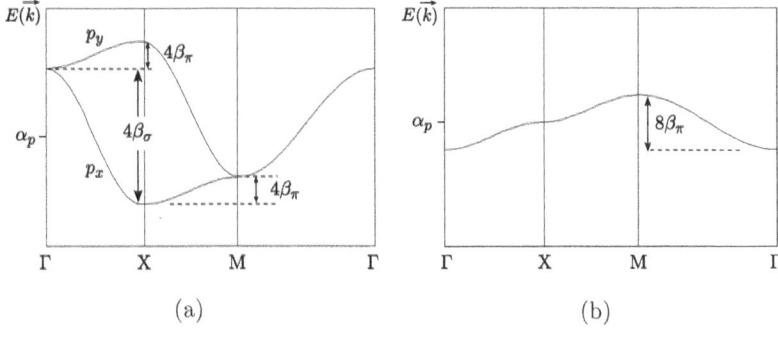

Fig. 9.18
$E(\vec{k})$ vs $\vec{k}$ dependence of: (a) the p_x and p_y bands and (b) the p_z bands of the $^2_\infty[A_n]$ system, assuming no s–p interaction but taking into account both σ- and π-type nearest-neighbour interactions.

9.2.3 π-type band structure of hexagonal graphene layers

In this section we will consider the case of hexagonal graphene layers as shown in Fig. 9.19a, where every filled circle is a carbon atom. When different graphene layers stack in an A-B-A- mode, the 3D structure of graphite is generated (see Fig. 9.19b). Since the different layers are tied together by relatively weak forces, many different types of electron donors and electron acceptors can be inserted between the layers to generate so-called intercalated graphite compounds: KC_8, KC_{24}, CaC_6, etc. [6] Graphite itself is a good electrical conductor but the doping resulting from the injection of electrons or holes into the graphite bands can strongly affect this conductivity. Very recently, success in isolating individual graphene layers has generated enormous interest in the remarkable properties of this 2D hexagonal system. [7] Here we will only consider the π-type bands of the layer since these are the bands that are responsible for the conductivity of the system. [8]

Fig. 9.19
(a) Hexagonal graphene layer where every filled circle is a carbon atom; (b) 3D graphite structure.

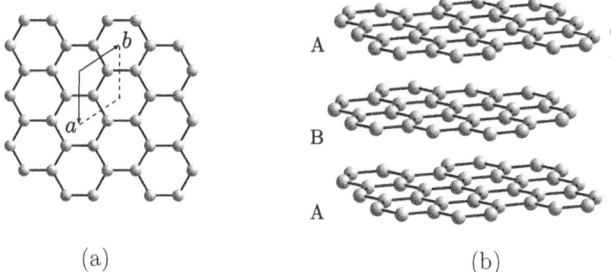

(a)   (b)

Brillouin zone

As shown in Fig. 9.19a, the repeat unit of a graphene layer is a pair of carbon atoms and the repeat vectors a and b ($= a$) can be chosen to make an angle of 120°. Thus the associated reciprocal lattice is also an hexagonal lattice, with repeat vectors a^* and b^* making an angle of 60°. The Brillouin zone of the system is an hexagon centred on a reciprocal lattice point (see Fig. 9.4b).

To find the irreducible part we must consider the symmetry properties of the system. It is easy to see in Fig. 9.19a that the layer contains all the symmetry elements of the D_{6h} symmetry group:

- C_6 axes perpendicular to the layer going through the centre of the hexagons
- a symmetry plane (σ_h) coinciding with the plane of the layer
- three symmetry planes perpendicular to the layer (σ_v) going through the centre of two opposite edges of the hexagon (i.e. along the directions a, b, and $-(a+b)$)
- three symmetry planes perpendicular to the layer (σ_d) going through two opposite carbon atoms of one hexagon
- an inversion centre (i) at the centre of the hexagon
- etc ...

The irreducible part may be determined by successively applying all symmetry operations to a general point of the Brillouin zone. Note that in contrast to the case of the square lattice, the a and b directions do not coincide with the a^* and b^* directions, so a symmetry element along, for instance, the a direction will be a symmetry element along the $(2a^* - b^*)$ direction. In Fig. 9.20 we have also drawn the a and b directions to facilitate the correct location of the

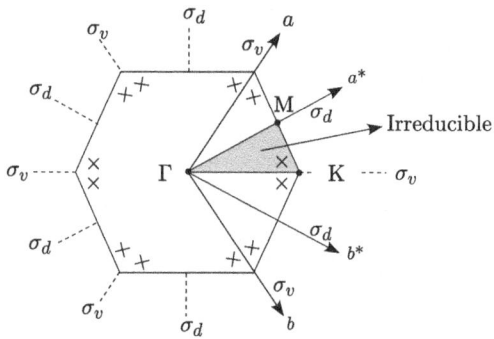

Fig. 9.20
Determination of the irreducible Brillouin zone for the 2D hexagonal graphene system.

2D model systems

symmetry elements. As a matter of fact, once the σ_v and σ_d planes are used (or the C_6 axis and the inversion centre, etc ...) the irreducible part cannot be further reduced. Thus the irreducible part of the Brillouin zone is the triangle with vertices at the Γ, K, and M points, i.e. 1/12 of the Brillouin zone. With the Γ, K, and M labels we refer to the k points (0,0), (1/3,1/3), and (1/2,0) in units of the reciprocal lattice vectors.

Obtaining the band structure

The repeat unit of the hexagonal layer contains two carbon atoms and consequently we must consider two p_z orbitals, which we will label $(p_z)_1$ and $(p_z)_2$. [8] Every one of these orbitals will generate a set of Bloch orbitals centred on each type of atom. Consequently, to obtain the π band structure of the graphene layers we need to solve a 2×2 secular determinant. When computing the different matrix elements of this determinant one needs to take care, due to the presence of two carbon sites in the unit cell and the non-orthogonality of the repeat vectors. Note that in this lattice, each carbon atom has three nearest neighbours of different type lying at $(2\vec{a}/3 + \vec{b}/3)$, $(-\vec{a}/3 + \vec{b}/3)$, and $(-\vec{a}/3 - 2\vec{b}/3)$. Consequently, using the usual approximations of the Hückel approach, the different elements of the secular determinant can be evaluated as:

$$h_{11}(\vec{k}) = \alpha = h_{22}(\vec{k})$$
$$h_{12}(\vec{k}) = \beta \left(e^{i\vec{k}(\frac{2}{3}\vec{a}+\frac{1}{3}\vec{b})} + e^{i\vec{k}(-\frac{1}{3}\vec{a}-\frac{2}{3}\vec{b})} + e^{i\vec{k}(-\frac{1}{3}\vec{a}+\frac{1}{3}\vec{b})} \right)$$
$$= \left(h_{21}(\vec{k}) \right)^* \quad (9.26)$$

The associated secular equation is easily solved and leads to the $E(\vec{k})$ vs $\vec{k}$ relationship:

$$E(\vec{k}) = \alpha \pm \beta \sqrt{[3 + 2\cos(2\pi(k_a + k_b)) + 2\cos(2\pi k_a) + 2\cos(2\pi k_b)]}$$
$$(9.27)$$

For the three high-symmetry points Γ, M, and K the energies are thus $\alpha \pm 3\beta$, $\alpha \pm \beta$, and $\alpha \pm 0\beta$. The band structure along the high-symmetry directions is shown in Fig. 9.21.

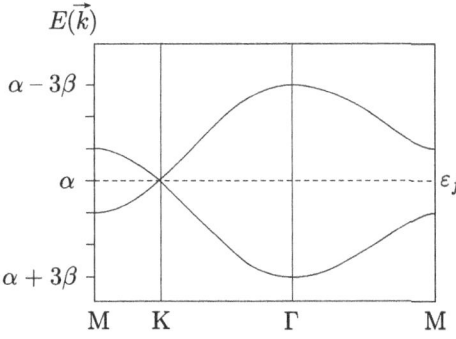

Fig. 9.21
π-type band structure of an hexagonal graphene layer.

2D and 3D systems

Analysis of the band structure

Although the crystal orbitals can be directly obtained from the previous approach, there is another way to tackle the problem, which is simpler and more informative. Let us assume that the basis orbitals used to build the band structure are not the individual $(p_z)_1$ and $(p_z)_2$ orbitals but the two linear combinations $\pi = \frac{1}{\sqrt{2}}((p_z)_1 + (p_z)_2)$ and $\pi^* = \frac{1}{\sqrt{2}}((p_z)_1 - (p_z)_2)$ (i.e. the π and π^* orbitals of ethylene; see orbitals 6 and 7 in Fig. 5.4). The different matrix elements of the 2×2 secular determinant are now:

$$h_{\pi\pi}(\vec{k}) = (\alpha + \beta) + \beta\cos(2\pi k_a) + \beta\cos(2\pi k_b)$$
$$h_{\pi^*\pi^*}(\vec{k}) = (\alpha - \beta) - \beta\cos(2\pi k_a) - \beta\cos(2\pi k_b)$$
$$h_{\pi\pi^*}(\vec{k}) = i\beta(\sin(2\pi k_a) + \sin(2\pi k_b))$$
$$= \left(h_{\pi\pi^*}(\vec{k})\right)^* \tag{9.28}$$

and solving the corresponding equation leads to the same $E(\vec{k})$ vs $\vec{k}$ dependence obtained above.

However, let us consider the problem in a simpler way. The crystal orbitals at the high-symmetry points Γ, M ($k_a = 1/2$, $k_b = 0$), and K ($k_a = 1/3$, $k_b = 1/3$) are easy to draw. Those at Γ are simply obtained by repeating in phase the π and π^* orbitals at all lattice points (see Fig. 9.22a). Every p_z orbital of $(\pi)_\Gamma$ makes three bonding interactions with its nearest neighbours. In contrast, every p_z orbital of $(\pi^*)_\Gamma$ makes three antibonding interactions with its nearest neighbours. Consequently, as far as we limit our evaluation of the energy to the nearest neighbours, the energy of the $(\pi)_\Gamma$ and $(\pi^*)_\Gamma$ Bloch orbitals is $\alpha + 3\beta$ and $\alpha - 3\beta$, respectively. Since the two Bloch orbitals are of different symmetry they cannot interact and consequently the two Bloch orbitals are really the two crystal orbitals for the Γ point.

For the M point the phase factor can be written as:

$$\exp\left(i2\pi(\frac{1}{2}m + 0p)\right)$$

so that the Bloch orbitals may be obtained by repeating the π and π^* orbitals out of phase along the a^* direction, but in phase along the b^* direction generating the two Bloch orbitals $(\pi)_M$ and $(\pi^*)_M$ of Fig. 9.22b. Every p_z orbital of $(\pi)_M$ makes two bonding and one antibonding interactions with its nearest-neighbours, whereas every p_z orbital of $(\pi^*)_M$ makes one bonding and two antibonding interactions. Consequently, the energy of the $(\pi)_M$ and $(\pi^*)_M$ crystal orbitals may be evaluated to be $\alpha + \beta$ and $\alpha - \beta$, respectively. Again, the two Bloch orbitals are of different symmetry and thus are the two crystal orbitals for the M point.

The Bloch orbitals for the K point can be obtained using the phase factor:

$$\exp\left(i2\pi(\frac{1}{3}m + \frac{1}{3}p)\right) = \exp\left(i2\pi\frac{1}{3}m\right)\cdot\exp\left(i2\pi\frac{1}{3}p\right)$$

Since the phase factors are not real we represent the two crystal orbitals as in Fig. 9.23a. The energy of these Bloch orbitals may seem less easy to evaluate

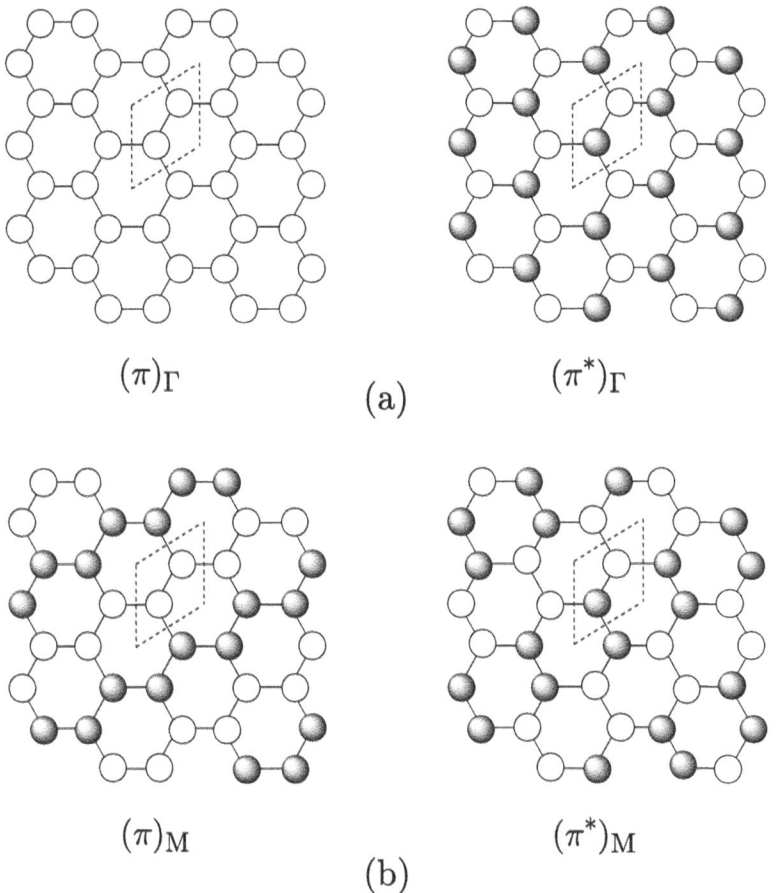

Fig. 9.22
Schematic representation of the Bloch orbitals BO(Γ) and BO(M) for the hexagonal graphene layer using the π and π^* basis orbitals.

than those of the Bloch orbitals at Γ and M. However, simple inspection of these orbitals makes it apparent that two linear combinations of these functions, i.e. $\frac{1}{\sqrt{2}}((\pi)_K + (\pi^*)_K)$ and $\frac{1}{\sqrt{2}}((\pi)_K - (\pi^*)_K)$, lead to two completely equivalent orbitals in which half of the positions have a zero phase factor (see Fig. 9.23b). These orbitals are purely non-bonding as far as the first nearest-neighbours are considered and their energy is thus α. However, since these two linear combinations are degenerate, it follows that the initial $(\pi)_K$ and $(\pi^*)_K$ orbitals must also be degenerate and non-bonding. Note that the orbitals of Fig. 9.23b are simply the Bloch functions associated with the individual $(p_z)_1$ and $(p_z)_2$ orbitals.

Thus, it has been possible to obtain the main features of the band structure of the hexagonal graphene planes without carrying out any type of calculation. Note that the actual energy of the two crystal orbitals at the K point depends on the method of evaluation but the fact that they are degenerate does not. It is simply a consequence of the phase factors for the K point and the symmetry of the lattice, so it is a result that is obtained independent of the approach used to calculate the $E(\vec{k})$ vs $\vec{k}$ dependence. The touching of the two bands at the K point qualifies the graphene planes as a semimetal. The two bands can only touch at this point since elsewhere the symmetry is lowered and the two bands

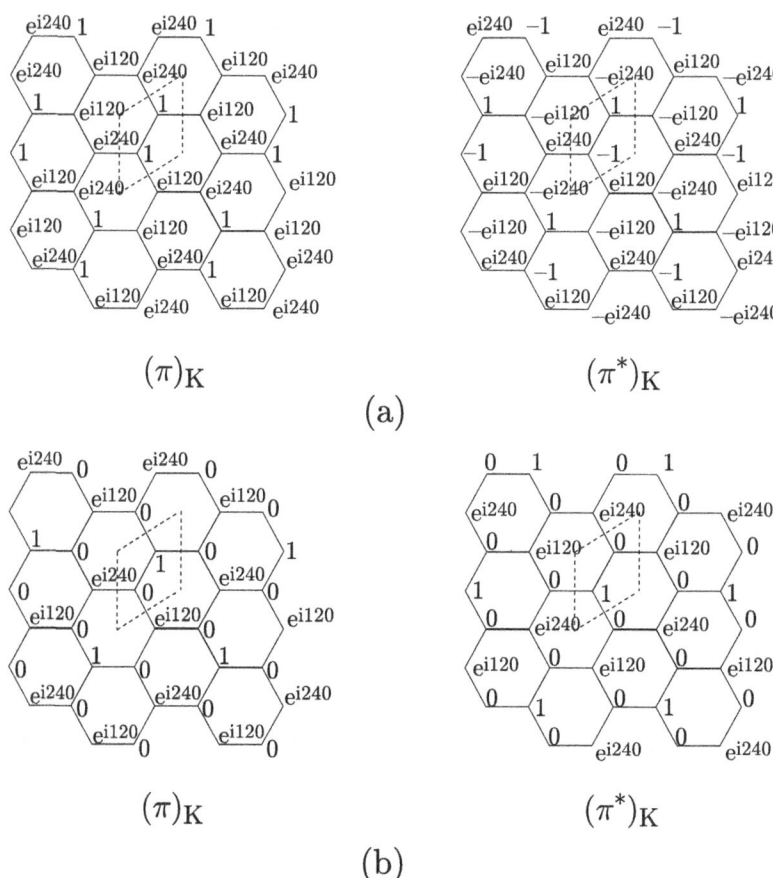

Fig. 9.23
Two different representations of the crystal orbitals CO(K) for a hexagonal graphene layer.

can interact. It is this feature of the band structure that lies behind the great interest in the study of single graphene layers. In graphite, where these planes are stacked, there are weak interlayer interactions, which are responsible for the fact that the Fermi level no longer cuts the bands precisely at the K point, and small electron and hole pockets are created around this point.

The preceding discussion clearly suggests a simple way to modify the transport properties of these hexagonal layers. Since the touching of the two bands at the K point results from the complete equivalence of the two atomic positions of the repeat unit, we simply need to destroy such equivalence to open a gap at the Fermi level, thus making the system behave as a semiconductor. By changing the electronegativity of one of the two sites with respect to the other (i.e. changing α to separate α_1 and α_2) one of the two CO(K) crystal orbitals is stabilised whereas the other is destabilised, leading to the appearance of a gap. This is simply what happens in hexagonal boron nitride, a solid built from the stacking of graphene-like layers in which half of the positions have been substituted by boron atoms and the remaining half by nitrogen atoms as shown in Fig. 9.24a. The band structure can be calculated exactly in the same way as for graphene. Since there is just one type of transfer integral, the only change in the secular determinant is that the diagonal elements are now (α_B–

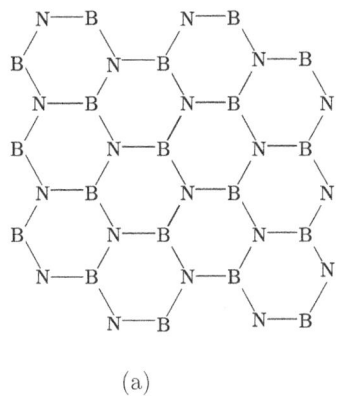
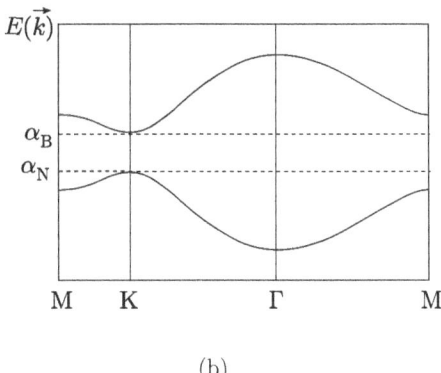

Fig. 9.24
(a) BN layers in hexagonal boron nitride; (b) schematic π-type band structure for a single layer of hexagonal boron nitride.

$E(\vec{k}))$ and $(\alpha_N - E(\vec{k}))$. Solving the secular determinant leads to the $E(\vec{k})$ vs $\vec{k}$ dependence:

$$E(\vec{k}) = \frac{1}{2}(\alpha_N + \alpha_B) \pm \frac{1}{2}\sqrt{(\alpha_B - \alpha_N)^2 + 4\beta^2(A(\vec{k}))}$$

$$A(\vec{k}) = 3 + 2\cos(2\pi(k_a + k_b)) + 2\cos(2\pi k_a) + 2\cos(2\pi k_b) \quad (9.29)$$

The band structure that may be calculated on the basis of this formula is shown in Fig. 9.24b and, as expected, clearly shows the presence of a band gap.

In conclusion, the electronic structure of 2D or 3D periodic solids is not conceptually different from that of the apparently simpler 1D systems. The main complication arises from dealing with a 2D or 3D Brillouin zone. In the next chapter we will look at some ways to condense the information of the full Brillouin zone in simple chemical terms. However, analysis of the relevant parts of the band structure usually provides a clear link between the details of the crystal and electronic structures. Let us note that Hückel type approaches such as those briefly considered in this chapter often provide straightforward but firm qualitative schemes to interpret the results of actual calculations in complex solids (see Chapter 11 for some cases).

Exercises

(9.1) The lattice vectors of a body-centred cubic lattice are $\vec{a} = \frac{a}{2}(-\vec{u} + \vec{v} + \vec{w})$, $\vec{b} = \frac{a}{2}(\vec{u} - \vec{v} + \vec{w})$ and $\vec{c} = \frac{a}{2}(\vec{u} + \vec{v} - \vec{w})$ where a is the side in the associated simple cubic lattice: $\vec{a} = a\vec{u}$, $\vec{b} = a\vec{v}$ and $\vec{c} = a\vec{w}$. Build the reciprocal lattice of the body-centred cubic lattice.

(9.2) Determine the Brillouin zone of the body-centred cubic direct lattice discussed in Exercise (9.1).

(9.3) Obtain the irreducible part of the Brillouin zone for a tetragonal lattice.

(9.4) Work out the irreducible Brillouin zone for the three different models (on top (a), bridging or twofold (b) and capping or threefold (c)) of the interaction of a CH$_4$ molecule with a hexagonal lattice, assuming that the repeat unit contains nine atoms of the hexagonal lattice and one CH$_4$ molecule.

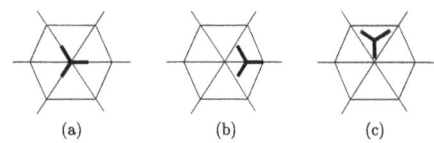

(a)　(b)　(c)

(9.5) Determine the different matrix elements of the 3×3 determinant needed to calculate the $E(\vec{k})$ vs $\vec{k}$ dependence of the s, p_x, and p_y bands of a square lattice taking into account the s–p interaction.

(9.6) Obtain the $E(\vec{k})$ vs $\vec{k}$ dependence for the bcc and fcc lattices with one s orbital per site according to a Hückel approach.

(9.7) Obtain the $E(\vec{k})$ vs $\vec{k}$ dependence for a simple cubic lattice with three p orbitals per site taking into account the σ and π interactions with first- and second-nearest neighbours.

References

1. N. W. Ashcroft and N. D. Mermin, *Solid State Physics*, Holt, Rinehart and Winston, Philadelphia, 1976.
2. R. M. Martin, *Electronic Structure: Basic Theory and Practical Methods*, Cambridge University Press, Cambridge, 2004.
3. J. K. Burdett, *Chemical Bonding in Solids*, Oxford University Press, New York, 1995.
4. S. I. Altmann, *Band Theory of Solids: An Introduction from the Point of View of Symmetry*, Oxford University Press, Oxford, 1991.
5. M. Tinkham, *Group Theory and Quantum Mechanics*, McGraw-Hill, New York, 1964.
6. M. S. Dresselhaus and G. Dresselhaus, *Adv. Phys.*, 51, 1, 2002.
7. A. K. Geim and A. H. MacDonald, *Physics Today*, 60, 35, 2007.
8. P. R. Wallace, *Phys. Rev. B*, 71, 622, 1947.

Density of states

10

In Chapters 3–8 we saw how to build the band structure of 1D systems from the viewpoint of orbital interactions. Except for the use of translational symmetry, the rules behind this process are not very different from those used to build the molecular orbitals of a molecule. This approach is extremely useful in understanding the electronic structure and physical properties of a large number of 1D solids. However, many solids are structurally complex and have large unit cells. In this case there will be many orbitals to consider in building the crystal orbitals, and the task of understanding the band structure of the material can be quite involved. In addition, complex unit cells usually have a very low symmetry, so that there are almost no restrictions on the mixing of orbitals. At this point it is useful to recall that, even when symmetry properties exist, the symmetry inside the Brillouin zone is usually quite low and consequently, since there will be many band crossings avoided, a certain orbital can be found spread among many bands. The orbital interaction analysis of the band structure can become too cumbersome in such cases and a simpler analytical tool must be devised. Very simply speaking, since the problem is that we have too much data when looking at the band structure of these systems, we must find a way to condense this information so that, even if we lose many details, we can easily extract an useful view of the electronic structure. It is obvious that this need for a simpler tool will be even more pronounced when dealing with 2D and 3D systems. In such cases the Brillouin zone will be a surface or a volume and consequently for each band there will be many different lines of the Brillouin zone to be considered. And, of course, most interesting solids are 2D or 3D! The simplest analytical tool devised for this purpose is the density of states (DOS). [1, 2]

10.1 Calculation and analysis of the density of states

10.1.1 Density of states

The DOS of a system, $n(e)$, is defined as the number of states in an energy interval between e and $e + \delta e$. Thus $n(e)$ is non-zero in the allowed energy

Fig. 10.1
π-type electronic structure of the cyclobutadiene molecule according to the Hückel approach: (a) molecular orbital diagram and (b) density of states.

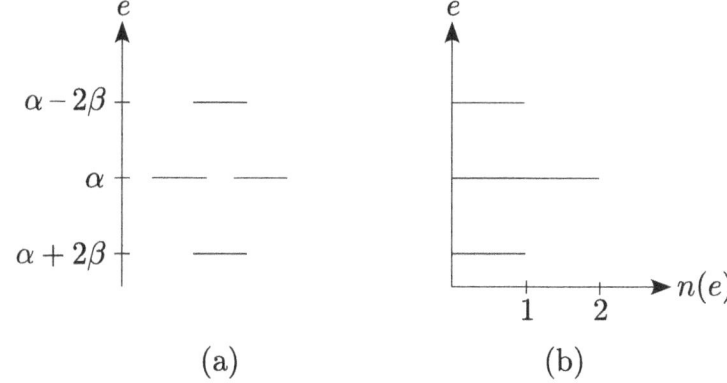

region of a band and vanishes in the forbidden energy region. The concept of the DOS can be easily illustrated by considering the example of the cyclobutadiene molecule (Fig. 2.1). As shown in Fig. 10.1a, the π-type orbitals of this molecule are, according to a Hückel approach: a) a non-degenerate bonding orbital at $\alpha + 2\beta$, b) a set of two non-bonding degenerate orbitals at α, and c) a non-degenerate antibonding orbital at $\alpha - 2\beta$ (see Fig. 2.10). Thus the number of π states at energy α is twice the number of π states at $\alpha + 2\beta$ and $\alpha - 2\beta$ and the π-type DOS of cyclobutadiene can be represented as in Fig. 10.1b.

For the simple case of an infinite chain of hydrogen atoms, as discussed in Section 3.2, the dispersion relation of the band according to a Hückel approach is given by a cosine function of k (see Fig. 10.2a and eqn. (3.20)). Since the k-space is a set of equally spaced k points, the DOS is proportional to the inverse of the slope of the energy versus k curve.

$$n(e) \propto \left(\frac{de(k)}{dk}\right)^{-1} \tag{10.1}$$

Thus, $n(e)$ will be minimum at the centre of the band and will tend to infinity at the borders, i.e. at $k = 0$ and $k = \pi/a$ (see Fig. 10.2b). The last feature is known as a van Hove singularity and is physically meaningless because the integrated DOS of the whole band must be equal to the number of electrons per unit cell associated with the band, i.e. two if the band is completely filled.[1] Since $n(e)$ is zero outside the band, the typical shape of the DOS for a 1D band is that shown in Fig. 10.2c. Thus the wider the band the lower the DOS will be. In other words, strong interactions along the chain are associated with a large energy dispersion and thus with lower DOS values. Conversely, very sharp peaks in the DOS are associated with flat bands and thus with electrons which tend to be localised.

The shape of the DOS curves for 2D and 3D systems is quite different. Although the detailed shape strongly depends upon the nature of the lattice, high DOS values generally occur around the centre of the band. In practice, to determine the DOS of a system the k-dependent Hamiltonian is diagonalised for a fine mesh of k-points in the Brillouin zone. This leads to an histogram of energy values, which is then numerically smoothed using a set of appropriate functions, for example Gaussians, leading to the DOS of the material. The finer

---

[1] Let us recall that, for simplicity, throughout this book we have assumed that the molecular orbitals for the α (or 'spin-up') and β (or 'spin-down') electrons are identical. In other words, we assume that a molecular orbital can accommodate two electrons and a band can accommodate two electrons per cell.

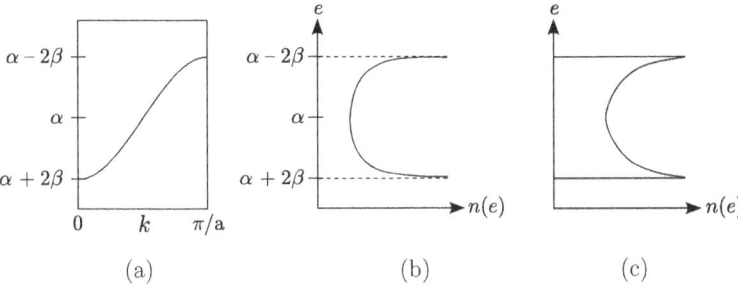

Fig. 10.2
(a) Band structure and (b) density of states for an infinite chain of hydrogen atoms according to a Hückel approach; (c) DOS for a typical 1D band.

the mesh of k points used, the more precise will be the DOS. The electronic structure of solids can be discussed solely on the basis of their DOS values. The DOS of a real solid, i.e. a multi-band system, can have a quite complex shape. Thus it is important to develop a qualitative way to analyse this type of plot. In this chapter we illustrate some chemically appealing ways to do it.

10.1.2 Projected density of states

The $n(e)$ values associated with a given band, i, should satisfy the normalisation condition:

$$\int_{-\infty}^{\infty} n_i(e)de = 1 \qquad (10.2)$$

Consequently, the DOS plot counts the number of levels available for a certain structure. Since a band can accommodate up to two electrons (with different spin) per unit cell, integration of the DOS (taking into account the occupation of each level) up to the Fermi level gives the total number of electrons of the system. Since the COs have been written as a linear combination of BOs associated with the different AOs of the unit cell, the DOS curves can be analysed in terms of contributions per AO of the unit cell. In other words, once the DOS plot has been calculated, we can project the contribution of certain AOs or fragment orbitals (FOs) of interest for the analysis.

The easiest way to illustrate this process is by considering a simple two-centre MO:

$$\Psi = c_1 \chi_1 + c_2 \chi_2 \qquad (10.3)$$

The electron distribution in this MO is given by the normalisation condition:

$$1 = c_1^2 + c_2^2 + 2c_1 c_2 S_{12} \qquad (10.4)$$

where S_{12} is the overlap integral between the AOs χ_1 and χ_2. Although it is clear that c_1^2 and c_2^2 represent the fraction of the electron described by this orbital, which on average is found near the centres 1 and 2, respectively, and consequently should be assigned to χ_1 and χ_2, it is not clear how to assign the term $2c_1 c_2 S_{12}$, which is an *overlap density*. The simplest solution has been suggested by Mulliken. [3] In the Mulliken population analysis, this term is equally shared between centres 1 and 2. Thus centre 1 is assigned a total of $c_1^2 + c_1 c_2 S_{12}$ electrons, which is the *gross population* of centre 1, and centre 2 a total of $c_2^2 + c_1 c_2 S_{12}$ electrons, which is the *gross population* of centre 2. These contributions should now be multiplied by the occupation number of this MO. This type of analysis can be performed for the different orbitals of a molecule or a solid. Consequently, a local DOS can be projected out from the total DOS. These projected DOS can be those of an AO, an FO, an atom, or a group of atoms that are convenient for the analysis. The projected DOS can be used in different ways. For instance, one can be interested in knowing with which atom or group of atoms of a complex structure the electrons in a certain energy range are associated. Typically, this question can be an important one when trying to correlate the structure and properties of complex materials. For instance, let us consider the 2D purple bronze KMo_6O_{17}. [4] This material has been the subject of many studies because it is a 2D metal and exhibits a resistivity anomaly at 120 K, which is the result of a metal-to-metal transition (see Chapter 11). The structure of KMo_6O_{17} is schematically shown in Fig. 10.3a. This bronze contains $Mo_6O_{17}^-$ layers made up of condensed MoO_6 octahedra and MoO_4 tetrahedra. Between these layers are found the K^+ cations. The electrostatic interactions between these K^+ cations and the outer oxygen atoms of the covalently bonded $Mo_6O_{17}^-$ layers hold the solid together. The layers of Fig. 10.3a contain only three different types of Mo atoms: a) the Mo atoms of the MoO_4 tetrahedra (denoted Mo_I in Fig. 10.3a), b) the Mo atoms of the MoO_6 octahedra in the two outer octahedral sublayers (denoted Mo_{II} in Fig. 10.3a), and c) the Mo atoms of the MoO_6 octahedra in the two inner octahedral sublayers (denoted Mo_{III} in Fig. 10.3a). With the usual oxidation states K^+ and O^{2-}, it turns out that three electrons per formula unit remain to fill the lower d-block levels of the $Mo_6O_{17}^-$ layers. Since there are three electrons to occupy these levels but six Mo atoms per formula unit, it is of interest to

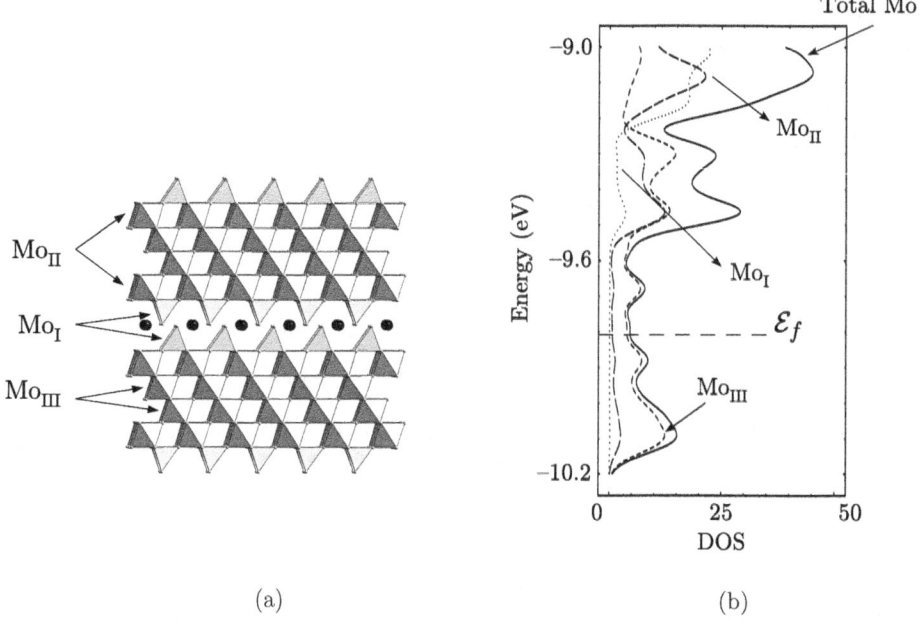

(a)   (b)

Fig. 10.3
(a) Perspective view of the crystal structure of KMo_6O_{17}, with the three different types of molybdenum atom labelled. (b) Projected DOS of the Mo atoms in KMo_6O_{17} as well as the individual contributions of Mo_I, Mo_{II} and Mo_{III}.

know if the electrons responsible for the metallic properties of this material (i.e. the electrons near the Fermi level) are delocalised all over the $Mo_6O_{17}^-$ layers or if they are confined to some of the sublayers. This information may be relevant in understanding the physical properties of the material. The projection of the Mo contribution to the DOS of the $Mo_6O_{17}^-$ layer as well as the individual contributions of Mo_I, Mo_{II} and Mo_{III} are shown in Fig. 10.3b. These results clearly show that the conducting electrons are almost completely confined into the inner sublayers of the $Mo_{III}O_6$ octahedra. Thus, they are quite well screened from the random potentials generated by possible alkali non-stoichiometry. This is a simple example of how the use of projected DOS can be useful in understanding key features of the electronic structure of complex materials. Other uses of the projected DOS will be discussed in the next sections.

10.1.3 Crystal orbital overlap population

In essence, the use of projected DOS allows an understanding of the composition of levels in a certain energy range, i.e. where in the structure these levels are located. This technique does not allow us to know if the associated orbitals are bonding, non-bonding or antibonding between a given pair of atoms (or orbitals). In molecular chemistry, this question is usually answered by looking at the Mulliken *overlap population*. [3] The overlap population between two orbitals 1 and 2 is related to the overlap density term $2c_1c_2S_{12}$ of the normalisation equation and is a measure of the strength of the bonding between the two orbitals (see eqn (10.4)). If the overlap integral is taken as positive, then this quantity is positive when the combination between the two orbitals is bonding and negative when the combination is antibonding. The Mulliken overlap population for two atoms is the addition of the terms $2c_1c_2S_{12}$ associated with all pairs of atomic orbitals of the two atoms over all MOs, multiplied by the occupation (0, 1, or 2) of these MOs. Exactly as for the local DOS, the quantities $2c_1c_2S_{12}$ of interest can be evaluated for the different energy levels of a solid and projected out from the total DOS. This type of DOS projections are thus overlap-population-weighted DOS curves and are usually known as *crystal orbital overlap population* (COOP) curves. [5, 6, 2] Integration of these curves up to the Fermi level leads to the Mulliken overlap population for the two atoms.[2]

To illustrate the use of the COOP curves let us consider those corresponding to the 1,2 and 1,3 π-type interactions in ideal *trans*-polyacetylene (see Fig. 10.4a). The π-type band structure of this simple system was discussed in detail in Chapter 5 and the important crystal orbitals are schematically shown in Fig. 10.4b. All the 1,2 interactions are bonding at the bottom of the lower band but antibonding at the top of the upper band. The levels at the top of the lower band and bottom of the upper band are degenerate and essentially non-bonding. Thus the COOP curve for the 1,2 π-type interactions will be as schematically shown in Fig. 10.5a. The 1,3 interactions are bonding, both at the top of the upper band and bottom of the lower band. However, they are antibonding at the top of the lower band and bottom of the upper band. Since these interactions are weaker than the 1,2 ones, the corresponding COOP

[2] A closely related tool for the analysis of bonding in solids are the so-called *crystal orbital Hamilton population* (COHP) curves. [2]

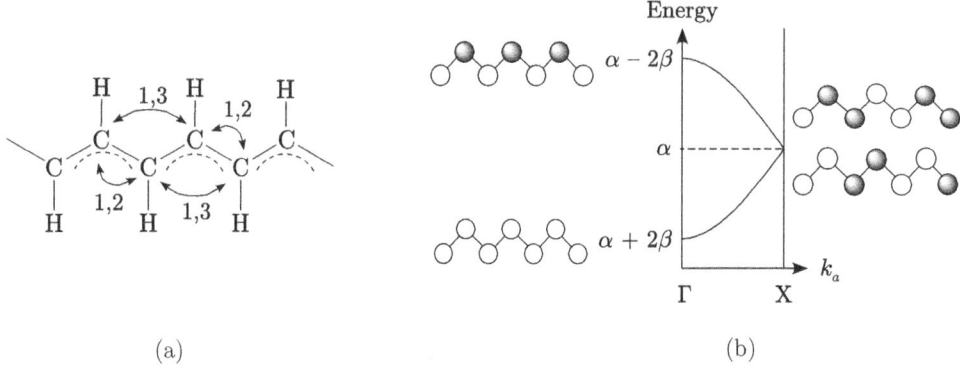

Fig. 10.4
(a) Regular *trans*-polyacetylene chain in which the direct 1,2 and indirect 1,3 interactions are labelled; (b) π-type band structure of regular *trans*-polyacetylene.

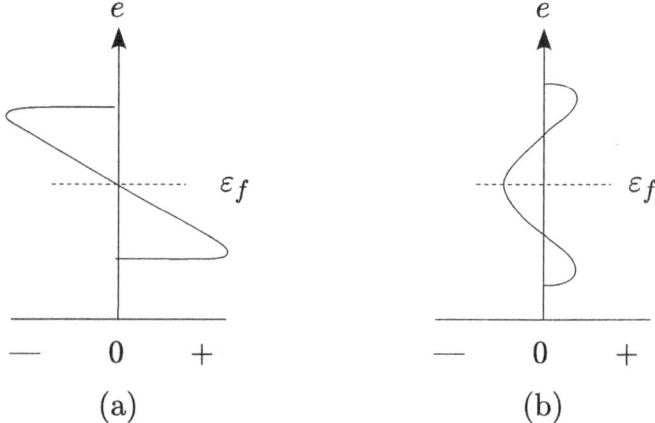

Fig. 10.5
Schematic COOP curves for the 1,2 (a) and 1,3 (b) interactions for the π system of a regular *trans*-polyacetylene chain.

curve should be as schematically shown in Fig. 10.5b. According to these curves and assuming that ideal *trans*-polyacetylene could be doped, keeping this ideal structure, doping with donors/acceptors (i.e. raising/lowering the Fermi level) should lead to an increase/decrease of the C–C bond length because antibonding/bonding levels in the COOP curve of Fig. 10.5a will be filled/emptied. In contrast, doping with donors/acceptors should lead to an increase/decrease of the C–C–C angle because the antibonding levels of Fig. 10.5b would be filled/emptied. The COOP curves are a very useful tool in rationalising structural changes as a function of the electron count.

10.2 Combined use of DOS and COOP: electronic structure of the MPS$_3$ layered phases

As an example of use of a combination of the projected DOS and COOP curves to understand the electronic structure of a material, let us consider the case of the MPS$_3$ phases (M = Mn, Fe, Co, Ni, Cd). These materials have been

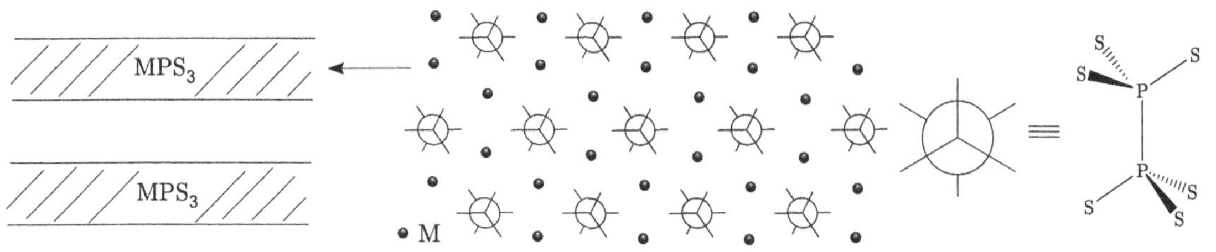

Combined use of DOS and COOP

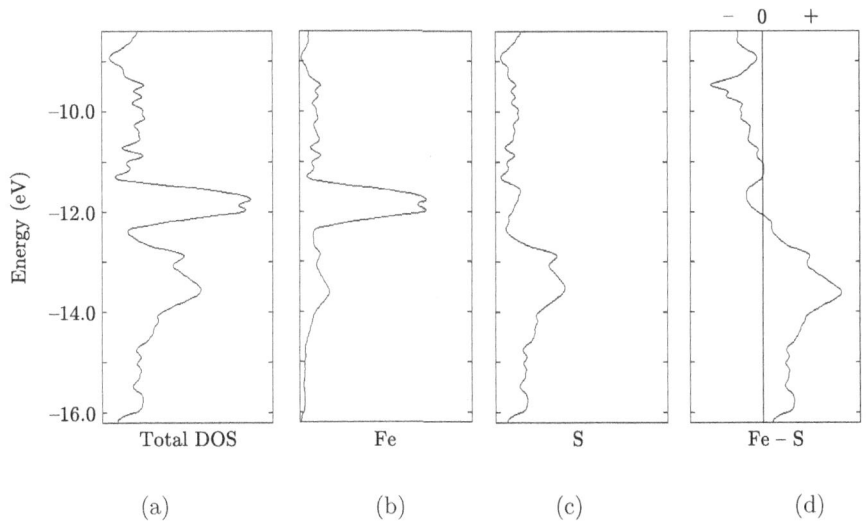

Fig. 10.6
Schematic representation of the crystal structure of the MPS$_3$ phases.

extensively studied because of their intercalation properties. As schematically shown in Fig. 10.6, the MPS$_3$ phases are layered materials with van der Waals gaps between the MPS$_3$ slabs. [7] These slabs are made up of $P_2S_6^{4-}$ ethane-like groups and M^{2+} transition metal cations. The M^{2+} cations sit on the octahedral holes created by six sulphur atoms of three different $P_2S_6^{4-}$ groups.

The total DOS of an FePS$_3$ layer in the region of the transition metal levels is shown in Fig. 10.7a. [8] The projected DOSs of the Fe and S atoms are shown in Figs 10.7b and 10.7c, respectively. The projection of the Fe orbitals shows three contributions: a strong peak centred at −11.9 eV and two broad contributions centred at −13.8 and −10.2 eV. Whereas there are important and quite broad contributions of the sulphur orbitals centred at −13.8 and −10.2 eV, the contribution at −11.9 is noticeably smaller. This immediately suggests that the peak at −10.2 eV in the total DOS contains the Fe–S $e_g{}^*$ levels, the peak at −11.9 eV contains the Fe t_{2g} levels, and the lowest part of the peak at −13.8 eV contains the Fe–S e_g levels. This is confirmed by the Fe–S COOP curve in Fig. 10.7d. This plot tells us that the peak at −10.2 eV contains the Fe–S antibonding states, as it should for the $e_g{}^*$ levels, whereas

Fig. 10.7
Total DOS (a), projected DOS for Fe (b), and S (c) as well as Fe–S COOP curve (d) calculated for FePS$_3$.

that at −13.8 eV contains Fe–S bonding states (i.e. those providing cohesion to the layers), as it should for the e_g levels. The Fe–S COOP curve exhibits a small global antibonding contribution in the region around −11.9 eV, i.e. the region of the t_{2g}, mostly non-bonding, levels. Comparison of the Fe and S contributions in the two regions of the e_g-type peaks shows that whereas the two contributions are comparable for the e_g* peak, the S contribution is much larger for the e_g peak. This means that there is a peak that contains the S lone pairs not interacting with Fe, which lies a little higher in energy than the e_g levels, but partially overlapping them. The curves in Fig. 10.7 make it clear that the interaction of the Fe orbitals with those of non-nearest neighbour atoms leads to a noticeable spread of the levels, so that the system can no longer be considered as made up simply of ions. Nevertheless, these interactions are not strong enough to significantly change the orbital splitting due to the local field of the nearest-neighbour atoms. This is not necessarily true in most solids, especially when there is extensive metal–metal bonding. However, combined use of the projected DOS and COOP (or COHP) curves usually leads to a detailed and yet relatively straightforward description of the electronic structure of these solids too.

10.3 Step-by-step determination of the density of states: the $(Pt(NH_3)_4Cl)^{2+}$ chain

In the previous sections, very simple examples were purposely chosen but, at least in principle, more complex systems can be studied along the same lines. As noted above, one of the advantages of using the DOS is to have a great deal of information summarised in a simple plot. However, a superficial look at the DOS of a complex system can quite often lead to the feeling that it is hopeless to try to understand the detailed shape of the plot. And yet, such an understanding is needed if we want to truly understand the physics and chemistry of the material. For instance, the use of projected DOS tells us that the conducting electrons in KMo_6O_{17} are confined to the inner octahedral layers but what the calculation does not tell us is why this is so. If we want to understand this result, so that we know where the conducting electrons are in similar oxides without having to carry out a new DOS calculation each time, we must understand the gross features of the DOS plot. Quite often, a convenient way to do it is to try to roughly sketch the DOS of the material. To do this we need some ideas to help us sketch the different parts of the DOS of the material from the features of the crystal structure. As we show in this and the following sections, these ideas are nearly the same as those used to understand how the geometrical and electronic structures of molecules are related.

The first system we consider in some detail is the $[Pt^{II}L_4 \cdot Pt^{IV}L_4Cl_2]^{4+}$ (L = $EtNH_2$) Wolfram red salt, which, as shown in Fig. 10.8a, contains linear arrays of alternating $Pt^{II}L_4$ and $Pt^{IV}L_4Cl_2$ units. [9] We may consider the chain in Fig. 10.8a to be the result of a Peierls distortion of the ideal $[Pt^{III}L_4Cl_2]^{4+}$ chain shown in Fig. 10.8b, and only the region dominated by the transition metal d levels will be analysed. [10] To build the DOS of the chain in Fig. 10.8b we can consider it as being made up of a series of octahedra sharing the apical chlorine atoms or as a series of square-planar PtL_4 units interacting with

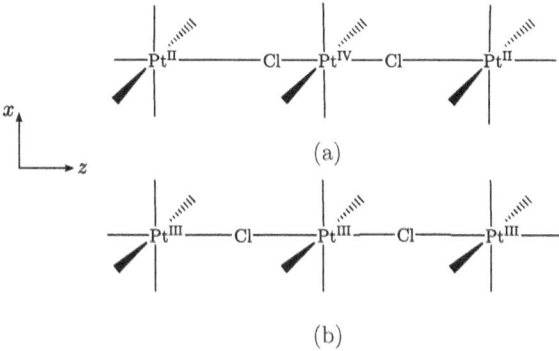

Fig. 10.8
Schematic representation of dimerised (a) and uniform (b) [PtL$_4$Cl]$^{2+}$ (L = EtNH$_2$) chains of the Wolfram red salt.

chlorine atoms. Although both approaches would lead to the same result, the difference in the donor character of the two types of ligand (Cl and EtNH$_2$) makes the second approach more useful.

The typical orbital splitting for an ML$_4$ fragment is depicted in Fig. 10.9a, i.e. a set of three almost degenerate orbitals (xz, yz, and xy), an orbital somewhat higher in energy (z^2), and finally another orbital ($x^2 - y^2$), which is at considerably higher energy, because it is strongly destabilised by σ-type antibonding interactions with the ligands. These orbitals will interact with the s and p orbitals of the bridging chlorines to generate the levels of the regular chain. Since the s and p levels of chlorine are considerably lower in energy than the metal levels, they will mix with them in an antibonding way and the mixing will not be very strong. In other words, every level of Fig. 10.9a will lead to a band of mainly metal character. The question is how wide these bands will be. The z^2 orbital can make σ-type interactions with the s and z orbitals of chlorine, the (xz, yz) pair can make π-type interactions with the (x, y) pair of orbitals of Cl and the $x^2 - y^2$ and xy orbitals do not interact with any orbital of chlorine because of their δ-type symmetry with respect to the chain axis. Thus, in view of the relative strength of the σ, π, and δ interactions, the levels of the Fig. 10.9a will be broadened into bands as shown schematically in Fig. 10.9b.

What should be remembered before sketching the transition metal projected DOS is that the integrated values for every band should be proportional to the density associated with every type of orbital. For instance, both the xy and $x^2 - y^2$ orbitals make δ-type interactions with the chlorines and consequently their partial DOS will be a very narrow peak. However, the xy orbital makes π-type interactions with the EtNH$_2$ ligands whereas the $x^2 - y^2$ orbital makes σ-type interactions with the EtNH$_2$ ligands. Consequently, the mixing of the ligand orbitals will be stronger in the $x^2 - y^2$ orbital and, because of the normalisation condition, the peak representing the partial DOS of this orbital will be weaker than that of the xy orbital. Since the z^2 orbital is spread into a broad band, the partial DOS of this orbital will have the typical shape shown in Fig. 10.2c and consequently the maximum values of this contribution will be much weaker than those of the xy or $x^2 - y^2$ orbitals. Of course, there will be some contribution of the EtNH$_2$ ligands into the z^2 band, although it will be small and equally spread all along the band because it is not primarily affected by the phase change along the chain. It is not quite clear how broad

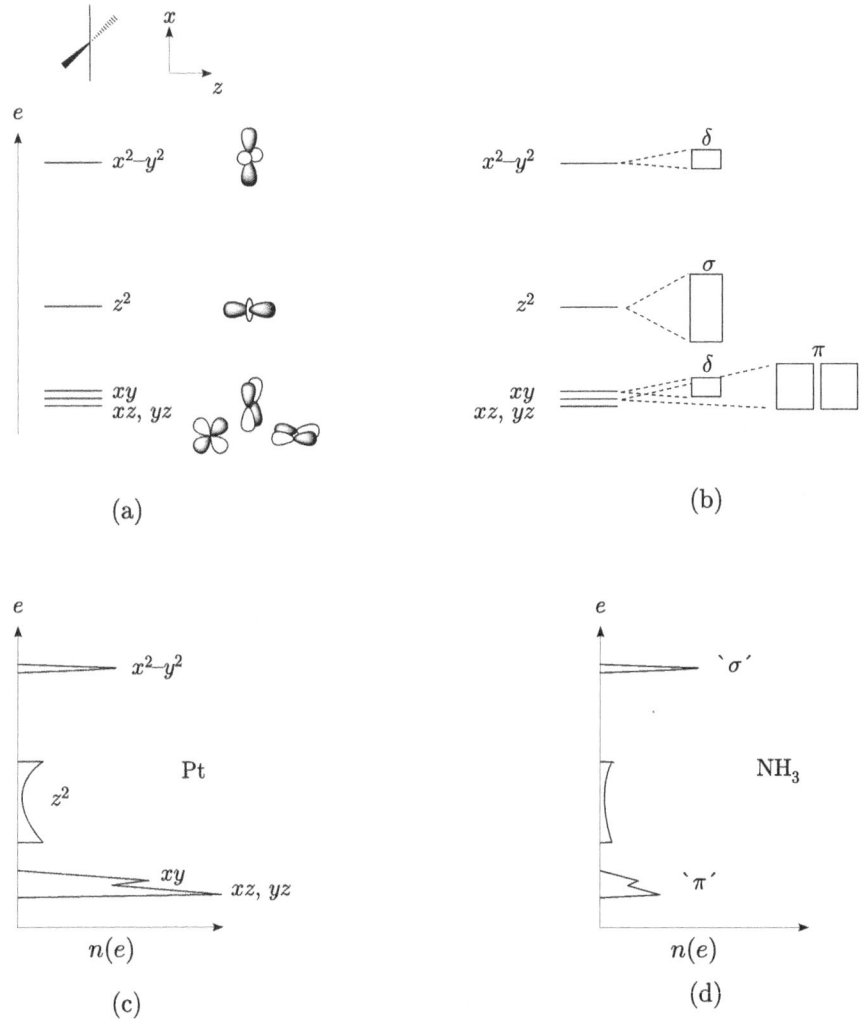

Fig. 10.9
(a) Metal-based molecular orbitals of an ML$_4$ fragment; (b) spread of the metal-based levels into bands for the chains of the Wolfram red salt. Sketched partial DOS for the platinum atom (c) and the ligands (d) in the regular chain of Fig. 10.8b.

[3] The situation could be different if by overlap or energy level proximity reasons the x and y orbitals of the bridging ligand could interact better with the xz and yz orbitals leading to sizeable π interactions along the chain. This is the case in d^0 compounds like ReCl$_4$N whose second-order Peierls distortion is discussed in Section 8.2.6 of Chapter 8. In that case the shape of the partial DOS would be better represented as in Fig. 10.2c.

the bands associated with the xz and yz orbitals will be. Since these orbitals make π-type interactions with the chlorine, which is a poor π donor, we can tentatively assume that the corresponding orbital mixing will be small and thus represent the partial DOS as a peak. Because there are two of these orbitals, this peak should be a somewhat less than twice as high but also a little bit broader than the peak of the xy orbital.[3] As a result of these elementary considerations we can sketch the partial DOS of the metal, ligands, and bridging chlorine and thus the total DOS. For instance, the result for the platinum and ligands is shown in Figs 10.9c and 10.9d, respectively. They compare quite well with those shown in Fig. 10.10, which are the results calculated for the chain in Fig. 10.8b, where for simplicity the ligands have been taken to be NH$_3$. Note

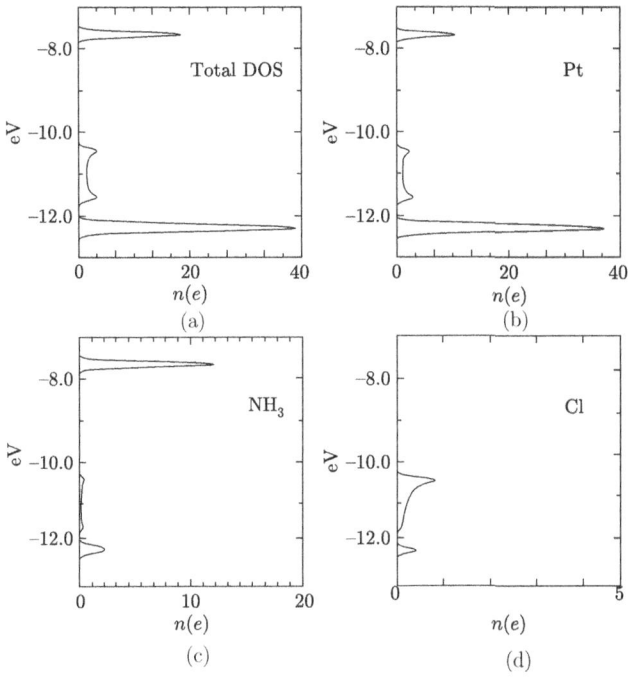

Fig. 10.10

Total DOS (a), projected DOS for Pt (b), the NH$_3$ ligands (c) and the bridging chlorine atoms (d) for the ideal chain of Fig. 10.8b, where the EtNH$_2$ ligands have been modelled with NH$_3$. Note the different scales for $n(e)$.

that the two peaks of the xy and yz contributions have merged into a single peak. The important point is that in the process of sketching the DOS we have gained a sound understanding of the electronic structure of the system.

Let us now consider the z^2 band more carefully. This is the band which is half-filled in the d^7 Wolfram salt and consequently, the one that is responsible for the dimerised structure and mixed valence of these systems. In the ideal chain of Fig. 10.8b, with C_{4v} symmetry, only the s and z orbitals of the bridging chlorine can mix with the platinum z^2 orbitals. However, the z orbital has a nodal plane perpendicular to the chain direction while the s orbital does not. Consequently the s and z orbitals cannot mix with the z^2 orbitals all along the band. The s orbital will mix into the z^2 band when the phase of two adjacent z^2 orbitals is similar in magnitude and sign (i.e. for wave vectors around $k = 0$) and will decrease elsewhere. For the z orbitals the mixing will be strong when the phase of two adjacent z^2 orbitals is similar in magnitude but opposite in sign (i.e. for wave vectors around $k = \pi/a$) and will decrease elsewhere. In addition, since the energies of the Cl s orbital and the Pt z^2 orbital are quite different, the antibonding mixing into the z^2 band will be very small. Consequently, the chlorine s contribution to the DOS in the region of the z^2 levels will be small and lie at the bottom of this region. In contrast, both in overlap and energy proximity terms the mixing of the Cl z orbital into the z^2 band will be stronger. Thus, the chlorine z contribution in the same region of the DOS will be quite more important, with a maximum at the top and decreasing towards the bottom of the partial DOS curve. All these simple predictions are in good agreement with the results of Fig. 10.11, which are those calculated for the [(NH$_3$)$_4$PtCl]$^{4+}$ chain. Note that in Fig. 10.11b the values at the upper part of the partial Pt z^2 DOS are smaller than those in the lower part. This is another consequence of

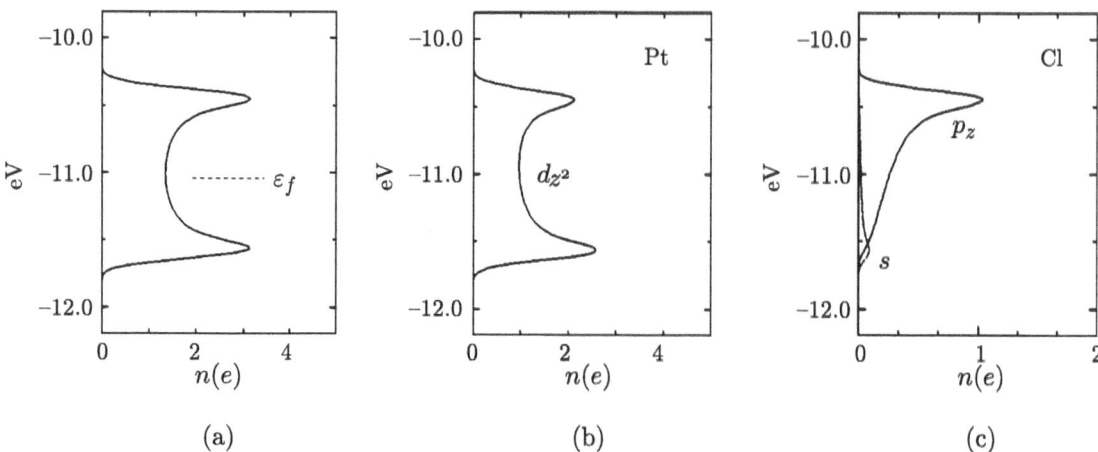

Fig. 10.11
DOS in the region of the z^2 bands for the uniform chain of Fig. 10.8b: (a) Total DOS, (b) projected DOS for the Pt z^2 orbital and (c) projected DOS for the z (continuous line) and s (dotted line) orbitals of the bridging chlorine. In (a) the dotted line refers to the Fermi level.

the stronger mixing of the Cl z orbital at the upper part of the band which, through the normalisation condition, decreases the Pt z^2 contribution.

Since the electron count for the platinum is d^7, the z^2 band is half-filled (the Fermi level is shown as a dotted line in Fig. 10.11a) and consequently, as discussed in Section 8.2.7, the chain is susceptible to a Peierls distortion towards a dimerised structure such as that of Fig. 10.8a. Without knowing any detail of the band structure we can say that when two chlorine atoms approach the Pt atoms (Fig. 10.8a), the z^2 levels of the Pt atoms with the shorter Pt–Cl bond lengths will be strongly destabilised while those of the other platinum atoms will be stabilised. Consequently, the DOS of Fig. 10.11a will split into two contributions. The lower one will mostly contain the z^2 levels of the Pt atom with long Pt–Cl bond lengths while the opposite will be true for the upper one. Since in the dimerised chain there are two Pt atoms per unit cell, there are two electrons to fill the z^2 levels and consequently only the lower contribution will be full. This means that under dimerisation, two different types of platinum atoms have been generated, Pt^{IV} and Pt^{II}, i.e. the

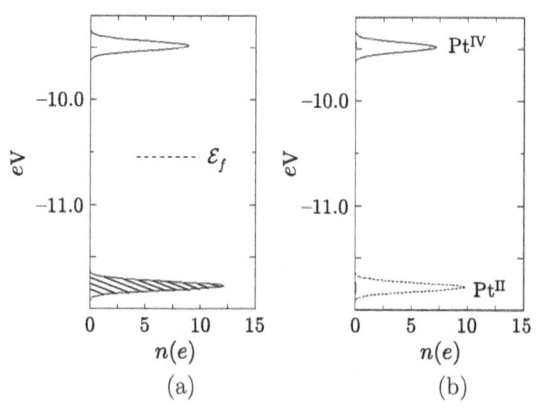

Fig. 10.12
DOS in the region of the z^2 bands for the dimerised chain of Fig. 10.8a: (a) total DOS, (b) projected DOS for Pt^{IV} (continuous line) and Pt^{II} (dotted line).

dimerisation leads to a mixed valence situation. This qualitative reasoning is supported by the calculated results in Fig. 10.12. That this dimerisation is a stabilising process can be seen by comparing Figs 10.11a and 10.12a, i.e. the upper filled levels are lower in energy in the distorted chain (Fig. 10.12a). Thus, the DOS plots of Figs 10.11 and 10.12 clearly tell us the chemical origin of the Peierls distortion in the Wolfram salt: creation of a mixed valence situation that is energetically preferred.

10.4 Density of states and fragment molecular orbital interaction analysis: application to the [(C₅H₅)M] chains

A very fruitful approach that is used by chemists to understand molecular structure and reactivity problems is that of the fragment molecular orbital (FMO) analysis of the wavefunction. [11] This method relies on the analysis of the molecular orbitals of a system in terms of the molecular orbitals of the chemically meaningful constituent fragments. It provides a simple yet very powerful way to relate the electronic and geometric structures of molecules. It is therefore a natural choice for chemists trying to understand why a certain solid has the physical properties or the crystal structure it has. The approach has been indeed very useful in understanding the chemistry and physics of solid state materials. A word of caution is, however, in order. Many solids can be seen as being formed through the condensation of an elementary building block, which by itself is a well-defined chemical unit. This building block thus provides a convenient way to describe the crystal structure of the material. However, because of the condensation it is not possible to perform an FMO analysis of the electronic structure. This problem is frequently found in solids and somewhat limits the usefulness of the FMO approach in the solid state. Nevertheless, there are many systems to which this approach is applicable. The ML_5 chain studied in the previous section, the MPS_3 phases (M = Fe, Co, Ni, Mn, Cd) [8] and the MMo_6X_8 Chevrel phases (M = Pb, Sn, Ba, Cu, Li and X = S, Se, Te) [12] are 1D, 2D and 3D examples, respectively. In this section we will study the $[(C_5H_5)M]$ chain (see Fig. 10.13a) to illustrate the FMO analysis of the DOS and we will compare this with a similar analysis of the $(C_5H_5)_2M$ molecule (see Fig. 10.13b).

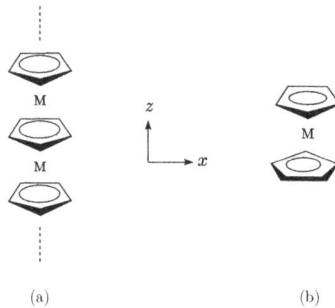

Fig. 10.13
Schematic representation of a $[(C_5H_5)M]$ chain (a) and a $(C_5H_5)_2M$ molecule (b). The coordinate axis used in the orbital analysis is shown.

The important orbital interactions to consider in both cases are those between the d orbitals of the metal and the π-type orbitals of C_5H_5 (see Fig. 10.14). Fig. 10.15 is a schematic interaction diagram for the $(C_5H_5)_2M$ molecule of Fig. 10.13b. Every π orbital of C_5H_5 generates two symmetry-adapted orbitals of the fragment $(C_5H_5)_2$, as shown on the left of Fig. 10.15. In the D_{5d} group of the molecule, the five d orbitals of the metal belong to the symmetries $a_{1g}(z^2)$, $e_{1g}(xz,yz)$, and $e_{2g}(x^2-y^2, xy)$. The metal z^2 orbital remains almost non-bonding, even if it has the correct symmetry to interact with one of the proper combinations of π_1 because its conical nodal surface points directly towards the π system of the C_5H_5 moieties. The pair of in-plane x^2-y^2 and xy (e_{2g}) orbitals are somewhat stabilised by one pair of combinations of the π_4 and π_5 orbitals. In contrast, the two out-of-plane xz

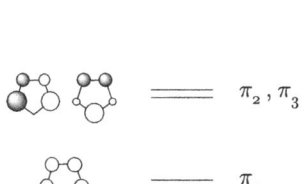

Fig. 10.14
Schematic representation of the π-type orbitals of a cypotentadienyl (C_5H_5) ligand.

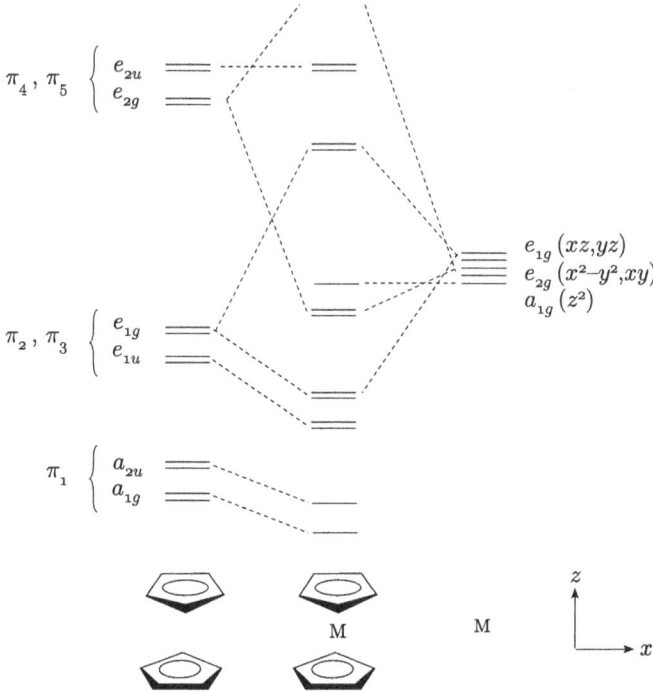

Fig. 10.15
Schematic interaction diagram for a $(C_5H_5)_2M$ molecule.

and yz (e_{1g}) orbitals interact very strongly with one pair of combinations of the π_2 and π_3 orbitals and are strongly destabilised. Thus, three metal orbitals have been left low in energy and the $(C_5H_5)_2M$ molecule will be very stable for a d^6 metal, i.e. $(C_5H_5)_2Fe$.

Let us now consider the DOS of the [$(C_5H_5)M$] chain (see Fig. 10.13a). [13] It should be noted that although the appropriate symmetry group at the centre and border of the Brillouin zone is D_{5h}, it is lower for a general k point inside the Brillouin zone (C_{5v}). Within this group the symmetry labels appropriate for the metal orbitals are $a_1(z^2)$, $e_1(xz, yz)$, and $e_2(x^2 - y^2, xy)$. Those for the π_1, (π_2, π_3), and (π_4, π_5) orbitals of C_5H_5 are a_1, e_1, and e_2, respectively. From now on we will refer to these orbitals as $\pi(a_1)$, $\pi(e_1)$, and $\pi(e_2)$. The calculated DOS for the infinite chain [$(C_5H_5)M$] as well as the projected DOS for the orbitals $\pi(e_1)$, $(x^2 - y^2, xy)$, z^2, (xz, yz), and $\pi(e_2)$ are shown in Fig. 10.16, where the $\pi(e_1)$ and $\pi(e_2)$ contributions are given in the same panel. Notice how the contributions of the interacting e_1 orbitals, $\pi(e_1)$ and (xz, yz), are split into two regions. The lower-energy orbital $\pi(e_1)$ is more heavily weighted in the lower-energy contribution and the higher-energy orbital (xz, yz) is more heavily represented in the upper-energy contribution. This is the equivalent of our understanding of the interaction of two non-degenerate orbitals in molecules. The contribution of these interacting orbitals extends over a considerable energy range because the two types of orbitals are well oriented to interact, leading to dispersive energy bands. As mentioned, the z^2 orbital does not interact well with the $\pi(a_1)$ ligand orbital. Consequently, the projected contribution of this orbital is very localised. The $\pi(a_1)$ DOS

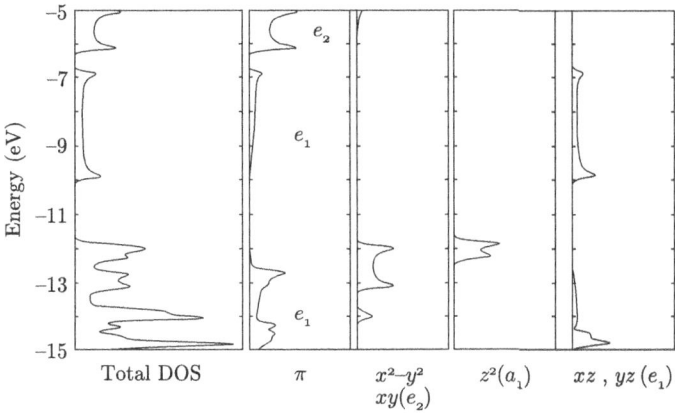

Fig. 10.16
DOS and projected DOS for the orbitals $\pi(e_1)$, $\pi(e_2)$, $(x^2 - y^2, xy)$, z^2, and (xz, yz) calculated for the $[(C_5H_5)M]$ chain where M = Fe.

lies at the bottom of Fig. 10.16, overlapping with other peaks. The $x^2 - y^2$ and xy orbitals (e_2) can interact with the ligand $\pi(e_2)$ ones and both types of orbital appear split into two contributions. However, the interactions are less effective than those of the xz and yz orbitals, because the $x^2 - y^2$ and xy orbitals are not directed towards the π orbitals of the ligand. As shown in Fig. 10.16, the mixing is weaker and the dispersion of the corresponding bands is smaller for the $(x^2 - y^2, xy)$ and $\pi(e_2)$ orbitals than it was for the (xz, yz) and $\pi(e_1)$ orbitals. With three metal bands left low in energy, the $[(C_5H_5)M]$ chain should be stable for a d^6 metal, i.e. $[(C_5H_5)Mn]$, and all levels should be filled up to the band gap around -11 eV. The main conclusions of this analysis are equivalent to those of the previous paragraph concerning the $(C_5H_5)_2M$ molecule. Thus, the use of projected FMO DOS is a very powerful tool for the analysis of the electronic structure of those complex solids to which this approach is applicable and to help in establishing a link between our understanding of the electronic structure of molecules and of solids.

The examples discussed up to now illustrate different ways to analyse the DOS of a material. For some of these cases, such as the ML_5 or $[(C_5H_5)M]$ chains, similar conclusions can be reached by looking at the band structure instead of the DOS. The analysis of the DOS is, however, simpler for KMo_6O_{17} or the MPS_3 phases, where many bands overlap. The choice to analyse the band structure or the DOS is largely dictated by the nature of the problem at hand. Usually, the DOS is the more convenient choice when looking at structural problems (especially for 2D or 3D materials – see later for several examples) or molecule–surface interactions. Examining the band structure can be a more appropriate choice when trying to understand the transport properties of a solid, because these properties are largely determined by a very limited number of bands, i.e. those very near the Fermi level. More often the two approaches are complementary. For instance, to understand the origin of the resistivity anomaly of KMo_6O_{17} we need to understand the detailed shape of the bands cut by the Fermi level, i.e. the lower t_{2g}-block bands. The DOS analysis leads us to concentrate on the inner octahedral double layer in order to qualitatively build these bands, and thus it considerably simplifies the problem. In a similar vein, when looking at a structural problem for a complex material, use of projected DOS and COOP curves usually allows a simple argument to be made based on selected contributions to the DOS. A more detailed understanding of the shape of these contributions and thus of the correlation between the fine details of the crystal and the electronic structures requires an analysis of the band structure in the energy range in which these contributions lie.

10.5 Transition metal diborides with the AlB$_2$ structure type: a 3D case study

The basic principles of the DOS analysis of the electronic structure of solids have already been discussed in the previous sections. Here we will consider an example of a real 3D material, so that the reader gains confidence and practice in the use of these tools. We believe that this example illustrates in simple terms the strategy that is followed in real research work. Any structure type calls for the rationalization of its main structural features in simple terms. The interested reader has a huge number of problems to consider with the help of these tools, along with computational code that enables their calculation.[4] Every structure type will demand a particular type of analysis and a catalogue of these is outwith the scope of this book. We urge the reader to practise the approach, carrying out their own DOS analyses either on materials already analysed in the literature or for other materials that are of interest to them. Let us remind ourselves that all the numerical results discussed in this book have been obtained using the extended Hückel approach, allowing them to be readily and quickly reproduced by the readers. However, the tools discussed in this chapter can equally well be obtained using more sophisticated first-principles approaches. [2]

We will consider here a particular structural aspect of transition metal diborides, many of which adopt the AlB$_2$ structure. [14] This structure can be regarded as built from graphite-like layers of boron in which the transition metal atoms occupy all the hexagonal prismatic sites (see Fig. 10.17a). This description is a convenient one because it is simple and describes the more relevant structural features, but it is just that: a simple geometrical description. Based on this description, one is tempted to assume that the transition metal transfers two electrons to the boron sheets so that these sheets become isoelectronic with those in graphite and the transition metal is thus M^{2+}. In practice, however, the stability of the structure suggests that there must be very sizeable transition-metal–boron bonds and the system can therefore more

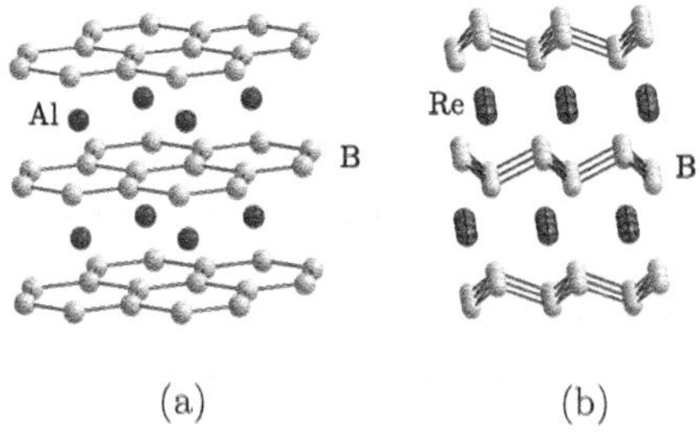

[4]For this purpose, an extended Hückel type approach is very convenient since it allows very fast calculation of the DOS, COOP, etc., for even very complex systems. Despite the qualitative nature of the approach, it is invaluable in providing simple explanations for structural trends.

Fig. 10.17
Crystal structure of AlB$_2$ (a) and ReB$_2$ (b).

usefully be described as a 3D network of covalently linked atoms. Transition metal diborides with the AlB_2 structure are known for the first-row elements, from scandium to manganese. For the second and third-row transition metal series a changeover in structure occurs between groups 6 and 7 and between 7 and 8. For instance, the structure of ReB_2 is shown in Fig. 10.17b. Although the RuB_2 structure is slightly different, it shares the main structural variations: puckering of the boron sheets and displacement of the transition metals in such a way that the metal–metal distances along the direction perpendicular to the sheets increases. Thus, an interesting line of inquiry is to understand why what seems to be a very stable structure has a tendency to lose the planarity of the boron sheets for transition metal atoms of groups 7 and 8. In addition, there are some structural trends of the AlB_2 structure that are interesting to address. For instance, going from scandium to manganese, the boron–boron distances decrease, whereas transition-metal–boron distances first decrease and then seem to slightly increase. Another interesting question is which of the two descriptions – graphite-like sheets kept together through electrostatic interactions with the transition metal cations, or a 3D network of covalent interactions – better captures the essence of the bonding in these phases.

In studying problems like this it is convenient to use a *rigid band* approach: the electronic structure of a given compound of the series is studied and the variation of the transition metal atom is simulated by appropriately changing the number of electrons of the system. Although simplistic, this approach provides simple guidelines for understanding how the electron filling of the bands affects the strength of the different types of bonds for a given structure. In any case, it is safe to say that the main *trends* obtained from this analysis are not sensitive to the specific compound used for the actual calculations. Let us use this approach to find a rationale of the puckering tendency noted above. [15]

In Fig. 10.18 we show the calculated DOS for a particular member of this family, TiB_2, as well as the projected DOS for the (xz, yz), $(x^2 - y^2, xy)$, and z^2 orbitals of the transition metal. The calculated Fermi levels for the different first-row transition metal diborides in the rigid band approximation are also shown. Fig. 10.19 shows the calculated COOP curves for the M–B and M–M interactions in the plane, the M–M interaction perpendicular to the boron planes, and the B–B interaction. The results shown are from a calculation on TiB_2.

Simple inspection of these curves suggests many useful ideas. For instance, note that the Fermi level for atoms of the groups 7 and 8 lies at a peak of the DOS. When the projected DOS of Fig. 10.18 are examined, it is clear that this peak is strongly based on the z^2 orbitals of the transition metal. This simple observation suggests that the noted instability of the MB_2 phases for these electron counts may be tied to these orbitals. Before pursuing this idea let us examine how this simple approach deals with the main structural trends. Looking at the COOP curves of Fig. 10.19, one can deduce that the overlap population for the B–B bonds must continuously increase with electron filling. This may appear surprising at first sight. One of the interesting features of transition metal borides is that the transition metal d levels lie between the boron s and p orbitals. When these levels broaden into the bands of the planar

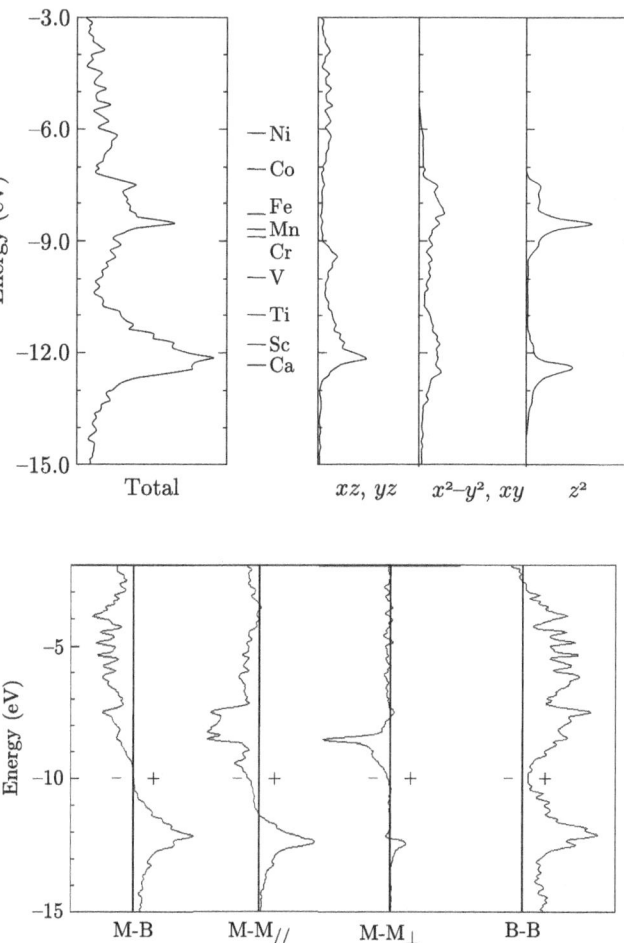

Fig. 10.18
Electronic DOS for MB$_2$ phases with the AlB$_2$ structure and the decomposition of the total DOS into contributions from the (xz, yz), $(x^2 - y^2, xy)$, and z^2 orbitals of the transition metal. The Fermi levels corresponding to various first-row metals in the rigid band approximation are also shown. The z^2 contribution in the right-hand panel has been multiplied by a factor of two.

Fig. 10.19
COOP curves for the four different types of interaction in MB$_2$ phases with AlB$_2$ structure.

sheets, the transition metal levels lie in a region of the energy band that is boron–boron bonding. As a consequence the boron–boron overlap population never becomes negative for the region of chemical interest. In other words, the B–B bonds should become shorter as we move from left to right in the transition metal series since we are filling more and more levels with a positive contribution to the overlap population. Despite the simplicity of the approach, this trend provides a good description of the experimental situation. The distances found experimentally are 1.816 Å (Sc), 1.748 Å (Ti), 1.727 Å (V), 1.714 Å (Cr), and 1.736 Å (Mn). If we now look at the transition-metal–boron COOP curve we see that the associated overlap population should increase from ScB$_2$ to VB$_2$ and then start to decrease for CrB$_2$, since antibonding states would then be filled. Thus according to this rigid band approach the transition-metal–boron bonds should first decrease and later should increase again. This seems also to describe the experimental observations very well, the M–B distances being as follows: 2.53 Å (Sc), 2.38 Å (Ti), 2.30 Å (V, Cr), and 2.31 Å (Mn). Since the calculations in Figs 10.18 and 10.19 were carried out for TiB$_2$ it is understandable that they describe the situation better for transition metals on the left of the series than for those on the right. The calculated values of the metal–boron overlap populations are of the order of 0.20, which is large and confirms that there are strong covalent interactions between the transition metal and boron atoms. Other features, such as the trend in heats of formation for diborides of the first transition row, are also well accounted for by this kind of simple procedure. [15] It is quite remarkable that such a simple approach reflects so well the structural features of these phases.

For our purpose, the important feature of Figs 10.18 and 10.19 is, however, that the Fermi level corresponding to transition metals of groups 7 and 8 hits a peak of the DOS and that this peak is strongly associated with the z^2 orbitals of the transition metal. When looking at the COOP curves for the metal–metal interactions along the direction perpendicular to the boron sheets, it is quite clear that the peak of the DOS is associated with antibonding metal–metal interactions. It therefore seems quite understandable that for atoms of groups 7 and 8 there is a tendency to undergo a distortion that eliminates these destabilising interactions. Thus a very simple calculation has given an invaluable rationalisation of one of the important structural features of these phases. At this point one may be tempted to think that the problem is solved. After all, the z^2 orbitals point directly towards those of the next transition metal layer and the fact that there are two peaks in the z^2 projection (see Fig. 10.18), the lower one being associated with bonding levels and the upper one with antibonding ones (see the corresponding COOP curve in Fig. 10.19), may be taken as an indication of the formation of direct bonding and antibonding levels. However, this would in fact be incorrect because we are dealing with an infinite system and thus the direct interaction of z^2 orbitals should lead to a relatively wide band, with the z^2 orbitals spread out over a wide energy range. This being the case, the change from bonding to antibonding should be very gradual – a sudden changeover would not be expected. If we want to have a clear view of the origin of the structural change we must then understand why the z^2 contributions appear as two very sharp peaks (i.e. originating from narrow bands). Without understanding this feature we only have a partial explanation: we must turn our attention to the band structure.

In Fig. 10.20a we show the band structure calculated along the line $\Gamma \to$ A of the hexagonal Brillouin zone appropriate for the AlB$_2$ structure (see Fig. 10.21). This is the direction in reciprocal space (c^*) where energetically interesting and important interactions occur for these structures and we must therefore explore the character of the crystal orbitals along this line. These are the interactions with the boron nearest-neighbour ring orbitals and the associated through-ring metal–metal interaction. At the Γ point, orbitals that are symmetric with respect to the plane of the transition metal sheets cannot mix with the in-phase combination of p_z orbitals of boron appropriate for this k point. As a result, the higher-energy crystal orbital labelled a_1 in Fig. 10.20a is simply made up of the boron π-type orbitals stabilised with some admixture of the p_z orbital of the transition metal. At the top of the line, the A point, this orbital has become an out-of-phase combination along the c-direction of transition metal s-d hybrids. The sense of the hybridization is such that the hybrid is well directed along c. The amount of z^2 character increases along the

Density of states

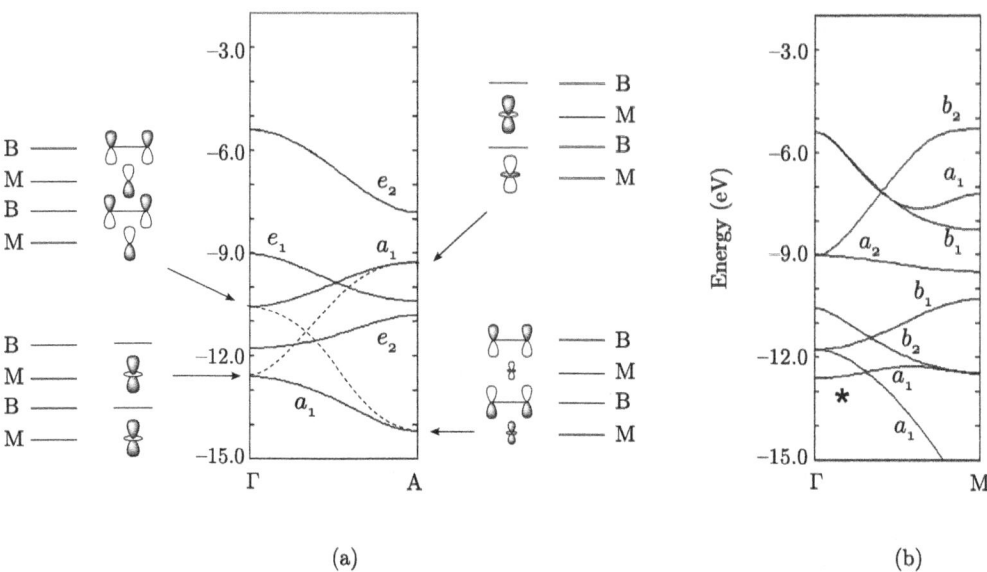

(a)   (b)

Fig. 10.20
Band structure calculated for the MB_2 systems along the (a) $\Gamma \to A$ and (b) $\Gamma \to M$ directions of the hexagonal Brillouin zone (see Fig. 10.21). The asterisk refers to the crystal orbital at Γ which is strongly based on the z^2 orbital.

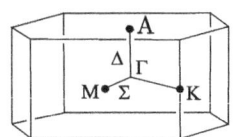

Fig. 10.21
Brillouin zone appropriate for the hexagonal AlB_2 structure.

$\Gamma \to A$ line and reaches a maximum at the A point. At this point, because of the enforced phase relationships from cell to cell in this direction, the crystal orbital is antibonding between the metal atoms perpendicular to the sheets. The energy of the crystal orbital is therefore controlled by these direct metal–metal antibonding interactions. Note that the energy of this crystal orbital coincides with the energy around the upper peak in the z^2 projection in Fig. 10.18. The lower energy z^2 band of Fig. 10.20a is purely metal, and located at Γ. The enforced cell-to-cell phase relationships at this point mean that the orbital is bonding between the metal atoms along the c-direction. Note that the energy of this crystal orbital coincides with the energy around the lower peak in the z^2 projection in Figure 10.18. The z^2 character in this orbital decreases on moving along the $\Gamma \to A$ line and in fact at A there is a large component of boron p_z orbitals. What we are describing here is a strongly avoided band crossing between the two bands, one based on the z^2 orbitals of the transition metal and another based on π-type levels of the boron sheets. This is indicated with dotted lines in Fig. 10.20a. In this figure we have also included symmetry labels appropriate for this line of the Brillouin zone. Thus on the basis of this crystal orbital analysis we can understand that the z^2 contributions may appear as two narrow and well-separated DOS peaks.

However, the proof is still not complete. We still need to see if the interactions along the perpendicular directions are very weak, thus leading to narrow bands. The band structure along one of these directions (the Σ line) is shown in Fig. 10.20b. The crystal orbital at Γ, which is strongly based on the z^2 orbital, is marked with an asterisk. Although near Γ, it mixes with what was the component of a doubly degenerate band at Γ, which strongly decreases in energy along Σ. The result is clear to see: the z^2 band occurs over a rather narrow energy range. This is also true for other lines perpendicular to the Δ line. Thus the rather sharp features associated with the z^2 projection of Fig. 10.18 are quite understandable.

Although a little elaborate, the crystal orbital analysis above substantiates our initial suspicion that metal–metal interactions perpendicular to the boron sheets lie at the origin of the structural change in transition metal diborides of groups 7 and 8. What this analysis has also made clear is that direct metal–metal interactions are by no means the key factor in explaining the origin of this structural change. The strongly avoided crossing and consequent localization of the z^2 contributions that were seen in the DOS and which were discussed above, are clearly telling us that the origin lies in the through-boron metal–metal interaction. In other words, a blend of metal–metal and metal–boron interactions lead to the structural changeover occurring in this structure.

The role of DOS analysis has been essential in understanding this problem. It has been a simple task to develop a crystal orbital analysis and a simple rationalisation, simply because DOS analysis through the different orbital projections and COOP curves tells us, in a very simple and efficient way, what to look for.

Exercises

(10.1) Suggest a general expression for the gross population of an atom A in: (a) a polyatomic molecule and (b) a 1D system with several atoms in the unit cell.

(10.2) Explain the origin of the dimerization of the NbX_4 chains discussed in Exercise (8.3) from the viewpoint of the DOS.

(10.3) There are many AB_2X_2 compounds, where A is typically a rare-earth, alkaline-earth or alkali element, B is a transition metal or main group element, and X is an atom of the groups 14 or 15 (and only occasionally 13), that exhibit the $ThCr_2Si_2$ structure shown below. One of the many interesting structural features exhibited by these phases is the variation in the X–X interlayer distance as a function of the electron count. The X–X distance ranges from those corresponding to no bond between the two atoms to those typical for a fully formed X–X bond. In general, X–X bonding is favoured for those phases containing a transition metal from the right-hand side of the periodic table. For instance the P–P distance in the series of compounds $CaFe_2P_2$, $CaCo_2P_2$, $CaNi_2P_2$, and $CaCu_{1.75}P_2$ is: 2.71, 2.45, 2.30, 2.25, respectively. Typical P–P single bond distances in molecules range from 2.19 to 2.26 Å. Provide an explanation of this structural feature on the basis of a DOS analysis for a typical compound of this series, such as $BaMn_2P_2$ (see reference [16] for computational details and reference [17] for structural details).

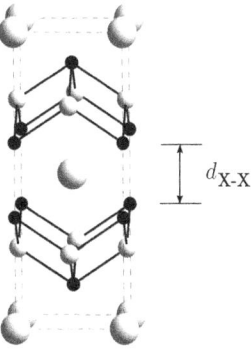

(10.4) Sketch the DOS for a ladder of hydrogen atoms where the interaction along the direction of the ladder is β and the interaction perpendicular to the layer is $k\beta$.

(10.5) To acquire some practice in the use of the DOS, PDOS and COOP tools in analysing the electronic structure of extended systems, the reader is advised to look at some of the following references: (i) J. Silvestre and R. Hoffmann, C_2H_n *Fragments on Metal Surfaces*, Langmuir, 1, 621 (1985), (ii) T. Hughbanks and R. Hoffmann, *Chains of Trans-Edge-Sharing Molybdenum Octahedra: Metal–Metal Bonding in Extended Systems*, J. Am. Chem. Soc., 105, 3528 (1983), (iii) J. K. Burdett and T. Hughbanks, *NbO and TiO: Structural and Electronic Stability of Structures Derived from Rock Salt*, J. Am. Chem. Soc., 106, 3101 (1984), (iv) S.-S. Shung and R. Hoffmann, *How Carbon Monoxide Bonds to Metal Surfaces*, J. Am. Chem. Soc., 107, 578 (1985) and (v) J. K. Burdett and B. A. Coddens, *Geometrical-Electronic Relationships in the Series PdP_2, PdPS, and PdS_2*, Inorg. Chem., 27, 418 (1988).

References

1. N. W. Ashcroft and N. D. Mermin, *Solid State Physics*, Holt, Rinehart and Winston, Philadelphia, 1976.
2. R. Dronskowski, *Computational Chemistry of Solid State Materials*, Wiley-VCH, Weinheim, 2005.
3. R. S. Mulliken, *J. Chem. Phys.*, 23, 1833, 1955.
4. *Low-Dimensional Electronic Properties of Molybdenum Bronzes and Oxides*, C. Schlenker. Ed., Kluwer, Dordrecht, 1989.
5. T. R. Hughbanks and R. Hoffmann, *J. Am. Chem. Soc.*, 105, 3528, 1983.
6. R. Hoffmann, *Solids and Surfaces: A Chemist's View of Bonding in Extended Structures*, VCH, New York, 1988.
7. F. Hulliger, *Structural Chemistry of Layer-Type Phases*, Reidel, Dordrecht, 1976, pp. 217–9.
8. For an analysis of the orbital interactions occurring in the MPS_3 phases, see: (a) M.-H. Whangbo, R. Brec, G. Ouvrard, and J. Rouxel, *Inorg. Chem.*, 24, 2459, 1985; (b) H. Mercier, Y. Mathey, and E. Canadell, *Inorg. Chem.*, 26, 963, 1987.
9. H. J. Keller, *Extended Linear Chain Compounds*, J. S. Miller, Ed., Plenum Press, New York, 1982, vol. 1, Chapter 8.
10. For an analysis of the orbital interactions occurring in the Wolfram red salt, see: M.-H. Whangbo and M. J. Foshee, *Inorg. Chem.*, 20, 113, 1981.
11. T. A. Albright, J. K. Burdett, and M.-H. Whangbo, *Orbital Interactions in Chemistry*, John Wiley, New York, 1985.
12. T. R. Hughbanks and R. Hoffmann, *J. Am. Chem. Soc.*, 105, 1150, 1983.
13. J. K. Burdett and E. Canadell, *Organometallics*, 4, 805, 1985.
14. A. F. Wells, *Structural Inorganic Chemistry*, 4th edn, Oxford University Press, Oxford, 1975.
15. J. K. Burdett, E. Canadell, and G. J. Miller, *J. Am. Chem. Soc.*, 108, 6561, 1986.
16. R. Hoffmann and C. Zheng, *J. Phys. Chem.*, 89, 4175, 1985.
17. A. Mewis, *Z. Naturforsch.*, 35B, 141, 1980.

Fermi surface and low-dimensional metals

The Fermi surface is an extremely useful concept which, when appropriately decoded, contains much information about the transport and structural properties of systems with partially filled bands. [1, 2] However, it does suffer from the disadvantage that it is a concept in reciprocal space whereas most chemists are used to reasoning in real space. This difference is the origin of the reluctance of many solid state chemists to consider this useful construct in their everyday work. As considered in detail in most of this book, extracting the chemical or physical meaning from crystal orbitals may be a little more lengthy but no more difficult than extracting useful information from the molecular orbitals of a discrete molecule or cluster. We will now try to show that the essential aspects of the Fermi surface of a given metal may be obtained in a relatively simple way using the orbital approach developed in previous chapters so that there is no reason for such reluctance on the part of chemists.

The Fermi surface has been a very powerful tool in the study of low-dimensional metals. [3] This important class of materials were predicted to exhibit very different behaviour from the usual, more isotropic, 3D metals. For some time this prediction could not be confirmed because most of these systems did not exist in nature and had to be synthesised. The preparation and study of many of these materials in the last few decades has fully confirmed the initial expectations. The Fermi surface has provided a very simple and versatile concept to understand the peculiar behaviour of many of these solids. Simple ideas like the relationship between the nature of the Fermi surface and electron counting, or the way in which the crystal structure and the Fermi surface are related, have been important tools in guiding and understanding research in this field.

Regardless of their structural complexity, only very few bands of any solid are partially filled. Consequently, in understanding the detailed shape of the Fermi surface we must focus on a small number of bands. This is what makes the study of the Fermi surface relatively straightforward. In contrast, low-dimensional metals quite frequently exhibit large and complex unit cells. Thus, in order to understand the behaviour of these systems, it is important to be able to single out the parts of their crystal structure that are essential for the description of their Fermi surface. Only after selection of this essential part of the complete structure is it possible to proceed to the qualitative building-up of the dispersion of the reduced number of bands of interest, i.e. the band structure around the Fermi level and the Fermi surface.

In this chapter we will illustrate this procedure for several classes of low-dimensional materials: transition metal oxides and chalcogenides as well as molecular conductors. These examples have been chosen not only to introduce the interesting and fast-developing field of low-dimensional metals but also to give the reader confidence in the qualitative construction of the detailed band structure of real complex materials. Thus, it should be a useful complement to the ideas developed in the previous chapter, which were based on the analysis of the full band structure through the density of states.

11.1 Notion of Fermi surface

Let us consider the case of a rectangular 2D system, with one s orbital and one electron per site, and where the interactions along the b-direction are weaker than those along the a-direction, i.e. $\beta_a < \beta_b \sim 0$ (see Fig. 11.1a). The band structure for this system can be obtained analytically in the nearest-neighbour approximation:

$$E(\vec{k}) = \alpha + 2\beta_a \cos(2\pi k_a) + 2\beta_b \cos(2\pi k_b) \qquad (11.1)$$

or qualitatively by simply considering the shape of the Bloch orbitals at the Γ, X, Y and M special points. This band structure is shown in Fig. 11.1b. Since there is one electron per site, the band is only half-filled and half of the k points of the rectangular Brillouin zone will correspond to filled levels and half to empty levels. As all k points of the Brillouin zone are equally allowed, the area of the Brillouin zone corresponding to the occupied states will be exactly half of the full Brillouin zone (see Fig. 11.1c, where the region of the filled levels is hatched). The Fermi surface is defined as the boundary dividing the region of the filled and empty k points and occurs as a surface (or series of surfaces) in a 3D representation, as a line (or a series of lines) in a 2D representation

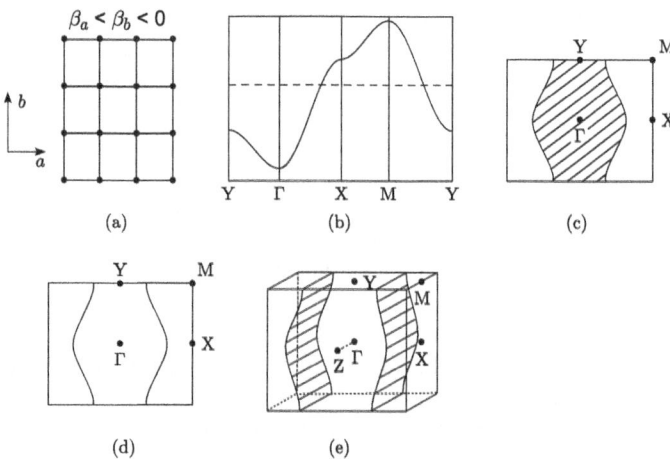

Fig. 11.1
(a) Schematic representation of a 2D rectangular network with stronger interactions along the a-direction; (b) band structure for the network in (a) where the dashed line refers to the Fermi level; (c) the region of the Brillouin zone corresponding to filled/empty levels is shown as hatched/non hatched; (d) Fermi surface of the system; (e) 3D representation of the Fermi surface.

and as a pair of points (or a series of pairs of points) in a 1D representation. In the present 2D case, the Fermi surface is the pair of warped lines of Fig. 11.1d which, in the full 3D representation, is a pair of warped planes as shown in Fig. 11.1e.

As shown in Fig. 11.2, depending on the relative value of the two βs, there are different possibilities for the band structure of this 2D model system with one electron per site and, of course, the shape of the Fermi surface also differs (see Fig. 11.3). Case (a) corresponds to a perfect 1D system and the Fermi surface would then be a pair of parallel lines in a 2D representation but a pair of parallel planes in the full 3D representation (see Fig. 11.4a). Cases (b) and (c) correspond to pseudo-1D systems. Since the interactions about one of the directions is no longer nil, the lines become warped (see Fig. 11.3b and c). The warping increases as the ratio between the absolute values of the βs (i.e. the relative strength of the interactions) decreases. Case (d) corresponds to the situation where the two interactions have the same strength. Under the nearest-neighbour approximation, the Fermi surface becomes a rhombus touching the X and Y points (see Fig. 11.3d; the rhombus becomes rounded near the region

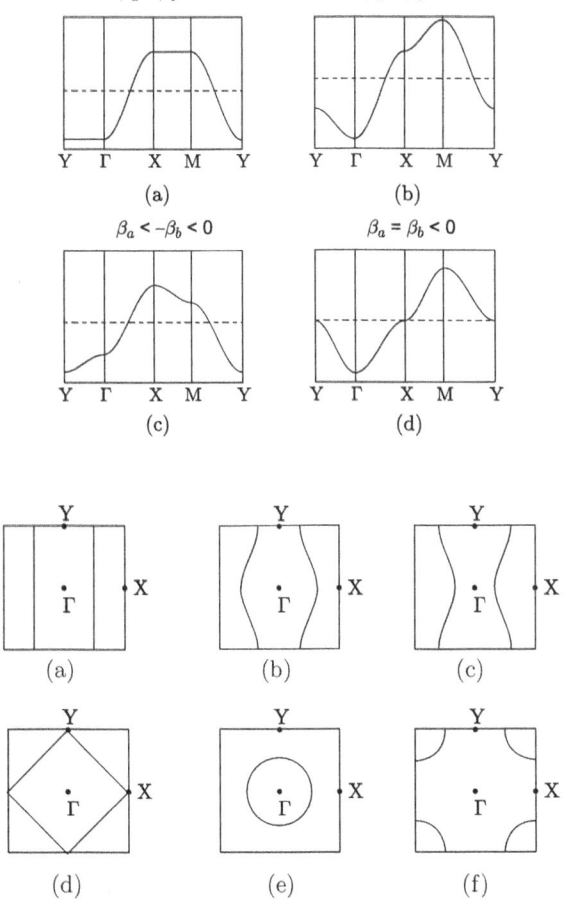

Fig. 11.2
(a)–(d) Different band structures for a 2D rectangular system with one orbital and one electron per site, depending on the strength of the two types of interaction.

Fig. 11.3
(a)-(f) Different Fermi surfaces for the 2D rectangular network depending on the strength of the two types of interaction and the filling of the band.

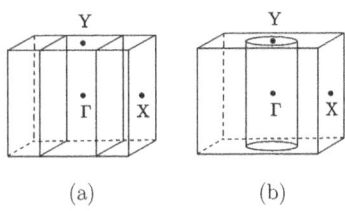

Fig. 11.4
3D representation of the Fermi surfaces of Fig. 1.3a (a) and 11.3e (b).

of X and Y in a non-nearest-neighbour approach). When the system is not exactly half filled, the Fermi surface for cases (a)–(c) essentially does not change in shape but the pairs of lines approach or separate if the number of electrons in the partially filled band decreases or increases, respectively. In case (d) the Fermi surface becomes the one shown in Figs 11.3e and 11.3f when there is a decrease or increase of electrons in the partially filled band, respectively. Although at first sight these two Fermi surfaces may appear different, their shape is identical because of the periodic nature of reciprocal space. The Fermi surface of Fig. 11.3e becomes a cylinder in the full 3D representation (see Fig. 11.4b). Fermi surfaces associated with less than half-filled bands or more than half-filled bands, such as those in Figs 11.3e and 11.3f respectively, are usually referred to as hole and electron Fermi surfaces.

The Fermi surface of a metal governs the conductivity of the system. As shown in Chapter 1, for the case of a 1D system in a free electron treatment, in the absence of an electric field the number of electrons moving towards the right is exactly the same as those moving towards the left (i.e. the number of electrons having positive values of k is the same as those having negative values of k) so that there is no net flux of electrical charge. When a constant and uniform electric field is applied, the electrons experience a force and move in the opposite direction. When the electrons reach a new stationary state because of the interaction with the underlying lattice, the electrons associated with a given value of k have acquired a new value of k equal to $k + \delta k$, where δk is proportional to the electric field (see Section 1.1.1). Thus all electrons experience an increase of their wave vector by the same amount. As schematically shown in Fig. 1.9a, this avoids the compensation in the number of electrons having positive and negative values of k. Thus, there is a net transport of electrons. Obviously, completely filled bands cannot contribute to the conductivity since it is not possible to have an imbalance between positive and negative values. Exactly the same argument holds for the 3D case (see Section 1.1.2), and we can therefore think about the application of a constant and uniform electric field as a rigid displacement of the Fermi surface that prevents a compensation of the number of electrons having positive and negative components of k in the direction of the applied field leading to transport of charge in that direction.

The analysis in Chapter 1 used the free electron model, but exactly the same argument holds for a *real* Fermi surface. In other words, application of an electric field along a given direction results in a translation of the Fermi surface as a *rigid body* along the opposite direction. For instance, application of an electric field along a in the case of a rectangular lattice with $\beta_a < \beta_b = 0$, i.e. the case of Fig. 11.3a, leads to a shift of the two lines (or planes in the 3D representation of Fig. 11.4a) along a and thus there is a net flow of charge along that direction. In contrast, if the electric field is applied along b there is no imbalance in the number of electrons having positive and negative components of k along this direction, and there is no transport of charge in that direction. Obviously, the same reasoning holds for direction c. This is a perfect 1D metal. In contrast, for a system with a closed Fermi surface like that shown in Fig. 11.3e (or in the 3D representation of Fig. 11.4b), for whatever direction of the applied field in the (a,b) plane, there will be a net shift of the circle (cylinder in the 3D representation), and the system will exhibit metallic behaviour for any direction of the (a,b) plane. This is an example of a 2D metal. Again, no transport will occur when the field is applied along c.

The situation represented in Figs 11.3b and 11.3c is intermediate and corresponds to a pseudo-1D metal, with the best conductivity along the a-direction. Thus all Fermi surfaces represented in Fig. 11.3 are associated with different degrees of anisotropic metallic behaviour. A Fermi surface with spherical shape would be an example of a perfectly isotropic 3D metal. As shown in Fig. 11.5 for copper, a typical 3D metal, most Fermi surfaces corresponding to real materials can have shapes that are not as ideal as those discussed above. The ways in which the real Fermi surface differs from these ideal ones may have important consequences, and details of the Fermi surface may be the source of interesting anomalies or unexpected properties. This is why it is important to be able to correlate the crystal structure and the shape of the Fermi surface. Let us also mention that experimental studies, most frequently based on magnetoresistance or photoemission measurements, can bring very valuable information concerning the Fermi surface.

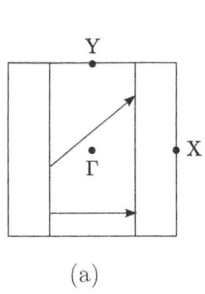

Fig. 11.5
3D Fermi surface for copper.

11.2 Nesting vector and electronic instabilities in low-dimensional metals

The Fermi surface is thus an important tool in understanding the resistivity (and even more importantly, the resistivity under a magnetic field) of metallic systems. The nature of the Fermi surface is also at the origin of numerous distortions and low-temperature structural modulations in low-dimensional materials, which in turn noticeably affect their transport properties. When a piece of a Fermi surface can be translated by a vector $\vec{q}$ and superimposed onto another piece of the Fermi surface, this Fermi surface is said to be nested by the vector $\vec{q}$. For instance, the Fermi surface for the perfect 1D system in Fig. 11.3a is nested by an infinite number of vectors (see Fig. 11.6a), all of them having the same component along a^*, i.e. the direction of the chains.

In looking for nesting vectors of a given Fermi surface it is often useful to take into account more than just one Brillouin zone. The Fermi surface of Fig. 11.3b is an example. As seen in Fig. 11.6b, this Fermi surface is

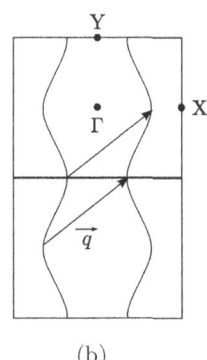

Fig. 11.6
(a) The Fermi surface of Fig. 11.3a showing two different nesting vectors. (b) The Fermi surface of Fig. 11.3b shown in a repeated zone scheme where the nesting vector is more easily seen.

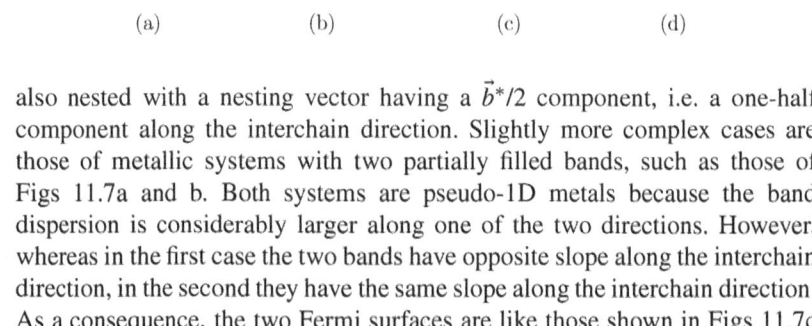

Fig. 11.7
(a) and (b) Band structure of two different pseudo-1D metallic systems; (c) and (d) Fermi surfaces associated with the band structures of (a) and (b), respectively, where the nesting vector is shown.

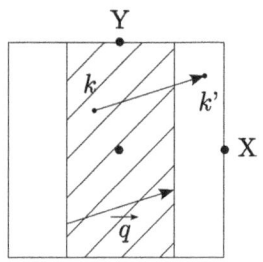

Fig. 11.8
2D representation of the Fermi surface for a 1D system, where the region of the filled wave vectors is hatched and a particular nesting vector $\vec{q}$ is shown.

also nested with a nesting vector having a $\vec{b}^*/2$ component, i.e. a one-half component along the interchain direction. Slightly more complex cases are those of metallic systems with two partially filled bands, such as those of Figs 11.7a and b. Both systems are pseudo-1D metals because the band dispersion is considerably larger along one of the two directions. However, whereas in the first case the two bands have opposite slope along the interchain direction, in the second they have the same slope along the interchain direction. As a consequence, the two Fermi surfaces are like those shown in Figs 11.7c and d, respectively. Although these Fermi surfaces contain four open lines, a single nesting vector can nest the full Fermi surface in both cases.

The importance of the Fermi surface nesting concept resides in the valuable insight it gives concerning the possibility and nature of electronic instabilities that a metallic system can exhibit. A metallic state predicted by one-electron band theory (i.e. neglecting electron repulsions) is not stable when the Fermi surface is nested, and may lead to a metal-to-insulator transition under the effect of some perturbation. Let us consider again the case of a 1D half-filled band the Fermi surface of which is shown in Fig. 11.8 in a 2D representation. The hatched/non-hatched regions of this figure are associated with full and empty band levels. The $\vec{q}$ nesting vector shown is one of the infinite number of possible nesting vectors and relates any pair of crystal orbitals associated with wave vectors $\vec{k}$ and $\vec{k}'$, i.e. $\vec{q} = \vec{k}-\vec{k}'$. The crystal orbitals $CO^0(\vec{k})$ and $CO^0(\vec{k}')$ are eigenfunctions of the unperturbed zero-order Hamiltonian, $\hat{H}^0$. If a certain perturbation, $\hat{V}$, is introduced into the system, the initial orbitals will no longer be eigenfunctions of the new Hamiltonian:

$$\hat{H} = \hat{H}^0 + \hat{V} \qquad (11.2)$$

and they will mix giving rise to the new crystal orbitals $CO(\vec{k})$ and $CO(\vec{k}')$. These new orbitals may be written in a non-normalised form as:

$$CO(\vec{k}) = CO^0(\vec{k}) + \gamma \exp(-i\phi_\gamma)CO^0(\vec{k}')$$
$$CO(\vec{k}') = \gamma \exp(i\phi_\gamma)CO^0(\vec{k}) - CO^0(\vec{k}') \qquad (11.3)$$

The consequences of such a mixing have already been examined in detail for the case of a purely 1D system in Section 4.2, and the detailed formal treatment will not be repeated here. The mixing leads to a modulation of the original electronic distribution associated with the two crystal orbitals governed by the term $\cos(\vec{q} \cdot \vec{r})$ in such a way that a maximum in the electronic density associated with $CO(\vec{k})$ corresponds to a minimum in the electronic density associated with $CO(\vec{k}')$. Since the energy mixing is more favourable as the energy difference between the initial states decreases, it will be maximum when $CO^0(\vec{k})$ and $CO^0(\vec{k}')$ are degenerate, i.e. they are on the Fermi surface. This orbital mixing lifts the degeneracy when $CO^0(\vec{k})$ and $CO^0(\vec{k}')$ are in the Fermi surface, and increases their energy difference when they are not. The energy consequences of the orbital mixing are thus maxima for states at the Fermi surface and quickly decrease for pairs of states that are not at the Fermi surface. A very important consequence of this mixing is that a gap opens for states that were originally at the Fermi surface. In the case of the Fermi surface of Fig. 11.8, since the nesting vector relates all pairs of states of the Fermi surface, a gap opens everywhere and the Fermi surface disappears. The consequence is that the system loses the initial metallic properties to become a semiconductor or insulator. In general we will speak of a metal-to-insulator transition, whatever the magnitude of the gap that is opened.

The kinds of perturbation that lead to orbital mixing as discussed above are essentially electron–phonon coupling and electronic repulsions. Assume that the orbital mixings associated with the $\vec{q}$ nesting vector ($= \vec{k}-\vec{k}'$) are carried out for all occupied $\vec{k}$ wave vectors of the Brillouin zone to obtain the new $CO(\vec{k})$ and $CO(\vec{k}')$ crystal orbitals. In the usual metallic state, the $CO^0(\vec{k})$ crystal orbitals are doubly filled and, as illustrated in Fig. 11.9a for the 1D system, each site has an identical charge and no magnetic moment. When the new $CO(\vec{k})$ crystal orbital is also doubly filled, a charge density wave (CDW) state arises where, as shown in Fig. 11.9b, there are no local magnetic moments and the charge density on different sites changes, in wave-like behaviour. If electron repulsions are important, so that the $CO(\vec{k})$ and $CO(\vec{k}')$ crystal orbitals are each singly occupied with spin-up and spin-down electrons respectively, a spin density wave (SDW) state results, which, as shown in Fig. 11.9c, is associated with a local magnetic moment but identical charge density at the different sites. Thus, for a metallic state with nested Fermi surface, when electronic interactions are not strong there will be a CDW instability, which will lead to the destruction of the Fermi surface and a structural CDW modulation through electron–phonon coupling. In contrast, when the electron repulsions are strong, there will be an SDW instability, which will also lead to the destruction of the Fermi surface, but without any need for (detectable) structural distortion.

When there is Fermi surface nesting, favourable orbital mixing can be achieved for a large region of the Brillouin zone; the associated electronic stabilisation is large and can overcome the cost in elastic energy needed to distort the lattice. This explains why metallic systems with nested Fermi surfaces generally undergo phase transitions associated with the opening of

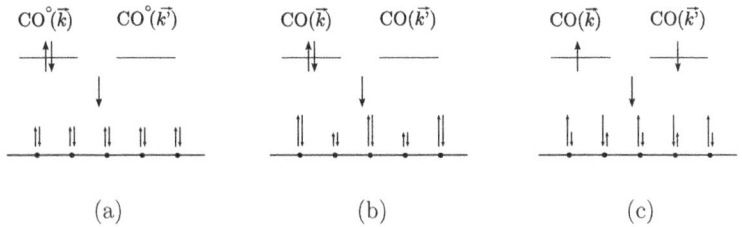

Fig. 11.9
Schematic representation of the occupation, charge, and spin density for the different sites of a 1D system: (a) metallic state, (b) CDW state, and (c) SDW state.

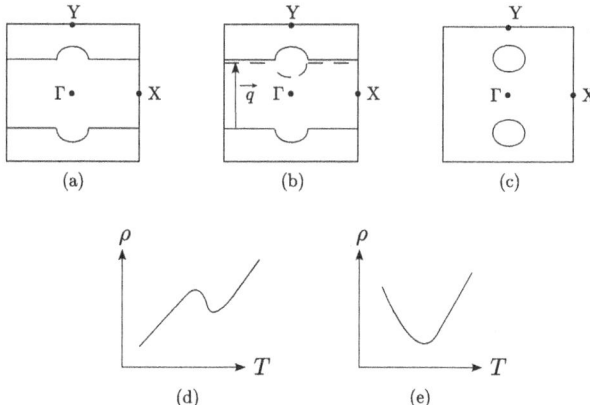

Fig. 11.10
(a)–(c) Schematic illustration of how a partial nesting of the Fermi surface can lead to a distortion while leaving Fermi surface pockets afterwards. Expected behaviour of the resistivity vs temperature curve in the case of the metal-to-metal (d) and metal-to-insulator (e) transitions as discussed in the text.

a gap at the Fermi level, i.e. a metal-to-insulator transition. Although we have so far assumed that the Fermi surface nesting is complete, this is not necessarily so. Sometimes even an incomplete Fermi surface nesting may afford enough electronic stabilisation and lead to a partial destruction of the Fermi surface. Figure 11.10a shows a schematic Fermi surface in which some parts have different curvatures and cannot be nested with the same nesting vector. However, a relatively large portion is well nested by a $\vec{q}$ nesting vector (Fig. 11.10b), which will lead to the destruction of the flat portions and the unnested portions will be left as small Fermi surface pockets (Fig. 11.10c). In this case some Fermi surface remains after the transition and thus the system will keep its metallic character. However, since an important part of the Fermi surface (i.e. electrons at the Fermi level) has been destroyed, the resistivity will experience a sudden increase at the moment of the transition and will then go down again (i.e. a metal-to-metal transition) as schematically shown in Fig. 11.10d, in contrast with the behaviour for a complete Fermi surface destruction (i.e. a metal-to-insulator transition) schematically shown in Fig. 11.10e.

These nesting ideas will be very useful in understanding the nature of many resistivity anomalies and the structural distortions exhibited by some low-dimensional metals, as well as in suggesting how to stabilise others. For instance, they suggest that a way to stabilise 1D metallic systems is to induce interchain contacts such that the nesting is degraded and the driving force for bringing about the distortion becomes too small. This type of idea has been very useful in the search for superconducting molecular solids, a subject which will be considered at the end of this chapter.

11.3 Monoclinic TaS$_3$ versus NbSe$_3$

NbSe$_3$, as well as blue bronzes and Bechgaard salts, is one of the best studied low-dimensional metals. More than thirty years after being reported by Meerschaut and Rouxel [4] it continues to be the focus of much attention. NbSe$_3$ was the first material for which detailed studies of a pseudo-1D metal experiencing two successive CDWs but retaining metallic behaviour down to the lowest temperatures could be carried out. Interestingly, monoclinic TaS$_3$, which is isostructural and also metallic at room temperature, becomes insulating after the two CDWs, thus offering an interesting contrast. [5] In this chapter we will show that a careful analysis of the crystal structure makes possible a qualitative understanding of some of the key aspects of the band structure and Fermi surface of these two low-dimensional solids. [6]

11.3.1 Crystal structure and electron counting

Both NbSe$_3$ and TaS$_3$ have a monoclinic crystal structure built from trigonal prismatic MX$_3$ chains condensed into MX$_3$ layers. TaS$_3$ also exists in a similar orthorhombic variety, which we will not consider here. From now on we will refer to monoclinic TaS$_3$ simply as TaS$_3$. The crystal structure of TaS$_3$ is shown in Fig. 11.11a. The repeat unit of the solid contains six TaS$_3$ units, which generate trigonal prismatic chains along the b-direction. Every one of these chains is displaced by $b/2$ along the chain direction with respect to their two neighbouring chains, so that every Ta atom is coordinated to the six sulphur atoms of its own chain and two extra S atoms of adjacent chains. Consequently, as shown in Fig. 11.11b, the coordination environment of every Ta atom is a bicapped trigonal prism. These two extra Ta–S contacts have bond lengths compatible with a Ta–S bond, so that there is a 2D network of Ta–S bonds. These layers are kept together through interlayer van der Waals interactions, leading to the 3D crystal structure. From a purely structural point of view this material can be considered a 2D solid.

From the six trigonal prismatic chains of the repeat unit, only three are independent. Thus we will refer to these chains as type I, type II, and type III. The key structural aspect of TaS$_3$ is the existence of S–S bonds in the chains of type I and type III. As shown in Fig. 11.11a, these two types of chain exhibit one S–S contact compatible with an S–S bond. In contrast, there is no short S–S contact in the type II chain. This feature is one of the keys to understanding the physical properties of this material. As far as electron counting is concerned, the three S atoms of chains II must be considered to be S^{2-}. However, as schematically shown in Fig. 11.12, the presence of one S–S bond lowers one of the sulphur 3p-based orbitals to keep the two bonding electrons, but also raises up another orbital, leading to the S–S antibonding counterpart. In terms

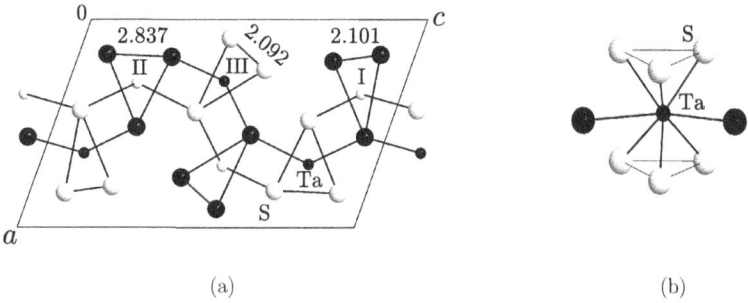

Fig. 11.11
(a) Projection of the crystal structure of monoclinic TaS$_3$ along the b-direction. Filled and empty circles refer to atoms at $y = 1/4$ and $y = 3/4$, respectively.
(b) Coordination environment of the Ta atoms in monoclinic TaS$_3$.

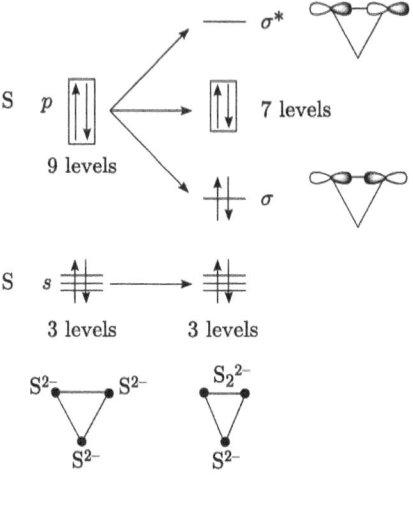

Fig. 11.12
Qualitative diagram showing the two different electron countings for the sulphur atoms of the three chains in TaS$_3$.

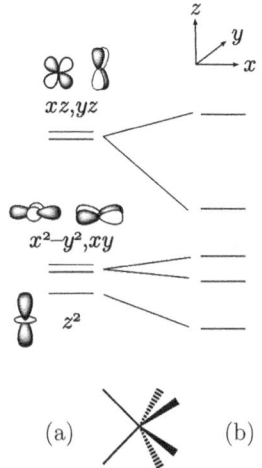

Fig. 11.13
Qualitative diagram showing the d orbital splitting in a trigonal prismatic environment with: (a) equilateral triangles and (b) isosceles triangles.

of electron counting, the three S atoms of chains I and III must be considered to be S^{2-} + S$_2^{2-}$. Thus the system can be formulated as 2 × [Ta$_{II}$(S^{2-})$_3$ + Ta$_I$(S^{2-})(S$_2^{2-}$) + Ta$_{III}$(S^{2-})(S$_2^{2-}$)].

The previous formulation leads to the important conclusion that the Ta atoms of chain of type II are formally d^0 whereas those in chains of types I and III are formally $d^{1/2}$. In other words, chains of type II should not participate in the conduction process and each pair of chains of types I and III should share one electron per repeat unit. From a very naive viewpoint we could assume that the metallic properties of this solid must be due to a series of double prismatic chains of type I and III, separated by 'insulating' chains of type II. However, Fig. 11.11 clearly shows that the different prismatic chains are connected, so to progress in our understanding of the material we must look in more detail at the electronic structure. The analysis of the crystal structure of NbSe$_3$ leads to identical conclusions.

11.3.2 Qualitative band structure

The d orbitals of a Ta atom in a trigonal prismatic environment are those schematically shown in Fig. 11.13. The very low number of electrons in the Ta-based bands and the very substantial splitting between the two groups of orbitals considerably simplify the construction of the relevant band structure. It is clear that we only need to focus on the set of (z^2, x^2-y^2, and xy) orbitals. Although only the lowest of these levels may be partially filled, the splitting between the three orbitals is small enough to require us to retain all of them in our analysis. However, an important structural feature has not yet been considered at this point: the capping of the Ta atoms by two S atoms of adjacent chains. As shown in Fig. 11.14, the Ta x^2-y^2 and xy orbitals are well oriented to interact with the S p_x or S p_y orbitals of adjacent chains depending on the phase factors of the different k points. Thus the x^2-y^2 and xy orbitals will

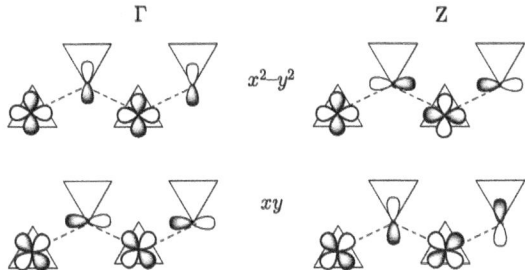

Fig. 11.14
Crystal orbitals based on the Ta x^2-y^2 and Ta xy orbitals at the Γ (left) and Z (right) points where Γ and Z refer to (0, 0, 0) and (0, 0, $c^*/2$) (For simplicity, we use the triangles in these and subsequent crystal orbital drawings to help visualise the crystal orbitals. These triangles are not meant to imply S–S bonds).

lead to the capping Ta–S antibonding orbitals and will be ejected to higher energies.[1] In contrast, the Ta z^2 orbital, which primarily concentrates in the lobe along the chain direction, makes only weak interactions with the S orbitals of the adjacent chains (see Figure 11.15). In addition, for the very same reason the z^2 orbitals did not interact with the π-type orbitals of the cyclopentadienyl ligands in the [M(C$_5$H$_5$)] chains (see Section 10.4), i.e. that its conical nodal surface hits the C p_z orbitals of the upper and lower pentagons of the ligands, neither do they interact appreciably with the sulphur orbitals of their own chain. Consequently, they are kept low in energy and they are the only ones that need to be retained. Since the electron counting analysis showed that only chains of type I and III can bear the conduction electrons, i.e. can be the main components of the partially field bands, only the four Ta z^2 orbitals of chains I and III must be retained to build the qualitative band structure of TaS$_3$.

The Ta z^2 orbitals are well oriented to strongly interact with those of the Ta atoms of the same chain, but they make only very weak interactions with the S p orbitals of adjacent chains, which are strongly involved in the creation of the capping Ta–S σ-type bonding orbitals. Thus, as shown in Fig. 11.15, the metal–metal interactions are all bonding at Γ, but antibonding at Y, so that the energy will increase from Γ to Y. Thus the qualitative band structure associated with the Ta z^2 orbitals of chains I and III will be as shown in Fig. 11.16a. Since there are six chains per repeat unit, four of which are of type I or III, the qualitative band structure for this system will be as shown in Fig. 11.16b. The weak separation among the bands will depend on the weak interchain interactions. The only detail left to complete this qualitative band structure – but an important one – is the positioning of the Fermi level. Since the Ta atoms of chains I and III are $d^{1/2}$ and there are a total of four of them per repeat unit, there is a total of two electrons to fill these bands. If the interchain interactions were nil so that these bands were fully degenerate, these bands would be one-quarter filled and the Fermi level would be found at $k_f = 0.125$. So long as the interchain interactions are weak though not nil, the position of the Fermi level

[1] Of course, two filled lower-lying bands, with a strong sulphur p character, are at the same time stabilised leading to the capping Ta–S bonding bands.

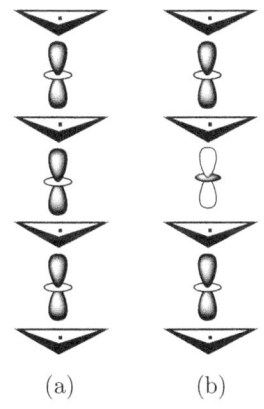

Fig. 11.15
Crystal orbitals based on the Ta z^2 orbital at the Γ (a) and Y (b) points where Γ and Y refer to (0, 0, 0) and (0, $b^*/2$, 0).

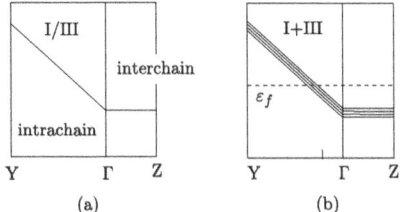

Fig. 11.16
(a) Qualitative band structure for the Ta z^2 band of one isolated bicapped trigonal chain of type I or III. (b) Qualitative band structure for the bottom part of the transition metal bands of TaS$_3$.

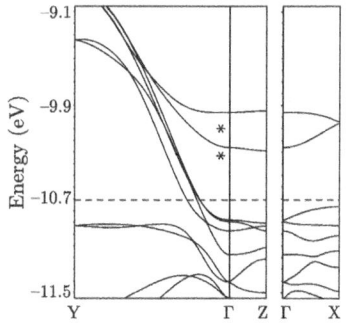

Fig. 11.17
Calculated extended Hückel band structure for TaS_3 where the dashed line refers to the Fermi level and $\Gamma = (0, 0, 0)$, $X = (a^*/2, 0, 0)$, $Y = (0, b^*/2, 0)$, and $Z = (0, 0, c^*/2)$.

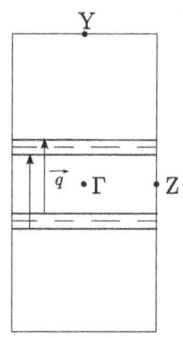

Fig. 11.18
Qualitative Fermi surface associated with a pair of chains of either type I or type III in TaS_3: the dashed lines refer to the Fermi surface for nil interchain interactions and the straight lines refer to the Fermi surface including the interchain interaction. The vector $\vec{q}$ nesting the full Fermi surface is shown.

will still be valid: the bands will be slightly shifted from this ideal position, with slightly lower or higher k_f values, as shown schematically in Fig. 11.16b.

The calculated band structure for TaS_3 around the Fermi level is shown in Fig. 11.17. The Fermi level cuts four bands that are dispersive along Γ to Y and very weakly dispersive along Γ to Z and Γ to X. These four bands are heavily based on the Ta z^2 orbitals of chains of type I and III. The two upper bands (marked with an asterisk), which are empty all along the Brillouin zone, are based on the Ta z^2 orbitals of chains type II. Thus, the calculated band structure completely substantiates the qualitative analysis developed above (note that an avoided crossing between the two bunches of bands around -9.65 eV may be a little confusing at first sight).

11.3.3 Qualitative Fermi surface: differences between $NbSe_3$ and TaS_3

Having thus substantiated the qualitative view of Fig. 11.16b we can now proceed to the construction of the qualitative Fermi surface. Every band will lead to a Fermi surface like that of Fig. 11.3a, corresponding to a perfect 1D metal. The full Fermi surface will be composed of four pairs of lines perpendicular to the Γ to Y direction located at around $\pm 0.125 \, \vec{b}^*$. If we assume that the interchain interactions are nil, it is useful to separate this full Fermi surface into two partial Fermi surfaces, one being associated with the pair of chains of type I and the other with the pair of chains of type III, each containing two pairs of lines (see Fig. 11.18 where the dashed lines, which correspond to this non-interacting situation, are really the superposition of two identical lines). Since the two chains of a given type are directly connected, there must be some kind of interchain interaction so that the two bands will weakly interact and be somewhat shifted. As shown in Fig. 11.18, the two pairs of lines of the Fermi surface (full lines) will be somewhat shifted with respect to the position where the interaction was nil (dashed line). Because of the inversion symmetry of the Fermi surface around the Γ point, the complete Fermi surface of Fig. 11.18 is nested by the same nesting vector. Consequently, with only two nesting vectors, every one being associated with a given type of chain, the full Fermi surface of TaS_3 can be destroyed.

As in any 1D system, the nesting vector is given by $2k_f$, so that in the present case, if the interchain interactions were nil, the nesting vector would be $0.25 \, \vec{b}^*$. In other words, there would be a tetramerisation in every chain of type I and III. However, in the real material, the interchain interactions are not nil and in addition they must be of differing strength for the two types of chain. This means that the displacement of the two pairs of straight lines in Fig. 11.18 with respect to the dashed line of the ideal case must not be the same. The two nesting vectors $\vec{q}_I$ and $\vec{q}_{III}$, associated with the prismatic chains I and III, respectively, will be slightly different. However, as in any 1D system, $2k_f$ gives the number of electrons in the partially filled band. Since the total number of electrons in partially filled bands per repeat unit is two (i.e. the equivalent of a full Brillouin zone) and two chains should undergo a distortion associated with the nesting vector $\vec{q}_I$ and two chains should undergo a distortion associated with the nesting vector $\vec{q}_{III}$, it must follow that:

$$2[(2k_f)_\mathrm{I}]\,\vec{b}^* + 2[(2k_f)_\mathrm{III}]\,\vec{b}^* = 2[\vec{q}_\mathrm{I} + \vec{q}_\mathrm{III}] = \vec{b}^* \qquad (11.4)$$

or in simple terms,

$$\vec{q}_\mathrm{I} + \vec{q}_\mathrm{III} = 0.50\,\vec{b}^* \qquad (11.5)$$

Thus our simple approach leads to the prediction that TaS_3 should be a metal and undergo two different CDW instabilities, each one being mostly associated with either type I or type III chains of the structure. These will lead to the progressive destruction of the Fermi surface. Thus the resistivity curve should exhibit metallic behaviour until, at some temperature, a bump associated with the first CDW occurs. Later, after displaying metallic behaviour again, the system will become insulating. It also leads to the prediction that the two distortions should be incommensurate, although not far from a tetramerisation, and should be such that the simple relationship $\vec{q}_\mathrm{I} + \vec{q}_\mathrm{III} = 0.50\,\vec{b}^*$ is obeyed.

All of these expectations are in excellent agreement with the experimental observations. [7] TaS_3 is a metal at room temperature, which experiences two different structural modulations at 240 K and 160 K, after which it becomes an insulator. The first transition is associated with chains of type III and the second with chains of type I. The values of $2k_f$ are 0.243 and 0.254, respectively, which obey the relationship outline above, to within the bounds of experimental error.

Since TaS_3 and $NbSe_3$ are isostructural and share all structural features, imposing the electron count analysed before they should also exhibit the same qualitative electronic structure. However, the selenium atoms have more diffuse orbitals which must lead to stronger interchain interactions, which in these materials mostly occur via direct Se...Se contacts of different chains. One should therefore expect stronger interchain interactions in $NbSe_3$, which will produce more warping to the pseudo-planar Fermi surfaces. As a result there will be some degradation of the nesting, although the main conclusions of the previous analysis should still apply. This is again in excellent agreement with experimental observations. [7] $NbSe_3$ is a metal at room temperature, which experiences two different structural modulations at 145 K and 59 K, after which it keeps its metallic character. This clearly indicates that, because of the increased warping, the sheets of the Fermi surface are incompletely nested so that some pockets remain after the transition. Again, the first transition is associated with chains of type III and the second with chains of type I. The values of $2k_f$ are 0.241 and 0.259, again in agreement with the expected relationship. Note that the two transitions now occur at lower temperatures, a consequence of the smaller stabilisation brought about by the distortions produced by the degradation of the nesting.

11.4 Molybdenum bronzes

Molybdenum bronzes have been the subject of great and continued interest, both from experimental and theoretical scientists. This is because most of them are low-dimensional metals and exhibit interesting physical properties related to their electronic instabilities. [8, 9, 10] Red bronzes $A_{0.33}MoO_3$ (A = Li, K,

Rb, Cs, Tl) are semiconductors while blue bronzes $A_{0.3}MoO_3$ (A = K, Rb, Tl) and purple bronzes $A_{0.9}Mo_6O_{17}$ (A = Li, Na, K) are metals at room temperature. As mentioned above, the physical behaviour of some of these materials, such as the blue bronzes and the 2D purple bronzes $A_{0.9}Mo_6O_{17}$ (A = Na, K), has been very puzzling to scientists for many years and some, such as the 3D purple bronze $Li_{0.9}Mo_6O_{17}$, are still not well understood. All these bronzes contain molybdenum–oxygen layers made up of edge- and corner-sharing MoO_6 octahedra, with large and structurally complex unit cells. Qualitative analyses of their electronic structures have played an important role in unravelling their peculiar behaviour. [3] In the following section, we will consider some of these materials and show that the qualitative methods developed so far may offer simple yet useful views of the electronic structure of these complex materials.

11.4.1 Octahedral distortions and t_{2g} level splitting in MoO_6 octahedra

The t_{2g}-block levels of a regular MoO_6 octahedron have π antibonding combinations between the metal $4d$ orbitals and the oxygen $2p$ orbitals (see Fig. 2.20). Therefore, an Mo–O shortening raises any t_{2g}-block level if this level has an antibonding combination of Mo $4d$ and O $2p$ orbitals along the shortened Mo–O bond. A distortion in which one Mo–O bond is shortened (see Fig. 11.19a) leaves one t_{2g} level unaffected, i.e. $x^2 - y^2$, which is the δ orbital with respect to the shortened Mo–O bond axis, but raises the remaining two levels, i.e. the xz and yz which are the π orbitals with respect to the shortened Mo–O bond axis. On the other hand, all three t_{2g}-block levels are raised by a distortion in which two or more bonds are shortened (see Fig. 11.19b). Simple inspection of the nature and extent of the distortions in the MoO_6 octahedra allows the prediction of which of them would keep the d electrons and what kind of t_{2g}-block bands the oxide under consideration is likely to have. Shortening of an octahedral Mo–O bond is usually accompanied with the lengthening of the opposite bond. In the extended structures considered in this section, this generally leads to a long-range alternation of short and long Mo–O bonds. In the discussion in this section we will use the term Mo–O bond alternation both for the extended structures and for isolated octahedra.

According to these simple ideas, if in a given structure there are octahedra with strong O...Mo–O bond alternation in either one or two directions, only

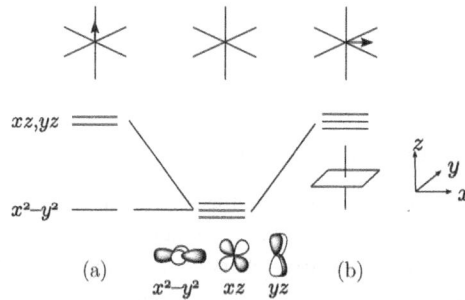

Fig. 11.19
Schematic diagram showing how the t_{2g}-block levels of a regular MoO_6 octahedron are split when: (a) one Mo–O bond is shortened (and the opposite bond is lengthened) and (b) two or more bonds are shortened (and the opposite ones are lengthened).

the t_{2g}-block orbital of the first class of octahedra that is δ with respect to the shortened bond, will lead to lower lying energy levels. If in another type of crystal structure all octahedra have more than one direction of strong alternation, there can be two situations: (i) the extent of the distortion is very different for different octahedra, and (ii) the extent of the distortion does not vary very much for the different octahedra. In case (i) the t_{2g} orbitals of the octahedra with smaller distortion are those leading to lower-lying t_{2g}-block levels. In case (ii) the t_{2g} orbitals of all octahedra can contribute to lower-lying t_{2g}-block levels.

11.4.2 MoO$_5$ chain with corner-sharing octahedra: counting of 2p oxygen antibonding contributions

In building qualitative band structures for these molybdenum bronzes, a simple yet very powerful technique is the counting of the number of oxygen $2p$ orbitals which mix into the t_{2g}-block bands. Since this approach will be extensively used in this chapter (and can be used for a large number of other transition metal oxides and bronzes [3]) we will now introduce the basic concepts and notation involved.

We will refer to the axial and equatorial oxygen atoms as O_{ax} and O_{eq}, respectively. With the system of axes shown in Fig. 11.19, the t_{2g} levels arise from the $x^2 - y^2$, xz, and yz orbitals. The $x^2 - y^2$ orbital does not interact with the O_{ax} $2p$ orbitals (Fig. 11.20a) but it does with those of the O_{eq} atoms (Fig. 11.20b). If the antibonding interaction in each Mo–O bond is denoted by the symbol (Y_t), where the subscript t indicates that the oxygen of the Mo–O bond is a terminal ligand, then the $x^2 - y^2$ level has 4(Y_t) interactions. The xz and yz orbitals each make 2(Y_t) interactions with the $2p$ orbitals of the O_{ax} atoms (see Fig. 11.20c). With the $2p$ orbitals of the O_{eq} atoms, the xz and yz orbitals interact as shown in Fig. 11.20d. Unlike the situation in Figs 11.20b and c, no Mo–O_{eq} bond in Fig. 11.20d is in the d orbital plane, and the Mo–O_{eq} therefore has a weaker antibonding character. The presence of such antibonding interactions in an Mo–O_{eq} bond may be denoted (y_t). Then the xz and yz levels each have 2(Y_t) + 4(y_t) interactions. These two levels are degenerate with the $x^2 - y^2$ level, so that 2(y_t) antibonding interactions are equal in magnitude to one (Y_t) antibonding interaction [$(Y_t) = 2(y_t)$].

Let us now consider the corner-sharing MoO$_5$ octahedral chain of Fig. 11.21a. For simplicity, except otherwise stated, all O $2p$ orbitals in the diagrams will be suppressed except those of the bridging oxygen atoms. The Bloch orbitals of the Mo xz orbital at the centre (Γ) and border (Z) of the 1D Brillouin zone are shown in Fig. 11.21b. Since the metal–metal distance is long, direct metal–metal interactions are nil. As shown by the dotted line in Fig. 11.21d, in the absence of Mo–O interactions the xz band

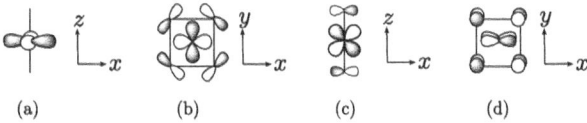

(a) (b) (c) (d)

Fig. 11.20
(a)–(d) Different types of Mo–O antibonding interactions for the t_{2g} orbitals of an MoO$_6$ octahedron.

Fig. 11.21
((a) MoO$_5$ corner-sharing octahedral chain. Bloch (b) and crystal (c) orbitals associated with the xz orbital of Mo in the MoO$_5$ chain. (d) Qualitative dispersion curve for the Bloch and crystal orbitals based on the xz orbital.

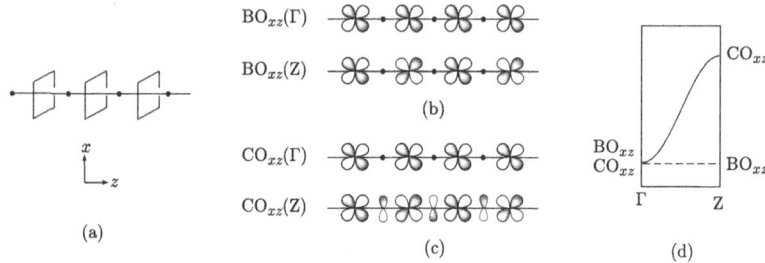

will be dispersionless. At Γ the oxygen $2p$ orbitals cannot mix into the xz Bloch orbital but mixing is possible at Z, where the resulting crystal orbital (Fig. 11.21c) has the maximum antibonding interactions and the band has the shape displayed in Fig. 11.21d (continuous line). For the yz orbital exactly the same arguments hold, so that the two bands are degenerate. The $x^2 - y^2$ orbitals of Mo are δ orbitals with respect to the chain axis (Fig. 11.22a) and therefore they do not interact with any $2p$ orbital of the bridging oxygen ligands. Thus the Bloch and crystal orbitals coincide and the associated band is dispersionless (Fig. 11.22b).

As we did for the levels of a regular MoO$_6$ octahedron, the occurrence of an antibonding interaction between the molybdenum $4d$ orbitals and the *bridging* oxygen $2p$ orbitals will be denoted by the symbol (Y$_b$). The absence of an oxygen $2p$ orbital at the bridging position may then be represented by the symbol (N) and by a dot in the orbital diagrams. Figure 11.23 shows the calculated dispersion relations for the t_{2g}-block bands of the ideal MoO$_5$ chain. As expected, the $x^2 - y^2$ band, which is characterised by 4(Y$_t$) interactions at both Γ and Z, is practically dispersionless. The degenerate xz and yz bands are characterised by 2(Y$_t$) interactions at Γ but 2(Y$_t$) + 2(Y$_b$) interactions at Z. In fact, since an oxygen in a bridging position implicates two antibonding interactions, we can further simplify the estimation of the relative energies by simply counting the number of oxygen $2p$ *contributions per repeat unit*

Fig. 11.22
(a) Bloch and crystal orbitals associated with the $x^2 - y^2$ orbital of Mo in the MoO$_5$ chain. (b) Qualitative dispersion curve for the Bloch and crystal orbitals based on the $x^2 - y^2$ orbital.

Fig. 11.23
Calculated t_{2g}-block bands of the ideal MoO$_5$ chain (extended Hückel type calculations with Mo–O distances of 1.96 Å). The number of O $2p$ contributions per repeat unit at Γ and Z are indicated.

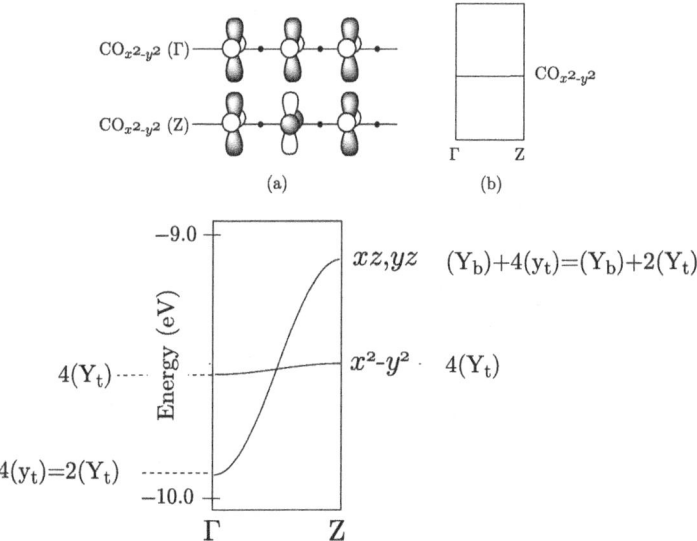

in a given crystal orbital. In that case, the degenerate xz and yz bands are characterised by $4(y_t) = 2(Y_t)$ interactions at Γ but $4(y_t) + (Y_b) = 2(Y_t) + (Y_b)$ interactions at Z, while the $x^2 - y^2$ band is characterised by $4(Y_t)$ interactions at both Γ and Z. Since the flat $x^2 - y^2$ band lies approximately in the middle of the xz and yz bands, it turns out that $(Y_b) \cong 4(Y_t)$.[2] With this and the relationship we saw above, namely $(Y_t) = 2(y_t)$ (and consequently, $(Y_b) = 2(y_b)$), it is possible to construct qualitative band structures for most of these molybdenum bronzes. In the following sections we will further simplify our notation and use $(Y_b) = Y$ as the reference in our qualitative estimation of crystal orbital energies.

[2] Of course, if we count the number of Mo–O antibonding interactions the relation is $(Y_b) \cong 2(Y_t)$.

11.4.3 $A_{0.33}MoO_3$ (A = K, Rb, Cs, Tl) 2D red bronzes: metallic or insulating?

The red bronzes $A_{0.33}MoO_3$ (A = Li, K, Rb, Cs, Tl) belong to two different structural types. $A_{0.33}MoO_3$ (A = K, Rb, Cs, Tl) consist of isolated layers of composition MoO_3, with A^+ cations between these layers. Hence they can be referred to as 2D red bronzes [11] to distinguish them from $Li_{0.33}MoO_3$ which has a complex 3D structure. [12] Simply on electron count grounds it is not clear if these solids should be metallic or semiconducting. In the absence of cations one would predict semiconducting or insulator behaviour. However, whether the extra electrons brought about by the cations will completely or partially fill some of the lower-lying transition metal-based bands is something that completely depends on the details of the crystal structure (number of formula units in the unit cell, specific distortions of the different octahedra, etc). In this section we will build the qualitative band structure of these 2D red bronzes to answer this question. [3, 13]

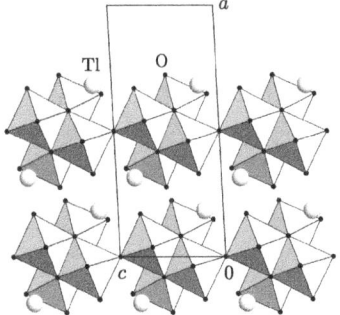

Fig. 11.24
Top view of the crystal structure of $TlMoO_3$.

Crystal structure analysis

A top view of the crystal structure of $Tl_{0.33}MoO_3$ is shown in Fig. 11.24. In analysing how the band structure of these materials is related to the crystal structure it is useful to construct the layers in terms of the Mo_6O_{26} clusters shown in Fig. 11.25a. Upon sharing the corners of the Mo_6O_{26} clusters, the Mo_6O_{24} chain (Fig. 11.25b) is obtained. A projection view of this chain along the chain direction can be represented as shown in Fig. 11.25c. Finally, the Mo_6O_{18} layer (i.e. the MoO_3 layer) is obtained by sharing the corners and edges of the Mo_6O_{24} chains as shown in Fig. 11.25d (successive Mo_6O_{24} chains are shown as shaded and non-shaded). The Mo_6O_{24} chains run along the b-direction and the MoO_3 layers alternate with layers of the A^+ cations along the a-direction.

The repeat unit of the layers contains six octahedra (i.e. Mo_6O_{18}, see Figs 11.25b and d). Hence there are only two electrons to fill the bottom portion of the t_{2g}-block bands. The different Mo–O bond lengths found in $TlMoO_3$ are shown in Fig. 11.26. The hump octahedra have two strong O...Mo–O bond alternations while the chain octahedra (i.e. those of the central Mo_4O_{18} chain; see Fig. 11.25e) have only one strong O...Mo–O bond alternation. Thus, the t_{2g}-block levels of the hump octahedra are high in energy. As far as the

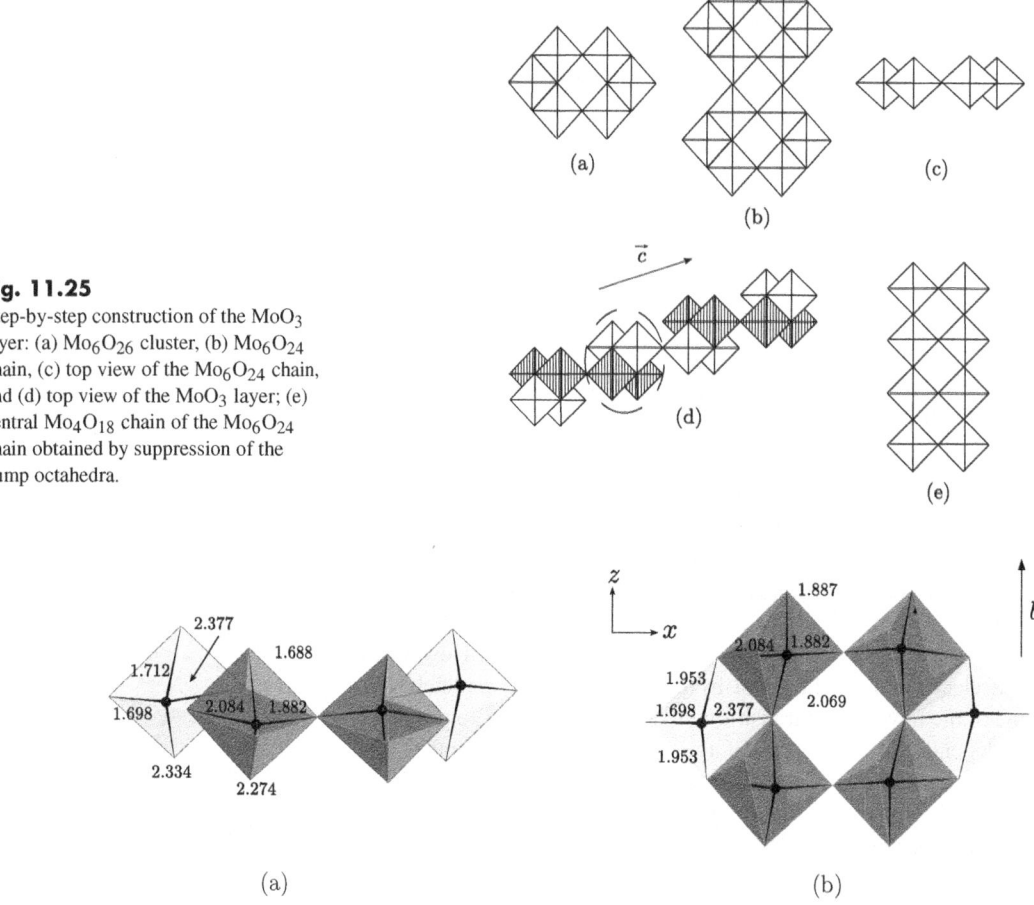

Fig. 11.25
Step-by-step construction of the MoO$_3$ layer: (a) Mo$_6$O$_{26}$ cluster, (b) Mo$_6$O$_{24}$ chain, (c) top view of the Mo$_6$O$_{24}$ chain, and (d) top view of the MoO$_3$ layer; (e) central Mo$_4$O$_{18}$ chain of the Mo$_6$O$_{24}$ chain obtained by suppression of the hump octahedra.

Fig. 11.26
(a) and (b): Different Mo–O bond lengths found in the crystal structure of Tl$_{0.33}$MoO$_3$. [40]

lower t_{2g}-block bands are concerned, we do not need to consider the Mo$_6$O$_{24}$ chain but the Mo$_4$O$_{18}$ chain. The strong alternation in the chain octahedra occurs in a direction perpendicular to the chain so that only the d orbital of each chain octahedron, which is δ with respect to this direction, will remain low in energy. This orbital, which is xz, with the axes shown in Fig. 11.26, makes π-type interactions with the $2p$ oxygen orbitals along the chain axis. Thus it will be sufficient to consider the xz bands of the Mo$_4$O$_{18}$ chain in describing the low-lying t_{2g}-block bands of 2D red bronzes.

Ideal Mo$_4$O$_{18}$ chain

The cluster orbitals relevant for the construction of the lower t_{2g}-block bands of these chains (i.e. those based on the Mo xz orbitals) and their energy ordering are shown in Fig. 11.27. By appropriately repeating these four cluster orbitals in-phase or out-of-phase, the crystal orbitals at the centre (Γ) and

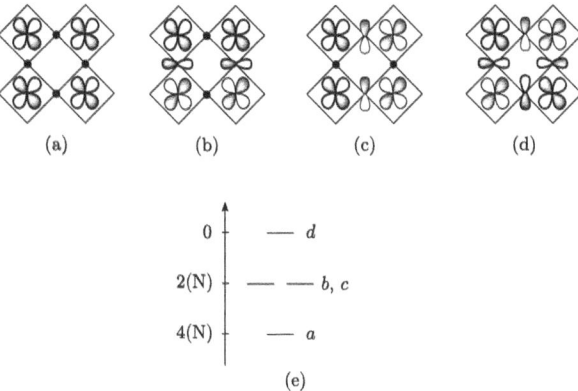

Fig. 11.27
Cluster orbitals relevant for the construction of the lower t_{2g}-block bands of the ideal Mo_4O_{18} chain (a)–(d) and their energy ordering (e).

border (Y) of the 1D Brillouin zone may be constructed. Their relative energy may be evaluated by counting the number of O $2p$ contributions and using the previously described relationships (see end of section 11.4.2). Finally, a qualitative band structure as shown in Fig. 11.28 can be obtained. For instance, band a at Γ has the nodal pattern of Fig. 11.29a; bands a and b at Y have those of Figs 11.29b and c; and band c at Γ has that of Fig. 11.29d. We should remind ourselves that a dot at the shared oxygen atom's position is used to indicate that no O $2p$ orbital is found at that position. In order to obtain the qualitative band structure of Fig. 11.28, the only thing we need to do is to count the number of O $2p$ contributions in the shared oxygen positions of the qualitative crystal orbitals, since those in the terminal oxygen atoms are exactly the same for all crystal orbitals. For instance, there are zero contributions for band a at Γ, two for bands a and b at Y and band c at Γ, so that all three are degenerate. We can evaluate the energy of all crystal orbitals in a similar way.

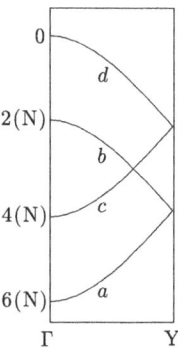

Fig. 11.28
Qualitative band structure for the ideal Mo_4O_{18} chain.

Real Mo_4O_{18} chains

Now we must consider how the distortions in the real Mo_4O_{18} chains modify the qualitative band structure of Fig. 11.28. As can be seen in Fig. 11.26, there is a bond-length alternation in a pattern Mo...O...Mo–O–Mo along the real chain (i.e. with bond lengths of 2.069, 2.069, 1.887, 1.887 Å). This lifts the degeneracy between bands a and b at Y (see Figs 11.29b and c, respectively). Since the longer distances are associated with the presence of

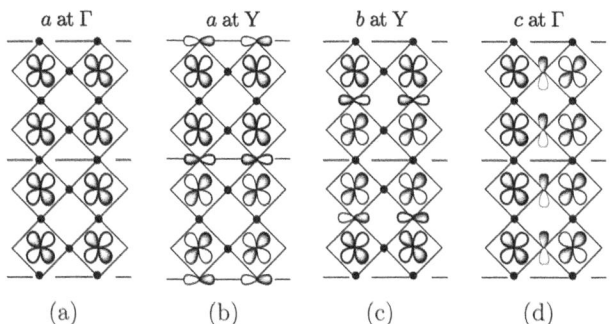

Fig. 11.29
Nodal patterns of different crystal orbitals for the ideal Mo_4O_{18} chain: (a) crystal orbital a at Γ, (b) crystal orbital a at Y, (c) crystal orbital b at Γ, and (d) crystal orbital c at Γ.

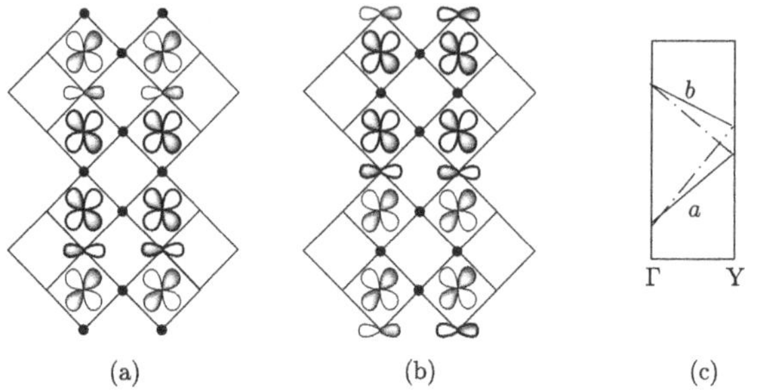

Fig. 11.30
Hump octahedra and crystal orbitals: lower (a) and higher (b) energy crystal orbitals originating from the degenerate crystal orbitals of the ideal Mo_4O_{18} chain (Figs 11.29b and 11.29c). The occurrence of an avoided crossing between bands a and b is schematically shown in (c).

the hump octahedra, the lower-energy orbital must be that of Fig. 11.30a and the higher-energy orbital must be that of Fig. 11.30b. The crystal orbital of Fig. 11.30a is built from the cluster orbital of Fig. 11.27b, which repeats out-of-phase so that it corresponds to band b of the ideal chain. The crystal orbital of Fig. 11.30b is built from the cluster orbital of Fig. 11.27a, which repeats out-of-phase so that it corresponds to band a of the ideal chain. Therefore in the real chain there must be an avoided crossing between bands a and b, as illustrated in Fig. 11.30c.

In the Mo–O–Mo linkages of Fig. 11.26a, the short Mo–O bonds (1.882 Å) that are perpendicular to the direction of the chain are comparable to the short Mo–O bonds (1.887 Å) of Fig. 11.26b along the chain. Thus the extent of antibonding in Figs 11.29b and 11.29d are comparable. Therefore, when the Mo_4O_{18} chain distorts from the ideal to the real structure, band a at Y (which because of the avoided crossing is given by crystal orbital b) is lowered, while both bands b at Y (which because of the avoided crossing is given by crystal orbital a) and band c at Γ are raised. A band gap should open along the chain direction making the real chains semiconducting. The calculated band structure for a real Mo_4O_{18} chain is shown in Fig. 11.31a and completely substantiates this analysis.

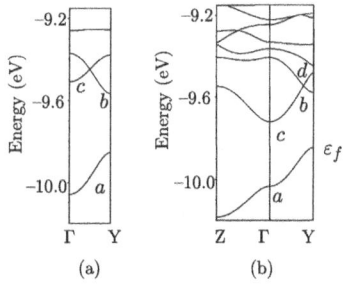

Fig. 11.31
Bottom portion of the extended Hückel t_{2g}-block bands calculated for: (a) the real Mo_4O_{18} chain, and (b) the real $Mo_6O_{18}^{2-}$ layers of $Tl_{0.33}MoO_3$, where $\Gamma = (0, 0)$, $Y = (b^*/2, 0)$ and $Z = (0, c^*/2)$.

Interchain interactions

The calculated dispersion relations for the bottom portion of the t_{2g}-block bands of the real Mo_6O_{18} layer of $Tl_{0.33}MoO_3$ are shown in Fig. 11.31b. The lowest band, a, is separated by an indirect band gap from the next band, c. Therefore, and in agreement with the previous discussion, we conclude that the 2D red bronzes are not metallic. However, we still have not considered the interchain interactions, which in principle could have closed the gap of the real chains again. Could the effect of the interchain interactions have been guessed?

To examine this point we must concentrate on the encircled unit of Fig. 11.25d. Essentially, the transmission of the electronic interactions along the c-direction should occur either through the oxygen $2p$ orbital of the edge-sharing unit or through the $4d$ orbitals of the Mo atom in the hump octahedron. When looking from the top, the crystal orbitals of Figs. 11.29a, b, and c may be represented as in Fig. 11.32a. The orbitals of Fig. 11.29d can be represented as in Fig. 11.32b, where the terminal oxygen orbitals have now been added. As shown in Figs 11.32c and d, with such a representation, bands a and c of the layer at Γ can be obtained by repeating Figs 11.32a and b in phase along *c*. Since the 2*p* orbital of each external oxygen atom of one Mo_6O_{18} chain is located along the nodal plane of the *xz* orbital of the adjacent Mo_6O_{18} chain, no strong interaction is expected. The same reasoning holds for all crystal orbitals built from the Mo *xz* orbitals. Hence, no strong direct interchain interactions are expected in the Mo_6O_{18} layer.

Can the hump octahedra influence the interchain interactions? The orbital patterns of Figs 11.32c and d around the hump octahedra are as shown in Figs 11.33a and b, respectively. The *d*-block levels of the hump octahedra are high in energy so that, if they mix into these crystal orbitals, they will mix in a bonding way. As illustrated in Figs 11.33c and d, the local symmetry in Fig. 11.33a is such that it does not allow such a mixing whereas that of Fig. 11.33b does. When this mixing occurs, the participation of the inner oxygen atoms in Figs 11.33a and b decreases, leading to an energy lowering. Thus depending on the orbital patterns at the zone of the hump octahedra, there may or may not be a stabilisation due to interchain interactions.

The relative orbital phases between adjacent chains must change from in-phase to out-of-phase and vice versa along the interchain direction (i.e. from Γ to Z). Therefore, bands with the orbital pattern of Fig. 11.33a at Γ will gradually pick the orbital pattern of Fig. 11.33b and will be stabilised. The opposite will be the case for bands with crystal orbitals of the pattern of Fig. 11.33b at Γ. This is schematically shown in Figs 11.34a and b. On the basis of these results we can draw the schematic band diagram shown in Fig. 11.35, which suggests that whether there is a band gap between bands *a* and *c* or not depends critically on the relative strengths of the intrachain distortion effect (ΔE_{intra}) and the interchain interaction effect (ΔE_{inter}).

The Mo–O distances associated with the antibonding interactions in band *a* at Y and band *c* at Γ are substantially different. For an isolated real Mo_4O_{18} chain, the Mo–O bonds engaged in antibonding interactions are 2.069 and 2.084 Å for band *a* at Y, but 1.882 and 2.084 Å for band *c* at Γ. Due

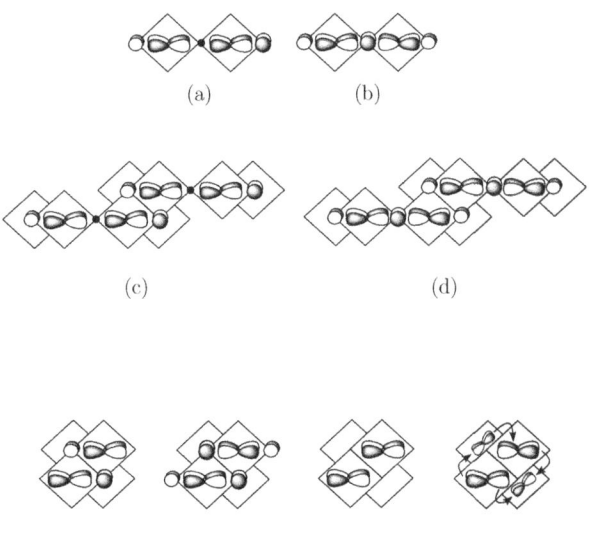

Fig. 11.32

Top view of the crystal orbitals of Fig. 11.29a–c (a) and Fig. 11.29d (b). Top view of the crystal orbitals a (c) and c (d) of the Mo_6O_{18} layer at Γ.

Fig. 11.33

Orbital patterns around the hump octahedra in the crystal orbitals of Fig. 11.32c (a) and 11.34d (b). (c) and (d) are schematic diagrams showing that the hump *d* orbitals may stabilise crystal orbitals with the orbital pattern in (b) but not those with the orbital pattern in (a).

Fig. 11.34
Correlation of the band shape of an Mo_6O_{18} layer band (solid lines) with those of the corresponding chain band (dashed lines) for the cases where the interchain orbital patterns at Γ are those of Fig. 11.33a (a) and Fig. 11.33b (b), respectively.

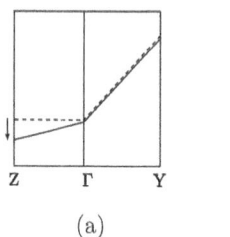
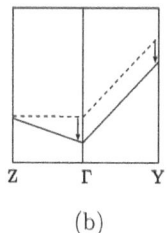

(a) (b)

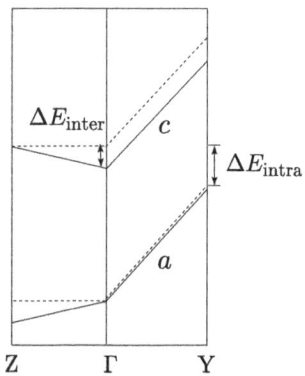

Fig. 11.35
Comparison of bands a and c for the Mo_6O_{18} layer (solid lines) with those of the Mo_6O_{18} chain (dashed lines).

to interchain interactions, the antibonding interaction of the 2.084 Å Mo–O bond is removed in band c at Γ but retained in band a at Y in the Mo_4O_{16} layer. Although this makes the band gap smaller, band c at Γ would still lie above band a at Y in the Mo_6O_{18} layer because the intrachain antibonding of band c at Γ is associated with a much shorter Mo–O bond (i.e. 1.882 Å). Comparison between the band structures of real Mo_4O_{18} chain and a $Mo_6O_{18}^{2-}$ layer (Fig. 11.31) shows that the strength of the intrachain interaction is about twice that of the interchain interaction. Ultimately, this leads to a band gap between bands a and c for the layers of 2D $Tl_{0.33}MoO_3$, as well as all other 2D red bronzes, leading to semiconducting properties.

In the end, the analysis of the correlation between the crystal and electronic structure of the 2D red bronzes has been relatively long and involved. However, none of the arguments used is complex if the details of the crystal structure are taken into account. Although most of the aspects discussed have been elaborated without calling for actual computations, this is not the way in which the orbital analysis is most useful. When studying structurally complex materials, orbital arguments are invaluable tools to *understand* the results of actual computations. This is the most effective way – to elaborate simple arguments, which can then be easily tested by additional computations or to suggest other materials to which similar ideas should be applicable and thus help in finding general trends. For instance, most of the arguments developed in this section are directly applicable to other transition metal oxides and bronzes.[3]

[3] Note that although in this section we have considered molybdenum as the transition metal atom, the orbital arguments are completely general and are practically independent of the nature of the transition metal.

11.4.4 $A_{0.3}MoO_3$ (A = K, Rb, Tl) blue bronzes: 2D solids with pseudo-1D behaviour

Blue bronzes $A_{0.3}MoO_3$ (A = K, Rb, Tl) are layered materials that exhibit a low-temperature metal-to-semiconductor transition. The electrical conductivity in the plane of the layers is quite anisotropic so that they are really pseudo-1D metals at room temperature. A number of experimental techniques have provided evidence that the metal-to-insulator transition is due to a CDW with a low-temperature 0.75 $\vec{b}^*$ wave vector component, where b is the direction with better conductivity. The discovery of non-linear conductivity due to sliding of the CDW in the potassium compound raised enormous interest in these bronzes, which are now among the most carefully studied low-dimensional materials. [8, 9]

Here we will consider how the essential feature of the electronic structure of this material – a CDW with a 0.75 $\vec{b}^*$ wave vector component that can completely destroy the Fermi surface – may be accounted for in terms of an orbital approach. However, since most of the analysis can be carried out using arguments very similar to those used for the 2D red bronzes, a large part of the analysis will be left as an exercise for the reader.

Crystal structure analysis

Blue bronzes contain MoO_3 layers in between which the A^+ cations reside. These MoO_3 layers are related to those of 2D red bronzes examined above. The basic structural unit of blue bronzes is not a double-chain Mo_4O_{18} but a quadruple-chain Mo_8O_{34} (Fig. 11.36a). When hump octahedra are added to this chain hump–quadruple-chain $Mo_{10}O_{40}$ is generated (Fig. 11.36b). These chains differ from those in the 2D red bronzes (Fig. 11.25b) in that the hump octahedra have been added to this chain in a zigzag way. Thus the resulting $Mo_{10}O_{40}$ chain has a twofold screw rotation axis along the chain direction. The $Mo_{10}O_{30}$ layer (i.e. the MoO_3 layer) of the blue bronzes (Figs 11.36c and d) is obtained by sharing the corners and edges of the $Mo_{10}O_{40}$ chains. The hump octahedra are shaded in Fig. 11.36 to help visualise the structure of the layer. In the C-centred monoclinic cell of the blue bronzes, each layer is contained in a plane defined by the two orthogonal vectors $\vec{b}$ and $2\vec{d}$, where $2\vec{d} = 2\vec{c} + \vec{a}$. As shown in Fig. 11.36d, each layer can be described with a primitive cell defined either by the two orthogonal vectors $\vec{b}$ and $2\vec{d}$ or the two nonorthogonal vectors $\vec{b}'$ ($= \vec{b}$) and $\vec{d}'$, where $\vec{d}' = \vec{d} + \vec{b}/2$. From the computational viewpoint it is more convenient to use the smaller cell defined by the $\vec{b}'$ and $\vec{d}'$ vectors.

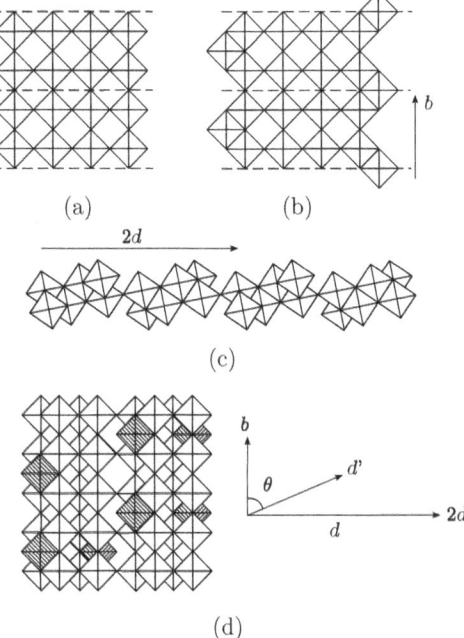

Fig. 11.36
Step by step construction of the MoO_3 layer in the blue bronzes: (a) Mo_8O_{34} chain, (b) $Mo_{10}O_{40}$ chain, (c) top view of the $Mo_{10}O_{30}$ layer, and (d) view of the $Mo_{10}O_{30}$ layer along a direction perpendicular to the layer (the hump octahedra have been shaded).

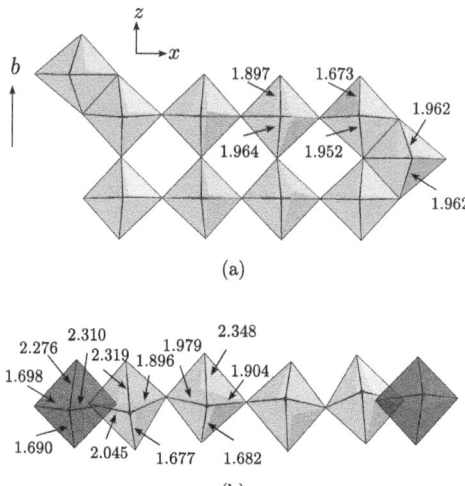

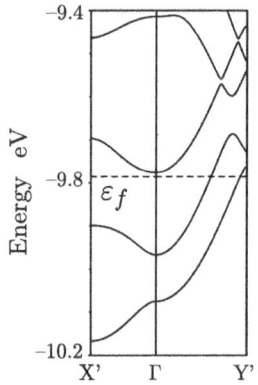

Fig. 11.37
(a) and (b): Different Mo–O bond lengths found in the crystal structure of $K_{0.3}MoO_3$ (only those distances non-equivalent by symmetry are shown).

Fig. 11.38
(a) Bottom portion of the extended Hückel t_{2g}-block bands calculated for the $Mo_{10}O_{30}^{3-}$ layers of the $K_{0.3}MoO_3$ blue bronze along the $\Gamma \rightarrow X'$ and $\Gamma \rightarrow Y'$ directions of the Brillouin zone defined by the vectors $\vec{b}'$ and $\vec{d}'$ (see Fig. 11.36d). The dashed line refers to the Fermi level and $\Gamma = (0, 0)$, $X' = (d'^*/2, 0)$ and $Y' = (0, b'^*/2)$.

Band structure and Fermi surface

The repeat unit of the $Mo_{10}O_{30}$ layer contains ten MoO_6 octahedra. With the formal oxidation states of Mo^{+6}, O^{2-} and A^+, there are three electrons to fill the t_{2g}-block bands of this layer. Thus, there are 30 t_{2g}-block bands but only the bottom portion is filled. The analysis may be considerably simplified again by considering the local distortions of the different octahedra in the layer. The different Mo–O bond lengths found for the $K_{0.3}MoO_3$ blue bronze are shown in Fig. 11.37 (only those distances non-equivalent by symmetry are shown). [14] The calculated extended Hückel band structure for the $Mo_{10}O_{30}^{3-}$ layers of the $K_{0.3}MoO_3$ blue bronze is given in Fig. 11.38. The construction of a qualitative band structure for the $Mo_{10}O_{30}^{3-}$ layers to explain these results is left as an exercise for the reader (Exercise (11.3)).

After deriving a qualitative band structure (see Exercise (11.3)) we are now ready to consider the nature of the CDW in this material. Given the qualitative (or the real) band structure for the $Mo_{10}O_{30}^{3-}$ layers it is clear that the three electrons will partially fill the two lower t_{2g} bands. Let us assume for the moment that the interchain interactions are nil. Using the rectangular representation of the Brillouin zone (see Exercise (11.3)), we must draw two series of lines perpendicular to the b^*-direction. Since the energy of the two bands rises from Γ to Y, the region corresponding to the filled k points is that around Γ (see Figs 11.39a and b). Based on electron counting considerations we can say that the addition of the areas corresponding to the filled states in

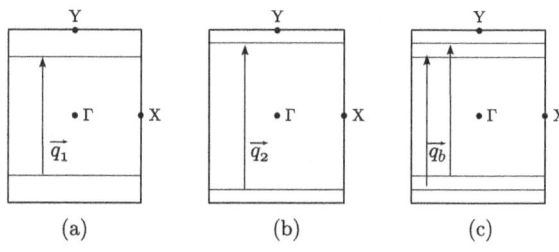

Fig. 11.39
(a)–(c) Qualitative Fermi surfaces associated with the two partially filled bands of the blue bronze without taking into account interchain interactions.

Fig. 11.39a and b must be equal to one and a half times the area of the rectangle (i.e. three electrons), so that $\vec{q}_1 + \vec{q}_2 = 3\vec{b}*/2$. The four lines of the combined Fermi surface can be completely nested by a single nesting vector $\vec{q}_b = \vec{q}_1/2 + \vec{q}_2/2$ (see Fig. 11.39c) which, taking into account the previous equation, is $\vec{q}_b = 3\vec{b}*/4$. Thus, if the interchain interactions are nil, the simple qualitative treatment leads to the prediction of a metal-to-insulator transition associated with a structural modulation with wave vector component of 0.75 along the chain's direction.

Now we must take into account the interchain interactions. There are essentially two possibilities: (i) the two partially filled bands have opposite slopes and (ii) the two partially filled bands have the same slope. In the first case the two pairs of lines will have opposite warping, and the combined Fermi surface will be like that of Fig. 11.7c. In the second case the two pairs of lines will have the same type of warping and the combined Fermi surface will be like that of Fig. 11.7d. As shown by the qualitative derivation of the band structure, we are in the first case and thus the qualitative Fermi surface should be like that of Fig. 11.40, which is completely nested by the vector $\vec{q}_b = 3\vec{b}/4$. The qualitative Fermi surface thus provides a simple explanation for the origin of the metal-to-insulator transition in the blue bronze.

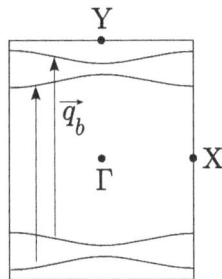

Fig. 11.40
Qualitative combined Fermi surface of the blue bronze taking into account the interchain interactions.

11.4.5 Looking for 1D systems where there seem to be none: the concept of hidden nesting

For 1D metallic systems it is usually straightforward to recognise the nesting vectors of their Fermi surface. However, there are more complex systems with several partially filled bands that lead to apparently complicated Fermi surfaces where the nesting vectors may be difficult to recognise. The band structures of complex materials sometimes exhibit avoided crossings in the vicinity of the Fermi level. These avoided crossings in band structure generally lead to avoided crossings in the associated Fermi surfaces. Therefore, a low-dimensional metal with several Fermi surfaces may give rise to apparently unnested Fermi surfaces, although their *intended* surfaces (i.e. those expected in the absence of the avoided crossings) are nested. In such cases the nesting is *hidden* by the avoided crossings. Such hidden nested Fermi surfaces can lead to CDWs, thereby destroying the nested portion of the full, combined Fermi surface and being responsible for resistivity anomalies and structural distortions. [16]

In this section we will look at one of these systems: the purple bronze KMo_6O_{17}.[4] We have already briefly considered this material in Section 10.2. The potassium purple bronze is a 2D metal above 120 K, at which temperature it undergoes a CDW phase transition, although in contrast to the blue bronze it remains metallic after the transition. Diffuse X-ray scattering as well as electron diffraction studies show the occurrence of superlattice spots at $\vec{a}*/2$, $\vec{b}*/2$ and $(\vec{a}*-\vec{b}*)/2$, i.e. there are structural modulations with wave vectors $\vec{a}/2$, $\vec{b}/2$ and $(\vec{a}-\vec{b})/2$. [17] Thus, it appeared that the low-temperature behaviour of this bronze was related to the existence of the associated nesting vectors in the Fermi surface. However, the calculated extended Hückel band structure

[4] Although the composition of the 2D purple bronzes has been traditionally considered to be $A_{0.9}Mo_6O_{17}$ (A= K, Na, Tl) it was later shown for the potassium compound to be KMo_6O_{17}.

Fig. 11.41
Bottom portion of the extended Hückel t_{2g}-block bands calculated for the $Mo_6O_{17}^-$ layers of KMo_6O_{17}. The dashed line refers to the Fermi level and $\Gamma = (0, 0)$, $X = (a^*/2, 0)$, and $L = (a^*/3, b^*/3)$.

Fig. 11.42
Fermi surfaces for the three partially filled bands of Fig. 11.41.

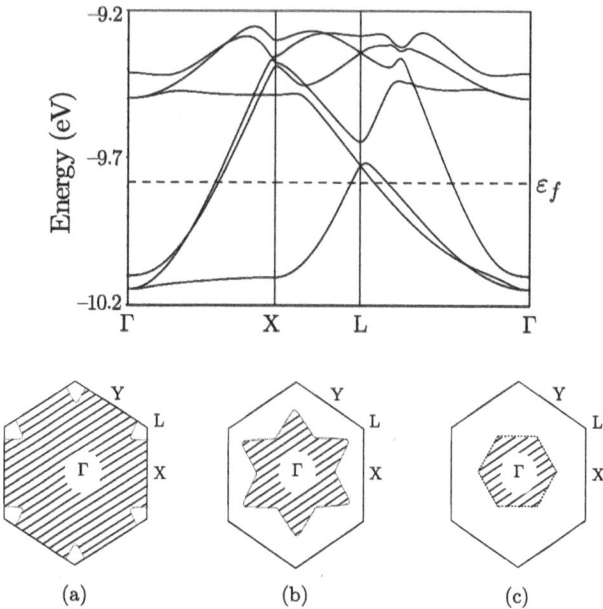

(Fig. 11.42a) shows the existence of three partially filled bands leading to three Fermi surfaces (Fig. 11.42b) that apparently do not exhibit such nesting vectors.

Crystal structure analysis

As shown in Fig. 10.3a, the bronze is made up of Mo_6O_{17} hexagonal layers, which are perpendicular to the c-direction. [18] With the formal oxidation states of Mo^{+6}, O^{2-}, and K^+, there are three electrons per formula unit to fill the t_{2g}-block bands of the Mo_6O_{17} layer. As shown in Fig. 10.3b, which outlines the contribution of the different molybdenum atoms to the DOS, only one of the three different types of molybdenum atoms – those labelled Mo_{III} – contribute appreciably to the levels around the Fermi level and below. These molybdenum atoms are those of the two inner layers of octahedra of the Mo_6O_{17} layer. Thus, in building the qualitative band structure (and Fermi surfaces) of this bronze we only need to take into account the small sublayer of the $Mo_{III}O_6$ octahedra.

We could have reached the same conclusion by examining the Mo–O distances associated with the $Mo_{II}O_6$ and $Mo_{III}O_6$ octahedra and the Mo_IO_6 tetrahedra in the crystal structure. Because of the smaller coordination number, the Mo–O distances in the Mo_IO_4 tetrahedra are short (1.759 Å on average) and the Mo_I d levels will be quite high in energy. The $Mo_{II}O_6$ and $Mo_{III}O_6$ octahedra both have three short and three long Mo–O distances. However the difference between the short and long distances is larger for $Mo_{II}O_6$ (2.101/1.771 Å) than for $Mo_{III}O_6$ (2.028/1.869 Å). Hence, only the t_{2g} levels of the $Mo_{III}O_6$ octahedra should make an important contribution to the bottom d-block bands, consistent with the DOS analysis.

The two inner sublayers associated with the $Mo_{III}O_6$ octahedra make the hexagonal Mo_2O_9 layer shown in Fig. 11.43a. This layer is generated from a

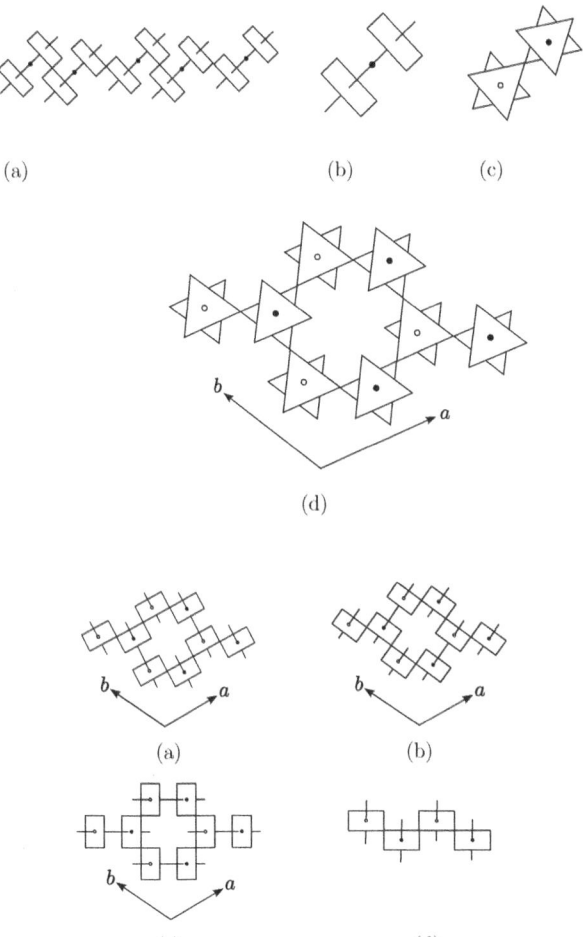

Fig. 11.43
(a) Mo_2O_9 layer associated with the $Mo_{III}O_6$ octahedra of Fig. 10.3a; (b)–(c) two alternative ways to represent the double octahedral Mo_2O_{11} unit; (d) alternative representation of the Mo_2O_9 layer.

Fig. 11.44
(a)–(c): three different but equivalent representations of the hexagonal Mo_2O_9 layer; (d) Mo_2O_{10} zigzag chain.

double octahedral Mo_2O_{11} unit (Fig. 11.43b) by sharing four oxygen atoms. The best way to understand the hexagonal nature of the layer is by using the alternative octahedral representation shown in Fig. 11.43c. Using this, the Mo_2O_9 layer, when viewed along the direction perpendicular to the layer, can be redrawn as in Fig. 11.43d. Because of the threefold rotational axis of an ideal octahedron, if we now redraw this layer according to the first octahedral representation, we must realise that there are three different but equivalent ways to do it (see Fig. 11.44). Thus we can describe the inner Mo_2O_9 layer as being built from parallel octahedral Mo_2O_{10} zigzag chains (Fig. 11.44d) parallel to the a-, b- and $(a+b)$-directions. Recognition of this feature is crucial in understanding the nature of the Fermi surface of this material. [3]

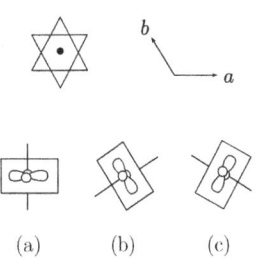

Fig. 11.45
Perspective view of the three t_{2g} orbitals of an MoO_6 octahedron in the Mo_2O_9 layers. The d orbital containing planes are aligned along the a-, b- and $(a+b)$-directions of the Mo_2O_9 layer in (a), (b) and (c), respectively.

Band structure

As shown in Fig. 11.45, the three t_{2g} orbitals of an octahedron are such that the d orbital-containing planes can be chosen to be aligned along the a-, b- and $(a+b)$-directions of the Mo_2O_9 layer, respectively. Every t_{2g} orbital thus makes

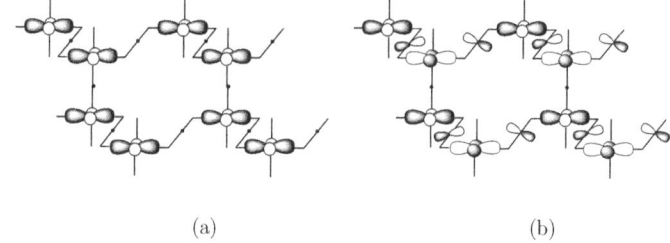

Fig. 11.46
Crystal orbitals at Γ, based on one of the three Mo t_{2g} orbitals.

δ-type interactions with those of the neighbouring parallel chains but π-type interactions along the zigzag chain. This can be easily realised by looking at the t_{2g}-based crystal orbitals for some representative k points. For instance, the two crystal orbitals – there are two Mo atoms per repeat unit of the Mo_2O_9 layer – at Γ associated with one of the t_{2g} orbitals are shown in Figs 11.46a and b, where a simplified representation of the zigzag chains has been used. When the phase is changed along the direction of the zigzag chains, oxygen $2p$ contributions in half of the shared oxygen atoms of the zigzag chain are created in the crystal orbital of Fig. 11.46a, but they are destroyed in the crystal orbital of Fig. 11.46b. However no Mo–O antibonding interactions are created or destroyed in any of the two orbitals when the phase is changed along the inter-zigzag chain direction because of the δ character of the t_{2g} orbital with respect to the interchain Mo–O–Mo bridges.[5] Hence, both crystal orbitals will exhibit a strong dispersion along the direction of the zigzag chains but a nil dispersion along the inter-zigzag chain direction. Exactly the same reasoning can be used for the other two t_{2g} orbitals and the appropriate zigzag chain direction.

[5] Readers interested in a detailed derivation of the six t_{2g}-based bands of an Mo_2O_9 layer can look at Section 14 of reference [3].

Thus, there will be three low-lying bands based on each of the three t_{2g} orbitals, these bands being dispersive along the associated intrachain direction but dispersionless along the interchain direction. In addition, due to the local orthogonality of the t_{2g} orbitals, the three chain bands resulting from the three t_{2g} orbitals are practically independent of one another to a first approximation. Thus, for qualitative purposes, the three bottom t_{2g}-block bands of KMo_6O_{17} can be approximated by the three independent 1D bands resulting from the three t_{2g} orbitals.

Fermi surface and hidden nesting

The bottom t_{2g}-block bands of KMo_6O_{17} must be filled with three electrons. These bands can be seen as the superposition of three almost independent half-filled 1D bands. As a result, the full ideal Fermi surface of KMo_6O_{17} can be obtained by the superposition of the three Fermi surfaces associated with these bands. In a 2D representation, the Fermi surface of an ideal 1D system is given by two parallel lines perpendicular to the chain direction. Note that for the present purpose the two representations of Fig. 11.47 are equivalent. The hexagonal Brillouin zone of Fig. 11.47a can be redrawn in the form of the rectangular one of Fig. 11.47b, which has exactly the same area. Note that in Fig. 11.47b, $\Gamma \to X'$ is perpendicular to $\Gamma \to Y$ and is therefore parallel to the

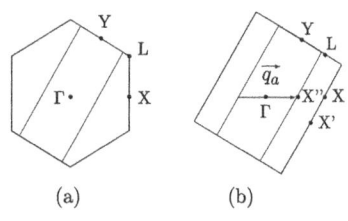

Fig. 11.47
Representation of one hidden 1D Fermi surface of KMo_6O_{17} in (a) the first Brillouin zone, and (b) a rectangular zone with the same area as the first Brillouin zone.

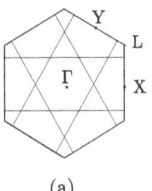 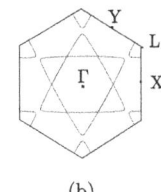 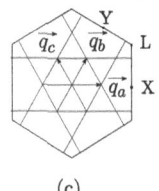

Fig. 11.48
(a) Qualitative Fermi surface for KMo_6O_{17} as a combination of hidden 1D Fermi surfaces. (b) Fermi surface for KMo_6O_{17} where the weak hybridisation of the hidden Fermi surfaces has been taken into account. (c) Expected nesting vectors.

a-direction. In other words, the two lines of the Fermi surface of Fig. 11.47b are perpendicular to the a-direction and therefore such a Fermi surface is associated with the Mo_2O_{10} zigzag chain running along the a-direction.

The full ideal Fermi surface can thus be obtained by superposing three such ideal Fermi surfaces with pairs of lines perpendicular to the a-, b- and ($a+b$)-directions. Such ideal full Fermi surface is shown in Fig. 11.48a. In the real Mo_2O_9 layer the interaction between the three different types of chain is not exactly nil and one should expect a weak interaction that will result in the opening of small gaps at the crossing points of the lines in Fig. 11.48a, thus leading to the expected Fermi surface of Fig. 11.48b. This Fermi surface is the same as that obtained by superposing the three calculated Fermi surfaces of Fig. 11.42b. [3]

There is a last point that should be considered: what nesting vectors are thought to be responsible for the CDW instabilities? For an ideal 1D Fermi surface there is an infinite number of possible nesting vectors (Fig. 11.3a). Is there any reason to assume that some of them will be more effective in stabilising the system? The extent of electronic instability is enhanced when the area of the nested Fermi surface increases. The appropriate nesting vectors for a Fermi surface such as that of Fig. 11.48a are those allowing the simultaneous nesting of more than one set of 1D Fermi surfaces. These nesting vectors are shown in Fig. 11.48. The vectors $\vec{q}_a$, $\vec{q}_b$ and $\vec{q}_c$ each nest two sets of 1D Fermi surfaces and they are parallel to the a^*- and b^*- and (a^*-b^*)-directions, respectively. Since the band is half-filled and $\Gamma \rightarrow X'$ in Fig. 11.47b is $\sqrt{3}a^*/4$, it follows that the $\vec{q}_a$ nesting vector is given by $\vec{a}^*/2$. Likewise the $\vec{q}_b$ and $\vec{q}_c$ nesting vectors are given by $\vec{b}^*/2$ and $(\vec{a}^*-\vec{b}^*)/2$, respectively. [3]

The KMo_6O_{17} bronze is thus a system with a 2D Fermi surface resulting from the weak hybridisation of three 1D Fermi surfaces. KMo_6O_{17} is thus a 2D metal, which at low temperature exhibits the behaviour of 1D metals. Since the nesting of the real Fermi surfaces is not as complete as those of the ideal hidden Fermi surfaces because of the weak hybridisation, after the CDW phase transition some small pieces of the Fermi surface will remain as small electron and hole pockets in the region of the avoided crossings. Thus the system keeps its metallic behaviour after the 120 K transition although the number of carriers has been considerably lowered.

The concept of hidden nesting is not a peculiarity of the KMo_6O_{17} bronze. It has also been found necessary to explain the nature of resistivity anomalies and structural modulations of a number of transition metal oxides, chalcogenides, bronzes, etc. Note that the three different hidden 1D systems of KMo_6O_{17} are identical, but in other cases, as for instance in η- and γ-Mo_4O_{11} or the

Fermi surface and low-dimensional metals

monophosphate tungsten bronzes, they are different. This leads to different instabilities and to a seemingly complex low-temperature behaviour. Using orbital interaction analysis to look for hidden nesting in the real calculated Fermi surfaces is a useful way to rationalise many of these observations.

11.5 Low-dimensional molecular conductors

We conclude our analysis of the electronic structure of low-dimensional systems by considering some *molecular* solids exhibiting metallic conductivity. Because of their molecular nature, the interactions leading to the partially filled bands are considerably weaker than those of most of the systems examined so far in this book. This feature, which lies at the origin of the interesting physical behaviour of these solids (see below), makes it more difficult to predict their band structure and Fermi surface solely on the basis of an analysis of the crystal structure as discussed above. In addition, these systems usually exhibit complex crystal structures. However, their band structures are conceptually simple. The only difference with the inorganic solids examined so far in this chapter lies in the strategy used to analyse their band structures.

The search for molecular solids exhibiting high conductivity dates back to the 1950s. However, it was only in 1973, with the successful marriage of TTF and TCNQ (see Fig. 11.49 for illustration of the different acronyms for molecular species discussed in this section) that the challenge of preparing a metallic solid exclusively built from closed-shell molecular units was

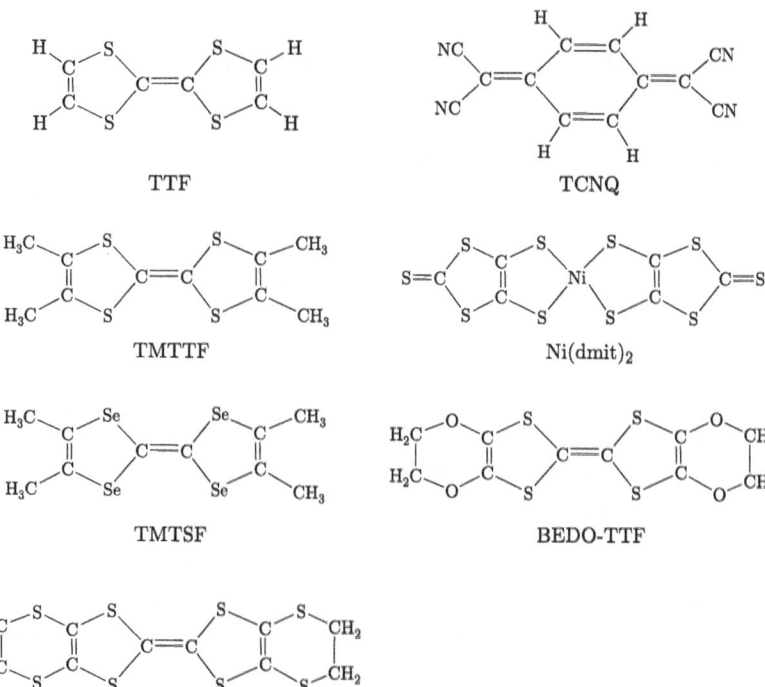

Fig. 11.49
Some molecules commonly used in the preparation of molecular metals discussed in this chapter.

realised. [19] TTF–TCNQ exhibits a crystal structure built up from parallel segregated stacks of donors (TTF) and acceptors (TCNQ). Due to the π–π interactions along the stacks and the weak interactions between molecular chains, the electrical conductivity is highly anisotropic. TTF–TCNQ is thus a pseudo-1D metal and exhibits a series of CDW transitions, which successively destroy the metallic conductivity of the TCNQ and TTF chains. [20] The discovery of TTF–TCNQ and its remarkable low temperature behaviour triggered an enormous interest in the physics of molecular conductors, which soon led to the discovery of superconductivity in the molecular solid $(TMTSF)_2PF_6$. [21] Since then, a quite impressive number of molecular conductors and superconductors has been prepared using different donor molecules, examples including TMTSF, BEDT–TTF, and BEDO–TTF. Molecular acceptors like $Ni(dmit)_2$ have also been used. A wealth of new fundamental phenomena has also been unveiled. [22, 23, 24]

Although these solids are metals for which a delocalised one-electron band description is in principle appropriate, a word of caution is in order. Throughout this book we have assumed that a delocalised description of the electrons is valid. This is the case when the bandwidth (W) is large compared with the on-site repulsion (U). When $U > W$, the electrons prefer to be localised on lattice sites because the on-site repulsion necessarily associated with the delocalisation of electrons overrides the stabilisation generated by such delocalisation. [25] In that case conductivity is activated because electron hopping from one site to the other leads to a situation in which two electrons reside on a single site, thereby causing on-site repulsion (we will discuss this problem in the next chapter). Thus, if the dispersion of the partially filled bands is small compared with the on-site repulsion, the system will prefer to have one electron localised in each site of the lattice and it will be a semiconductor.

In molecular conductors two facts often conspire to make U and W comparable. First, the molecular nature of the solids leads to weak intermolecular interactions and thus to a small band dispersion. Second, the need to use systems with relatively extended π orbitals so as to force strong enough interactions along the stacks, leads also to small values of U because the electrons (or holes) that remain localised in one molecule of the stack are, in fact, quite delocalised within the molecule. Thus, U and W are not very different and these solids lie at the border of the validity range of the localised and delocalised descriptions of their electronic structure, thus leading to a competition between delocalised (i.e. metallic) and localised (i.e. semiconducting) electronic states. In this case, in addition to the usual electronic instabilities of the low-dimensional metals, at a certain temperature the metallic system can undergo a different metal-to-semiconductor transition as a result of the occurrence of an electronic localisation. In short, there are several ground states relatively close in energy, and very rich phase diagrams result. Relatively small differences in temperature, pressure, or chemical composition can have a pronounced effect on the physical properties of molecular metals. [22, 23] In the rest of this chapter we will focus on the delocalised description of molecular conductors so as to be able to describe the large body of experimental information related to the metallic state of these systems.

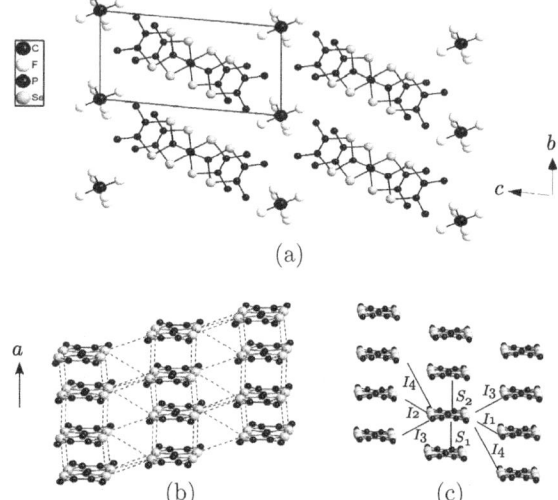

Fig. 11.50
(TMTSF)$_2$PF$_6$: (a) crystal structure, (b) donor layers where the Se...Se contacts shorter than the sum of the van der Waals radii are shown as dotted lines, and (c) donor layers where the six different transfer integrals used in the analytical treatments of the electronic structure are defined. Hydrogen atoms are not shown for clarity.

11.5.1 An archetypal molecular metal: (TMTSF)$_2$PF$_6$

(TMTSF)$_2$PF$_6$ was the first molecular metal for which a superconducting state could be stabilised (under pressure). The crystal structure of this salt (Fig. 11.50) contains layers of TMTSF donors (see Fig. 11.49) separated by layers of isolated PF$_6$ anions. Since these PF$_6$ anions are isolated, their energy levels cannot lead to energy bands. In contrast, there are many donor...donor interactions (mainly through short Se...Se contacts (see Fig. 11.50b)), which can lead to the formation of energy bands based on the TMTSF π-type molecular orbitals. As mentioned above, because of the molecular nature of charge transfer salts, the intermolecular interactions leading to the spread of the molecular energy levels into bands are weaker than the intramolecular chemical bonding forces that determine the molecular energy levels of the donor and the acceptor. Consequently, the highest occupied energy band of this salt (which in fact will be partially filled as a result of the charge transfer) should be adequately described by considering just the HOMO of TMTSF.[6] Because of the stoichiometry, one electron every two TMTSF molecules is transferred to PF$_6$, so that the TMTSF donors have an average charge of +1/2. This means that the HOMO bands should be partially filled and thus, if the HOMO of TMTSF is involved in HOMO...HOMO interactions through the crystal leading to the creation of dispersive HOMO bands, the (TMTSF)$_2$PF$_6$ salt could be metallic.

The HOMO of TMTSF is shown in Fig. 11.51a. It is a π-type orbital with strong contributions from the selenium $4p$ orbitals. Thus, the overlap between HOMOs of adjacent TMTSF donors along the chain direction of the crystal (i.e. the a-direction) will be important. Since the unit cell of the salt contains two TMTSF donors, two HOMO combinations can be formed and every one of them will lead to an energy band through interactions between unit cells. These two bands should be filled with only three electrons because of the charge transfer and thus, at least one of them should be partially filled. The

---

[6] A number of molecular conductors, the so-called *two band* systems, are an exception and need the explicit consideration of two molecular orbitals of the same chemical species to correctly account for their electronic structures. [26]

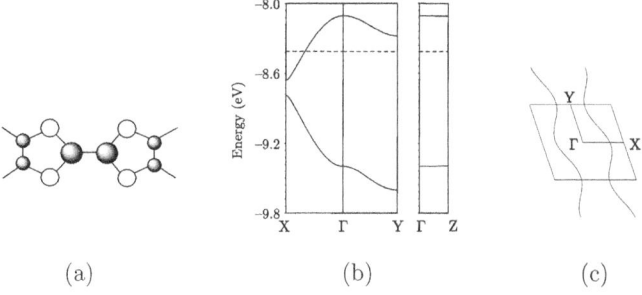

Fig. 11.51
HOMO of TMTSF (a), band structure (b), and Fermi surface (c) calculated for $(TMTSF)_2PF_6$. The dashed line in (b) denotes the Fermi level. $\Gamma = (0, 0, 0)$, $X = (a^*/2, 0, 0)$, $Y = (0, b^*/2, 0)$, and $Z = (0, 0, c^*/2)$.

calculated extended Hückel band structure in the region of the Fermi level is shown in Fig. 11.51b, where the dashed line refers to the Fermi level. The two bands shown there are almost completely built from the HOMO of the TMTSF donors and they are well separated from the other filled or unfilled energy bands, as expected. Since there are three electrons per unit cell to fill the two bands of Fig. 11.51b, the lower band is completely filled and the upper one is half filled. What the band structure of Fig. 11.51b is also telling us is something about the strength of the HOMO...HOMO interactions along the crystal. For instance, the band dispersion along the chain direction (i.e. when going from Γ to X) is clearly stronger than the band dispersion along the interchain direction of the TMTSF layers (i.e. along the Γ to Y direction). The band dispersion along the interlayer direction (i.e. along Γ to Z) is practically nil. These observations are easily understandable when taking into account the π-type nature of the HOMO and the details of the crystal structure.

Since the dispersion of the partially filled band is quite sizable, $(TMTSF)_2PF_6$ should be a pseudo-1D metal with higher conductivity along the chain direction. This is in agreement with the calculated Fermi surface shown in Fig. 11.51c. This Fermi surface is open and contains warped lines approximately perpendicular to the chain direction. In addition, this Fermi surface is quite well nested. As a consequence, $(TMTSF)_2PF_6$ loses its metallic character at 12 K due to a spin density wave instability. The metallic character can, however, be restored under pressure of 9 kbar, and the salt becomes superconducting at 0.9 K. [21]

11.5.2 Chemically modifying the electronic structure of molecular conductors

The two main parameters that chemists try to manipulate in the search for new molecular conductors are the band filling and the transfer integrals. Finding simple and novel strategies towards this end lies at the heart of the chemist's artistry in this area and since the very beginning has been one of the strongest driving forces in the field. For instance, it was the wish to circumvent the usual activated conductivity associated with 1D systems with one electron per site (i.e. formally a half-filled 1D band in a one-electron description, for which a localised description is usually more pertinent; see next chapter) that led to the tendency to avoid 1:1 stoichiometry when the counter-ion was a singly charged species (i.e. controlling the band filling). The synthesis of the celebrated BEDT–TTF (see Fig. 11.49) had its origin in the attempt to avoid the well known CDW or SDW instabilities of 1D systems by using a donor the shape of which tends to preclude the chain-stacking typical of planar molecules while at the same time increasing the dimensionality of the HOMO...HOMO electronic interactions by increasing the number of sulphur atoms in the donor (i.e. controlling the transfer integrals or intermolecular interactions).

As an illustrative example of how the conductivity regime can be changed by modifying the filling of the bands let us compare two compounds formulated as $(TMTSF)_3[Ti_2F_8(C_2O_4)]$ and $(TMTSF)_3[Ta_2F_{11}]$, both with essentially identical stacks but with different anion charges. [27, 28] As shown in Fig. 11.52a the donor stacks of the first are built from trimerised chains of the TMTSF donor. Since the repeat unit of the donor network contains three donors, the band structure near the Fermi level (see Fig. 11.53a) contains three bands almost completely based on the donor's HOMO. Because of the anion charge, i.e. $[Ti_2F_8(C_2O_4)]^{2-}$, two holes must be housed on these bands so that the upper one is empty. Thus as a consequence of the band gap separating the filled and empty bands, the salt exhibits activated conductivity. The $Ta_2F_{11}^-$ and $[Ti_2F_8(C_2O_4)]^{2-}$ anions have a similar shape but different charge. The $(TMTSF)_3[Ta_2F_{11}]$ salt is indeed structurally very similar (see Fig. 11.53b) and, as a consequence, the band structures of the two salts are almost identical

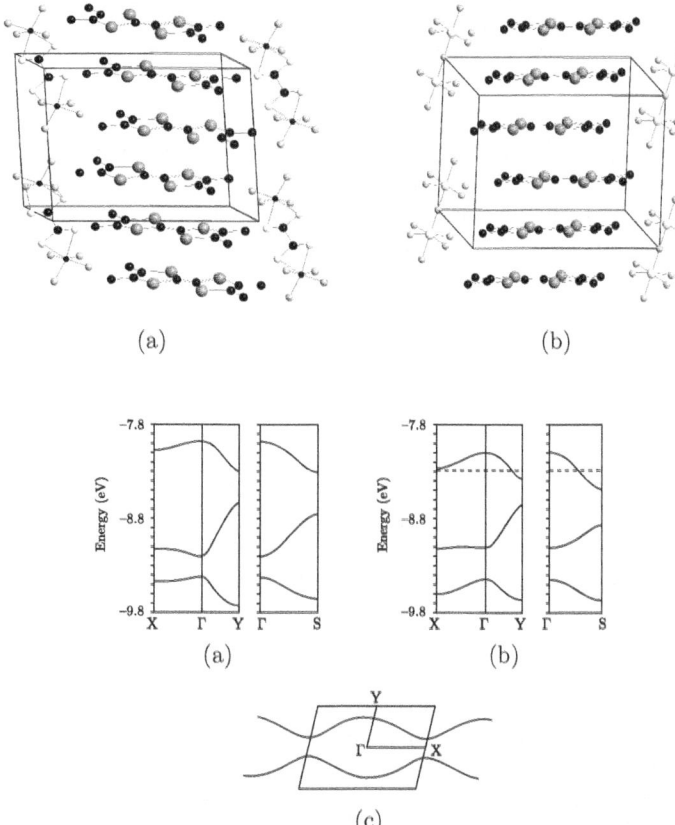

Fig. 11.52
Crystal structure of: (a) $(TMTSF)_3[Ti_2F_8(C_2O_4)]$ and (b) $(TMTSF)_3[Ta_2F_{11}]$. Hydrogen atoms are not shown.

Fig. 11.53
Band structures calculated for the donor lattices of $(TMTSF)_3[Ti_2F_8(C_2O_4)]$ (a) and $(TMTSF)_3Ta_2F_{11}$ (b). The Fermi surface of the latter is shown in (c). The dashed line in (b) refers to the Fermi level. $\Gamma = (0,0)$, $X = (a^*/2, 0)$, $Y = (0, b^*/2)$, and $S = (a^*/2, -b^*/2s)$.

(compare Fig. 11.53a and b). Note the larger dispersion of the bands along the b^*-direction (i.e. from Γ to Y) so that these salts must be considered pseudo-1D conductors. However, for (TMTSF)$_3$[Ta$_2$F$_{11}$] the HOMO bands must bear only one hole so that the upper band in Fig. 11.53b must be half-filled and, from the viewpoint of one-electron band theory, the system should exhibit metallic behaviour in contrast with the semiconducting one of (TMTSF)$_3$[Ti$_2$F$_8$(C$_2$O$_4$)]. These simple ideas are in agreement with reported conductivity measurements. Since the band dispersion of the upper band is larger along the donor stacks, the Fermi level does not cut this band along the a^*-direction, and the Fermi surface is open (Fig. 11.53c). Note however that the warping of the lines is quite large, a result of the existence of several transverse interstack Se...Se contacts. Thus knowledge of the structure and conductivity type of a given salt suggests that, using an anion with relatively similar shape and chemical features so as to keep the general structure of the salt, but a different charge, one can change the band filling and the transport properties. This strategy has been very fruitful.

The second approach chemists use in order to modify the transport properties of molecular conductors is to induce changes in some or all of the transfer integrals that control the band-structure topology. In other words, the kind of network associated with the conducting species is kept but variations in the strength of the intermolecular interactions within this lattice are introduced. There are several ways to do this, by using counter-ions of the same charge but slightly different shape, by changing the size or symmetry of the guest molecules that are often included in the crystal structure of molecular conductors, by modifying the shape or the nature of selected atoms in the donor, etc. As a simple example let us consider the case of the three salts (NMe$_4$)[Ni(dmit)$_2$]$_2$, (NHMe$_3$)[Ni(dmit)$_2$]$_2$, and (NH$_2$Me$_2$)[Ni(dmit)$_2$]$_2$. [29] The conducting layers of these three salts are structurally very similar and are built from Ni(dmit)$_2$ stacks and have two alternating types of intermolecular overlaps. As shown in Fig. 11.54, one of these overlap modes, the so-called intra mode associated with the shorter S...S contacts along the stacks, is kept

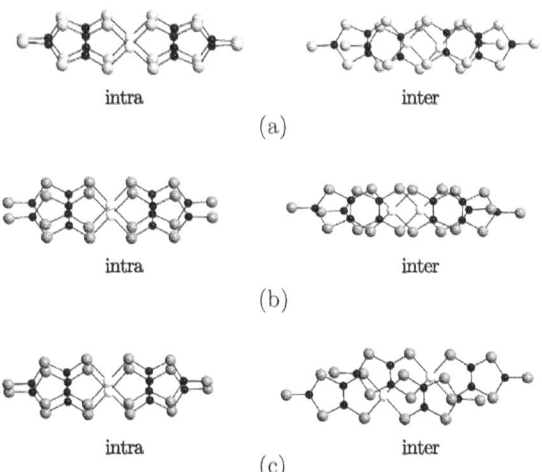

Fig. 11.54
Intermolecular overlap modes along the chain stacks in the salts (NMe$_4$)[Ni(dmit)$_2$]$_2$ (a), (NHMe$_3$)[Ni(dmit)$_2$]$_2$ (b), and (NH$_2$Me$_2$)[Ni(dmit)$_2$]$_2$ (c). Hydrogen atoms are not shown. The inter and intra labels refer to the overlap between and within the [Ni(dmit)$_2$]$_2$ units.

almost unchanged in the three salts so that these layers may be considered to be built from the same type of [Ni(dmit)$_2$]$_2$ units. Note that, in contrast with the preceding example, the Ni(dmit)$_2$ building block of the conducting layers is an acceptor (i.e. the partially filled bands will originate from the LUMO of the molecular unit).

Changing the shape of the cation from NMe$_4$ to NHMe$_3$ (see Figs 11.54a and b) does not appreciably change the overlap mode between these units (inter mode) along the stacks. However, further substitution of hydrogen for a methyl group (i.e. using the NH$_2$Me$_2$ cation) leads to a salt in which the overlap between these units has noticeably changed (see Fig. 11.54c). Simply speaking, the less voluminous doubly substituted cation has moved slightly differently the [Ni(dmit)$_2$]$_2$ units with respect to each other. All of these salts have very small interstack transfer integrals so that they are 1D conductors. However, the two different transfer integrals in the NMe$_4$ and NHMe$_3$ salts along the stacks are large and very similar, so the LUMO...LUMO interactions along the chains are in fact very uniform. This leads to very sizable band dispersions. Taking into account the stoichiometry, the lowest of the two LUMO bands is only half-filled and thus, because of the band dispersion, the two salts are metallic. Although the interplanar spacing is quite similar in the three salts, it is clear from Fig. 11.54 that the overlap between [Ni(dmit)$_2$]$_2$ units is different in the NH$_2$Me$_2$ salt – a consequence of the different cation shapes. As a matter of fact, the associated transfer integral becomes four times smaller and results in very flat LUMO bands. As a consequence, we are in the typical situation where an insulating localised state is more stable than the delocalised metallic state and the salt exhibits activated conductivity. From a more chemical point of view, it can be said that the electron in the lowest combination of the LUMOs of the two Ni(dmit)$_2$ molecules forming a [Ni(dmit)$_2$]$_2$ unit, Ψ_{LUMO}^+, can delocalise along the chain because of the favourable interunit mode of overlap in the (NMe$_4$)[Ni(dmit)$_2$]$_2$ and (NHMe$_3$)[Ni(dmit)$_2$]$_2$ salts. However, in the (NH$_2$Me$_2$)[Ni(dmit)$_2$]$_2$ salt it remains localised in each unit because of the less favourable mode of overlap between the [Ni(dmit)$_2$]$_2$ units. Thus, when one electron hops from one [Ni(dmit)$_2$]$_2$ unit to the next one, an energy repulsion equal to U is generated, and this competes with the hopping (see Chapter 12). Consequently, the conductivity is activated for this salt. Changing the size of the counter-ion, modifying the anion–cation hydrogen-bonding interactions, and using different solvent molecules to fill the structural holes are common strategies used to modify the properties of molecular conductors.

11.5.3 Structurally complex materials with simple band structures

As the reader can guess, after looking at the crystal structures of the molecular conductors discussed above, the crystal packing of these solids results from a very subtle balance between several contributions, such as hydrogen bonding, π–π interactions, cation–anion electrostatic interactions, etc. This makes the prediction of the final outcome difficult but at the same time makes the structure easily amenable to modification. The control of the conductivity of molecular conductors still relies on quite naïve and approximate ideas. The main reason for this is that the factors that determine the structure of these materials are not the same as those that determine their transport properties. To a large extent, conductivity is determined by the nature of the states near the Fermi level, which originates from π-type orbitals of the donor and/or acceptor. The way in which these orbitals overlap in the solid state and either spread into energy bands or remain as localised states is, of course, imposed by the crystal packing. Except for the $\pi-\pi$ interactions, none of the main contributors to the energy balance leading to the final crystal packing contributes to the dispersion of the π-type based bands near the Fermi level. In order to build a molecular conductor with predetermined transport properties, we must first know the nature of the π-type molecular level that is going to lead to the bands near the Fermi level. Then we must analyse what type of crystal packing will result in the appropriate overlap between the π levels of adjacent molecules so that the required band structure topology is obtained. Finally, the molecules must be directed to that crystal structure by functionalising or modifying their shape, using counter-ions of different shapes, etc. Thus a first step towards a real control of the transport properties of molecular conductors is the ability to rationalise in simple, yet precise, terms why apparently slight differences between two structurally related phases can be at the origin of a vastly different conductivity behaviour. Only if we are able to understand what kind of changes we must induce in the electronic structure of a given structural family in order to achieve the intended conductivity behaviour, as very naïvely discussed in the previous section, can we think of a way to direct the structure in an appropriate way.

Despite the inherent difficulty of such approach, it must be said that molecular conductors have relatively simple band structures. As mentioned before, the intermolecular interactions leading to the broadening of the molecular energy levels into bands in molecular solids are weaker than the intramolecular chemical bonding forces that determine the molecular energy levels. In practise, most molecular conductors, like $(TMTSF)_2PF_6$ (Fig. 11.50), are layered solids made of a molecular donor with an extended π-type HOMO that can spread into bands through interaction with those of the neighbouring donors, and anions with quite localised acceptor levels. From the electronic viewpoint the donor and acceptor layers are practically independent and the main role of the anions is to create holes in the HOMO-based bands of the donor layers. Thus to a very good approximation the electronic structure of a family of molecular conductors can be understood by considering the shape of the HOMO (or the LUMO when the molecule leading to the conducting layers is an acceptor and the counterion a donor) and using very basic symmetry and overlap ideas to evaluate the different transfer integrals, i.e. the β integrals describing the strength of the interaction between two HOMOs (or two LUMOs for acceptors) in adjacent molecules of the layer. In the field of molecular conductors such transfer integrals are usually referred to as the transfer or hopping integral t, and to avoid any confusion we will use the same labelling in this section. Once the values of these transfer integrals are known it is a simple matter to derive the band structure of the system and test the effect of each type of interaction (i.e. transfer integral) on the band structure.

To understand how the HOMOs combine to form the conduction bands of the solid as a result of the molecular interactions (t_j), we consider a simple example, the (BEDO–TTF)$_2.4$I$_3$ phase, [30] the donor sublattice of which is illustrated in Fig. 11.55. Layers of this type are said to exhibit a β'' packing motif. [31] There is a substantial number of different salts with this type of packing, which we will consider in detail below. This layer has one molecule per repeat unit, so it is a particularly simple case. As indicated by the stoichiometric formula, BEDO–TTF has been oxidised (BEDO–TTF$^{+0.417}$) and this results in a partially occupied HOMO band in the solid. Let us call α the energy of the BEDO HOMO. Every molecule of this layer interacts with six neighbours: along the step-chains in the $\pm a$-direction with transfer integral t_{sc} = 0.146 eV, along the π-stacks in the $\pm b$-direction with $t_{\pi s}$ = –0.049 eV, and along the π-chains in the $\pm (a - b)$-direction with $t_{\pi c}$ = –0.142 eV. With this information in mind we can write a simple solution for the band dispersion, $E(\vec{k})$. Since every HOMO interacts with another HOMO along $\pm a$ (t_{sc}), along $\pm b$ ($t_{\pi s}$), and along $\pm (a - b)$ ($t_{\pi c}$), the energy of the Bloch function associated with the HOMO can be simply written as:

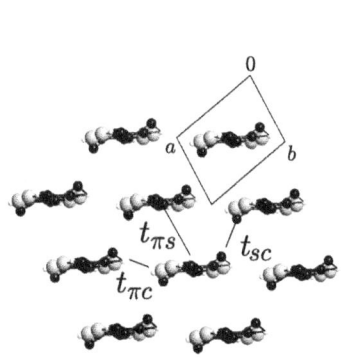

Fig. 11.55
Structure of the β'' layer of the BEDO–TTF molecules in the (BEDO–TTF)$_2.4$I$_3$ salt. The different transfer integrals needed for the tight binding treatment are indicated. Hydrogen atoms are not shown.

$$E(\vec{k}) = \alpha + 2[t_{sc}\cos(2\pi k_a) + t_{\pi s}\cos(2\pi k_b) + t_{\pi c}\cos(2\pi (k_a - k_b))]$$

(11.6)

The band dispersion curve obtained from this equation (Fig. 11.56a) compares very well with that obtained from a full extended Hückel calculation. [32] The advantage of having an analytical formula is that we can understand all details of the band structure in a very simple way. For instance, given the relatively small value of $t_{\pi s}$ and the large value of t_{sc} one may be surprised by the small dispersion along the $\Gamma \rightarrow$ X direction but the large one along $\Gamma \rightarrow$ Y. However, this is a system with three different interactions per site, and simply by looking at eqn (11.3), it becomes clear that the weak dispersion along $\Gamma \rightarrow$ X results from cancellation between t_{sc} and $t_{\pi c}$, while the large dispersion along $\Gamma \rightarrow$ Y is a consequence of cooperation between $t_{\pi s}$ and $t_{\pi c}$. It is also clear from this equation that the highest energy region will lie around Y and that the energy will go down along any line starting from this k point. The Fermi surface will therefore most likely be a closed line centred on Y.

The calculated Fermi surfaces according to eqn (11.3) and the full extended Hückel calculation are shown in Figs 11.56b and c, respectively. Both calculations provide almost identical ellipses centred around Y, as expected. As

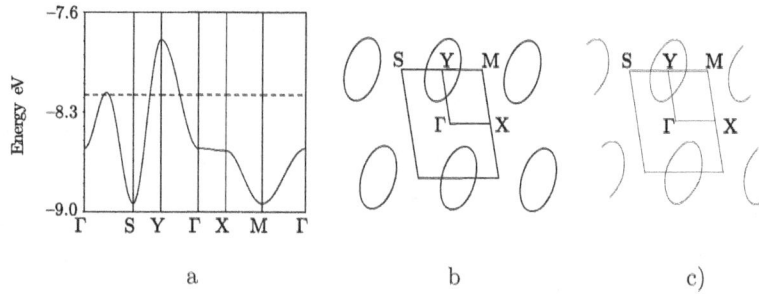

Fig. 11.56
Electronic structure of the β'' layer of (BEDO–TTF)$_2.4$I$_3$ salt: band structure (a) and Fermi surface (b) obtained by application of eqn (11.3); (c) Fermi surface obtained from the extended Hückel calculation.

far as the Fermi surface is concerned, the agreement between the two types of calculation and even with more sophisticated first-principles DFT-type calculations, is almost invariably excellent. Cases in which there are clear differences with the experimentally based Fermi surfaces are rare. There may be some differences between the two types of calculation regarding the lowest part of the band structure because in the full extended Hückel calculation molecular orbitals slightly lower in energy are included in the calculation (but not in the simple HOMO-based calculation) and the associated bands may interact and thus move the lowest part of the HOMO band somewhat. However, these changes have hardly any influence on the Fermi surface of the material, which is what determines the conductivity behaviour. This is because the Fermi level is located at energies sufficiently higher than the bottom of the band. Thus one may conclude that the simple one-orbital approach provides an excellent description of the Fermi surface of the material.

To extend this simple one-orbital picture to materials where the unit cell contains two molecules is usually not difficult. For instance, as shown in Fig. 11.50c, the donor layers of $(TMTSF)_2PF_6$ contain six different types of HOMO...HOMO interaction: two of them occur along the stacks in the a-direction, S_1 and S_2, and four are associated with interstack interactions, I_1 to I_4. The energy dispersion of the two bands can be obtained following standard procedures (i.e. evaluating the matrix elements of the 2×2 secular determinant and solving the associated second-order equation) and leads to the equation:

$$E(\vec{k}) = \alpha + 2[t_{I_3} \cos(2\pi k_b) + t_{I_4} \cos(2\pi (k_a - k_b))]$$
$$\pm \mid t_{S_2} + t_{S_1} \exp(2i\pi k_a) + t_{I_2} \exp(i2\pi k_b)$$
$$+ t_{I_1} \exp(i2\pi (k_a - k_b)) \mid \qquad (11.7)$$

which allows an in-depth discussion of the dependence of the band structure, Fermi surface, and the nature of the possible nesting vectors etc. for salts of this family. [33]

Extension of this analytical one-orbital approach to systems with more molecules in the repeat unit soon becomes too cumbersome. In these cases, it is more practical to directly build the Bloch orbitals associated with the HOMO of each molecule in the repeat unit, then solve the corresponding secular equations for different values of the k wave vector. This is a very simple procedure once the values of the different transfer integrals are specified. Note that when this is done, the energy of the HOMO (or LUMO for the acceptors) of the symmetry-inequivalent molecules of the repeat unit may be different. This type of calculation is extremely fast and allows a detailed analysis of the relationship between the different transfer integrals or the relative energy of the HOMOs of the symmetry inequivalent molecules with the band structure and Fermi surface of the salt. So long as the transfer integrals and orbital energies chosen are realistic, these calculations are not only a very successful tool for understanding the results of more sophisticated calculations, but also provide a predictive description of the Fermi surface.

Finally, let us briefly comment on the estimation of the different parameters used in these calculations. If there is just one type of symmetry-inequivalent molecule, the energy is not really needed since it can be taken as the origin of the energies. If there is more than one symmetry-inequivalent molecule, the energy differences are needed, and these can always be obtained from a simple molecular calculation.

There are several models in the literature for estimating the transfer integrals, although two of these are more frequently used. The simplest one involves making the transfer integral proportional to the corresponding overlap integral, i.e. $t_{ij} = k S_{ij}$, where k is a constant that is often taken as the energy of the HOMO and S_{ij} is the overlap integral between the ith and j^{th} molecular orbital (HOMO for donors or LUMO for acceptors) on neighbouring molecules.

The more widely used model is the so-called dimer splitting approximation (DSA). [33] This involves calculating the energy splitting of the appropriate molecular orbital for each nearest-neighbour pair. The transfer integral is taken to be half the dimer splitting. Care should be taken in choosing the appropriate sign for the transfer integrals. Since we are dealing with π-type orbitals, if the transfer integral is positive (overlap integral negative), the orbitals of the two molecules will combine with opposite sign in the lower energy (i.e. bonding) orbital of the dimer. Conversely, they will combine with the same sign if the transfer integral is negative (overlap integral positive). Strictly speaking, the DSA is only valid if the two interacting molecules are identical, so that the two interacting molecular orbitals have the same energy. If this is not the case, so long as the initial energy difference is not very large, subtraction of such initial energy differences from the dimer splitting should be carried out. Although it is not frequently the case, if the initial separation is large, the transfer integrals must be calculated directly from the molecular orbital energies and the Hamiltonian and overlap matrix elements between the two orbitals (i.e. $t_{ij} = H_{ij} - \alpha S_{ij}$, where $\alpha = \frac{\alpha_i - \alpha_j}{2}$ and H_{ij} is the Hamiltonian matrix element between the two HOMOs). [32] With such an implementation, this very simple one-orbital model provides a fast and reliable framework by which the properties of molecular conductors may be intuitively rationalised.

11.5.4 A case study: 1D vs 2D character of the carriers in some α phases of BEDT-TTF

In this section we illustrate the usefulness of the one-orbital simulation approach by considering just a case study: how an apparently innocuous structural detail can lie behind the beautiful physics of the α-(BEDT-TTF)$_2$MHg(XCN)$_4$ (M = Tl, K, Rb, NH$_4$; X = S, Se) molecular metals. [34, 35] These phases are 2D metals and have produced many intriguing results. Their room-temperature crystal and band structures as well as Fermi surfaces are very similar for all salts of this series. However, their low-temperature physical behaviour is remarkably different. All of these salts exhibit Shubnikov–de Haas oscillations associated with closed 2D sections of the Fermi surface. At the same time, some of the salts exhibit a low-temperature resistivity anomaly, which originates from a CDW instability due to the presence of nested 1D parts of the Fermi surface. [36] Thus, despite being 2D metals, these salts must have two different types of charge carrier with different dimensionality.

A perspective view of one BEDT–TTF layer in these salts is shown in Fig. 11.57. This layer is constructed from two non-equivalent stacks parallel to the c axis. One of them (I) contains only one type of donor, I, while the other (II) contains two different donors, types II and III, which alternate. Within these donor layers there are many short S...S intermolecular contacts. Assuming the usual oxidation states of M^+, Hg^{2+}, and $(SCN)^-$, the average charge per BEDT–TTF donor molecule is +1/2. The non-integral charge of the BEDT–TTF donors and the existence of short S...S contacts between the different donors of the layer are at the origin of the metallic conductivity of these salts.

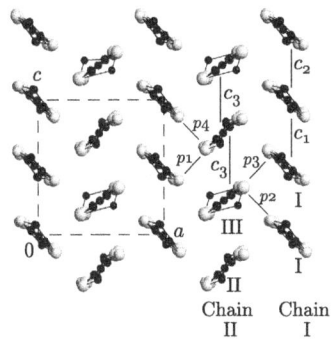

Fig. 11.57
Donor layers in the crystal structure of the α-(BEDT–TTF)$_2$MHg(XCN)$_4$ family of salts. The different types of chains, donors and interactions are labelled. Hydrogen atoms are not shown.

The calculated extended Hückel band structure near the Fermi level for a representative example of these salts, i.e. α-(BEDT–TTF)$_2$ KHg(SCN)$_4$, is shown in Fig. 11.58a. As expected, the four bands in this figure are almost exclusively made of the HOMO of the four BEDT–TTF molecules of the unit cell and hence from now on we will refer to these four bands as the HOMO bands. With the average oxidation state of (BEDT–TTF)$^{+1/2}$, there are six electrons to fill these bands. Since the third and fourth ones (from the bottom) overlap, the system should be metallic and the Fermi surface should exhibit both electron and hole contributions. The calculated Fermi surface is shown in Fig. 11.58b. The hole contribution (i.e. the part of the Fermi surface arising from the almost filled band) is closed, whereas the electron contribution (i.e. the part of the Fermi surface arising from the almost empty band) is open and parallel to the c^* direction. These results are very similar to those for all other salts of this family. They correctly describe the main aspects of the electronic structure of these salts, i.e. their metallic character and the existence of both 1D and 2D types of carrier.

We can now proceed to a simulation study to enquire why the Fermi surface of these salts possess both 1D and 2D portions. Is there actually a structural motif causing this, i.e. are there actually chains in the BEDT–TTF slabs of

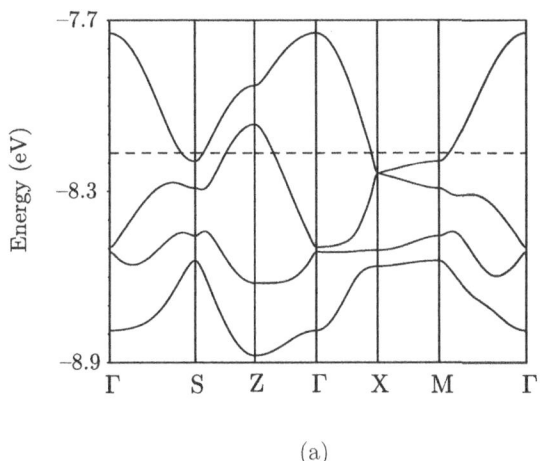

(a)

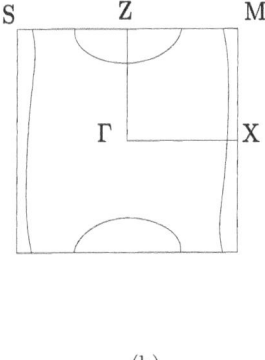

(b)

Fig. 11.58
(a) Band structure and (b) Fermi surface calculated for the donor lattice of α-(BEDT–TTF)$_2$KHg(SCN)$_4$. The dashed line in (a) refers to the Fermi level. $\Gamma = (0,0)$, $X = (a^*/2, 0)$, $Z = (0, c^*/2)$, $M = (a^*/2, c^*/2)$, and $S = (a^*/2, -c^*/2)$.

Fig. 11.57? If yes, where are such chains and why is there a closed 2D orbit in the Fermi surface as well? Let us remind ourselves that the open part of the Fermi surface consists of two warped lines along a direction perpendicular to the a-direction. Thus, if there are chains in the BEDT–TTF slabs, they should be along the a-direction. However, Fig. 11.57 suggests the existence of chains along the c-direction.

A typical set of transfer integrals for the seven types of interaction defined in Fig. 11.57 is the following (in eV): $c_1 = -0.012$, $c_2 = 0.140$, $c_3 = 0.074$, $p_1 = 0.129$, $p_2 = 0.116$, $p_3 = 0.162$, and $p_4 = 0.161$. Three of the transfer integrals (c_1 to c_3) will be referred to as intrachain transfer integrals and the remaining four (p_1 to p_4) as interchain transfer integrals. Of the c-type interaction energies, only c_2 is sizable in chain type I. Although from just a visual inspection this chain appears to be just that – a chain – in terms of the HOMO...HOMO interactions it is a series of weakly interacting dimers. From now on we will call them c_2 dimers. Now consider Fig. 11.57 in the light of this fact. It can be seen that the integrals p_2 and p_3 are different from their counterparts p_1 and p_4 in their orientation with respect to this dimer. In fact, we can see that the different c_2 dimers interact through donors type III (interactions p_2 and p_3) to form a chain along the a-direction. Thus, at least conceptually, it is tempting to consider the lattice of the α-(BEDT–TTF)$_2$MHg(XCN)$_4$ phases as being made up of chains of donor types III and I (c_2 dimers), which communicate through donors of type II.

To further explore this idea we can now carry out a series of model calculations where taking the previous transfer integrals as the basis, the values of the p_1, p_4 and c_3 transfer integrals will be gradually decreased while keeping the other integrals constant. [37] In such a way the suggested chains are gradually isolated within the lattice. The band structure of the lattice will be calculated as a function of the parameter γ, which defines the different values of the p_1, p_4 and c_3 transfer integrals (t') through the relationship $t' = \gamma t$, where t refers to the real values given above. Thus for $\gamma = 0$ the chains are completely isolated, whereas for $\gamma = 1$ the real 2D lattice of interactions is recovered. The results are shown in Fig. 11.59.

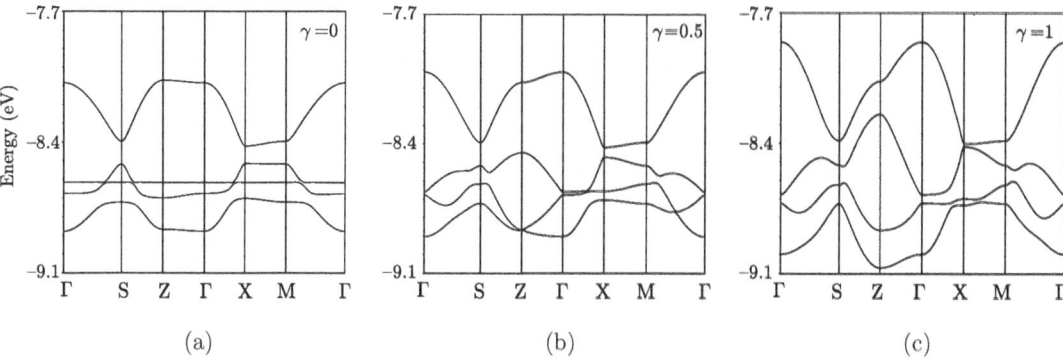

Fig. 11.59
Model band structure for the α-(BEDT–TTF)$_2$MHg(XCN)$_4$ phases as a function of the parameter γ.

For $\gamma = 1$ the band structure is very similar to the results of the full calculation (Fig. 11.58a). For $\gamma = 0$ the band structure consists of a flat band, that of the non-interacting $HOMO_{II}$ levels, and a set of three bands characteristic of the chain along a. It is important to understand the topology of these three bands. Looking at them carefully, it is easily seen that in fact they result from the interaction of a flat band (the second band at Γ) and two dispersive ones. When analysing the orbital character of these bands it can be seen that the major components of the lower/upper dispersive bands are the bonding/antibonding combinations of the HOMOs of the c_2 dimer. The bonding/antibonding levels of the different c_2 dimers interact through the $HOMO_{III}$ and thus acquire dispersion (of course, leading to bands with opposite slope) along the $\Gamma \rightarrow X$ (i.e. a^*) direction. Since the $HOMO_{III}$ levels are energetically in between the bonding and antibonding HOMO levels of the c_2 dimer, they interact in a stabilising way with the antibonding level but in a destabilising way with the bonding one, leading to a quite flat band.

Thus, the nature of the upper dispersive band and the completely flat one of Fig. 11.59a ($\gamma = 0$) are easy to understand. It is these two bands that are going to lead to the two partially filled bands of the α-(BEDT–TTF)$_2$ MHg(XCN)$_4$ phases. This can be clearly seen in Fig. 11.59. When we gradually switch the p_1, p_4, and c_3 interactions (i.e. γ changes from 0 to 1), the bands of the two subsystems mix. The originally flat band, being in the middle of the diagram, mixes with all other bands and loses any trace of its original shape, leading to a band dispersive in all directions. The original upper dispersive band is pushed up in energy but being the upper one, interacts largely with the original flat band and changes its shape much less. The important point, however, is that the shape is practically unaltered at the bottom of the band. Thus there is a strong memory of the chain of c_2 dimers and donors of type III in the bottom of the upper band of the α-(BEDT–TTF)$_2$ MHg(XCN)$_4$ phases. Thus even though there is a 2D slab of donors in this system, the electrons at the bottom of the upper band experience significant 1D interactions and contribute with 1D electron pockets to the Fermi surface.

The third band originates from the $HOMO_{II}$ levels and its strong dispersion along all directions is caused by the switching of the p_1, p_4, and c_3 interactions. Hence, this band can be described as arising from the cross-linking of the chains of the c_2 dimers and donors of type III through donors of type II. Therefore the following physical model can be suggested to describe the electronic structure of the α-(BEDT–TTF)$_2$ MHg(XCN)$_4$ phases. The electron pockets of the Fermi surface arise from the chains of c_2 dimers and donors of type III along the a-direction. The hole pockets of the Fermi surface originate from the coupling of such chains through donors of type II. Thus the carriers of the Fermi surface of these phases reflect different aspects of the lattice. This model view of the donor lattice of the α-(BEDT–TTF)$_2$ MHg(XCN)$_4$ phases, at least as far as the HOMO...HOMO interactions near the Fermi level are concerned, is schematically presented in Fig. 11.60.

With reference to this figure, the electron carriers almost only feel the interactions along the chains of donors in black, whereas the hole carriers feel all the interactions. This conclusion may appear strange at first sight. However, it is not. The reason again lies in the topology of the lattice. Because of the

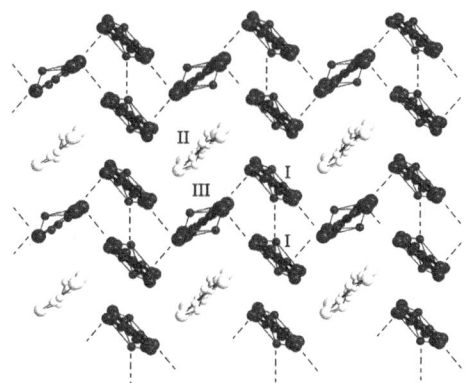

Fig. 11.60
Schematic representation of the donor lattice of the α-(BEDT–TTF)$_2$ MHg(XCN)$_4$ salts illustrating the nature of the chains leading to the 1D portion of the Fermi surface. The donors of the chains are shown with filled atoms and the intrachain interactions are marked with dotted lines. Hydrogen atoms are not shown.

geometrical nature of the p_1 and p_4 interactions, i.e. those which are largely responsible for the coupling of the chains, their contributions can add or can compensate depending on the phase of the different HOMOs. Whereas in the second case the 1D character survives, in the first one it completely disappears. The bottom of the fourth band lies in a region of the Brillouin zone, where such phases impose a near cancellation of the p_1 and p_4 interactions, but the opposite is true for the top of the third band. This is why we can say that the electrons and holes of these phases experience different aspects of the donor lattice.

This model is a useful conceptual framework, using which many experimental results concerning these phases can be discussed. [37] However, our interest here is in pointing out how sometimes apparently irrelevant structural details can be at the heart of the interesting physics of molecular conductors. The large difference in HOMO...HOMO transfer integrals in chains of type I is ultimately responsible for the shape of the Fermi surface and many of the interesting transport phenomena of these salts. This subtle feature is clearly captured by the simple one-orbital approach. The one-orbital simulations provide a simple yet very powerful technique to understand the results of more sophisticated calculations that would otherwise be difficult to analyse. They also are useful to elaborate simple models or to provide correlations between, on the one hand, crucial aspects of the band structure and Fermi surface and, on the other, structural details that determine the values of the transfer integrals. In molecular conductors it is difficult to introduce geometrical changes or to consider distortions from an idealised structure as is usually done when analysing the electronic structure of solids with a continuous network of chemical bonds. Here, the crucial bands result from a series of small interactions that are almost impossible to change without at the same time modifying other interactions of the lattice. This fact, as well as the intrinsic weakness of the interactions, makes it difficult to devise a strategy to correlate electronic and structural details. However, the one-orbital approach allows us to test the role of the different transfer integrals and thus, of the different interactions. Once the range of values of interest for the transfer integrals has been determined with these simulations, it is easy to translate the results into structural parameter terms by carrying out calculations for the appropriate dimeric units, which tell what combination of structural parameters lead to a given transfer integral value.

11.5.5 Electronic structure and folding: how to relate the band structure and Fermi surface of different salts of the same family

It is a commonplace to say that a very small change in structure can completely change the physical behaviour of molecular conductors. This is in fact an observation that cannot be denied. As a matter of fact, consider the β'' salts of BEDO–TTF and BEDT–TTF, of which there is a considerable number. The basic structure of the donor layer of these salts is always the same, namely the layers of Fig. 11.55, where every donor interacts with six neighbours – two along the step-chains, two along the π-chains, and two along the π-stacks. Magnetoresistance experiments, which give direct information concerning the Fermi surface, performed on three BEDO–TTF phases of this structural family, namely (BEDO–TTF)$_2$ Cl(H$_2$O)$_x$, (BEDO–TTF)$_2$ ReO$_4$·H$_2$O, and (BEDO–TTF)$_5$ [CsHg(SCN)$_4$]$_2$, have led to very different results concerning their Fermi surfaces. Of course, the different anions present in these salts induce small variations in the general structure of the β''-type layers, such that their repeat units contain one, four, and five donors, respectively. Thus instead of just one transfer integral for each type of interaction, as there is in the first salt, there are several for the second and third. However, the calculated transfer integrals in each type of interaction are not very different and the average value does not vary noticeably from salt to salt. Thus, one is tempted to conclude that, effectively, slight structural changes completely reorganise the electronic structure. However, this is at odds with our chemical intuition. The electronic structure of structurally related materials should exhibit variations over a general scenario but should not be completely different. To reconcile the observation of a rich variety of transport properties with the need for a common Fermi surface pattern for a series of salts of the same family it is helpful to use the concept of *folding*, [32] which we have already discussed in Section 3.4.1.

Let us consider what would happen to the band dispersion of Fig. 11.56a, generated through the use of eqn (11.6), if we had considered a unit cell with two BEDO–TTF donors in the b-direction (see Fig. 11.55). Since the repeat unit is double in size, the associated Brillouin zone (BZ) would have half the area of the original and is composed of vectors $\vec{k}' = k'_a \vec{a}'^* + k'_b \vec{b}'^*$ where $\vec{a}'^* = \vec{a}^*$ and $\vec{b}'^* = 1/2\vec{b}^*$, so that $k'_a = k_a$ and $k'_b = 2k_b$. The boundary conditions, $1/2 \geq k'_a, k'_b \geq -1/2$, lead to a translation of states, with $|k_b| \geq 1/4$ to the inside of the folded BZ. Although in this particular case, as for the (TMTSF)$_2$ PF$_6$ salt, it is possible to derive the analytical formula for $E(\vec{k})$, it becomes prohibitively complicated for large supercells. However, it is possible to use eqn (11.6) to generate the folded band structure by correlating the wave vectors of the folded and unfolded BZs. For example, the centre of the folded BZ, Γ' ($k'_a = 0$, $k'_b = 0$) corresponds to the points Γ ($k_a = 0$, $k_b = 0$) and Y ($k_a = 0$, $k_b = 1/2$), whereas, X' ($k'_a = 1/2$, $k'_b = 0$), Y' ($k'_a = 0$, $k'_b = 1/2$) and M' ($k'_a = 1/2$, $k'_b = 1/2$) correspond to ($k_a = 1/2$, $k_b = 0, 1/2$), ($k_a = 0$, $k_b = \pm 1/4$), and ($k_a = 1/2$, $k_b = \pm 1/4$), respectively. Following this procedure for all points in k-space results in the band dispersion curve of Fig. 11.61, which contains *exactly* the same information as the one-orbital band structure of Fig. 11.56a but condensed to a

Fig. 11.61
Folded band structure corresponding to Fig. 11.56a as a result of using a unit cell twice as large along b.

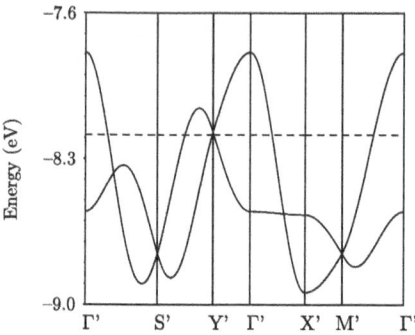

Fig. 11.62
Folding of the Fermi surface: the one-molecule Fermi surface obtained from eqn (11.6) and the translation vector needed for the folding process are shown on the left; the folded BZ with the translated Fermi surface are shown at the right.

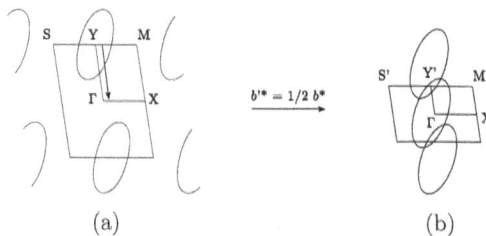

smaller BZ. The effect this has on the Fermi surface is shown in Fig. 11.62: one now obtains exactly the same Fermi surface as before, except that the ellipses are now touching in the new, smaller BZ.

One must think of folding as a renormalization of k-space resulting in a *translation* of parts of the band structure. For a unit cell extended n times in real space along one of the main axis directions (γ), i.e. $\gamma' = n\gamma$, it is necessary to apply the vectors $\gamma'^*, 2\gamma'^*, \ldots, (n-1)\gamma'^*$ to respectively translate the $n-1$ successive replicas of the new smaller BZ along the γ'^* direction (Fig. 11.62 is an illustration for the $n = 2$ case). By so doing, all the information contained in the n times larger initial BZ has been transferred into the new BZ (here the periodicity of k-space has been used). Thus, we can generate a first-order description of the band structure and/or Fermi surface adapted to the nature of the unit cell, whatever size it is, solely on the basis of the information contained in eqn (11.6).[7]

The result of the folding process leads to nothing new: the folded and unfolded band structures are physically indistinguishable. However, the counterion layer invariably enforces upon the organic layer an external potential with periodicity larger than the simple one-molecule packing motif. This external potential further perturbs the intermolecular interactions and results in orbital mixings in the folded BZ. These alter the band structure and thus the Fermi surface, resulting in the diversity of electrical properties of these materials. This is incorporated into the picture above by changes in both t_j and α_i. The transfer integrals may deviate in value due to structural modulations in the packing motif and, to a lesser extent, by variation in the molecular orbitals as induced by changes in the local molecular environment. Modifications of α may arise from electrons being preferentially localised upon certain sites, as well as from small distortions in the molecular geometry due to crystal

---

[7]Because of the counterions, the crystal structures are frequently described on the basis of axes that are not along the same directions as those of the parent single molecule system chosen. In general, it is desirable to preserve the unit cell of the reported crystal structures thus retaining consistent axes between physical measurements and the derived electronic structure. This restriction may result in the need to generate the band structure or Fermi surface for supercells that are not integer multiples of a particular, or any, single molecule cell to which one wishes to compare. Some care is needed in these cases (see some examples in ref. [32]).

packing effects. Overall, these changes are not independent of each other but they induce deviations with respect to the parent one-molecule lattices, which are small and thus the underlying one-orbital band structure is largely retained. The main effect of these changes is to open gaps at the crossing points of the Fermi surface in the folded one-orbital picture (see Fig. 11.62).

The above picture provides a simple hierarchy by which to understand the electronic structure and, more specifically, the Fermi surface of a family of molecular conductors and its effect upon electrical conductivity (see Fig. 11.63). First, the molecular structure of the organic molecule determines the nature of the molecular orbital that is going to lead to the conduction band. The crystal structure imposes the magnitude and type of transfer integrals within an organic slab. This determines the form of the energy dispersion curve and thus the Fermi surface of the parent one-molecule lattice. Second, the nature of the anion imposes a unit cell that is generally larger than that of the parent one-molecule lattice. The new Fermi surface is just the appropriately folded version of the parent one. Finally, the local environment causes variations in the intermolecular interactions, which lead to modifications of the electronic structure and the Fermi surface (i.e. opening of gaps as shown in Fig. 11.63). The effect this has on the properties of a material will depend

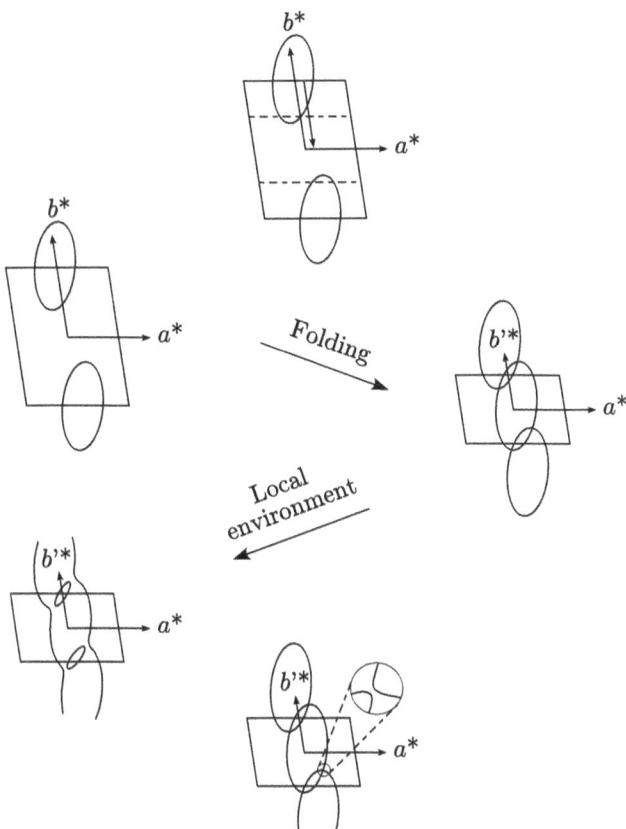

Fig. 11.63
Hierarchy of electronic structure considerations used in building the Fermi surface of a molecular conductor from that of the simple one-molecule case: (1) folding is used to generate the Fermi surface of systems with any unit cell size from that of the one-molecule parent system; (2) gaps in the folded Fermi surface arise due to orbital mixings caused by the structural perturbations induced by the counterion layer.

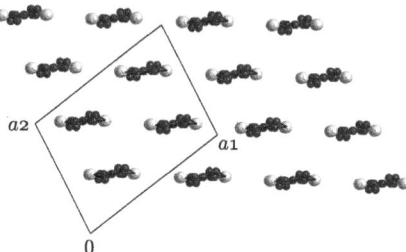

Fig. 11.64
β'' layer with four molecules (2 × 2) in the repeat unit. Hydrogen atoms are not shown.

on the magnitude of the perturbations as well as the property under consideration. For instance, the combination of folding and gap opening schematically shown in Fig. 11.63 will result in small changes in the conductivity but strong differences in the magnetoresistance of the parent and actual materials. The two Fermi surfaces are associated with an almost isotropic conduction in the plane of the layers, which will hardly indicate any difference between the two compounds. However, magnetoresistance experiments (i.e. measurement of the resistance in the presence of an applied magnetic field) will lead to quite different results. When a Fermi surface contains closed portions, the resistance of the material as a function of the inverse of the applied field usually exhibits oscillations (Shubnikov–de Haas oscillations) with frequency being related to the cross-section (percentage of the area of the BZ) of such closed portions perpendicular to the field direction. Thus, magnetoresistance measurements with a magnetic field applied perpendicular to the layers will lead to large frequency oscillations in the parent case but quite small ones in the second.

As an illustration of these ideas let us consider the case of β''-type donor layers of Fig. 11.64, with four molecules in the repeat unit (i.e. a 2×2 repeat unit). Examples of this kind of layer include the series of metallic salts β''-(BEDT–TTF)$_4$(guest)$_n$· [Re$_6$Q$_6$Cl$_8$] (Q = S, Se; guest = H$_2$O, dioxane, THF,....)[38] and (BEDO–TTF)$_2$· ReO$_4$· H$_2$O. [39] These salts have been extensively studied because of the tuning of the low-temperature behaviour in the former series by guest (solvent) molecules and the low-temperature phase transitions (including superconductivity) of the second. The repeat unit of these layers (Fig. 11.64) contains four donors, and the two repeat vectors coincide with the π-stacks and step-chain directions. We have used the labels a_1 and a_2 in this figure because the naming of the crystallographic axes in the original structural determinations is different in the two compounds: $(a_1/a_2) = (c/-a)$ for β''-(BEDT–TTF)$_4$(guest)$_n$· [Re$_6$Q$_6$Cl$_8$] but $(a_1/a_2) = (b/-a)$ for (BEDO–TTF)$_2$· ReO$_4$·H$_2$O. For any of these systems there will be four HOMO bands and, since the average charge of the donors is +1/2, these bands will be filled with six electrons. To generate the band structure we just need to use the analytical expression for the band dispersion (a simple adaptation of eqn (11.6) to the actual main axes directions in the systems under consideration) and the coordinates of the lines of the original unfolded BZ, which will become the four bands associated with a given line in the folded BZ. This is left as an exercise for the reader (see Exercise (11.7)).

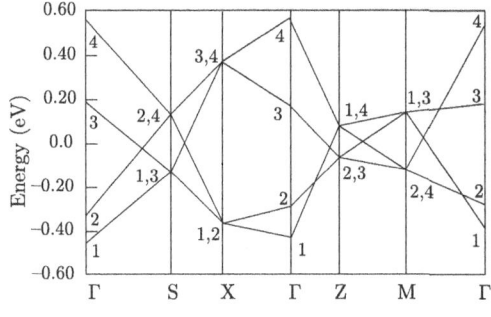

Fig. 11.65
Qualitative band structure for the β'' layers of β''-(BEDT–TTF)$_4$(guest)$_n$·[Re$_6$Q$_6$Cl$_8$] (Q = S, Se; guest = H$_2$O, dioxane, THF,....). $\Gamma = (0,0)$, X = $(a_1^*/2, 0)$, Y = $(0, a_2^*/2)$, M = $(a_1^*/2, a_2^*/2)$, and S = $(a_1^*/2, -a_2^*/2)$.

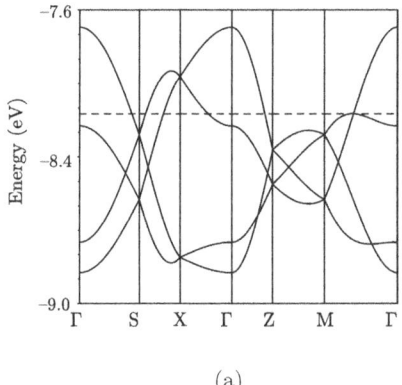
(a)

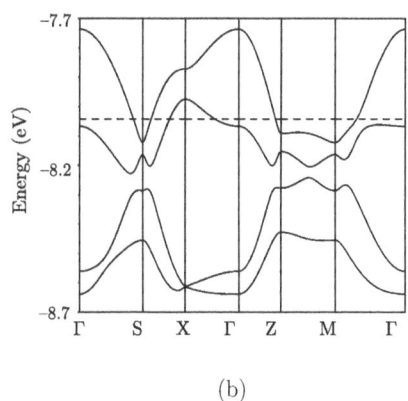
(b)

Fig. 11.66
β''-(BEDT–TTF)$_4$(1-4-dioxane)$_n$·[Re$_6$Q$_6$Cl$_8$]: (a) simulated band structure using the average transfer integrals and the same energy for the HOMOs, and (b) extended Hückel band structure. $\Gamma = (0,0)$, X = $(a^*/2, 0)$, Z = $(0, c^*/2)$, M = $(a^*/2, c^*/2)$, and S = $(a^*/2, -c^*/2)$.

Average transfer integrals along the step chains, π-stacks and π-chains calculated for β''-(BEDT–TTF)$_4$(1,4-dioxane)·[Re$_6$S$_6$Cl$_8$] are t_{sc} = +0.185 eV, $t_{\pi c}$ = −0.035 eV and $t_{\pi s}$ = −0.066 eV. Using the results of the analysis developed in Exercise (11.7) and these transfer integrals, the qualitative band structure of Fig. 11.65 can be drawn. Fig. 11.66a shows the simulated band structure calculated with the analytical treatment, while Fig. 11.66b shows the full calculated band structure using the extended Hückel approach. The simple model does indeed reproduce the band structure from the full calculation very well. We thus have a model (see Fig. 11.65) that captures, in a simple way, the essence of the electronic structure of this kind of layer. Since the different transfer integrals are associated with specific intermolecular interactions and we know how the different types of transfer integral influence the shape of the band structure (and thus the Fermi surface), we can understand the correlation between the details of the crystal and the electronic structures for any of these layered materials.

With the previous results in mind it is not difficult to predict 2D metallic behaviour for salts of this family. However, the two-dimensionality goes much deeper. What the present folding-based approach is telling us is that the four

HOMO bands are really just one 2D HOMO band with gaps opened at specific points in the BZ caused by the lower symmetry of the crystal structure. As a matter of fact, the pattern of overlapping ellipses because of folding is clearly seen in the extended Hückel Fermi surface calculated for β''-(BEDT–TTF)$_4$(1,4-dioxane)· [Re$_6$S$_6$Cl$_8$] (see Fig. 11.67) once the avoided crossings are taken into account. It is now very easy to see how the actual band structures (like that of Fig. 11.66b) can be understood on the basis of the ideal band structure (Fig. 11.65) when the details of the real lattice topology are taken into account. Firstly, this process lowers the symmetries of all the bands and causes all of the crossings to become avoided crossings. Second, the degeneracies at X and Z are removed since t_{sc} and $t_{\pi s}$ are average transfer integrals. At Z, for example, it is the dissymmetry in the step-chain interactions (there are four different interactions of this type in real β''-(BEDT–TTF)$_4$(1,4-dioxane)· [Re$_6$S$_6$Cl$_8$] with transfer integrals equal to +0.195, +0.189, +0.230 and +0.120 eV) that is the major contributor to the lifting of the degeneracies. At X, the major contributors to the lifting of the degeneracies are the smaller interactions, which are much more similar to each other (there are two different interactions of this type with transfer integrals equal to –0.045 and –0.022 eV), and the energy splitting will be much smaller. At M, it is again the dissymmetry in the step-chain interactions that is the leading factor breaking the degeneracies. Most of the details of the band structure of any of these systems can thus be correlated, in a similar vein, with those of the crystal structure, once the transfer integrals are calculated.

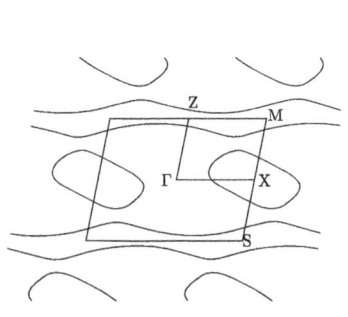

Fig. 11.67
Fermi surface for β''-(BEDT–TTF)$_4$(1-4-dioxane)$_n$·[Re$_6$Q$_6$Cl$_8$] calculated with the extended Hückel method.

The advantage of such a treatment is that it greatly facilitates the comparison between different systems. The average values for the three types of transfer integrals used for β''-(BEDT–TTF)$_4$(1,4-dioxane)· [Re$_6$S$_6$Cl$_8$] (t_{sc} = +0.185, $t_{\pi c}$ = –0.066 and $t_{\pi s}$ = –0.035 eV) are quite typical for many BEDT–TTF salts with β'' type donor layers. If these values are compared with those typical for BEDO–TTF salts (for instance, t_{sc} = +0.146, $t_{\pi c}$ = –0.142, and $t_{\pi s}$ = –0.049 eV for (BEDO–TTF)$_{2.4}$I$_3$), it is clear that the more noticeable difference is that, whereas the interactions along the step-chains and π-chains have a comparable strength for the BEDO–TTF salt, the difference is very large for the BEDT–TTF salt. The reason for the strongly decreased value of the interaction along the π-chains is simply the smaller size of the oxygen atoms in the six-member rings of BEDO–TTF with respect to the sulfur atoms in BEDT–TTF, which allows for closer lateral contacts. Simple simulations in which the absolute value of $t_{\pi c}$ increases show that the ellipse rotates in such a way that in the four-fold folded BZ the short axis of the ellipse (see Fig. 11.67) slightly turns in a clockwise direction. The relatively small rotation can, however, strongly influence the shape of the final Fermi surface depending on the superposition pattern imposed by the particular type of folding (i.e. by the shape of the unit cell). However, the pattern of superposing ellipses always emerges, spotlighting the common structural link between these phases. The rich variety in Fermi surfaces can be traced back to the structural details of the lattice through the use of qualitative band structures such as that of Fig. 11.65.

Exercises

(11.1) Show that the Fermi surface of a half-filled square 2D system in the nearest-neighbour approximation is a square.

(11.2) Describe the effect on the band structure and conductivity of red bronzes if the hump octahedra are added at the two sides of the Mo_4O_{18} chain (see Fig. 11.25e) in a zigzag way instead of the parallel manner discussed in the text.

(11.3) Justify the shape of the partially filled bands in Fig. 11.38 on the basis of the structural information of Fig. 11.37 and an orbital analysis similar to that employed for the 2D red bronzes.

(11.4) $Sr_3V_2O_7$ (see figure below) is a metallic system, the crystal structure of which is built from V_2O_7 double octahedral layers, where the black and white atoms refer to the metal and oxygen, respectively. [15] Build the qualitative band structure of these double layers and predict the dimensionality of the partially filled bands.

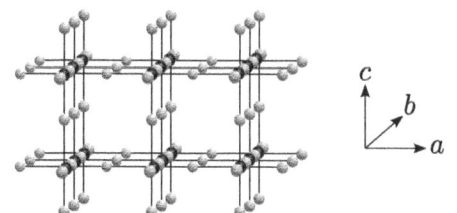

(11.5) Discuss how the different HOMO...HOMO interactions influence the band structure of the $(TMTSF)_2X$ family of salts using eqn (11.7) and the set of transfer integrals $t_{S_1} = +0.395$ eV, $t_{S_2} = +0.334$ eV, $t_{I_1} = -0.009$ eV, $t_{I_2} = -0.036$ eV, $t_{I_3} = +0.041$ eV and $t_{I_4} = +0.010$ eV as a reference.

(11.6) The so-called θ-phases contain donor layers as schematically shown below, where the thick lines represent donor molecules viewed along the long molecular axis. There are only two different types of interaction: c and p. Derive an analytical solution for the band dispersion, $E(\vec{k})$ of the two bands.

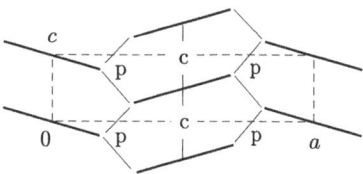

(11.7) Using eqn (11.6) and the concept of folding, build a qualitative band structure for the donor layers of β''-$(BEDT–TTF)_4(guest)_n \cdot [Re_6Q_6Cl_8]$ (Q = S, Se; guest = H_2O, dioxane, THF,...) and $(BEDO–TTF)_2 \cdot ReO_4 \cdot H_2O$.

References

1. N. W. Ashcroft and N. D. Mermin, *Solid State Physics*, Holt, Rinehart and Winston, Philadelphia, 1976.
2. J. Ziman, *Electrons in Metals: A Short Guide to the Fermi Surface*, Taylor and Francis, London, 1963.
3. E. Canadell and M.-H. Whangbo, *Chem. Rev.*, 91, 965, 1991.
4. A. Meerschaut and J. Rouxel, *J. Less-Common. Met.* 39, 197, 1975.
5. A. Meerschaut and J. Rouxel in *Crystal Chemistry and Properties of Materials with Quasi-One-Dimensional Structures*, J. Rouxel, Ed., Reidel, Dordrecht, The Netherlands, 1986, p. 205.
6. For some theoretical studies of the electronic structure of $NbSe_3$ and TaS_3 focusing on the qualitative aspects of the band structure see: (a) R. Hoffmann, S. Shaik, J. C. Scott, M.-H. Whangbo and M. J. Foshee, *J. Solid State Chem.*, 34, 263, 1980; (b) M.-H. Whangbo in *Crystal Chemistry and Properties of Materials with Quasi-One-Dimensional Structures*, J. Rouxel, Ed., Reidel, Dordrecht, The Netherlands, 1986, p. 27; (c) E. Canadell, I. E.-I. Rachidi, J. P. Pouget, P. Gressier, A. Meerschaut, J. Rouxel, D. Jung, M. Evain and M.-H. Whangbo, *Inorg. Chem.*, 29, 1401, 1990.

7. R. Moret and J. P. Pouget in *Crystal Chemistry and Properties of Materials with Quasi-One-Dimensional Structures*, J. Rouxel, Ed., Reidel, Dordrecht, The Netherlands, 1986, p. 87.
8. C. Schlenker, Ed., *Low-Dimensional Electronic Properties of Molybdenum Bronzes and Oxides*, Kluwer Academic Publishers, Dordrecht, 1989.
9. C. Schlenker, J. Dumas, M. Greenblatt and S. van Smaalen, Eds., *Physics and Chemistry of Low-Dimensional Inorganic Conductors*, NATO Advanced Study Institute, Series B: Physics, vol. 354, Plenum Publishing, New York, 1996.
10. M. Greenblatt, *Chem. Rev.*, 88, 31, 1988.
11. (a) N. C. Stephenson and A. D. Wadsley, *Acta Crystallogr.*, 18, 241, 1965; (b) P. P. Tsai, J. A. Potenza and M. Greenblatt, *J. Solid State Chem.*, 69, 329, 1987; (c) J. M. Réau, C. Fouassier and P. Hagenmuller, *Bull. Chem. Soc. Fr.*, 8 2884, 1971.
12. P. P. Tsai, J. A. Potenza, M. Greenblatt and H. Schugar, *J. Solid State Chem.*, 64, 47, 1984.
13. M.-H. Whangbo, M. Evain, E. Canadell and M. Ganne, *Inorg. Chem.*, 28, 267, 1989.
14. M. Ghedira, J. Chenavas, M. Marezio and J. Marcus, *J. Solid State Chem.*, 57, 300, 1985.
15. N. Suzuki, T. Noritake, N. Yamamoto and T. Hioki, *Mat. Res. Bull.*, 26, 1, 1991.
16. M.-H. Whangbo, E. Canadell, P. Foury and J.P. Pouget, *Science*, 96, 252, 1991.
17. C. Schlenker, J. Dumas, C. Escribe-Filippini and C. Guyot in C. Schlenker, Ed., *Low-Dimensional Electronic Properties of Molybdenum Bronzes and Oxides*, Kluwer Academic Publishers, Dordrecht, 1989, p.159.
18. H. Vincent, M. Ghedira, J. Marcus, J. Mercier and C. Schlenker, *J. Solid State Chem.*, 47, 113, 1983.
19. J. Ferraris, D. O. Cowan, V. V. Walatka and J. H. Perlstein, *J. Am. Chem. Soc.*, 95, 948, 1973; L. B. Coleman, M. J. Cohen, D. J. Sandman, F. G. Yamagishi, A. F. Garito and A. J. Heeger, *Solid State Commun.*, 12, 1125, 1973.
20. D. Jérome and H. Schulz, *Adv. Phys.*, 31, 299, 1982.
21. D. Jérome, A. Mazaud, M. Ribault and K. Bechgaard, *J. Phys. Lett. (France)*, 41, L-95, 1980.
22. J. M. Williams, J. R. Ferraro, R. J. Thorn, K. D. Carlson, U. Geiser, H. H. Wang, A. M. Kini and M.-H. Whangbo, *Organic Superconductors*, Prentice-Hall, 1992.
23. T. Ishiguro, K. Yamaji and G. Saito, *Organic Superconductors*, 2nd edition, Springer Verlag, Berlin, 1996.
24. See the different reviews in the special issue concerning molecular conductors: *Chem. Rev.*, 104, 4887, 2004.
25. T. Giamarchi, *Chem. Rev.*, 104, 5037, 2004.
26. E. Canadell, I. E.-I. Rachidi, S. Ravy, J. P. Pouget, L. Brossard and J. P. Legros, *J. Phys. (France)* 50, 2967, 1989; E. Canadell, S. Ravy, J. P. Pouget and L. Brossard, *Solid State Commun.*, 75, 633, 1990; E. Canadell, *New. J. Chem.*, 21, 1147, 1997.
27. A. Pénicaud, P. Batail, K. Bechgaard and J. Sala-Pala, *Synt. Met.*, 22, 201, 1988.
28. C. Lenoir, K. Boubekeur, P. Batail, E. Canadell, P. Auban, O. Traetteberg and D. Jérome, *Synt. Met.*, 41-43, 1939, 1991.
29. B. Pommarède, B. Garreau, I. Malfant, L. Valade, P. Cassoux, J. P. Legros, A. Audouard, L. Brossard, J.-P. Ulmet, M.-L. Doublet and E. Canadell, *Inorg. Chem.*, 33, 3401, 1994.
30. F. Wudl, H. Yamochi, T. Suzuki, H. Isotalo, C. Fite, H. Kasmai, K. Liou, G. Srdanov, P. Coppens, K. Maly and A. Frost-Jensen, *J. Am. Chem. Soc.*, 112, 2461, 1990.
31. T. Mori, *Bull. Chem. Soc. Jpn.*, 71, 2509, 1998; T. Mori, H. Mori and S. Tanaka, *Bull. Chem. Soc. Jpn.*, 72, 179, 1999; T. Mori, *Bull. Chem. Soc. Jpn.*, 72, 2011, 1999.
32. R. Rousseau, M. Gener and E. Canadell, *Adv. Funct. Mater.*, 14, 201, 2004.
33. P. M. Grant, *Phys. Rev. B*, 26, 6888, 1982; P. M. Grant, *J. Phys.* (Paris), 44, C3-847, 1983; L. Ducasse, M. Abderrabba, J. Hoarau, M. Pesquer, B. Gallois and J. Gaultier, *J. Phys.C.: Solid State Phys.*, 19, 3805, 1986.

34. H. Mori, S. Tanaka, M. Oshima, G. Saito, T. Mori, Y. Maruyama and H. Inokuchi, *Bull. Chem. Soc. Jpn.*, 63, 2183, 1990.
35. R. P. Shibaeva, L. P. Rozenberg, N. D. Kushch and E. B. Yagubskii, *Crystallogr. Rep.*, 39, 747, 1994.
36. M. V. Kartsovnik and V. N. Laukhin, *J. Phys. I*, 6, 1753, 1996; J. Wosnitza, *Fermi Surfaces of Low-Dimensional Organic Metals and Superconductors*, Springer Verlag, 1996; J. Singleton, *Rep. Prog. Phys.*, 63, 1111, 2000.
37. R. Rousseau, M.-L. Doublet, E. Canadell, R. P. Shibaeva, S. S. Khasanov, L. P. Rozenberg, N. D. Kushch and E. B. Yagubskii, *J. Phys. I (France)*, 6, 1527, 1996.
38. A. Deluzet, R. Rousseau, C. Guilbaud, I. Granger, K. Boubekeur, P. Batail, E. Canadell, P. Auban-Senzier and D. Jérome, *Chem. Eur. J.*, 8, 3884, 2002.
39. S. Kahlich, D. Schweitzer, I. Heinen, S. E. Lan, B. Nuber, H. J. Keller, K. Winzer and H. W. Helberg, *Solid State Commun.*, 8, 191, 1991; L. I. Buravov, A. G. Khomenko, N. D. Kushch, V. N. Laukhin, A. I. Schegolev, E. B. Yagubskii, L. P. Rozenberg and R. P. Shibaeva, *J. Phys. I*, 2, 59, 1992.
40. M. Ganne, M. Dion and A. Boumaza, *C. R. Acad. Sci.*, Ser. 2, 302, 635, 1986.

12 Electron repulsion

In the previous chapters we saw that semi-empirical methods such as the extended Hückel model are very powerful ways to describe the electronic structure of a wide variety of solids. At this level of approximation, electrons are free to move in the infinite solid. Using this qualitative picture, we established a general methodology allowing us (i) to interpret the electronic properties of various organic and inorganic solids, (ii) to correlate these properties to the crystal structure and the local chemical bonds of the systems and (iii) to predict some of the electronic instabilities that are likely to occur in 1D metals. For some solids however, the hypothesis of non-interacting electrons is no longer valid,[1] and a localised description of the electrons is necessary to properly account for their electronic structure. This is the case for the so-called *(strongly) correlated systems* the electronic structure of which is at least partially governed by the interaction between electrons. In this chapter, we will thus increase the level of theory to account for the electronic repulsion U_{ee}, firstly in an effective manner, starting from the extended Hückel model ($U_{ee} = 0$) and going to the Hubbard [1] ($U_{ee} \neq 0$) and Heisenberg [2] ($U_{ee} \to +\infty$) models, and secondly in a more quantitative manner using the exact electronic Hamiltonian of the many-body problem, i.e. a system of n interacting particles.

[1] In the Hückel model, also called the tight-binding model, the electrons are considered to be non-interacting particles.

12.1 From the Hückel model to the Hubbard model

The basics of electron repulsion can be understood with the simple case of the H_2 molecule. In this section, we will use this example to go step by step from the Hückel (non-interacting electrons) to the Hubbard (interacting electrons) model, in order to formally introduce electron repulsion and its effect on the H_2 energy and wavefunction.

12.1.1 The delocalised picture of H_2

In the Hückel model and the LCAO approach,[2] the interaction diagram of H_2 consists of two molecular orbitals (MOs) coming from the bonding and antibonding interactions of the two $1s$ orbitals of H_a and H_b (see Fig. 12.1).

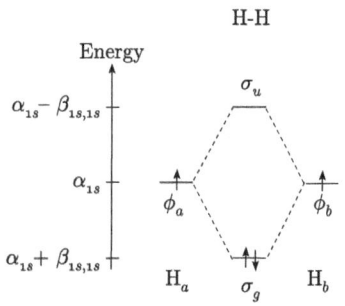

Fig. 12.1
Molecular orbital diagram of H_2 in the Hückel approximation.

[2] LCAO is the linear combination of atomic orbitals.

The Hubbard model

The energy difference between the σ_g and the σ_u molecular orbitals represents the strength of the orbital interaction, which is here directly proportional to the Hückel $\beta_{1s,1s}$ parameter.[3] At the equilibrium H–H distance, the two electrons of the system lie in the bonding MO (σ_g), with opposite spins so that the total electronic energy of H_2:

$$2 \times e(\sigma_g) = 2(\alpha_{1s} + \beta_{1s,1s}) \qquad (12.1)$$

is lower than the energy of two isolated H atoms:

$$2 \times e(1s) = 2\alpha_{1s} \qquad (12.2)$$

Electrons being indistinguishable fermions, the ground-state electronic wavefunction of H_2 must be antisymmetric with respect to the exchange of two electrons and is thus written as a Slater determinant:

$$\Psi_{GS}^{LCAO} = |\sigma_g \bar{\sigma}_g| = \frac{1}{\sqrt{2}} \begin{vmatrix} \sigma_g(1) & \bar{\sigma}_g(1) \\ \sigma_g(2) & \bar{\sigma}_g(2) \end{vmatrix}$$

$$= \frac{1}{\sqrt{2}} \{\sigma_g(1)\bar{\sigma}_g(2) - \sigma_g(2)\bar{\sigma}_g(1)\} \qquad (12.3)$$

where (1) and (2) refer to the electron label and σ_g and $\bar{\sigma}_g$ refer to the occupation of the σ_g orbital by an electron with spin up ($\uparrow$) and spin down ($\downarrow$), respectively.

In this representation, the two electrons of H_2 are fully delocalised over the two H–H atoms, so that the average electron density per atom (see Fig. 12.2) corresponds to half an electron with spin up and half an electron with spin down.

This average picture illustrates that each electron spends half its time on H_a and half its time on H_b, independently of where the second electron is located (but not independently of its spin, so as to satisfy the Pauli exclusion principle). In a more instantaneous or local picture, this can be represented by four different electronic configurations describing all possible distributions of two electrons on two sites with opposite spins (see Fig. 12.3).[4] In the Hückel model, these four configurations (electron distributions) have strictly the same energy ($e^0 = 2\alpha_{1s}$) and the same probability of occurrence ($\frac{1}{4}$) so that the average number of electrons per site is easy to obtain:

$$\langle n_a^\uparrow \rangle = \left(\tfrac{1}{4} \times 1\right) + \left(\tfrac{1}{4} \times 0\right) + \left(\tfrac{1}{4} \times 1\right) + \left(\tfrac{1}{4} \times 0\right) = \tfrac{1}{2}$$
$$\langle n_a^\downarrow \rangle = \left(\tfrac{1}{4} \times 0\right) + \left(\tfrac{1}{4} \times 1\right) + \left(\tfrac{1}{4} \times 1\right) + \left(\tfrac{1}{4} \times 0\right) = \tfrac{1}{2}$$
$$\langle n_b^\uparrow \rangle = \left(\tfrac{1}{4} \times 0\right) + \left(\tfrac{1}{4} \times 1\right) + \left(\tfrac{1}{4} \times 0\right) + \left(\tfrac{1}{4} \times 1\right) = \tfrac{1}{2} \qquad (12.4)$$
$$\langle n_b^\downarrow \rangle = \left(\tfrac{1}{4} \times 1\right) + \left(\tfrac{1}{4} \times 0\right) + \left(\tfrac{1}{4} \times 0\right) + \left(\tfrac{1}{4} \times 1\right) = \tfrac{1}{2}$$

[3] Note that in the extended Hückel approach, $\beta_{1s,1s}$ is directly proportional to the overlap integral between $1s_a$ and $1s_b$ ($S_{ab} = \langle \phi_a | \phi_b \rangle$), as shown in eqn (2.9).

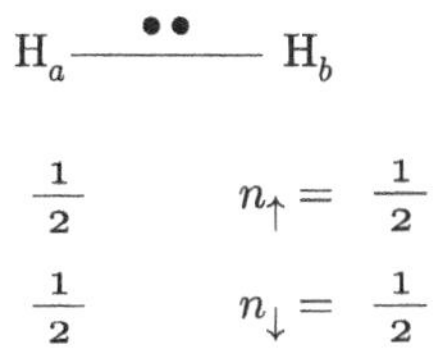

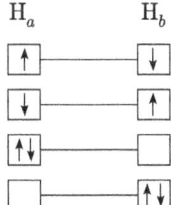

Fig. 12.2
Lewis representation of the two electrons delocalised over the H–H bond and average electronic population per H atom in the fully delocalised picture.

Fig. 12.3
Electron configurations required to describe the fully delocalised half-filled H_2 system (with the assumption that the high-spin configurations can be neglected).

[4] In neglecting the high-spin configurations, i.e. two electrons located on H_a and H_b with equivalent spins, we implicitly assume that the total spin of the H_2 molecule is zero which is true for the ground state.

To demonstrate this, let us write the four electronic configurations of Fig. 12.3 as the following basis vectors:

$$\begin{aligned}
|{\overset{\uparrow\downarrow}{a\,b}}\rangle &= |\phi_a \uparrow_a \phi_b \downarrow_b\rangle \equiv |a\bar{b}\rangle \\
|{\overset{\downarrow\uparrow}{a\,b}}\rangle &= |\phi_a \downarrow_a \phi_b \uparrow_b\rangle \equiv |\bar{a}b\rangle \\
|{\overset{\uparrow\downarrow}{a\,a}}\rangle &= |\phi_a \uparrow_a \phi_a \downarrow_a\rangle \equiv |a\bar{a}\rangle \\
|{\overset{\uparrow\downarrow}{b\,b}}\rangle &= |\phi_b \uparrow_b \phi_b \downarrow_b\rangle \equiv |b\bar{b}\rangle
\end{aligned} \qquad (12.5)$$

The first two basis vectors (hereafter termed the *neutral* states) represent the configurations associated with two electrons on different sites, i.e. H•–H•, while the two latter (hereafter termed the *ionic* states) represent the configurations associated with two electrons on the same site, i.e. H$^-$–H$^+$ and H$^+$–H$^-$. These vectors represent two-electron wavefunctions and are Slater determinants.[5]

[5] A Slater determinant is, by definition, an eigenfunction of the $\hat{S}_z$ spin operator but not always an eigenfunction of the $\hat{S}^2$ operator as we will see below.

In the Hückel model, the energy of these vectors simply corresponds to twice the energy of an electron located in a $1s$ orbital:

$$\begin{aligned}
\langle a\bar{b} | \hat{H}(t) | a\bar{b}\rangle &= 2\alpha_{1s} = e^0 \\
\langle \bar{a}b | \hat{H}(t) | \bar{a}b\rangle &= 2\alpha_{1s} = e^0 \\
\langle a\bar{a} | \hat{H}(t) | a\bar{a}\rangle &= 2\alpha_{1s} = e^0 \\
\langle b\bar{b} | \hat{H}(t) | b\bar{b}\rangle &= 2\alpha_{1s} = e^0
\end{aligned} \qquad (12.6)$$

As illustrated in Fig. 12.4, the ability of the system to delocalise the two electrons over the H–H bond can be intuitively understood as the ability of each electron to hop from one site to another. In the local basis set given above, this means that the neutral states $|a\bar{b}\rangle$ and $|\bar{a}b\rangle$ are coupled with the ionic states $|a\bar{a}\rangle$ and $|b\bar{b}\rangle$ through the so-called *hopping* integral t.

This hopping integral has exactly the same physical meaning as the β parameter in the Hückel approach (or the overlap integral between $\phi_a = 1s_a$ and $\phi_b = 1s_b$ in the extended Hückel approach). Its magnitude is directly correlated to the electron transfer between two adjacent sites which, in the case of H$_2$, is given by half the energy difference between the σ_g and the σ_u molecular orbitals:

$$t \equiv \beta_{1s,1s} = \frac{1}{2}\left\{e(\sigma_g) - e(\sigma_u)\right\}$$

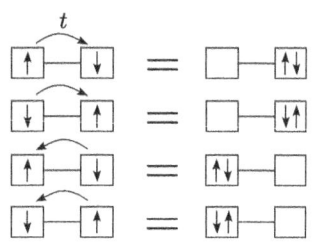

Fig. 12.4
Electron hopping from one site to another.

In a formal treatment, the hopping processes illustrated in Fig. 12.4 then become:

$$\begin{aligned}
\langle b\bar{b} | \hat{H}(t) | a\bar{b}\rangle &= +t \\
\langle b\bar{b} | \hat{H}(t) | \bar{a}b\rangle &= -t \\
\langle a\bar{a} | \hat{H}(t) | a\bar{b}\rangle &= +t \\
\langle a\bar{a} | \hat{H}(t) | \bar{a}b\rangle &= +t
\end{aligned} \qquad (12.7)$$

[6] In Fig. 12.4, the second and fourth processes lead to a distribution where electrons (1) and (2) need to be exchanged to retrieve the $|a\bar{a}\rangle$ or $|b\bar{b}\rangle$ distributions, as defined in eqn (12.5).

In these expressions, the negative sign stands for processes involving the exchange of the two electrons.[6]

The complete Hamiltonian matrix $\hat{H}(t)$ for the half-filled H_2 system can now be written in the local basis set of eqn (12.5) as:

$$\hat{H}(t) = \begin{array}{c} \\ \langle a\bar{b} | \\ \langle \bar{a}b | \\ \langle a\bar{a} | \\ \langle b\bar{b} | \end{array} \begin{array}{c} |a\bar{b}\rangle \ |\bar{a}b\rangle \ |a\bar{a}\rangle \ |b\bar{b}\rangle \\ \begin{pmatrix} e^0 & 0 & +t & +t \\ 0 & e^0 & -t & -t \\ +t & -t & e^0 & 0 \\ +t & -t & 0 & e^0 \end{pmatrix} \end{array} \quad (12.8)$$

This is called the 'tight-binding' model, and corresponds to a parametrised version of the Hückel model.[7] Following the general method given in Section 2.1.4, its diagonalization involves solving the secular determinant:

$$| \hat{H}(t) - \epsilon \hat{I} | = 0 \quad (12.9)$$

where I is the identity matrix.[8] We then get the eigenvalues:

$$\epsilon_\pm = e^0 \pm 2t$$
$$\epsilon = e^0 \quad (12.10)$$

where ϵ is doubly degenerate. Given that e^0 and t are negative quantities, the ground-state energy of H_2 is strictly equivalent to that obtained in eqn (12.2):

$$e_{GS}(t) = \epsilon_+ = e^0 + 2t = 2\left(\alpha_{1s} + \beta_{1s,1s}\right) \quad (12.11)$$

Its associated ground-state wavefunction, $\Psi_{GS}(t)$ is obtained by solving the linear equations:

$$\begin{pmatrix} e^0 - \epsilon_+ & 0 & +t & +t \\ 0 & e^0 - \epsilon_+ & -t & -t \\ +t & -t & e^0 - \epsilon_+ & 0 \\ +t & -t & 0 & e^0 - \epsilon_+ \end{pmatrix} \begin{pmatrix} c_1 \\ c_2 \\ c_3 \\ c_4 \end{pmatrix} = 0 \quad (12.12)$$

where the c_i ($i = 1, 4$) are the coefficients of the linear expansion of $\Psi_{GS}(t)$ over the neutral and ionic states ϕ_i:

$$\Psi_{GS}(t) = \sum_{i=1}^{4} c_i \phi_i \quad (12.13)$$

Using the normalisation condition:

$$\sum_{i=1}^{4} c_i^2 = 1$$

we obtain:

$$\Psi_{GS}(t) = \frac{1}{2}\left(|a\bar{b}\rangle - |\bar{a}b\rangle + |a\bar{a}\rangle + |b\bar{b}\rangle\right) \quad (12.14)$$

[7] In the Hückel model, the overlap integrals between the atomic orbitals are computed to determine the β integral, whilst in the tight-binding model, t is a parameter.

[8] This explicitly means that the basis set of eqn (12.5) is an orthonormalised basis set, i.e. $\langle ij | i'j' \rangle = \delta_{ii'}\delta_{jj'}$, where $\delta_{ll'} = 1$ when $l = l'$, and $\delta_{ll'} = 0$ when $l \neq l'$.

which can also be written as:

$$\Psi_{GS}(t) = \cos\left(\frac{\pi}{4}\right) \left\{ \frac{|a\bar{b}\rangle - |\bar{a}b\rangle}{\sqrt{2}} \right\} + \sin\left(\frac{\pi}{4}\right) \left\{ \frac{|a\bar{a}\rangle + |b\bar{b}\rangle}{\sqrt{2}} \right\}$$

(12.15)

This wavefunction confirms that the four electronic configurations of Fig. 12.4 (namely the neutral and the ionic states) contribute in an equivalent manner to the ground state H_2 with a probability of occurrence $c_i^2 = \frac{1}{4}$. $\Psi_{GS}(t)$ is a spin singlet. It is strictly equivalent to the wavefunction obtained in eqn (12.3) and is simply written in a more local basis set than the delocalised one chosen in the LCAO representation (see Exercise (12.1)):

$$\Psi_{GS}(t) = \frac{2}{\sqrt{2}} \cos\left(\frac{\pi}{4}\right) |\sigma_g \bar{\sigma}_g| = |\sigma_g \bar{\sigma}_g|$$

As we will see now, the local basis set is more convenient for the introduction of electron repulsion and assessing its effect on H_2 energy and wavefunction.

12.1.2 The localised picture of H_2

Let us now introduce the idea that electrons are not independent particles, but negatively charged particles interacting through the Coulomb potential imposed by each electron i on the other electrons j $(j \neq i)$. Because this (bi-electronic) coulombic interaction is repulsive and depends on the distance between the interacting charges, it is expected that $e^- - e^-$ repulsion will be stronger when electrons are located on the same site than when they are located on different sites. Let us call $U_{aa} = U_{bb} = U > 0$ the energy associated with the on-site $e^- - e^-$ repulsion and $V_{ab} = V_{ba} = V > 0$ the energy associated with the intersite $e^- - e^-$ repulsion. In the Hubbard model [1] the intersite repulsion V is assumed to be negligible compared to the on-site repulsion U.[9] In the local basis set of eqn (12.5) the Hubbard Hamiltonian matrix is thus:

$$\hat{H}(t,U) = \begin{matrix} & |a\bar{b}\rangle & |\bar{a}b\rangle & |a\bar{a}\rangle & |b\bar{b}\rangle \\ \langle a\bar{b}| \\ \langle \bar{a}b| \\ \langle a\bar{a}| \\ \langle b\bar{b}| \end{matrix} \begin{pmatrix} e^0 & 0 & +t & +t \\ 0 & e^0 & -t & -t \\ +t & -t & e^0 + U & 0 \\ +t & -t & 0 & e^0 + U \end{pmatrix}$$

(12.16)

where U appears as an energy penalty ($U > 0$) of the electron distributions associated with the double occupancy of a site. The resolution of the secular determinant of eqn (12.9) using $\hat{H}(t,U)$ instead of $\hat{H}(t)$ leads to the following eigenvalues:

$$\epsilon_\pm = e^0 + \frac{U \pm \sqrt{U^2 + 16t^2}}{2}$$

$$\epsilon = e^0$$

$$\epsilon' = e^0 + U$$

(12.17)

[9] In the extended Hubbard model, both the on-site U and the intersite V electronic repulsions are explicitly taken into account.

As in the delocalised picture, the energy spectrum of H_2 does not depend on the reference energy (e^0) since all eigenvalues are shifted by the same quantity e^0. Solving Exercise (12.2), we can show that the ground-state energy of H_2 is $\epsilon_{GS}(t, U) = \epsilon_-$ and its associated wavefunction is given by:

$$\Psi_{GS}(t, U) = \cos\theta \left\{ \frac{|a\bar{b}\rangle - |\bar{a}b\rangle}{\sqrt{2}} \right\} + \sin\theta \left\{ \frac{|a\bar{a}\rangle + |b\bar{b}\rangle}{\sqrt{2}} \right\}$$

(12.18)

where the ratio of the ionic over the neutral states is given by:

$$\tan\theta = \frac{\sin\theta}{\cos\theta} = \frac{|4t|}{U + \sqrt{U^2 + 16t^2}}$$

(12.19)

As expected, the ground-state energy and wavefunction of H_2 depend on U, i.e. on the strength of the on-site electron repulsion. Interestingly, it is not represented by one, but by two electronic configurations of H_2, when expressed in the $\{\sigma_g, \sigma_u\}$ basis set:

$$\Psi_{GS}(t, U) = \left(\frac{\cos\theta + \sin\theta}{\sqrt{2}} \right) |\sigma_g \bar{\sigma}_g| + \left(\frac{\sin\theta - \cos\theta}{\sqrt{2}} \right) |\sigma_u \bar{\sigma}_u|$$

(12.20)

Thus the effect of U is not simply to raise in energy the doubly occupied orbitals, but also to induce electronic excitations from the σ_g to the σ_u MO to minimise the effective on-site Coulomb energy $U_{\sigma_g \sigma_g}$. The total ground-state wavefunction of H_2 is then a linear combination of different electronic configurations when $U \neq 0$, as clearly shown in eqn (12.20). In contrast to the Hückel model, the neutral and the ionic states no longer contribute in an equivalent manner to the ground-state wavefunction of H_2. As shown in Fig. 12.5, $\tan\theta$ follows an hyperbolic decrease as a function of $U/4|t|$, showing that the contribution of the ionic states to $\Psi_{GS}(t, U)$ rapidly decreases with the increase of U. In other words, when the energy associated with the electron pairing on one given site (either H^-–H^+ or H^+–H^-) becomes significant, the two electrons tend to separate and to statistically distribute over both sites to minimise their total electronic energy ($H^\bullet$–$H^\bullet$). The ionic states being essential in the hopping process between H_a and H_b (see Fig. 12.4), a loss of electron delocalisation over the H–H bond is then naturally expected when on-site electron repulsion increases.

In the $U_{+\infty}$ limit, this leads to a situation where the two electrons are fully localised on each site (H_a and H_b). Indeed, in this limit, $\tan\theta$ tends to zero and the ground-state wavefunction of H_2 becomes:

$$\Psi_{GS}(t, U_{+\infty}) = \cos 0 \left\{ \frac{|a\bar{b}\rangle - |\bar{a}b\rangle}{\sqrt{2}} \right\} + \sin 0 \left\{ \frac{|a\bar{a}\rangle + |b\bar{b}\rangle}{\sqrt{2}} \right\}$$

$$= \frac{1}{\sqrt{2}} \left(|a\bar{b}\rangle - |\bar{a}b\rangle \right)$$

(12.21)

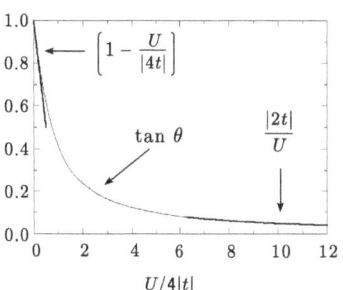

Fig. 12.5
$\tan\theta$ as a function of $U/4|t|$. The thick lines at small and large U values represent the asymptotic limits of $\tan\theta$ as defined in Exercise (12.3).

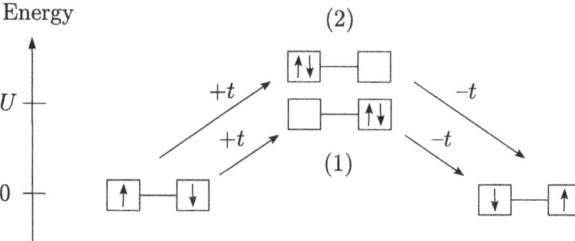

Fig. 12.6
Schematic representation of the coupling between the neutral states through the ionic states.

Interestingly, we see that the ground state of H_2 is now doubly degenerate since:

$$\lim_{U \to +\infty} \epsilon_{GS}(t, U) \approx e^0 = \epsilon_{ES}(t, U) \quad (12.22)$$

Solving the linear equations for this energy e^0 then leads to the second wavefunction:

$$\Psi_{ES}(t, U) = \frac{1}{\sqrt{2}} \left(|a\bar{b}\rangle + |\bar{a}b\rangle \right) \quad (12.23)$$

The two degenerate wavefunctions $\Psi_{GS}(t, U_{+\infty})$ and $\Psi_{ES}(t, U)$ both correspond to linear combinations of the neutral states with no contribution of the ionic states. This result emphasises that electron hopping between adjacent sites can no longer occur when the on-site repulsion is too high to allow double occupancy.[10] As shown in eqns (12.21) and (12.23), $\Psi_{GS}(t, U_{+\infty})$ and $\Psi_{ES}(t, U)$ simply differ in the sign of the linear expansion, which in fact reflects the spin state of the wavefunctions. After doing Exercise (12.4), the reader will be able to demonstrate that:

$$\Psi_{ES}(t, U) = \frac{1}{\sqrt{2}} \left\{ |\sigma_g \bar{\sigma}_u| + |\bar{\sigma}_g \sigma_u| \right\} \quad (12.24)$$

which corresponds to a spin triplet ($S = 1$). Actually, $\Psi_{ES}(t, U)$ is one of the three components of the triplet state of H_2. It corresponds to the spin projection $m_s = 0$ while the two high-spin electronic configurations neglected in this study (namely $|ab\rangle$ and $|\bar{a}\bar{b}\rangle$) correspond to the spin projections $m_s = \pm 1$.[11] As a consequence, in the limit of strong repulsions ($U \to +\infty$), the two electrons of H_2 fully localise on H_a and H_b, either with same spins (triplet) or with opposite spins (singlet) and the total energy of the system is $e^0 = 2\alpha_{1s}$. In other words, the two electrons of H_2 become localised spins and the Hubbard model turns out to be a spin Hamiltonian.

To demonstrate this, let us assume that $U > 4|t|$, so that the low-energy spectrum of H_2 is restricted to two states, $\Psi_{GS}(t, U)$ and $\Psi_{ES}(t, U)$, that are well separated in energy from the second excited state. Since the ionic states contribute nothing (or almost nothing) to these wavefunctions,[12] they can then be removed from the basis set. In doing so, we obtain a diagonal 2×2 matrix showing no explicit coupling between the localised spins. Yet we know from the Hubbard results that the two neutral states are indirectly coupled with each other through two-step processes involving the ionic states. Indeed, as illustrated in Fig. 12.6, $|a\bar{b}\rangle$ is coupled with $|\bar{a}b\rangle$ through:

[10] Note that in a less than half-filled system, electron hopping is still possible, even at large U, since it does not necessarily imply the creation of site double occupancy.

[11] These two states would have been degenerate with $\Psi_{ES}(t, U)$ if they were included in the development.

[12] Since $\tan\theta$ rapidly decreases with U, the ionic-state contribution to $\Psi_{GS}(t, U)$ rapidly becomes negligible over the neutral-state contribution.

1. an electron transfer from site a to site b (+t) *and* an electron transfer from site b to site a including the exchange of the two electrons ($-t$),
2. *or* an electron transfer from site b to site a (+t) *and* an electron transfer from site a to site b including the exchange of the two electrons ($-t$).

Using a second-order perturbative approach, the energy required for this process can be estimated as:

$$E^{(2)} = \sum_{\phi_I} \frac{\langle \phi_N | \hat{H}(t,U) | \phi_I \rangle \langle \phi_I | \hat{H}(t,U) | \phi_N \rangle}{\epsilon_{\phi_I} - \epsilon_{\phi_N}} \quad (12.25)$$

where ϕ_I and ϕ_N refer to the ionic and neutral states, respectively. We then obtain:

$$E^{(2)} = \frac{\langle a\bar{b} | \hat{H}(t,U) | b\bar{b} \rangle \langle b\bar{b} | \hat{H}(t,U) | \bar{a}b \rangle}{U}$$

$$+ \frac{\langle a\bar{b} | \hat{H}(t,U) | a\bar{a} \rangle \langle a\bar{a} | \hat{H}(t,U) | \bar{a}b \rangle}{U}$$

$$E^{(2)} = -\frac{2t^2}{U} \quad (12.26)$$

Therefore, if we remove the ionic states from the basis set of H$_2$, we have to account for their effect on the neutral state coupling in such a way that we accurately reproduce the hierarchy of the singlet/triplet energies obtained in eqn (12.17). This can be done by introducing a coupling constant between the neutral states. This coupling constant is the so-called *exchange interaction* parameter J. In the truncated basis set of the neutral states, the Hubbard matrix of eqn (12.16) thus becomes:

$$\hat{H}(U_{+\infty}) = \hat{H}(J) = \begin{array}{c} \\ \langle a\bar{b} | \\ \langle \bar{a}b | \end{array} \begin{pmatrix} |a\bar{b}\rangle & |\bar{a}b\rangle \\ e^0 + J & -J \\ -J & e^0 + J \end{pmatrix} \quad (12.27)$$

Its diagonalisation leads to two eigenvalues $\epsilon_0 = e^0$ and $\epsilon_J = e^0 + 2J$ and to the associated wavefunctions:

$$\Psi_{Tp} = \frac{1}{\sqrt{2}} (|a\bar{b}\rangle + |\bar{a}b\rangle) = \Psi_{ES}(t,U)$$

$$\Psi_{Sg} = \frac{1}{\sqrt{2}} (|a\bar{b}\rangle - |\bar{a}b\rangle) = \Psi_{GS}(t,U_{+\infty}) \quad (12.28)$$

A direct mapping using the Hubbard results for the half-filled H$_2$ molecule thus leads to:

$$J = \frac{1}{2} \{\epsilon_{ES}(t,U) - \epsilon_{GS}(t,U)\}$$

$$= \frac{U - \sqrt{U^2 + 16t^2}}{4} \quad (12.29)$$

Using a first-order Taylor expansion with $\epsilon = \frac{4|t|}{U}$, it is easy to show that $J = E^{(2)}$ in the $U_{+\infty}$ limit:

$$\lim_{\epsilon \to 0} J = \lim_{\epsilon \to 0} \frac{U}{4}\left(1 - \sqrt{1+\epsilon^2}\right) \approx \frac{|t|U}{4|t|}\left(\frac{\epsilon^2}{2}\right) = -\frac{2t^2}{U} \quad (12.30)$$

The exchange parameter J is negative in the whole range of U values so that the singlet state of H_2 is always favored over the triplet state, except at $U_{+\infty}$ where the two states are degenerate. At $U_{+\infty}$, the Hubbard Hamiltonian turns out to be the Hamiltonian of the $S = 1/2$ antiferromagnetic spin system.

In this section we have seen that the Hubbard model can be derived from the Hückel model by simply adding an energy penalization ($U \neq 0$) to the electronic states corresponding to the double occupation of a site. In other words, the Hubbard model states that the energy of a given orbital i is no longer α_i but α'_i defined as:

$$\alpha'_i = \alpha_i + q_i U \quad (12.31)$$

where $q_i = \langle n_i^\uparrow \rangle \langle n_i^\downarrow \rangle$ is the average number of paired electrons in orbital i. It also states that its ground-state electronic wavefunction is no longer represented by one but several electronic configurations, all associated with an equivalent spin state. The Hubbard model is a spinless Hamiltonian. In contrast to the Hückel model, the solutions of the Hubbard model (i.e. the eigenvalues and the eigenfunctions) depend not only on the U value but also on the number of electrons in the system (i.e. on electron filling).

Solving the half-filled Hubbard model leads to the following results:

- The eigenvalues (energies) do not depend on the reference energy e^0.
- The ground state is a spin singlet. It is written as a linear combination of the neutral and the ionic states, their relative contributions depending on U through eqn (12.19). This leads to an equivalent contribution of the neutral and ionic states when $U = 0$ (Hückel), a larger contribution of the neutral states when $U \neq 0$ (Hubbard), and an exclusive contribution of the neutral states when $U \to +\infty$ (Heisenberg).
- In the strong repulsion limit ($U_{+\infty}$) the ground state is a doubly degenerate state described by one spin singlet and one spin triplet. The two associated wavefunctions correspond to the symmetric (triplet) and antisymmetric (singlet) combinations of the neutral states with no contribution of the ionic states. The Hubbard model then turns out to be a spin model at $U_{+\infty}$, namely the Heisenberg model. In the Heisenberg model, the J parameter can be either positive (ferromagnetic exchange interaction) or negative (antiferromagnetic exchange interaction), in contrast to the $\hat{H}(t, J)$ model where it is always negative.

Note that the Hückel (tight-binding) and the Hubbard Hamiltonians are usually written in the formalism of second quantisation as: [4]

$$\hat{H}(t) = t \sum_{\langle i,j \rangle} \sum_\sigma \left(c_{i\sigma}^\dagger c_{j\sigma} + c_{j\sigma}^\dagger c_{i\sigma}\right) \quad (12.32)$$

and

$$\hat{H}(t, U) = \hat{H}(t) + \hat{H}(U)$$
$$= t \sum_{\langle i,j \rangle} \sum_\sigma \left(c_{i\sigma}^\dagger c_{j\sigma} + c_{j\sigma}^\dagger c_{i\sigma} \right) + U \sum_i \hat{n}_i^\uparrow \hat{n}_i^\downarrow \quad (12.33)$$

where $c_{i\sigma}^\dagger$ and $c_{i\sigma}$ are the second quantised operators acting as creator and annihilator of an electron with spin σ on site i, and $\hat{n}_i^\uparrow \hat{n}_i^\downarrow$ are electron count operators. In the limit of strong repulsion it becomes:

$$\hat{H}(t, J) = \hat{H}(t) + \hat{H}(J)$$
$$= t \sum_{\langle i,j \rangle} \sum_\sigma \left(c_{i\sigma}^\dagger a_{j\sigma} + c_{j\sigma}^\dagger a_{i\sigma} \right)$$
$$- J \sum_{\langle i,j \rangle} P \left(\hat{S}_i \hat{S}_j - \frac{1}{4} n_i n_j \right) P \quad (12.34)$$

where P is the projector over all configurations excluding the ionic states, written as $P = \Pi_i (1 - n_i^\uparrow n_i^\downarrow)$. Eventually, in the limit of $t \to 0$ and $U \to +\infty$ it restricts to the Heisenberg spin Hamiltonian:

$$\hat{H}(J) = -J \sum_{\langle i,j \rangle} \hat{S}_i \hat{S}_j \quad (12.35)$$

where $\hat{S}_i$ and $\hat{S}_j$ are the spin operators on sites i and j. Although seemingly complicated, the second quantisation formalism is a very elegant way of incorporating the antisymmetry property of the wavefunction into the algebraic properties of the operators, and is a relatively simple treatment of the many-body problem. The interested reader can refer to ref. [4] to become more familiar with this formalism.

The Hückel, Hubbard, $t - J$, and Heisenberg Hamiltonians are *phenomenological* models in the sense that they are parameterised. They are also *effective* models, since the role of the t, U, and J parameters is to account, in an effective manner, for the 'exact' interactions occurring in the real system.[13] In that sense, effective Hamiltonians are considered the simplest models for the study of the electronic and magnetic properties of a wide variety of systems. A fascinating observation here is that one might see in the Hubbard model a simplistic illustration of the wave-particle dualism of quantum physics. Indeed, when U is negligible or nil, the problem restricts to a $\hat{H}(t)$ spinless single-particle problem, which is exactly and analytically soluble. The solutions of this fully delocalised system are trivial (i.e. plane waves) and describe electrons that can be assimilated to 'waves'. On the other hand, when U is infinite, no more hopping can take place between adjacent sites and the spatial part of the electronic wavefunction is no longer needed to solve the $\hat{H}(J)$ spin problem. In other words, electrons are fully localised in space and become localised spins. The solutions of this fully localised system are also trivial, and now describe electrons that can be assimilated to 'particles'. In contrast, the $\hat{H}(t, U)$ Hubbard model leads to non-trivial solutions, which account

[13] These parameters can be either determined from experiments or computed using the sophisticated methods of quantum chemistry (see Section 12.2.1).

12.1.3 From the molecule to the solid state

Let us now consider the linear chain H_n described in Section 3.2. As shown in Chapter 3, in the Hückel approximation, the half-filled H_n chain is found to be metallic, just like any other chain whose unit cell contains an odd number of electrons. Its electronic structure is described by one 1s band (i.e. one crystal orbital $\Psi_{1s}(\vec{k})$) with energy given by eqn (3.22):

$$E(\vec{k}) = \alpha_{1s} + \frac{W}{2}\cos(2\pi k_a)$$

where k_a is the real projection of the $\vec{k}$ vector in the reciprocal lattice. Its bandwidth is $W = 4|\beta_{1s,1s}|$ (see Fig. 3.8). The Fermi level for this half-filled system refers to the dimensionless $\pm k_{a_f}$ values, so the total energy of H_n is obtained by integrating $E(k_a)$ over k_a from $k_a = -k_{a_f}$ to $+k_{a_f}$. Using the relations:

$$\vec{k} = k_a \vec{a}^*$$
$$\vec{a}^* \cdot \vec{a} = 2\pi$$

we obtain:

$$E(H_n) = \frac{a}{2\pi}\int_{-k_f}^{+k_f} n_{occ} E(\vec{k}) d\vec{k} = \int_{-k_{a_f}}^{+k_{a_f}} n_{occ} E(k_a) dk_a \quad (12.36)$$

where $\frac{a}{2\pi}$ is a normalisation factor and n_{occ} the number of electrons per electronic level.

As shown Fig. 12.7, in the Hückel model, the Fermi level of H_n corresponds to $\vec{k} = \pm\frac{\pi}{2a}$ (i.e. $k_{a_f} = \pm\frac{1}{4}$) and all crystal orbitals are doubly occupied ($n_{occ} = 2$). The total energy of the system thus corresponds to:

$$E^d(H_n) = 2 \times \int_{-k_{a_f}}^{+k_{a_f}} \left(\alpha_{1s} + \frac{W}{2}\cos(2\pi k_a)\right) dk_a$$

$$= 2 \times \left[\alpha_{1s} k_a - \frac{W}{4\pi}\sin(2\pi k_a)\right]_{-\frac{1}{4}}^{+\frac{1}{4}}$$

$$= \alpha_{1s} - \frac{W}{\pi} \quad (12.37)$$

If we now account for on-site electron repulsion, we need to consider that the energy of the 1s orbital is now $\alpha'_{1s} = \alpha_{1s} + q_{1s}U$, with $q_{1s} = \frac{1}{2} \cdot \frac{1}{2} = \frac{1}{4}$ for the delocalised half-filled system. As for the H_2 molecule, the effect of U is then to destabilise the total energy of the H_n chain by $\frac{U}{4}$ with respect to the non-interacting particle model, since:

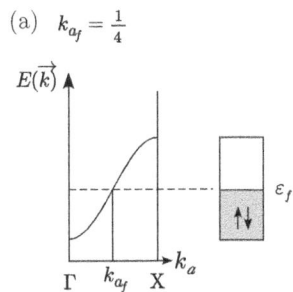

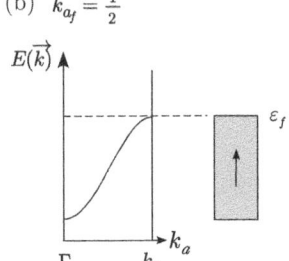

Fig. 12.7
Qualitative representation of the half-filled H_n electronic band structure (a) in the delocalised limit and (b) in the localised limit.

$$E^d(\mathrm{H}_n) = 2 \times \int_{-\frac{1}{4}}^{+\frac{1}{4}} \left(\alpha'_{1s} + \frac{W}{2} \cos(2\pi k_a) \right) dk_a$$

$$= 2 \times \left[\alpha_{1s} k_a + q_{1s} U k_a - \frac{W}{4\pi} \sin(2\pi k_a) \right]_{-\frac{1}{4}}^{+\frac{1}{4}}$$

$$= \alpha_{1s} + \frac{U}{4} - \frac{W}{\pi} \tag{12.38}$$

In the limit of strong repulsion (i.e. $U \to +\infty$), the system is expected to be fully localised and each crystal orbital simply occupied ($n_{occ} = 1$). The Fermi level of H_n then corresponds to the wave vectors $\vec{k} = \pm\frac{\pi}{a}$ (i.e. $k_a = \pm\frac{1}{2}$) so that the total energy of the localised half-filled chain becomes:

$$E^l(\mathrm{H}_n) = \int_{-\frac{1}{2}}^{+\frac{1}{2}} \left(\alpha_{1s} + \frac{W}{2} \cos(2\pi k_a) \right) dk_a$$

$$= \left[\alpha_{1s} k_a - \frac{W}{4\pi} \sin(2\pi k_a) \right]_{-\frac{1}{2}}^{+\frac{1}{2}}$$

$$= \alpha_{1s} \tag{12.39}$$

The energy difference between the localised and the delocalised states of the H_n chain is then:

$$E^l(\mathrm{H}_n) - E^d(\mathrm{H}_n) = \frac{W}{\pi} - \frac{U}{4} \tag{12.40}$$

This result is very close to the one previously obtained for the H_2 molecule.

It states that the H_n chain is metallic (delocalised) when the bandwidth $W = 4|t|$ is greater than $\frac{\pi}{4} U$, and insulating (localised) when W is smaller than $\frac{\pi}{4} U$. A half-filled system characterised by $W \sim U$ is therefore susceptible to a metal-to-insulator transition induced by electronic repulsion, namely a *Mott–Hubbard* transition. In contrast to first-order Peierls transitions, Mott–Hubbard transitions are second-order metal-to-insulator transitions, which are not restricted to 1D systems. They are not associated with a structural distortion but reflect a progressive localisation of the electrons, for example when the temperature is lowered. This electron localisation can be understood as resulting from a decrease in the screening of the on-site repulsion U by the flux of delocalised electrons when temperature is decreased. In other words, electrons feel an effective on-site repulsion that is progressively less efficiently screened as the temperature decreases.[14] This leads to a progressive trapping of electrons on the different sites of the system and to an energy band gap opening in the electronic structure of the system. The band gap in a Mott–Hubbard transition is proportional to U, in contrast to that of a Peierls transition, which is proportional to the amplitude of the *n*-merisation. Since the band gap corresponds to the energy required to promote one electron from the valence band (i.e. highest occupied band) to the conduction band (i.e. lowest unoccupied band), U is directly linked to the *ionisation potential* (IP) and *electron affinity* (EA) of the system, through:

[14] This is the case even if U does not explicitly depend on the temperature.

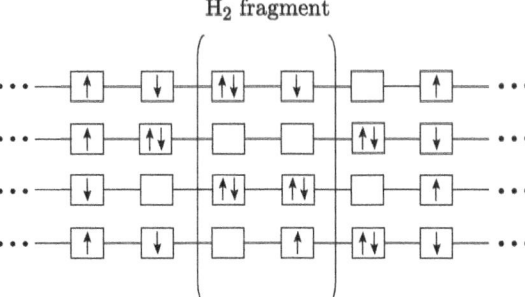

Fig. 12.8
Some of the electronic distributions associated with n electrons on n sites.

$$E_g \propto U = (E_n - E_{n-1}) - (E_{n+1} + E_n)$$
$$= \text{IP} - \text{EA} \qquad (12.41)$$

where E_n, E_{n+1}, and E_{n+1} refer to the energy of the system with n, $n-1$, and $n+1$ electrons. It should be noticed here that other types of second-order metal-to-insulator transitions exist, for example the Anderson transition in defective materials, where electrons are progressively trapped in the potential well created by defects in the structure as the temperature is decreased.

How do these results compare with the Hubbard model in the case of an infinite half-filled system?

Following the general procedure outlined in the previous section, the basis set for the H_n chain would consist of an infinite number of electronic states, each corresponding to a different distribution of the n electrons over the n sites of the H_n chain (see Fig. 12.8). As for the H_2 molecule, a given site i is thus alternately empty, singly occupied by an electron with spin ↑ or ↓, or doubly occupied, each situation having the same probability of occurrence in the free electron model, i.e. $\frac{1}{4}$. However, in contrast to the H_2 molecule, two adjacent sites can now be either fully empty or filled by one, two, three, or four electrons, independent of their spin. Indeed, in contrast with the H_2 molecule, which is a closed system from and to which no electron transfer can take place, an H_2 fragment belonging to H_n is able to exchange electrons with the rest of the chain, provided that the total number of electrons in the infinite chain ($n_e = n$) and the total spin ($S_{tot} = 0$) remain unchanged. As a consequence, solving the Hubbard model in the case of an infinite H_n chain involves diagonalising a $4^n \times 4^n$ matrix. This obviously requires use of numerical techniques, in particular when no analytical solution exists. While it is not the purpose of this book to detail these different techniques, it should be noted here that the infinite half-filled Hubbard model has been exactly solved by Lieb and Wu, [5] leading to the ground-state energy per site:

$$E_{GS}(t, U) = -\frac{4|t|}{\pi} + \frac{U}{4} - 0.017\frac{U^2}{|t|} \qquad (12.42)$$

which is very close to eqn (12.40) in the limit of weak repulsion and to the solution of the infinite $S = 1/2$ Heisenberg spin Hamiltonian in the limit of strong repulsion:

$$\lim_{U \to +\infty} E_{GS}(t, U) = -\frac{4t^2}{U} ln2 \qquad (12.43)$$

The infinite case of H_n thus compares well with the results obtained for the H_2 molecule, showing that the local parameters t and U are pertinent to describing the electronic properties of infinite systems. As already mentioned, the main effect of long-range interactions is to partially screen the local (short-range) interactions.

12.1.4 Application to one-band systems

To understand the effect of the on-site repulsion U on the electronic band structure of a real system, we will now consider the example of organic conductors. As shown in Chapter 11, organic conductors are low-dimensional molecular systems the electronic properties of which are generally supported by one orbital per site. In this respect, the one-band Hubbard model is perfectly adapted to the study of their electronic properties. The NH_xMe_{4-x} [Ni(dmit)$_2$]$_2$ series of compounds [7] exhibit various electronic behaviours depending on the size of the countercation $NH_xMe^+_{4-x}$, i.e. depending on x:

- NMe$_4$ [Ni(dmit)$_2$]$_2$ is metallic and exhibits a sample-dependent resistivity bump at around 90 K
- NHMe$_3$ [Ni(dmit)$_2$]$_2$ is a metal down to 220 K where it undergoes a metal-to-semiconductor transition
- NH$_2$Me$_2$ [Ni(dmit)$_2$]$_2$ is a semiconductor at room temperature with an energy band gap of 0.21 eV.

In these systems, the transport properties are governed by the Ni(dmit)$_2$ anionic slabs, i.e. by the LUMO orbital of the Ni(dmit)$_2$ acceptor molecules (see Fig. 12.9). The 1:2 stoichiometry implies that the LUMO band is formally quarter-filled since one electron is transferred from the NR_4^+ cation to a [Ni(dmit)$_2$]$_2$ dimeric unit. In previous chapters we learned that quarter-filled 1D metals are susceptible to a metal-to-insulator transition associated with a tetramerisation along the chains. In the present case, we will see that the Peierls instability may compete with the Mott–Hubbard instability. As illustrated in Fig. 12.10, each anionic slab of the NR_4 [Ni(dmit)$_2$]$_2$ series consists of tilted chains of Ni(dmit)$_2$ monomers. Based on extended Hückel molecular calculations, an analysis of the intermolecular interactions within these slabs confirms the 1D character of these systems, with negligible interchain intermolecular interactions compared to intrachain intermolecular interactions. Surprisingly, the analysis also reveals that the Ni(dmit)$_2$ chains are more or less dimerised (and not tetramerised) depending on the NR_4 cation size. While nearly equivalent intradimer β_1 interaction energies are found for all systems, various interdimer β_2 interaction energies are obtained depending on the system considered. Given that $\delta = (\beta_1 - \beta_2)/\beta_1$ represents a direct measure of the dimerisation amplitude along the chains, we find that NMe$_4$ [Ni(dmit)$_2$]$_2$ and NHMe$_3$ [Ni(dmit)$_2$]$_2$ are weakly dimerised ($\delta = 0.05$) and NH$_2$Me$_2$ [Ni(dmit)$_2$]$_2$ is strongly dimerised ($\delta = 0.75$).

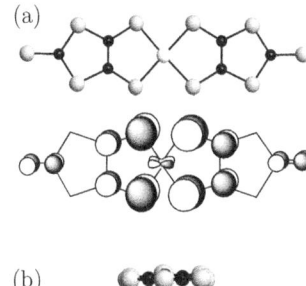
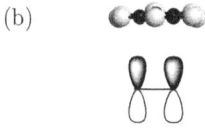

Fig. 12.9
Projection view of the Ni(dmit)$_2$ molecule and LUMO orbital (a) perpendicular to the molecule plane and (b) along the longidutinal axis of the molecule. Black, grey, and white circles refer to the C, S, and Ni atoms.

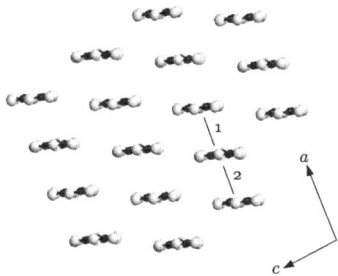

Fig. 12.10
Slab of Ni(dmit)$_2$ molecules in the NH_xMe_{4-x} [Ni(dmit)$_2$]$_2$ systems. Labels 1 and 2 refer to the intradimer and interdimer interactions, respectively.

Hubbard electronic band gaps

The quarter-filled $[Ni(dmit)_2]^{0.5-}$ chains of the NH_xMe_{4-x} $[Ni(dmit)_2]_2$ series have been studied in the Hubbard approximation. In the following, we will assume an equivalence between the Hückel resonance integrals β_i and the hopping integrals t_i. To solve the infinite chain problem, the so-called DMRG[15] numerical technique has been used. [8, 9] This technique allows one to extrapolate the properties of the infinite chain from successive size-increasing finite chain calculations. Various dimerisation amplitudes along the chains are considered to reproduce the physics of the whole series from $\delta = 0.05$ (almost regular chain) to $\delta = 0.75$ (strongly dimerised chain). Since there is no explicit reference to the crystal lattice coordinates in effective Hamiltonians, no band structures can be drawn from the results of the calculation. Instead, site energy as well as ground-state wavefunctions and band gaps can be computed. The energy band gaps are then computed using eqn (12.41) for various dimerisation degrees and various U values. As shown in Fig. 12.11, a band gap opens in the electronic structure of the $Ni(dmit)_2$ chains, which increases with U and δ. When accounting for the room temperature (RT) thermal activation energy e_a^{RT},[16] we find that neither NMe_4 $[Ni(dmit)_2]_2$ nor $NHMe_3$ $[Ni(dmit)_2]_2$ ($\delta = 0.05$) should undergo a metal-to-insulator transition induced by electron repulsion: their band gap Δ remains smaller than e_a^{RT} (or equivalently Δ/t_1 remains smaller than e_a^{RT}/t_1) in the whole range of U, from the Hückel limit ($U = 0$) to the strong repulsion limit ($U \longrightarrow +\infty$). This suggests another driving force for the transitions observed at low temperature for these two systems. Given the 1D character of their $Ni(dmit)_2$ slabs, it is likely that a Peierls distortion would be at the origin of these transitions. This will be discussed below. In contrast, NH_2Me_2 $[Ni(dmit)_2]_2$ ($\delta = 0.75$) shows completely different behaviour: except for very small U values, its band gap is larger than e_a^{RT}. An electronic localisation is thus expected along the $Ni(dmit)_2$ chains for this system, in perfect agreement with its semiconducting state at room temperature. Since experimental data are available for this system, the effective U value characterising the $Ni(dmit)_2$ molecule can easily be deduced from a direct mapping of the RT activation energy and the gap computed for $\delta = 0.75$. By simply adding the RT thermal energy e_a^{RT} to the $T = 0$ K computed gaps, Δ, U can be shown to be close to 1.0 eV for the $Ni(dmit)_2$ molecule.

[15] DMRG stands for Density Matrix Renormalisation Group.

[16] $e_a^{RT} \sim 25$ meV at T = 300 K.

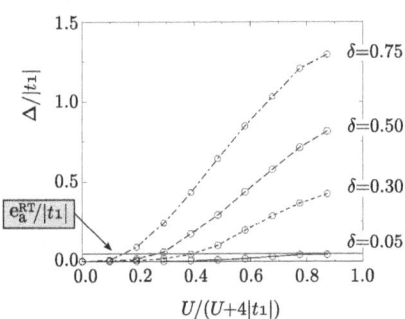

Fig. 12.11
Dimensionless energy band gaps Δ/t_1 computed in the Hubbard approximation using eqn (12.41) for Δ and plotted as a function of $U/(U + 4 | t_1 |)$. Note that t_1 and t_2 hopping integrals used in the Hubbard calculations correspond to the β_1 and β_2 intermolecular interaction energies extracted from the extended Hückel calculations.

How does this U value compare with the LUMO bandwidth $W = 4|t|$ in the different compounds?

In the Hückel approximation, the electronic band structure of the NH_2Me_2 [Ni(dmit)$_2$]$_2$ system ($\delta = 0.75$) is very similar to that obtained for the dimerised $(H_2)_n$ model system (see Chapter 2), while the electronic band structure for the $NHMe_3$ [Ni(dmit)$_2$]$_2$ system ($\delta = 0.05$) is very similar to that obtained for the regular $(H^{+0.5})_n$ model system (see Chapter 3). This is illustrated in Fig. 12.12, where the electronic structure of the Ni(dmit)$_2$ chains is schematically drawn from the intra- and interdimer orbital interactions β_i. For the weakly dimerised chain ($\delta = 0.05$), the intra- and interdimer interactions are very close, i.e. $\beta_1 \sim \beta_2 = \beta$, leading to one quarter-filled LUMO band, with a bandwidth W proportional to $2(|\beta_1| + |\beta_2|) \sim 4|\beta|$. For the dimerised chain ($\delta = 0.75$), the intradimer interaction is four times greater than the interdimer interaction, i.e. $\beta_1 = 4\beta_2$. This leads to the splitting of the LUMO band into one low-lying half-filled band coming from the ϕ_{LUMO}^+ orbital of the dimer and one high-lying empty band coming from the ϕ_{LUMO}^- orbital of the dimer. A large energy gap occurs in this electronic structure in response to the strong dimerisation amplitude along the chain.

We have seen that, in the Hubbard approximation, transport properties are primarily governed by the $U/4|t|$ ratio, where U is the on-site repulsion energy and t is the hopping integral between two adjacent sites. Given that the leading parameter corresponds to the interdimer interaction ($t = \beta_2$) for the strongly dimerised system and to the intermonomer interaction ($t = \beta_1$) for the weakly dimerised system, the $U/4|t|$ ratio is much larger for NH_2Me_2 [Ni(dmit)$_2$]$_2$ than for $NHMe_3$ [Ni(dmit)$_2$]$_2$: given that U and β_1 are equivalent in magnitude for all compounds of the NR_4 [Ni(dmit)$_2$]$_2$ series, the $U/4|t|$ ratio is close to 1 for $NHMe_3$ [Ni(dmit)$_2$]$_2$ and close to 4 for the NH_2Me_2 [Ni(dmit)$_2$]$_2$, in perfect agreement with the metallic (delocalised) and semiconducting (localised) properties of these two systems, respectively. This qualitative local picture is confirmed by the electronic band structures computed in the extended Hückel approximation for the three anionic slabs of the three NH_xMe_{4-x} [Ni(dmit)$_2$]$_2$ ($x = 0$, 1, and 2) compounds (see Figs 12.13 and 12.14). NMe_4 [Ni(dmit)$_2$]$_2$ and $NHMe_3$ [Ni(dmit)$_2$]$_2$ thus exhibit a quasi 1D electronic conductivity along the chains and should both undergo a metal-to-insulator Peierls-type transition at low temperature. The transition should correspond to a tetramerisation of the Ni(dmit)$_2$ monomers along the chain direction (quarter-filled band with $W = 2(|\beta_1| + |\beta_2|) = 4|\beta|$) or, more rigorously, to a dimerisation of the [Ni(dmit)$_2$]$_2$ weak dimers (half-filled band with $W = 2|\beta_2|$). This is confirmed by the Fermi surface computed for the $NHMe_3$ [Ni(dmit)$_2$]$_2$ system, for which a nearly perfect nesting vector is found at $q = a^*/2$.[17] The reason why NMe_4 [Ni(dmit)$_2$]$_2$ does not undergo a real metal-to-insulator transition at low temperature is related to its crystal structure. This system is built on two crystallographically independent Ni(dmit)$_2$ slabs (rather than one, as is the case for $NHMe_3$ [Ni(dmit)$_2$]$_2$), which present different orientations along the c-axis, probably due to the tetrahedral symmetry of the NMe_4^+ cations. This obviously decreases the driving force of the Peierls transition since the distortion must occur simultaneously in both slabs with a common nesting vector. [7]

[17] As detailed in ref. [7] the unit cell of $NHMe_3$ [Ni(dmit)$_2$]$_2$ consists on two monomers so that a tetramerisation along the chain is obtained by doubling the chain axis a.

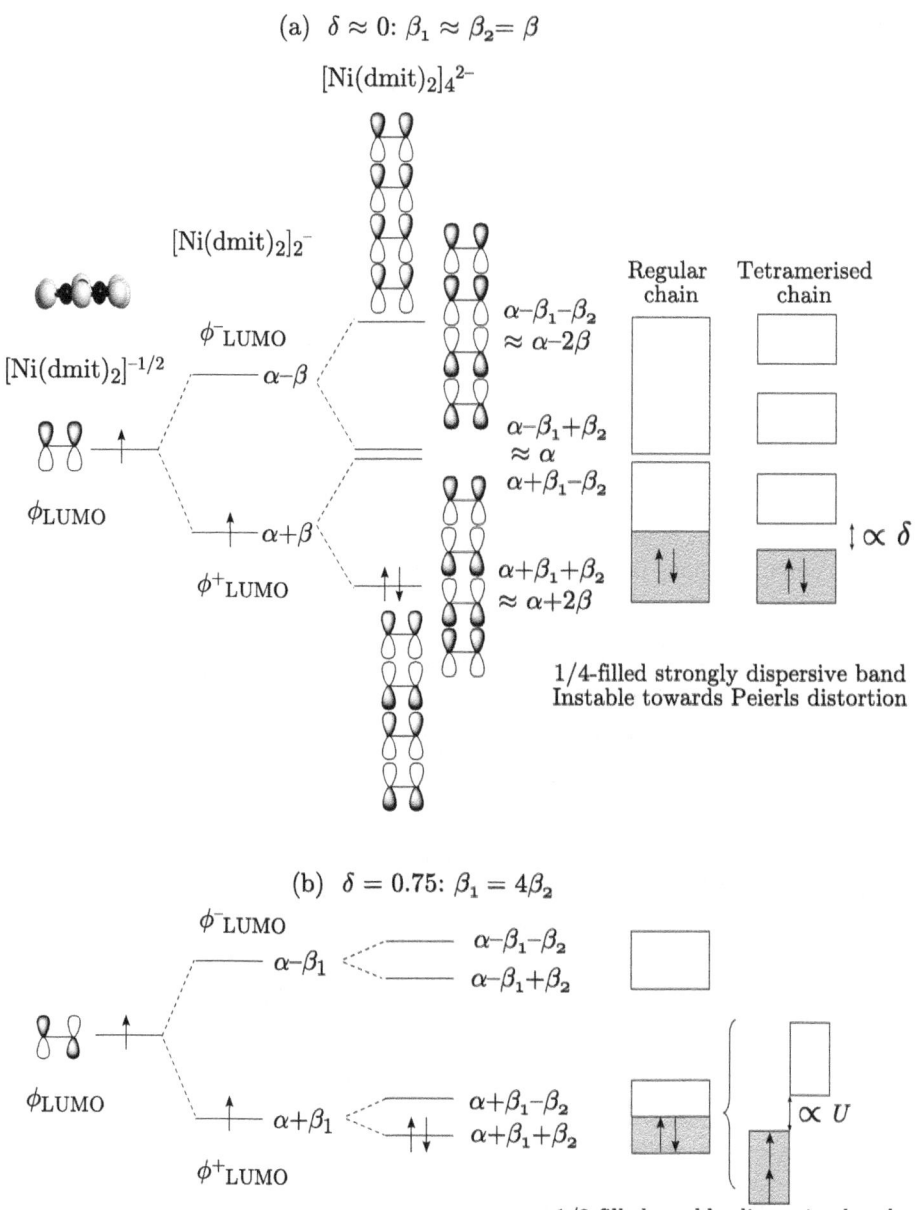

Fig. 12.12
Qualitative electronic band structure of the Ni(dmit)$_2$ chain for (a) the regular chain ($\delta = 0$) and (b) the dimerised chain ($\delta \neq 0$). The Ni(dmit)$_2$ molecule and its associated LUMO molecular orbital are projected along the longitudinal molecular axis. The hopping integrals t_1 and t_2 are assumed to be equivalent to the extended Hückel parameter β.

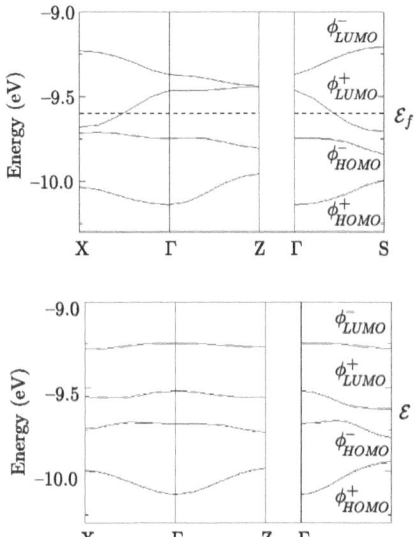

Fig. 12.13
Electronic band structure of the Ni(dmit)$_2$ slabs in NHMe$_3$ [Ni(dmit)$_2$]$_2$, computed in the extended Hückel approximation. The electronic band structure of NMe$_4$ [Ni(dmit)$_2$]$_2$ is very similar and not shown.

Fig. 12.14
Electronic band structure of the Ni(dmit)$_2$ slabs in NH$_2$Me$_2$ [Ni(dmit)$_2$]$_2$, computed in the extended Hückel approximation.

12.2 Mean-field approaches

While *phenomenological* and *effective* models are the simplest models for studying the transport properties of condensed matter, they rapidly become difficult to handle when the number of active states required to describe the system increases. This is the case for systems where the electronic properties are supported by more than one orbital per site, such as the transition metal (TM) oxides. As shown in Chapters 10 and 11, the electronic structure of TM-based compounds can be described by different overlapping bands, each associated with different bandwidths and occupancies. In practice, not only the basis set size but also the number of parameters required to describe the physics of these systems increases for these materials. Let us illustrate this with the simple case of the half-filled A$_2$ molecule in which A is described by two non-orthogonal orbitals. As illustrated in Fig. 12.15, seven parameters are required to describe this molecule: the energy difference between the two different orbitals ($\Delta e = e_1 - e_2$), three different on-site repulsion energies U_{11}, U_{22}, and U_{12}, and three different hopping integrals t_{11}, t_{22}, and t_{12}. Solving the Hubbard model for this system is then a tricky problem involving seven degrees of freedom, and obviously leads to a very complicated phase diagram depending on the magnitude of each parameter. In such multi-band systems, it is often more convenient to use first-principles quantum methods based on the so-called *mean-field approximation* to study their electronic structure.

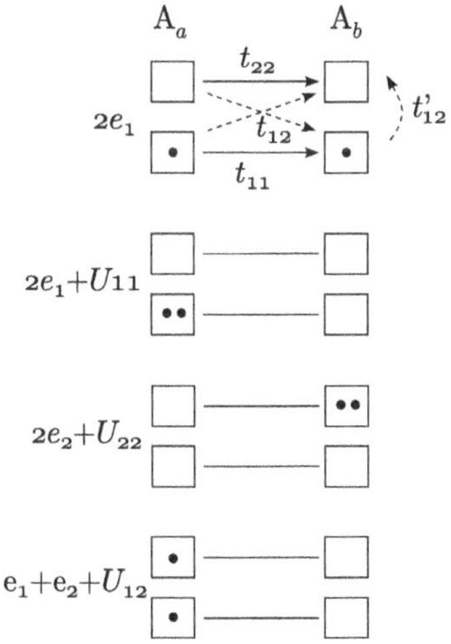

Fig. 12.15
Some different electronic configurations (of the 28 possible) for the A$_2$ molecule, showing the different intrasite and intersite hopping integrals t_{ij} and the different on-site repulsion energy. The reference energy of each configuration is given on the left of the figure.

12.2.1 The many-body problem

In the Born–Oppenheimer approximation, the time-independent non-relativistic[18] Hamiltonian of the n-interacting-particle system is written in atomic units as:

[18] This non-relativstic approximation means that we fully neglect the spin-orbit coupling.

Electron repulsion

$$\hat{H}^{el} = \sum_{i=1}^{n} -\frac{1}{2}\nabla_i^2 - \sum_{i=1}^{n}\sum_{\alpha}^{M} \frac{Z_A}{r_{i\alpha}} + \frac{1}{2}\sum_{i=1}^{n}\sum_{j \neq i}^{n} \frac{1}{r_{ij}}$$

$$= \sum_{i=1}^{n} \left(\hat{T}_i + \hat{V}_{i\alpha} + \frac{1}{2}\hat{V}_{ij} \right) \tag{12.44}$$

where $\hat{T}_i$ is the kinetic operator acting on electron i, and $\hat{V}_{i\alpha}$ and $\hat{V}_{ij}$ are the coulombic potentials felt by electron i due to the M immobile nuclei (attractive potential) and the $j = (n-1)$ mobile electrons (repulsive potential). Every electron i thus experiences a mean-field potential due to the M nuclei and the j other electrons of the system. The $\frac{1}{2}$ factor preceding $\hat{V}_{ij}$ prevents double-counting of the interaction between electrons i and j. The first two terms of $\hat{H}^{el}$ are one-electron operators, the solutions of which are analytical and therefore easy to compute. In contrast, the last term of $\hat{H}^{el}$ is a two-electron operator with unknown solutions. For this reason, the Schrödinger equation has no analytical solution for any system of n interacting particles where $n > 1$, and its resolution therefore requires approximations. Among these, the mean-field approach is an elegant numerical method, using the so-called *self-consistent field* (SCF) procedure and the *variational principle*[19] to solve the Schrödinger equation in such a way that the mean-field solutions (wavefunctions and energies) get as close as possible to the exact solutions. Both the Hartree–Fock [10] and the density functional theory [11] formalisms are based on this mean-field approach. However, they use different approximations to reach the final solutions. As we will see below, the former seeks the best approximation of the total electronic wavefunction ($\Psi_n(x_1, x_2, \ldots, x_n)$) using the exact expression of the total electronic Hamiltonian, while the latter seeks the exact ground-state electron density ($\rho(r)$) using an approximation of the V_{ij} two-electron potential energy. While the mathematical development of these two formalisms is beyond the scope of this book, it is interesting to summarise their main assumptions and results for the sake of comparison with the effective models discussed above. This will be done in the following sections.

[19] The variational principle states that the 'best' solution of the Schrödinger equation is the one that minimises the total energy of the system.

12.2.2 The Hartree–Fock method

Given the factored form of $\hat{H}^{el}$:

$$\hat{H}^{el} = \sum_{i=1}^{n} \left(\hat{T}_i + \hat{V}_{i\alpha} + \frac{1}{2}\hat{V}_{ij} \right)$$

$$= \sum_{i=1}^{n} \left(\hat{h}_i + \frac{1}{2}\sum_{j \neq i} \hat{V}_j \right) = \sum_{i=1}^{n} \hat{F}_i \tag{12.45}$$

where $\hat{F}_i$ is called the *Fock operator*, it is straightforward to determine that the n-interacting-particle problem could reduce to n one-electron problems if the solutions were known exactly. In this one-electron assumption, it is therefore tempting to assume that a good approximation of the n-particle wavefunction $\Psi_n(x_1, x_2, \ldots, x_n)$ is a product of one-electron wavefunctions, namely the *Hartree product*, written as:

$$\Psi_n(x_1, x_2, \ldots, x_n) = \Pi_i \phi_i(x_i)$$
$$= \phi_i(x_1)\phi_j(x_2)\ldots\phi_n(x_n) \quad (12.46)$$

In this expression, the $\phi_i(x_i)$ terms represent the n occupied spin orbitals of the system:

$$\phi_i(x_i) = \phi_i(r_i)\sigma_i \quad (12.47)$$

also written as:

$$\phi_i = \phi_i(r_i)\alpha_i$$
$$\bar{\phi}_i = \phi_i(r_i)\beta_i$$

and $r_i = x_i, y_i, z_i$, and σ_i stand for the electron spatial and spin coordinates, respectively. In the LCAO approximation, each spin orbital is written as a linear combination of atomic orbitals:

$$\phi_i(x_i) = \sum_j c_{ij} \chi_j(x_i) \quad (12.48)$$

where the c_{ij} coefficents are unknown. Following the general procedure given in Section 2.1, the determination of the c_{ij} coefficients involves solving the secular determinant:

$$| \underline{\underline{H}}^{el} - E\underline{\underline{S}} | = 0 \quad (12.49)$$

in the χ_j basis set. While this procedure is straightforward in the case of the Hückel Hamiltonian, here we face a mathematical issue arising from the two-electron $\hat{V}_j$ term of the Fock operator. As already mentioned, this operator reflects the repulsion exerted on each electron i by the j surrounding electrons. As shown in eqn (12.50), the form of this operator depends on its solutions ϕ_j ($j \neq i$):

$$\hat{V}_j \phi_i(x_1) = \int \phi_j^*(x_2)\phi_j(x_2)\frac{1}{r_{12}}\phi_i(x_1)dx_2 \quad (12.50)$$

This means that the secular determinant cannot be solved analytically and that the eigenvalues of the Schrödinger equation are functionals of ϕ_j, namely $E[\phi]$.[20] To circumvent this issue, the so-called *self consistent field* (SCF) iterative procedure has been proposed. This involves seeking the best approximation of Ψ_n, i.e. the one that minimises the total energy of the system, by varying the c_{ij} coefficients until a stable solution is found.[21] In practical terms, it involves: (i) starting with a trial set of c_{ij}^0, (ii) building the $\hat{V}_j^0$ operator with the c_{ij}^0 set, (iii) solving the Schrödinger equation to obtain a new set of c_{ij}^1 coefficients, (iv) using the c_{ij}^1 to build a new $\hat{V}_j^1$ potential, etc ... At each iterative step i, the total energy and wavefunction are compared to those obtained in step $i-1$, and a convergence criterion is applied to stop the procedure when the solutions become invariant upon further variations of the c_{ij}^i, i.e. when:

[20] The term 'functional' means that E is a function of the ϕ functions.

[21] This procedure is performed using so-called *Lagrange multipliers*, namely λ_{ij}, through the minimisation of:

$$\langle \Psi_n | \hat{H}^{el} | \Psi_n \rangle - \sum_{i=1}^{n} \lambda_{ij} \langle \phi_i | \phi_j \rangle$$

where the spin-orbitals ϕ are constrained to be orthonormal.

$$\frac{\partial E[\phi]}{\partial c_{ij}} = 0$$

In the present approximation of Ψ_n, this procedure would lead to the Hartree energy $E^H[\phi]$:

$$E^H[\phi] = \langle \Psi_n | \sum_{i=1}^{n} \hat{F}_i | \Psi_n \rangle$$

$$= \langle \phi_i(x_1) \ldots \phi_n(x_n) | \sum_{i=1}^{n} \hat{F}_i | \phi_i(x_1) \ldots \phi_n(x_n) \rangle$$

$$= \sum_{i=1}^{n} \left(e_i + \frac{1}{2} \sum_{j \neq i} J_{ij} \right) \quad (12.51)$$

where e_i and J_{ij} are one- and two-electron integrals written as:

$$e_i = \langle \phi_i(x_1) | \hat{h}_1 | \phi_i(x_1) \rangle$$

$$= \int dx_1 \, \phi_i^*(x_1) \hat{h}_1 \phi_i(x_1)$$

$$J_{ij} = \langle \phi_i(x_1) \phi_j(x_2) | \frac{1}{r_{12}} | \phi_i(x_1) \phi_j(x_2) \rangle$$

$$= \int \int dx_1 dx_2 \, \phi_i^*(x_1) \phi_i(x_1) \frac{1}{r_{12}} \phi_j^*(x_2) \phi_j(x_2) \quad (12.52)$$

Hence, in the Hartree approximation of the n-particle wavefunction, the total electronic energy of the system reduces to one orbital contribution (e_i) and one coulombic contribution (J_{ij}). J_{ij} is called the *Coulomb integral*, in reference to the classical Coulomb interaction between two charge distributions, here given by $\phi_i^* \phi_i$ and $\phi_j^* \phi_j$. Given that the Fock operator has no explicit dependence on spin, and that:

$$\langle \sigma_i | \sigma_j \rangle = \delta_{\sigma_i \sigma_j}$$

the integrals of eqn (12.52) become:

$$e_i = \langle \phi_i(r_1)\sigma_i | \hat{h}_1 | \phi_i(r_1)\sigma_i \rangle$$

$$= \langle \phi_i(r_1) | \hat{h}_1 | \phi_i(r_1) \rangle \langle \sigma_i | \sigma_i \rangle$$

$$= \int dr_1 \phi_i^*(r_1) \hat{h}_1 \phi_i(r_1)$$

$$J_{ij} = \langle \phi_i(r_1)\sigma_i \phi_j(r_2)\sigma_j | \frac{1}{r_{12}} | \phi_i(r_1)\sigma_i \phi_j(r_2)\sigma_j \rangle$$

$$= \langle \phi_i(r_1)\phi_j(r_2) | \frac{1}{r_{12}} | \phi_i(r_1)\phi_j(r_2) \rangle \langle \sigma_i | \sigma_i \rangle \langle \sigma_j | \sigma_j \rangle$$

$$= \int \int dr_1 dr_2 \phi_i^*(r_1)\phi_i(r_1) \frac{1}{r_{12}} \phi_j^*(r_2) \phi_j(r_2)$$

Interestingly, we note that J_{ij} is non-zero whatever the spin of the two interacting electrons. Since ϕ_i are spin orbitals, this implicitly means that two electrons can lie in the same orbital with equivalent spins, which obviously violates the Pauli exclusion principle that electrons must satisfy. A better approximation of the total n-electron wavefunction must therefore include the antisymmetry property required to satisfy this exclusion principle. This wavefunction is an antisymmetrised product of one-electron wavefunctions, the so-called *Slater determinant*:

$$\Psi_n(x_1, x_2, \ldots, x_n) = \frac{1}{\sqrt{n!}} \begin{vmatrix} \phi_1(x_1) & \phi_2(x_1) & \ldots & \phi_n(x_1) \\ \phi_1(x_2) & \phi_2(x_2) & \ldots & \phi_n(x_2) \\ \ldots & \ldots & \ldots & \ldots \\ \phi_1(x_n) & \phi_2(x_n) & \ldots & \phi_n(x_n) \end{vmatrix} \quad (12.53)$$

written in its contracted form as:

$$\Psi_n(x_1, x_2, \ldots, x_n) = |\phi_1(x_1)\phi_2(x_2) \ldots \phi_n(x_n)|$$

Using this wavefunction, the action of the Fock operator on the n-occupied ϕ_i now leads to the Hartree–Fock energy:

$$E^{HF}[\phi] = \langle \Psi_n | \sum_{i=1}^{n} \hat{F}_i | \Psi_n \rangle$$

$$= \sum_{i=1}^{n} \left\{ e_i + \frac{1}{2} \sum_{j \neq i} (J_{ij} - K_{ij}) \right\} \quad (12.54)$$

In this expression, a additional two-electron integral K_{ij} appears as a direct consequence of the antisymmetrisation of Ψ_n:

$$K_{ij} = \langle \phi_i(x_1)\phi_j(x_2) | \frac{1}{r_{12}} | \phi_j(x_1)\phi_i(x_2) \rangle$$

$$= \langle \phi_i(r_1)\sigma_i \phi_j(r_2)\sigma_j | \frac{1}{r_{12}} | \phi_j(r_1)\sigma_j \phi_i(r_2)\sigma_i \rangle$$

$$= \langle \phi_i(r_1)\phi_j(r_2) | \frac{1}{r_{12}} | \phi_j(r_1)\phi_i(r_2) \rangle \langle \sigma_i | \sigma_j \rangle \langle \sigma_j | \sigma_i \rangle$$

$$= \int\int dr_1 dr_2 \, \phi_i^*(r_1)\phi_j(r_1)\frac{1}{r_{12}}\phi_j^*(r_2)\phi_i(r_2).\delta_{\sigma_i\sigma_j}\delta_{\sigma_j\sigma_i}$$

This is called the *exchange integral* and its associated $\hat{K}_j$ operator is:

$$\hat{K}_j\phi_i(x_1) = \int dx_2 \, \phi_j^*(x_2)\phi_i(x_2)\frac{1}{r_{ij}}\phi_i(x_1) \quad (12.55)$$

In contrast to J_{ij}, the K_{ij} energy depends on the spin of the electrons through $\delta_{\sigma_i\sigma_j}$ and therefore has no classical analogue. Its physical effect is to correlate the motion of electrons with equivalent spins. This property is a very interesting and conterintuitive property of the Hartree–Fock method.

Indeed, whilst using a purely one-electron representation of the n-particle wavefunction, the Hartree–Fock method allows us to partially account for the

Electron repulsion

[22] We shall see at the end of this section that the so-called *configuration interaction* (CI) procedure can solve this shortcoming.

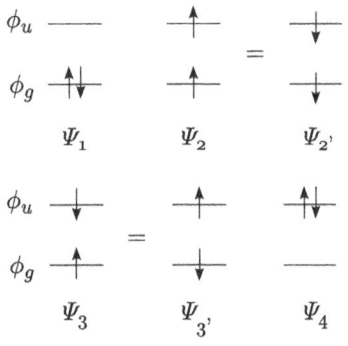

Fig. 12.16
Different electronic configurations of the H_2 molecule.

[23] Strictly speaking, the ϕ_g and ϕ_u orbital energies should not be equivalent when associated with different electron filling.

correlation between electrons with equivalent spins. In contrast, electrons with opposite spins are totally uncorrelated in the Hartree–Fock approximation.[22] This result has interesting consequences for the electronic structures of molecules and solids, which we now aim to illustrate in the simple case of the H_2 molecule, in order to compare them with the results obtained with the Hubbard approach in Section 12.1.2.

Limits of the Hartree–Fock method

Let us consider the H_2 molecule described in its minimal basis set by the two atomic orbitals χ_a and χ_b. For this two-electron system, the set of molecular spin orbitals consists of four orthogonal one-electron wavefunctions, i.e. the $\phi_g = \sigma_g$ and $\phi_u = \sigma_u$ MO of Fig. 12.1, both filled with one spin-up electron (α) or one spin-down electron (β):

$$\phi_g = \phi_g(r)\alpha(\sigma)$$
$$\bar{\phi}_g = \phi_g(r)\beta(\sigma)$$
$$\phi_u = \phi_u(r)\alpha(\sigma)$$
$$\bar{\phi}_u = \phi_u(r)\beta(\sigma) \quad (12.56)$$

From this basis set, one may build up six different Slater single-determinants:

$$\Psi_1 = |\phi_g \bar{\phi}_g|$$
$$\Psi_2 = |\phi_g \phi_u| \equiv |\bar{\phi}_g \bar{\phi}_u| = \Psi_2'$$
$$\Psi_3 = |\phi_g \bar{\phi}_u| \equiv |\bar{\phi}_g \phi_u| = \Psi_3'$$
$$\Psi_4 = |\phi_u \bar{\phi}_u|$$

Following eqn (12.54) and Exercise (12.4), the energy associated with these Slater determinants is:[23]

$$E_1^{HF} = 2e_{\phi_g} + J_{\phi_g \phi_g}$$
$$E_2^{HF} = e_{\phi_g} + e_{\phi_u} + J_{\phi_g \phi_u} - K_{\phi_g \phi_u}$$
$$E_3^{HF} = e_{\phi_g} + e_{\phi_u} + J_{\phi_g \phi_u}$$
$$E_4^{HF} = 2e_{\phi_u} + J_{\phi_u \phi_u}$$

Here J_{kk} and J_{kl} represent the Coulomb repulsion energy between two electrons lying in the same orbital and in two different orbitals, respectively, and correspond to an effective representation of the on-site (U_{ee}) and intersite (V_{ee}) electron–electron repulsion energies of the Hubbard model. In contrast, K_{kl} represents the exchange interaction between two electrons with equivalent spins, which is fully neglected in the spinless Hubbard model. Given that both J and K are positive energies, and assuming that their magnitude is significantly smaller than the energy difference $e_{\phi_u} - e_{\phi_g}$, then $E_1^{HF} = E_{GS}^{HF}$ should be the ground-state energy of H_2 in the Hartree–Fock approximation, with $\Psi_1 = \Psi_{GS}^{HF}$ its associated wavefunction. Following this line, $E_2^{HF} = E_{ES}^{HF}$ should be the first excited energy of H_2, with $\Psi_2 = \Psi_2' = \Psi_{ES}^{HF}$ its associated doubly degenerate wavefunctions. These two results actually differ from the Hubbard results.

- In the Hubbard model, we have shown that the ground-state electronic wavefunction of H$_2$ is written as a linear combination of the neutral and ionic states (see eqns (12.18) and (12.20)):

$$\Psi_{GS}(t, U) = \cos\theta \left(\frac{|a\bar{b}\rangle - |\bar{a}b\rangle}{\sqrt{2}} \right) + \sin\theta \left(\frac{|a\bar{a}\rangle + |b\bar{b}\rangle}{\sqrt{2}} \right)$$

with relative contributions depending on the on-site repulsion energy U, through the θ angle. In the $\{\phi_g, \phi_u\}$ basis set, this wavefunction thus corresponds to a linear combination of two Slater determinants describing two different electronic configurations of H$_2$ (see Exercise 12.1):

$$\Psi_{GS}(t, U) = \left(\frac{\cos\theta + \sin\theta}{\sqrt{2}} \right) |\phi_g\bar{\phi}_g| + \left(\frac{\sin\theta - \cos\theta}{\sqrt{2}} \right) |\phi_u\bar{\phi}_u|$$

The variational Hubbard solution is then a *multi-determinantal* representation of the wavefunction, while the Hartree–Fock method imposes a single-determinantal representation $|\phi_g\bar{\phi}_g|$. In fact, the multi-determinantal representation allows us to correlate the motion of electrons (here electrons with opposite spins), through the coupling of the $|\phi_g\bar{\phi}_g|$ and $|\phi_u\bar{\phi}_u|$ Slater determinants. In other words, electrons lying in the ϕ_g orbital can be promoted to the less contracted (or less 'correlated') ϕ_u antibonding MO to minimise the on-site repulsion in σ_g. The statistical weight of $|\phi_u\bar{\phi}_u|$ in $\Psi_{GS}(t, U)$ represents the probability of finding the two electrons in the ϕ_u orbital and obviously depends on the relative repulsion strength in the ϕ_g and ϕ_u orbitals and on the relative energy of these two orbitals.

- As for the excited state of H$_2$, we have shown that this is triply degenerate in the Hubbard model to represent the three different components of the spin triplet, i.e. $S = 1$ and $m_s = 0, \pm 1$.[24] The reason why the Hartree–Fock method does not reproduce this degeneracy arises from the non-antisymmetry of Ψ_3 and Ψ'_3. In fact, while Ψ_2 and Ψ'_2 are purely spin eigenfunctions (eigenfunctions of $\hat{S}^2$), Ψ_3 and Ψ'_3 are simply eigenfunctions of $\hat{S}_z$, but not of S^2. A multi-determinant representation of the wavefunction is then required to properly describe the third component of the spin triplet ($m_s = 0$). Indeed, while mixing the two Slater determinants Ψ_3 and Ψ'_3, we obtain the Ψ_{Tp} and Ψ_{Sg} wavefunctions:

$$\Psi_{Tp} = \frac{1}{\sqrt{2}} \{\Psi_3 + \Psi'_3\}$$

$$= \frac{1}{\sqrt{2}} \{|\phi_g\bar{\phi}_u| + |\bar{\phi}_g\phi_u|\}$$

$$= \frac{1}{\sqrt{2}} \{|a\bar{b}\rangle + |\bar{a}b\rangle\}$$

[24] Equation (12.23) corresponds to the $m_s = 0$ component of the triplet while the two other components $m_s = \pm 1$ are those which were neglected in the study, i.e. $|ab\rangle$ and $|\bar{a}\bar{b}\rangle$.

$$\Psi_{Sg} = \frac{1}{\sqrt{2}} \{\Psi_3 - \Psi_3'\}$$

$$= \frac{1}{\sqrt{2}} \{|\phi_g \bar{\phi}_u| - |\bar{\phi}_g \phi_u|\}$$

$$= \frac{1}{\sqrt{2}} \{|a\bar{a}\rangle - |b\bar{b}\rangle\} \quad (12.57)$$

which are now eigenfunctions of S^2 and represent the (open-shell) triplet and singlet states of H_2. The former is degenerate with the Ψ_2 and Ψ_2', while the latter is pushed up to higher energy by $2K_{\phi_g \phi_u}$. Hence because of the single-determinantal representation of the wavefunction, the energy of Ψ_3 is a mixing of the triplet and singlet spin energies. This error is known as the *spin contamination* error and occurs for each spin state that cannot be represented by one single determinant. In the case of magnetic systems, the two methods can lead to very different results. The spinless Hubbard method generally favours antiferromagnetic local interactions (singlet states) while Hartree–Fock generally favours ferromagnetic local interactions (triplet states).

Beyond the Hartree–Fock method

In the Hartree–Fock approximation, the use of a single determinant to represent the total wavefunction does not allow us to properly account for the correlation between electrons, although it permits us to correlate electrons with equivalent spin (through the exchange integral K_{ij}). The post-Hartree–Fock method was proposed to overcome this shortcoming. In principle, this method is very simple. It consists of writing the n-electron wavefunction as a linear combination of Slater determinants:

$$\Psi_n(x_1, \ldots, x_n) = \Psi^{HF}(x_1, \ldots, x_n) + \sum_K D_K \Psi^K(x_1, \ldots, x_n) \quad (12.58)$$

where Ψ^{HF} is the zero-order Hartree–Fock wavefunction and Ψ^K are a set of excited determinants, and then solving the Fock one-electron linear equations to get the variational D_K coefficients. This is called the *configuration interaction* (CI) procedure. Assuming that the correlation energy E^{corr} corresponds to:

$$E^{corr} = E^{exact} - E^{HF} \quad (12.59)$$

the variational principle states that if the basis set used to build the excited Ψ^K determinants is infinite,[25] then the eigenvalues and eigenfunctions of the n-interacting-particle system should be exact, i.e. $E^{CI} = E^{exact}$. In practice, the use of an infinite basis set is obviously not possible as it would require solution of an infinite number of linear equations. Since this post-Hartree–Fock method is very demanding in computational time, a compromise has always to be reached between the number of Slater determinants chosen for the calculation and the accuracy of the calculation. Mostly, the choice of determinants to be included in the CI procedure is dictated by the intuition of chemists regarding the system/property of interest. In the particular case of periodic systems, the

---

[25] In case of an infinite basis set, the Hartee-Fock energy E^{HF} in eqn (12.59) is known as the *Hartree–Fock limit*.

CI procedure cannot be straightforwardly implemented due to the Brillouin zone integration required to solve the problem.

Although there are perturbative alternatives to the variational CI, which explicitly include correlation effects in periodic calculations, these methods are still very demanding in computational time and do not receive much attention from the solid state community. From this point of view, the Hartree–Fock method is often preferred, even if it neglects correlation effects. This method is well adapted to the study of classical semiconductors, i.e. materials with electronic structures exhibiting a charge gap. In the case of metallic compounds with band structures exhibiting electronic state degeneracy at the Fermi level, the Hartree–Fock variational solutions are sometimes difficult to reach, in particular due to the single-determinant representation of the wavefunction and to the tendency of the Hartree–Fock method to favour triplet states over singlet states. As we will see in the next section, the density functional theory (DFT) formalism is much less time-demanding than the Hartree–Fock method and is well adapted to the study of metallic systems.

12.2.3 Density functional theory

As shown previously, the n-interacting electron wavefunction is a mathematical object of a bewildering complexity, which includes an amazing number of variational coefficients that must be computed to access the properties of the system. The electron density $\rho(r)$, meanwhile, appears as an object of incredible simplicity. Luckily, the first theorem of DFT states that the ground-state energy of the n-interacting-particle system can be deduced from its electron density. Provided that a unique correspondence exists between the external potential $v_R(r)$ [26] and the electron density, then the knowledge of $\rho(r)$ is sufficient to determine the ground-state properties of the n-particle system. The total energy of $\hat{H}^{el}$ is therefore a functional of ρ:

$$E^{\text{DFT}}[\rho] = F[\rho] + \int \rho(r) v_R(r) dr \qquad (12.60)$$

[26] Here v_R is the attractive potential due to the M immobile nuclei, which is determined, to a constant, from the electron density.

with:

$$F[\rho] = T_e[\rho] + V_{ee}[\rho] \qquad (12.61)$$

$F[\rho]$ is called the Hohenberg and Kohn functional. It is independent of $v_R(r)$ and is therefore an universal function, in the sense that it is valid for any n-electron system described by a Hamiltonian of type $\hat{H}^{el}$ (see eqn (12.44)). While containing all information required about the electrons, this function is unfortunately unknown. The goal pursued by DFT is to approach the exact shape of this function, in order to solve the n-interacting-particle problem through a minimisation procedure:

$$E[\rho] = \min_{\rho} \left\{ F[\rho] + \int \rho(r) v_R(r) dr \right\} \qquad (12.62)$$

Basically, the system of n-interacting particles is replaced by a fictitious system of n non-interacting particles, the wavefunction of which is a Slater determinant built on the natural Kohn–Sham orbitals $\phi_i(x_i)$.[27] The total energy

[27] Natural orbitals are orbitals that diagonalise the electron density.

Electron repulsion

of the exact system can now be written as:

$$E^{\text{DFT}}[\rho] = T_s[\rho] + (T_e[\rho] - T_s[\rho]) + \int \rho(r)v_R(r)dr + V_{ee}[\rho]$$

where $T_s[\rho]$ is the exact kinetic energy of the fictitious system:

$$T_s[\rho] = -\frac{1}{2}\sum_{i=1}^{n}\langle\phi_i(x_i) | \nabla_i^2 | \phi_i(x_i)\rangle \qquad (12.63)$$

and where $V_{ee}[\rho]$ is a global term accounting for all two-electron effects. These effects are missing in the fictitious system but are required to describe the exact system. From the previous section, we know that V_{ee} is divided into three main contributions:

- the classical Coulomb energy between two interacting electron distributions
- the non-classical exchange energy between two interacting spins
- the correlation energy between all interacting electrons.

Now, given that V_{ee} is a density functional, it is required to be written as a function of the electron density. This is done through the introduction of the density matrix γ_n:

$$\gamma_n(x_1,\ldots,x_n; x_1',\ldots,x_n') = \Psi_n(x_1,\ldots,x_n)\Psi_n^*(x_1',\ldots x_n')$$

$$\gamma_n(x_1,\ldots,x_n; x_1,\ldots,x_n) = \rho_n(x_1,\ldots,x_n) \qquad (12.64)$$

and its one- and two-particle reduced density matrices:

$$\gamma_1(x_1; x_1') = n\int\ldots\int dx_2\ldots dx_n$$

$$\Psi_n(x_1, x_2,\ldots,x_n)\Psi_n^*(x_1', x_2,\ldots x_n)$$

$$\gamma_1(x_1; x_1) = \rho_1(x_1) = \sum_{\sigma_1}\rho^\sigma(r_1)$$

$$\gamma_2(x_1, x_2; x_1', x_2') = \frac{n(n-1)}{2}\int\ldots\int dx_3\ldots dx_n$$

$$\Psi_n(x_1, x_2,\ldots,x_n)\Psi_n^*(x_1', x_2',\ldots x_n)$$

$$\gamma_2(x_1, x_2; x_1, x_2) = \rho(x_1, x_2) = \sum_{\sigma_1\sigma_2}\rho^{\sigma_1\sigma_2}(r_1, r_2) \qquad (12.65)$$

where γ_1 represents the probability of finding one electron in r_1 with spin σ_1, irrespective of the position of the $(n-1)$ other electrons, and γ_2 is the probability of finding an electron pair in r_1, r_2 with σ_1 and σ_2, irrespective of the position of the $(n-2)$ other electrons. In this representation, the total energy is written as:

$$E^{\text{DFT}}[\rho] = -\frac{1}{2}\int_{x_1=x_1'}\nabla^2\gamma_1(x_1, x_1')dx_1 + \int\rho(r)v_R(r)dr$$

$$+\frac{1}{2}\int\int\frac{1}{r_{12}}\rho(x_1, x_2)dx_1dx_2 \qquad (12.66)$$

Because electrons are correlated, $\rho(x_1, x_2)$ cannot reduce to the product of probabilities $\rho(x_1)\rho(x_2)$, but only to the conditional probability product:

$$\rho_2(x_1, x_2) = \rho(x_1)\rho(x_{2/1}) \tag{12.67}$$

In other words, the probability that one given electron i lies in an elemental volume around r_i with spin σ_i cannot be determined since it depends on the position and spin of the j other electrons, exactly as for the two-electron potential of the Fock operator (see eqns (12.50) and (12.55)). To circumvent this issue, the basic idea of DFT is to introduce the so-called *exchange-correlation hole,* which reflects the volume of 'anti-density' created around each electron i to exclude any other electron j. In the two-particle case, this means that electron 2 not only interacts with electron 1 but also with the hole created by electron 1 to exclude electron 2, i.e.:

$$\rho_2(x_1, x_2) = \rho(x_2)\left(\rho(x_1) + \rho^{xc}_{\text{hole}}(x_1)\right)$$
$$= \rho(x_1)\rho(x_2) + \rho(x_2)\rho^{xc}_{\text{hole}}(x_1) \tag{12.68}$$

The exclusion volume represented by ρ^{xc}_{hole} obviously depends on the spin of electrons i and j, since two electrons with equivalent or opposite spins do not repel each other the same way. The *Coulomb hole* describes the exclusion volume for electrons with opposite spins, while the *Fermi hole* describes the exclusion volume for electrons with equivalent spins. Combining eqns (12.65) and (12.68), V_{ee} becomes:

$$E^{\text{DFT}}[\rho] = T_s[\rho] + \int \rho(r)v_R(r)dr$$
$$+ \frac{1}{2}\int\int \frac{1}{r_{12}}\rho(r_1)\rho(r_2)dr_1 dr_2 + E_{xc}[\rho]$$
$$= T_s[\rho] + \int \rho(r)v_R(r)dr + J[\rho] + E_{xc}[\rho] \tag{12.69}$$

where $J[\rho]$ represents the classical Coulomb energy and $E_{xc}[\rho]$:

$$E_{xc}[\rho] = (V_{ee}[\rho] - J[\rho]) + (T_e[\rho] - T_s[\rho]) \tag{12.70}$$

contains all the remaining electronic effects, including the kinetic correction to $T_s[\rho]$. This term is the unknown energy functional that DFT seeks to approximate. Once it is known, the ground-state energy of the n-interacting-particle system can be determined by solving the linear equations:

$$\left(-\frac{1}{2}\Delta + v_{\text{eff}}(r)\right)\phi_i(r) = \epsilon_i \phi_i(r) \tag{12.71}$$

where $v_{\text{eff}}(r)$:

$$v_{\text{eff}}(r) = v_R(r) + \int \frac{\rho(r_1)}{|r - r_1|}dr_1 + \frac{\partial E_{xc}[\rho]}{\partial \rho(r)} \tag{12.72}$$

The search for the best $E_{xc}[\rho]$ functional is still a central topic in the DFT community. Several variants of this functional exist today. The common point for all these functionals is the uniform electron gas model, in which more or less sophisticated corrections are added. Although a thorough description

Electron repulsion

of these functionals is outwith the scope of this book, it is interesting to understand the basic idea on which the two most popular density functionals rely.

The local density approximation

The *local density approximation* (LDA) assumes that the exchange-correlation energy of one given particle located at r only depends on the electron density at this point:

$$E_{xc}^{LDA}[\rho] = \int \rho(r) \varepsilon_{xc}^{LDA}(\rho(r)) dr \qquad (12.73)$$

where $\varepsilon_{xc}(r)$ can be decomposed into:

$$\varepsilon_{xc}^{LDA}(r) = \varepsilon_{x}^{LDA}(r) + \varepsilon_{c}^{LDA}(r) \qquad (12.74)$$

The analytical form of the exchange energy is known from the uniform electron gas:

$$\varepsilon_{x}^{LDA}(r) = -\frac{3}{4}\left(\frac{3}{\pi}\right)^{\frac{1}{3}} \rho(r)^{\frac{4}{3}} \qquad (12.75)$$

while the expression for the correlation energy is deduced from parametric equations fitting the results of sophisticated post-Hartree–Fock CI calculations. As simple as it may seem, the 'local' approximation of DFT gives very satisfactory results for a large number of materials. It should be noted, however, that this success is partly attributed to an error cancellation between the exchange energy and the correlation energy. Since LDA is, in principle, exact for a uniform electron gas, it is well adapted to the study of systems with homogeneous electron densities, such as systems based on main group s, p elements. For systems exhibiting inhomogenous electron densities, it is necessary to account for the shape of $\rho(r)$ around r, i.e. for the density gradient.

The generalised gradient approximation

The *generalised gradient approximation* (GGA) introduces a 'semi-local' character of the electron density in the general formulation of E_{xc} so as to account for the density variations around r:

$$E_{xc}^{GGA}[\rho, \nabla\rho] = \int \rho(r) \varepsilon_{xc}^{GGA}(\rho(r), \nabla\rho(r)) dr \qquad (12.76)$$

This leads, for the exchange energy, to:

$$E_{x}^{GGA}[\rho, \nabla\rho] = \int F_x(s(r)) \rho(r)^{4/3} dr \qquad (12.77)$$

where F_x is a function of the reduced density gradient:

$$s(r) = \frac{|\nabla\rho(r)|}{6\pi^2 \rho(r)^{\frac{4}{3}}} \qquad (12.78)$$

As with LDA, the expression for the correlation energy comes from parametric equations fitting accurate atomic and molecular calculations.[28] A number of

---

[28] Note that although accounting for some kind of 'non-local' character of the electron density, the GGA exchange-correlation potential keeps a strictly local expression, in the mathematical sense.

similar approaches based on modifications of the GGA approach have been derived and are still being tested today. These all seek a better description of the exchange-correlation hole, using a combination of physical intuition, full CI calculations, and experimental data fits. These GGA functionals lead to better results than LDA for systems exhibiting sharp variations of the electron density, such as systems based on d- or f-elements.

Limits of the DFT method

In contrast to the Hartree–Fock method, the DFT method does not rely on the explicit calculation of the two-electron integrals. This feature makes DFT a considerably less time-consuming method than the Hartree–Fock method. However, this feature also leads to one of the main drawbacks of the DFT method, the so-called *self-interaction error*. The self-interaction error is a systematic error of DFT arising from the non-exact representation of the exchange energy, and leads to the unphysical interaction of an electron with itself. This error does not occur in the Hartree–Fock method since the Coulomb and exchange energies are built in such a way that they cancel 'self-interaction'. This is clear in the particular case of a single-particle system, where the expectation value of V_{ee} is strictly nil in the Hartree–Fock framework. Mostly negligible in the case of delocalised systems, the self-interaction error may become very large in the case of localised systems. This can be understood by the screening effect that has already been introduced for Mott–Hubbard transitions (Section 12.1.3): in delocalised compounds the interaction of an electron with itself is efficiently screened by the flux of other electrons, but this is not the case in localised systems.

To better describe systems with localised electrons, different self-interaction-corrected functionals have been developed. Among these, the most popular for studying periodic systems are the hybrid DFT/Hartree–Fock and DFT/Hubbard methods. The former takes advantage of the exact exchange energy of the Hartree–Fock method to design hybrid functionals, such as the so-called B3LYP functional for instance: [12]

$$E_{xc}^{B3LYP}[\rho, \{\phi_i\}] = E_{xc}^{LDA} + a_0(E_x^{HF} - E_x^{LDA}) + a_x(E_x^{GGA} - E_x^{LDA})$$
$$+ a_c(E_c^{GGA} - E_c^{LDA}) \quad (12.79)$$

In this expression, the coefficients a_0, a_x, and a_c are optimised to reproduce a set of experimental data such as the atomisation energies, ionisation potentials, proton affinities, and total energies. [29] The latter introduces an effective on-site Coulomb and exchange parameter $U_{eff} = U - J$ on the subset of correlated orbitals, so as to better describe the repulsion effects in these orbitals. It is called the DFT+U method [13] and is used to partially correct the self-interaction error in strongly correlated systems. It is written as:

$$E^{DFT+U}\left[\rho, \{n^\sigma\}\right] = E^{DFT}[\rho] + E^U\left[\{n^{I\sigma}\}\right]$$
$$= E^{DFT}[\rho] + E_{Hub}\left[\{n^{I\sigma}\}\right] - E_{dc}\left[\{n^{I\sigma}\}\right] \quad (12.80)$$

[29] Up to now, various flavours of hybrid functional have been reported in the literature by tuning either the a_0, a_x, and a_c coefficients and/or the LDA or GGA functionals for the DFT terms of this expansion.

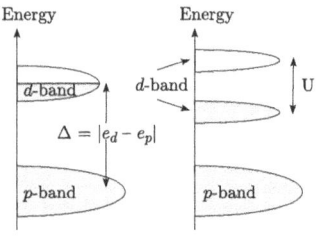

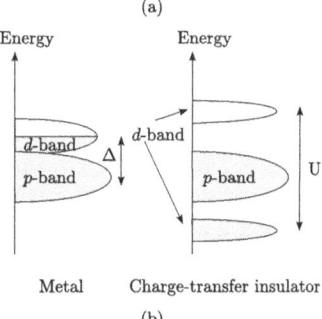

Fig. 12.17
Schematic representation of the effect of U_{eff} on the electronic structure of strongly correlated systems in the case of (a) Mott–Hubbard insulators and (b) charge transfer insulators. In both pictures, the left and right parts correspond to the electronic structure obtained in the DFT framework and in the DFT+U framework, respectively.

where ρ is the total density and $\{n^{I\sigma}\}$ the spin occupation matrix of the subset of strongly correlated orbitals I (e.g. d or f). In this expression E_{Hub} represents the 'correct' on-site repulsion energy, as defined in the Hubbard and Hartree–Fock models. Since $E^{\text{DFT}}[\rho]$ also contains a term intended to account for the repulsion energy, the so-called $E_{dc}[\{n^{I\sigma}\}]$ term has to be subtracted to avoid double counting:

$$E_{dc}\left[\left\{n^{I\sigma}\right\}\right] = \frac{U}{2}n^{I\sigma}(n^{I\sigma}-1) - \frac{J}{2}\left[n^{I\uparrow}\left(n^{I\uparrow}-1\right) + n^{I\downarrow}\left(n^{I\downarrow}-1\right)\right] \quad (12.81)$$

E^U is then written as:

$$E^U\left[\left\{n^{I\sigma}\right\}\right] = \frac{U_{\text{eff}}}{2}\sum_{I,\sigma}\left[n^{I\sigma} - n^{I\sigma}n^{I\sigma}\right] \quad (12.82)$$

This expression shows that the self-interaction correction applies only on spin-orbitals with occupations different from 0 or 1. In other words, the DFT+U method penalises fractional occupancies by U_{eff}, exactly as the Hubbard model penalises double occupancies. This may lead to non-zero magnetic moments in the strongly correlated orbitals and therefore to possible magnetic ordering in the structures of these materials. As schematically illustrated in Fig. 12.17, this U-correction opens a gap in the electronic structure of strongly correlated systems. This leads to either Mott–Hubbard insulators ($\Delta > U$) or charge-transfer insulators ($\Delta < U$) depending on the chemical nature of the system.

The failure of the conventional DFT method to reproduce the electronic structure of strongly correlated systems is very well known in the case of transition metal oxides (TMO). As an example, the cubic CoO phase is an antiferromagnetic insulator that is found to be metallic in the conventional DFT–LDA and DFT–GGA frameworks. As shown in Fig. 12.18, the density of states of this system is gapless and shows a majority contribution of the Co($3d$) orbitals around the Fermi level. Using the DFT+U formalism (or equivalently the hybrid DFT/HF formalism), a band gap opens in this electronic structure. Given that the experimental gap is known for this system, a direct mapping of the experimental and computational gaps can then be used to extract the range of acceptable U_{eff} values required to describe the ground-state properties of this oxide (see Fig. 12.19). This corresponds to U_{eff} ranging from 3.0 to 5.0 eV. Considering this value in DFT+U calculations now leads to the DOS plot of

Fig. 12.18
Electronic structure of the cubic CoO phase computed for the antiferromagnetic magnetic structure in (a) the DFT framework and (b) the DFT+U framework. In both plots, the black areas correspond to the projection of the total density of states over the metallic Co ($3d$) orbitals and the horizontal dotted line stands for the Fermi level.

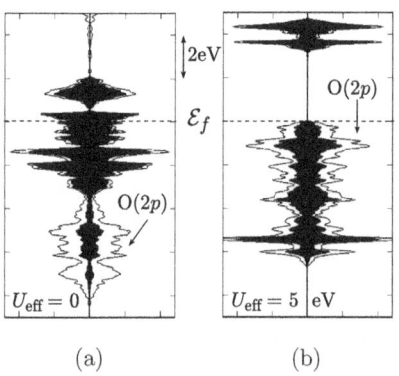

Fig. 12.18b, where the metallic d-band of CoO splits into one low-lying filled band and one high-lying empty band. Interestingly, a significant number of oxygen p-levels contribute to the total density of states around the Fermi level. This is typical for systems with behaviour between the Mott–Hubbard and the charge-transfer regimes.

12.3 Conclusion

In this chapter, we have seen that the repulsion between correlated electrons can take different mathematical forms depending on the Hamiltonian and wavefunction chosen to describe the n-interacting particles system.

- Despite their apparent simplicity, the Hubbard and extended Hubbard models offer an effective representation of electron repulsion through on-site and intersite soundable parameters. While lacking as regards the exchange interaction, the Hubbard Hamiltonian includes the correlation effects through its multi-reference representation. As a consequence, provided that all states involved in the property of interest have been taken into account in the Hilbert space, the calculation should lead to very accurate solutions. It should be noted here that whilst the Hubbard parameters t, U, V are phenomenological parameters, they can be extracted from very accurate first-principles calculations using *ab-initio* methods.
- The Hartree–Fock method uses a single-determinant wavefunction to describe the n-interacting electron problem and solve the Schrödinger equation in a mean-field approach. The expectation values of the two-electron $\hat{V}_{ij}$ operator decompose into one coulombic contribution J_{ij} and one exchange energy K_{ij}. The Coulomb term J_{ij} corresponds to the on-site U_{ee} and intersite V_{ee} repulsion terms of the Hubbard model depending on whether the two electrons lie in the same orbital or in two different orbitals. In contrast to the Hubbard model, the Hartree–Fock method lacks the correlation energy E_{corr}, which can be (at least partially) recovered through so-called post Hartree–Fock methods, including the configuration interaction procedure and/or a multi-determinantal representation of the n-interacting-particle wavefunction.
- Density functional theory evaluates the total ground-state energy of the n-interacting-particle system as a functional of the electron density. In contrast to the Hartree–Fock method, the DFT method accounts for the correlation energy through the introduction of a semi-empirical exchange-correlation energy. The expectation values of the $\hat{V}_{ij}$ operator thus decompose into one coulombic contribution $J[\rho]$, one exchange contribution $E_x[\rho]$, and one correlation contribution $E_c[\rho]$. In this formalism, the non-exact representation of the exchange energy avoids the compensating for the Coulomb energy of the one-particle system. This leads to a systematic self-interaction error of the DFT, which can be partly corrected through the use of DFT/HF or DFT/Hubbard hybrid functionals.

While acknowledging the existence of these methods and their remarkable contribution to the understanding of many complex properties of molecular and extended systems, it is nevertheless true that the orbital concepts conveyed

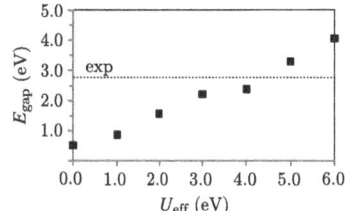

Fig. 12.19
Energy band gap computed for the CoO cubic phase as a function of the U_{eff} correction applied in the DFT+U framework. The horizontal dotted line refers to the experimental gap.

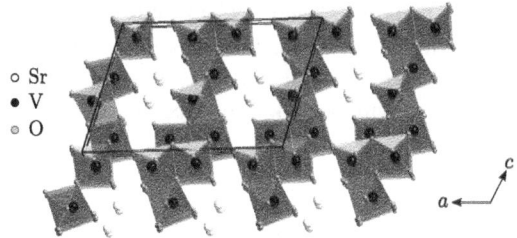

Fig. 12.20
Crystal structure of the β-AV$_6$O$_{15}$ magnetic oxide.

throughout this book may offer a powerful alternative to these sophisticated methods to rationalise the properties of complex systems. As a bottom line, we would like to conclude this chapter by illustrating this on the example of the β-SrV$_6$O$_{15}$ magnetic oxide. This system exhibits interesting magnetic properties that have long been discussed in the literature without being fully rationalised. As shown in Fig. 12.20, the structure of this material is very complex: it is built on 24 vanadium atoms per unit cell, with six being crystallographically independent. At first glance, the study of this system should require very sophisticated methods: (i) it contains d-electrons, which are known to be strongly correlated, especially for third-row elements, (ii) its magnetic properties imply complex metal–metal interactions, and (iii) its crystal packing in zigzag chains and ladders leads to tricky interactions and a large unit cell. However, a thorough look at the local distortions occurring in this structure is sufficient to entirely determine the leading magnetic interactions in this system and to interpret its antiferromagnetic behaviour. In this structure, all vanadium ions adopt a pyramidal environment due to the presence of a short vanadyl bond (V=O). Such an O_h to C_{4v} symmetry lowering splits the t_{2g}-block orbital degeneracy, as shown in Fig. 12.21.

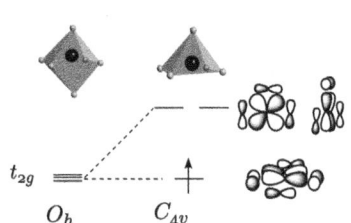

Fig. 12.21
t_{2g} orbital splitting due to the $O_h \longrightarrow C_{4v}$ symmetry lowering.

From the formal electron count Sr^{2+}(V$^{+5-1/3}$)$_6$(O^{-2})$_{15}$, one-third of an electron should then fill the lowest d orbital of each pyramid. Given that this orbital is perpendicular to the vanadyl bond, locating the vanadyl local axes in this structure is sufficient to determine the leading magnetic interactions. As shown in Fig. 12.22, π-type interactions occur along three ladders illustrated in light grey. In these ladders the metal–metal interactions are mediated by

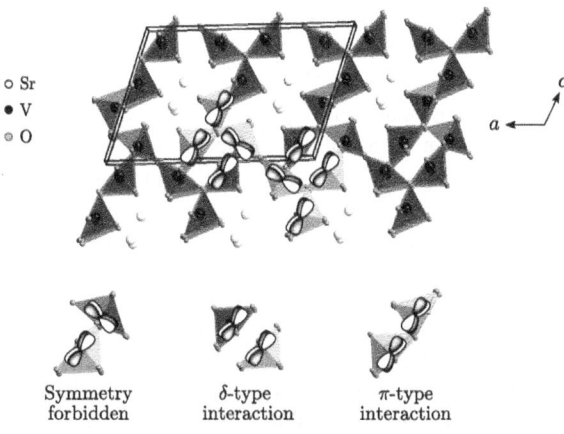

Fig. 12.22
Representation of the magnetic orbitals in the structure of β-AV$_6$O$_{15}$.

an oxygen atom, through a V–O–V angle close to 180°. Such interactions are known to be strongly antiferromagnetic. Given that all other interactions are either of δ-type or forbidden for symmetry reasons, then the three antiferromagnetic ladders correspond to the magnetic units of the β-SrV$_6$O$_{15}$ phase. Based on this very simple representation, the magnetic properties of the β-AV$_6$O$_{15}$ series of compounds have been fully explained. [14]

Exercises

(12.1) Re-write the ground-state wavefunction of H$_2$:

$$\Psi_{GS}(t) = \frac{1}{2}\left(|a\bar{b}\rangle - |\bar{a}b\rangle + |a\bar{a}\rangle + |b\bar{b}\rangle\right)$$

in the basis set formed by the two delocalised σ_g and σ_u molecular orbitals using the following notations:

$$\sigma_g = \frac{1}{\sqrt{2}}(\phi_a + \phi_b)$$

$$\bar{\sigma}_g = \frac{1}{\sqrt{2}}(\bar{\phi}_a + \bar{\phi}_b)$$

and show it is written as a closed-shell Slater determinant:

$$\Psi_{GS}^{LCAO} = \frac{1}{\sqrt{2}}\begin{vmatrix}\sigma_g(1) & \bar{\sigma}_g(1) \\ \sigma_g(2) & \bar{\sigma}_g(2)\end{vmatrix} = |\sigma_g\bar{\sigma}_g|$$

$$= \frac{1}{\sqrt{2}}\{\sigma_g(1)\bar{\sigma}_g(2) - \sigma_g(2)\bar{\sigma}_g(1)\}$$

where (1) and (2) refer to the electron labels and σ_g and $\bar{\sigma}_g$ to the ↑ and ↓ spins of the electrons.

(12.2) Assuming that the reference energy $e^0 = 0$, show that the energy spectrum of H$_2$ is strictly equivalent to the ones obtained with $e^0 = 2\alpha_{1s}$. Solve the linear equations of the Hubbard matrix for the ground-state energy $\epsilon_{GS}(t, U)$ of eqn (12.17) and show it is independent of e^0.

(12.3) As shown eqn (12.19) and Fig. 12.5, $\tan\theta$ gives the relative contribution of the ionic and neutral states to the ground-state wavefunction of H$_2$ in the Hubbard model. Show that the asymptotes of $\tan\theta$ can be written as a function of $U/4|t|$ using a first-order Taylor expansion, and deduce from the results that the θ domain from the limit of weak repulsion ($U \to 0$) to the limit of strong repulsion ($U \to +\infty$) is $\theta \in \left[\frac{\pi}{4}, 0\right]$.
Note: The first-order Taylor expansion of $(1 + \epsilon)^\alpha$ is $(1 - \alpha\epsilon)$.

(12.4) Using the same procedure as in Exercise 12.1:

(a) Show that the ground-state wavefunction of H$_2$ in the $U_{+\infty}$ limit $\Psi_{GS}(t, U_{+\infty})$ is a spin singlet written as:

$$\Psi_{GS}(t, U_{+\infty}) = \frac{1}{\sqrt{2}}\left(|\sigma_g\bar{\sigma}_g| - |\bar{\sigma}_u\sigma_u|\right)$$

(b) Show that it is degenerate with the first-excited state of H$_2$, $\Psi_{ES}(t, U)$.
(c) Write this first-excited state as a function of the σ_g and σ_u molecular orbitals and show it is a spin triplet.

(12.5) Using the total energy of the n-interacting electrons given in eqn (12.54), evaluate the energies of the different Slater determinants:

$$\Psi_1 = |\phi_g\bar{\phi}_g| \quad \Psi_2 = |\phi_g\phi_u|$$
$$\Psi_3 = |\phi_g\bar{\phi}_u| \quad \Psi_4 = |\phi_u\bar{\phi}_u|$$

for the H$_2$ molecule, and deduce from the results that the total energies of the closed-shell (CS) and open-shell (OS) systems, described by:

$$\Psi_n^{CS} = |\phi_1(x_1)\bar{\phi}_1(x_2)\ldots\phi_{n/2}(x_{n-1})\bar{\phi}_{n/2}(x_n)|$$

$$\Psi_n^{OS} = |\phi_1(x_1)\phi_2(x_2)\ldots\phi_{n-1}(x_{n-1})\phi_n(x_n)|$$

are written as:

$$E^{CS} = \sum_{i=1}^{n/2}\left\{2e_i + \frac{1}{2}\sum_{j\neq i}^{n/2}(2J_{ij} - K_{ij})\right\}$$

$$E^{OS} = \sum_{i=1}^{n}\left\{e_i + \frac{1}{2}\sum_{j\neq i}^{n}(J_{ij} - K_{ij})\right\}$$

References

1. J. Hubbard, *Proc. Roy. Soc. London A*, 277, 237, 1964.
2. R. J. Baxter, in *Exactly Solved Models in Statistical Mechanics*, London, Academic Press, 1982.
3. M. Wolfsberg and L. J. Helmholz, *J. Chem. Phys.*, 20, 837, 1952.
4. A. Szabo and N. S. Ostlund in *Modern Quantum Chemistry: Introduction to Advanced Electronic Structure Theory*, Macmillan Publishing Co. Inc., New York, 1982, 89.
5. E. H. Lieb and F. Y. Wu, *Physica A*, 321, 1–27, 2003.
6. R. S. Mulliken, *J. Chem. Phys.*, 23, 1833, 1955.
7. B. Pomarède, B. Garreau, I. Malfant, L. Valade, P. Cassoux, J.-P. Legros, A. Audouard, J.-P. Ulmet, M.-L. Doublet and E. Canadell, *Inorg. Chem.*, 33, 3401, 1994.
8. S. R. White, *Phys. Rev. Lett.*, 69, 2863, 1992; S. R. White, *Phys. Rev. B*, 48, 10345, 1993.
9. M.-L. Doublet and M.-B. Lepetit, *J. Chem. Phys.*, 110, 1767, 1999.
10. D. R. Hartree, *Proc. Cambridge Phil. Soc.*, 24, 89, 111, 1928; V. Fock, *Z. Physik*, 61, 126, 1930. And for a more recent review see: F. Jensen, *Introduction to Computational Chemistry*, Wiley, Chichester, 1999.
11. P. Hohenberg and W. Kohn, *Phys. Rev.*, 136, 1964, B864; W. Kohn and L. J. Sham, *Phys. Rev.*, 140, A1133, 1965; R. G. Parr and W. Yang, in *Density-Functional Theory of Atoms and Molecules*, Oxford University Press, New York, 1989.
12. A. D. Becke, *J. Chem. Phys.*, 98, 5648, 1993; C. Lee, W. Yang, and R. G. Parr, *Phys. Rev. B*, 785, 37, 1998.
13. V. I. Anisimov, J. Zaanen, and O. K. Andersen, *Phys. Rev. B*, 44, 943, 1991; A. I. Liechtenstein, V. I. Anisimov, and J. Zaanen, *Phys. Rev. B*, 52, R5467–R5470, 1995; S. L. Dudarev, G. A. Botton, S. Y. Savrasov, C. J. Humphreys, and A. P. Sutton, *Phys. Rev. B*, 57, 1505–1509, 1998.
14. M.-L. Doublet and M.-B. Lepetit, *Phys. Rev. B*, 71, 75–119, 2005.

Solutions for exercises

Chapter 1

(1.1) For an electron trapped in an infinite potential well, the potential may be drawn as shown in Fig. 1.

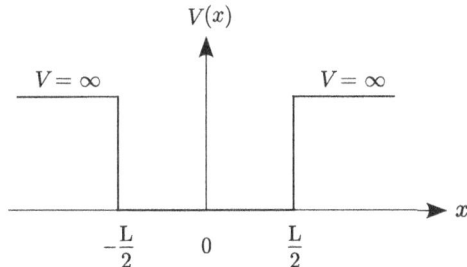

Fig. 1
Potential for an electron trapped in an infinite potential well.

Replacing the following boundary conditions,

$$\Psi\left(-\frac{L}{2}\right) = \Psi\left(\frac{L}{2}\right) = 0$$

which simply preclude the presence of the electron in the zones where the potential is infinite, for those in eqn (1.6), we can easily obtain the even solutions denoted $\Psi_n^e(x)(n = 1, \ldots)$ as well as the odd solutions noted $\Psi_n^o(x)(n = 0, 1, \ldots.)$ with energies E_n^e and E_n^o, respectively:

$$\Psi_n^e(x) = \sqrt{\frac{2}{L}} \cos(k_n^e x) = \sqrt{\frac{2}{L}} \cos\left(\frac{2(n+\frac{1}{2})\pi x}{L}\right)$$

associated with energy

$$E_n^e = \frac{\hbar^2 (k_n^e)^2}{2m} = \frac{h^2 (n+\frac{1}{2})^2}{2mL^2}$$

and

$$\Psi_n^o(x) = \sqrt{\frac{2}{L}} \sin(k_n^o x) = \sqrt{\frac{2}{L}} \sin\left(\frac{2n\pi x}{L}\right)$$

associated with energy

$$E_n^o = \frac{\hbar^2 (k_n^o)^2}{2m} = \frac{h^2 n^2}{2mL^2}$$

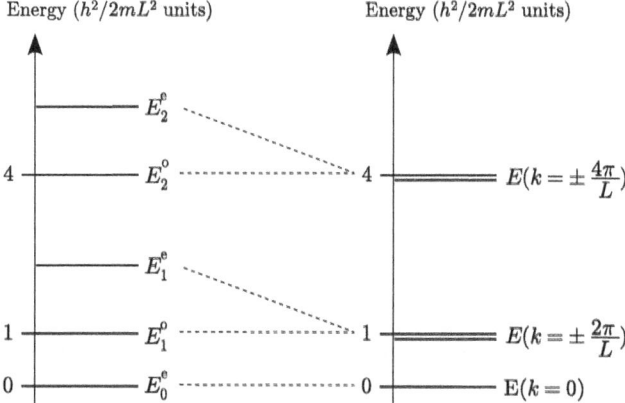

Fig. 2
Correspondence between the spectra obtained by means of the boundary conditions of this exercise (left) and those used in Chapter 1 (right).

If we now compare the energy spectrum obtained in this way with that of Fig. 1.4 (see Fig.2), we can see that the boundary conditions used in this exercise lead to a degeneracy lift of the energy spectrum obtained in the text. If L is very large, the energy levels are very near to each other. As a result, the spectra obtained converge towards the same continuous spectra when L tends to infinity. In particular, the density of states is independent of the particular nature of the boundary conditions.

Chapter 2

(2.1) The molecular orbitals Ψ for the π system of ethylene result from the interaction of the $2p_z$ orbitals of the two carbon atoms with indices (a) and (b) and may be written as:

$$\Psi = c_a 2p_{z_a} + c_b 2p_{z_b}$$

The secular equations are of the form:

$$\begin{cases} c_a(\alpha - E) + c_b(\beta - ES) = 0 \\ c_a(\beta - ES) + c_b(\alpha - E) = 0 \end{cases}$$

and are associated with the secular determinant:

$$\begin{vmatrix} \alpha - E & \beta - ES \\ \beta - ES & \alpha - E \end{vmatrix} = 0$$

where $\alpha = <2p_{z_a}|\hat{h}|2p_{z_a}>$ and $\beta = <2p_{z_a}|\hat{h}|2p_{z_b}>$. Here α and β are negative.

Solution of this equation leads to the allowed energies ($E_\pm$):

$$E_\pm = \alpha \pm \frac{\beta - \alpha S}{1 \pm S}$$

associated with the molecular orbitals

$$\Psi_\pm = N_\pm(2p_{z_a} \pm 2p_{z_b})$$

A Hückel-type calculation implies that $S = 0$ so that the allowed energies are $E_\pm = \alpha \pm \beta$.

Extended Hückel and Hückel-type calculations lead to the molecular orbital diagrams of Fig. 3.

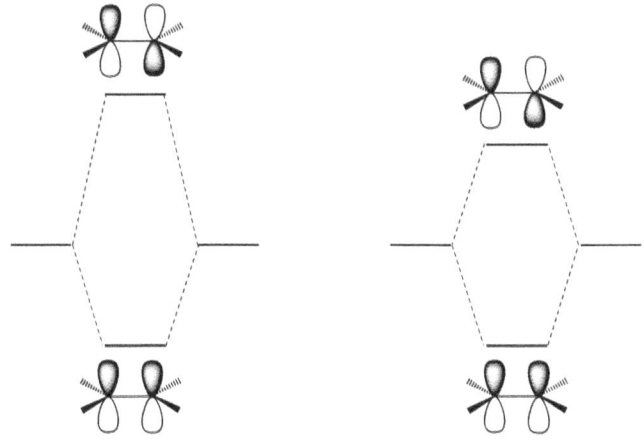

Fig. 3
Extended Hückel (left) and Hückel (right) molecular orbital diagrams for ethylene.

(2.2) Solution of the secular equations according to the Hückel approach leads to the molecular orbital diagram of Fig. 4

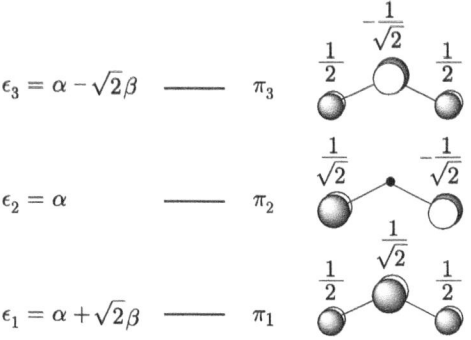

$\epsilon_3 = \alpha - \sqrt{2}\beta$ —— π_3

$\epsilon_2 = \alpha$ —— π_2

$\epsilon_1 = \alpha + \sqrt{2}\beta$ —— π_1

Fig. 4
Hückel molecular orbital diagram for planar allyl.

(2.3) The 48 symmetry operations are the following (see Fig. 5):
  (1) the identity E,
  (2) the inversion i,

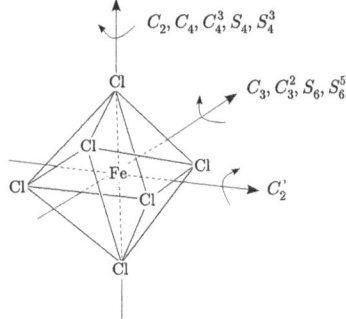

Fig. 5
Representation of the symmetry operations in the (FeCl$_6$)$^{4-}$ transition metal complex.

(3) the four axes C_3, C_3^2, S_6, and S_6^5 perpendicular to the centres of the triangular faces (i.e. 16 symmetry operations),

(4) the three axes C_2, C_4, C_4^3, S_4 and S_4^3 going through pairs of opposite chlorine atoms (i.e. 15 symmetry operations),

(5) the six axes C_2' going through the centre of the lines joining two opposite chlorine atom pairs,

(6) the three symmetry planes σ_h containing four chlorine atoms, and

(7) the six symmetry planes σ_d leaving invariant two chlorine atoms.

(2.4) The symmetry group appropriate for an octahedron deformed along the vertical axis is the D$_{4h}$ group.

(2.5) In the D_{4h} group, the rotations $\{C'_{2a}, C'_{2b}\}$ are physically equivalent and thus belong to the same class. Note, however, that these rotations are different from the $\{C''_{2a}, C''_{2b}\}$ rotations, which form a different class, as does C_2. Thus, although all these rotations are two-fold rotations, they can be classified into three different types of rotation. For exactly the same reason, the symmetry planes may be classified into three different classes, $\{\sigma_{xy}\}$, $\{\sigma_{v_a}, \sigma_{v_b}\}$, and $\{\sigma_{d_a}, \sigma_{d_b}\}$, which leave invariant 4, 2, and 0 chlorine atoms, respectively.

(2.6) The different sets of atomic orbitals of the cyclobutadiene molecule constituting bases of a representation of the D_{4h} group are, for instance, the four 1s orbitals of the hydrogen atoms or the four 2s orbitals of the carbon atoms....

(2.7) H$_2$O is a neutral ligand with a closed shell, so the oxidation state of copper in Cu(H$_2$O)$_6^{2+}$ is +2. In contrast, the CN ligand does not possess a closed shell, so we must formally consider it a CN$^-$ anion. Consequently, the oxidation state of platinum in the Pt(CN)$_4^{2-}$ complex is formally +2. Thus the mainly d levels of these systems are filled with 9 and 8 d electrons, respectively. In other words, copper and platinum in these complexes may be described as being Cu(II): 3d^9 and Pt(II): 5d^8, respectively.

(2.8) The characters of the representation (Γ_π) associated with the 12 π-type orbitals shown in Fig. 2.18 are shown in the table below. Using the table of characters of the O_h group it is possible to write:

O_h	E	8C_3	6C_2'	6C_4	3C_2	i	8S_6	6σ_d	6S_4	3σ_h
Γ_π	12	0	0	0	-4	0	0	0	0	0

$$\Gamma_\pi = T_{1g} + T_{1u} + T_{2g} + T_{2u}$$

Chapter 3

(3.1) If the number of cells (n) is odd, we may label cell M_i of the system by means of an integer i which can have any value in the interval $[-n'+0.5, n'-0.5]$, where $n' = n/2$.

(1) Following the same procedure as in Fig. 3.2 of Chapter 3 we will connect cells $M_{n'-0.5}$ and $M_{-n'+0.5}$.

(2) If we calculate the character of the representation $\Gamma_{\phi_j} = \{(\phi_j)_{-n'+0.5} \ldots (\phi_j)_0, \ldots (\phi_j)_{n'-0.5}\}$ made of the n ϕ_j orbitals that are equivalent by rotational symmetry, a nil character is obtained for any rotation other than the identity.

C_n (n odd)	E	C_n	C_n^{-1}	...	C_n^m	C_n^{-m}	...	$C_n^{n'-0.5}$	$C_n^{-n'+0.5}$
Γ_{ϕ_j}	n	0	0	...	0	0	...	0	0

Projecting Γ_ϕ thus leads to:

$$\Gamma_{\phi_j} = \sum_{\ell=-n'+0.5}^{n'-0.5} \Gamma_\ell$$

In other words, a decomposition identical to that obtained in eqn (3.1). Determination of the linear combination of the ϕ_j orbitals that is a basis for the irreducible representation Γ_ℓ may be carried out by application of the projector P_{Γ_ℓ} (see eqn (2.14)) over the orbital of the reference cell $(\varphi_j)_0$. Since a C_n^m rotation transforms $(\varphi_j)_0$ into $(\varphi_j)_m$, we obtain:

$$(\Psi_j)_\ell = \frac{1}{\sqrt{n}} \sum_{m=-n'+0.5}^{n'-0.5} \exp\left(i\frac{2\pi m\ell}{n}\right)(\varphi_j)_m$$

which is a completely equivalent expression to that in (eqn 3.2). The only difference is the number of Bloch orbitals generated: it is an odd number in this case, but even in the one considered in the text. Among these orbitals, only the one corresponding to ℓ equal to zero is non-degenerate whereas all others are pairs of degenerate orbitals $((\Psi_j)_\ell$ and $(\Psi_j)_{-\ell})$. The physical properties of a solid thus having a large number of cells are obviously independent of the even or odd number of cells or of the existence of one or two non-degenerate levels.

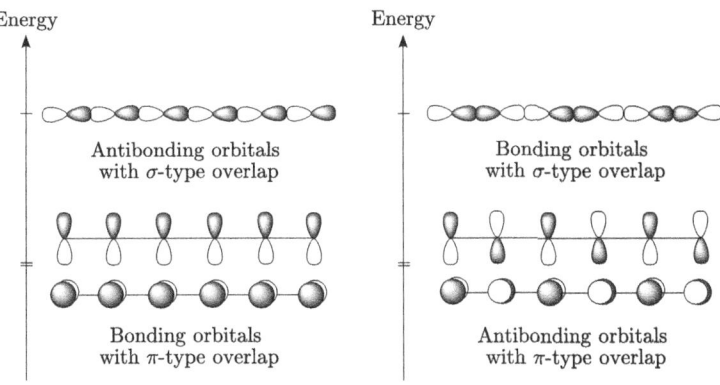

Fig. 6
Bloch orbitals at the Γ (left) and X (right) points for the $2p$ orbitals of the A_n chain.

(3.2) The Bloch orbitals of the A_n chain with just $2p$ orbitals are shown in Fig. 6.

(3.3) Since the wavelength λ is given by $2\pi/k$, its value at the Γ point is infinite but $2a$ at the X point. This means that the phase factor, $\exp(2\pi i k_a)$, is constant at Γ for all crystal orbitals whereas it possesses a periodicity of $2a$ at the X point.

(3.4) The wavelength associated with the $k_a = 1/4$ and $k_a = -1/4$ points is $4a$. This can be clearly verified by looking at Figs 3.11 and 3.13 since the coefficients are successively:

... 1 0 –1 0 1 0 –1 0 1 0 –1 0 ... on Fig. 3.11

and

... 1 1 –1 –1 1 1 –1 –1 1 1 –1 –1 ... on Fig. 3.13.

(3.5) Both functions, $S(\pm 1/4)$ and $A(\pm 1/4)$, are antisymmetric with respect to a translation by $2\vec{a}$. In contrast, a translation by a vector $\vec{a}$ transforms function $S(\pm 1/4)$ into $A(\pm 1/4)$ and vice versa. These functions are not individually well adapted to a translation by a vector $\vec{a}$ but they are to a $2\vec{a}$ translation.

(3.6) In Section 2.3, the molecular orbitals for the π system of cyclobutadiene were obtained. Following exactly the same procedure, the molecular orbitals of H_4 may be obtained. In this case, the $1s$ orbitals of the hydrogen atoms play the same role as the $2p_z$ orbitals of the carbon atoms in cyclobutadiene. Fig. 7 is then obtained. Alternatively, the general

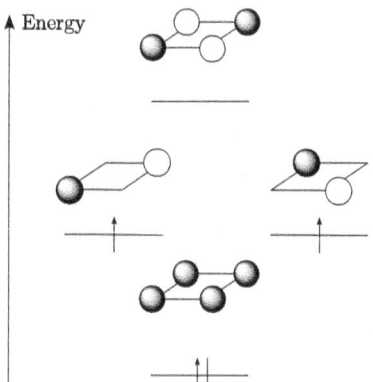

Fig. 7
Molecular orbital diagram of H_4 using the procedure developed in Chapter 2.

formula (3.15) may be used to obtain the orbital diagram of Fig. 8 (note that the two central orbitals are complex).

These two diagrams are completely equivalent. In the first diagram, the doubly degenerate levels are real functions, which may be labelled as ψ_1 and ψ_2, whereas in the second figure the same levels are complex functions given by the linear combinations $\psi_1 + i\psi_2$ and $\psi_1 - i\psi_2$.

(3.7) Let us first consider the $(H^{0.5+})_n$ chain. Every cell contributes half an electron to the 1s band so that the band is a quarter filled and the Fermi wave vector is given by $k_a = (0.5/2) \cdot 0.5 = 1/8$. In the case of the $(H^{0.3+})_n$ chain, the Fermi wave vector is given by $k_a = (0.7/2) \cdot 0.5 = 0.175$.

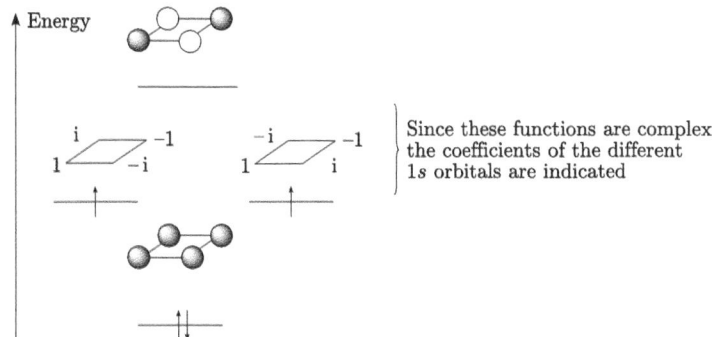

Fig. 8
Molecular orbital diagram of H$_4$ using eqn (3.15).

(3.8) The Fermi levels ($\epsilon_{f_1}, \epsilon_{f_2}, \epsilon_{f_3}, \epsilon_{f_4}$) for the filling of these bands with 1, 2, 3, and 4 electrons are shown in Fig. 9. They correspond to the filling of the equivalent of 1/2, 1, 3/2, and 2 bands, respectively.

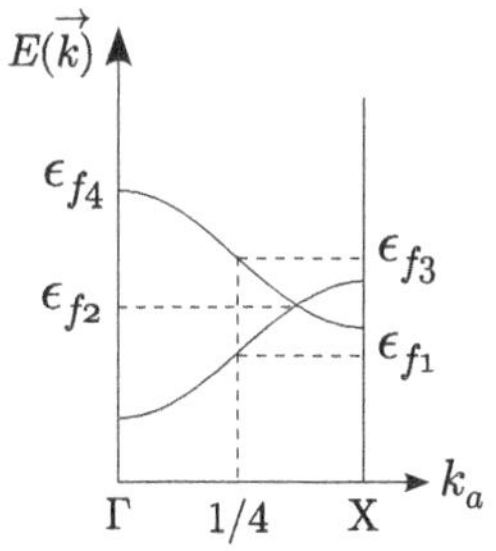

(3.9) Detailed analysis of the electron distribution in the H$_n$ system ($n' = n/2$, $n'' = n/4$).

(1) According to eqn (3.6), the possible values for k_a are 0, $\pm\frac{1}{n}$, $\pm\frac{2}{n}$, ..., $\pm\frac{n'-1}{n}$, $\frac{1}{2}$. Consequently, to consider the point associated with $k_a = 1/4$ it is necessary that 1/4 can be written as $\frac{m}{n}$, i.e. n must be a multiple of 4.

(2) If n is a multiple of 4, the $(n'' - 1)$ allowed levels characterised by values of k_a equal to $\pm\frac{1}{n}, \pm\frac{2}{n}, \ldots, \pm\frac{n''-1}{n}$ are filled with 4 electrons whereas the level with $k_a = 0$ is filled with 2 electrons. The two last electrons partially fill the doubly degenerate levels characterised by $k_a = \pm 1/4$.

(3) If n is even but not a multiple of 4, the $(n'' - 0.5)$ allowed levels characterised by values of k_a equal to $\pm\frac{1}{n}, \pm\frac{2}{n}, \ldots, \pm\frac{n''-1}{2n}$ are filled with 4 electrons whereas the level with $k_a = 0$ is filled with 2 electrons. The Fermi level is thus characterised by the k_a value of $(n' - 1)/2n$.

(4) Since n is very large and tends to infinity, $(n' - 1)/2n$ tends to 1/4 so that the Fermi level is characterised by the k_a value of $\pm 1/4$.

Fig. 9
Different Fermi levels for a two-band system.

Chapter 4

(4.1) The symmetry operation rendering the S($\pm 1/8$) and A($\pm 1/8$) functions equivalent is a translation by a $2\vec{a}$ vector.

(4.2) The band structures for the regular and distorted systems are shown in Figs 10 and 11. The tetramerisation makes possible the interaction of the Ψ_{1s} and Ψ_{2s} orbitals as well as the interaction of the Ψ_{1a} and Ψ_{2a} orbitals at both the Γ and Z points. This leads to the new orbitals φ_{1s}, φ_{2s}, φ_{1a}, and φ_{2a}. At Γ the situation is relatively simple: the interaction between Bloch orbitals and the subsequent distortion results in the stabilisation of the φ_{1s} and φ_{1a} orbitals as well as the destabilisation of the φ_{2s} and φ_{2a} orbitals. At Z, the main result of the distortion is the degeneracy

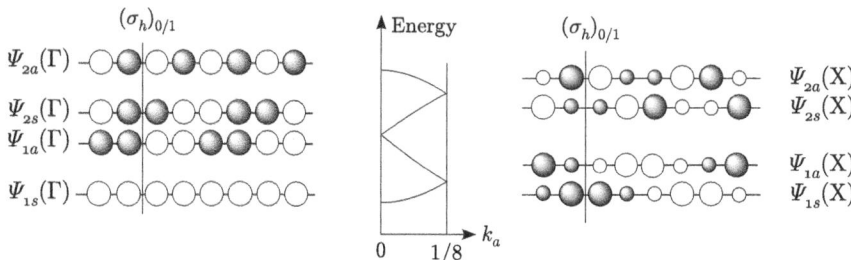

Fig. 10
Band structure of the regular system.

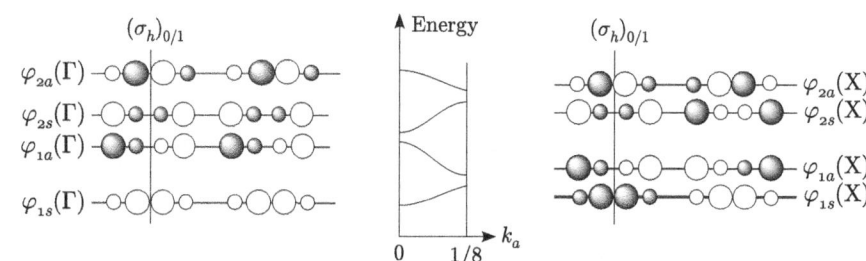

Fig. 11
Band structure of the distorted system.

lift of the orbitals since the distortion reinforces or diminishes either bonding or antibonding interactions.

Chapter 5

(5.1) Let us schematically represent the interaction between $\frac{1}{\sqrt{n}}\pi_0$ and $\mathrm{BO}_\pi(\vec{k})$ (see Fig. 12).

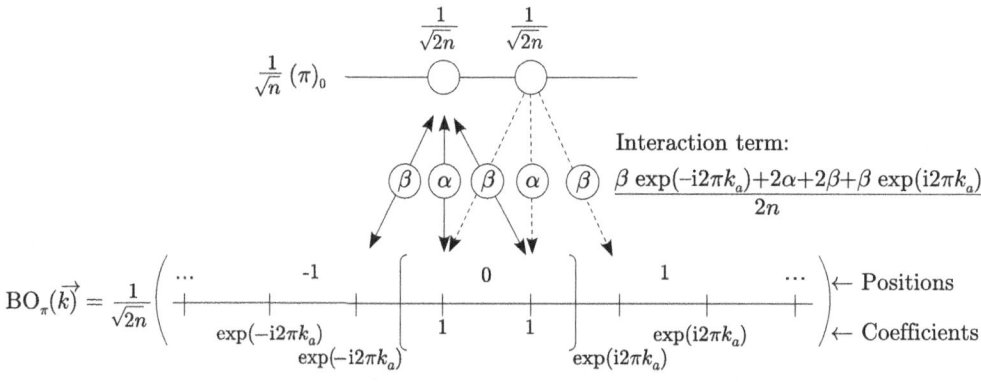

Fig. 12
Schematic evaluation of the interaction term between $\frac{1}{\sqrt{n}}\pi_0$ and $\mathrm{BO}_\pi(\vec{k})$.

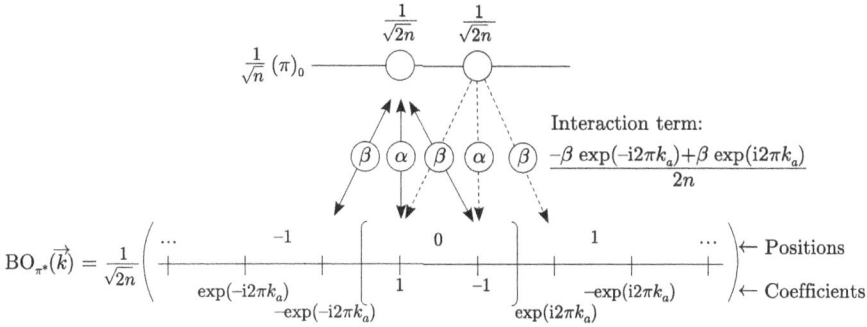

Fig. 13
Schematic evaluation of the interaction term between $\frac{1}{\sqrt{n}}\pi_0$ and $BO_{\pi^*}(\vec{k})$.

Note: To simplify the scheme we have adopted a top view of the $2p_y$ orbitals (i.e. we have only depicted the lobe pointing towards the positive values of the y-axis, which consequently looks like a circle).

Using the procedure developed in Section 3.2.2 we obtain:

$$h_{\pi,\pi}(\vec{k}) = n\langle \frac{1}{\sqrt{n}}\pi_0 \mid \hat{h} \mid BO_\pi(\vec{k})\rangle$$
$$= \alpha + \beta(1 + \cos(2\pi k_a))$$
$$= \alpha + 2\beta \cos^2(\pi k_a)$$

If we now use the same representation for the interaction between $\frac{1}{\sqrt{n}}\pi_0$ and $BO_{\pi^*}(\vec{k})$ (see Fig. 13), we obtain:

$$h_{\pi,\pi^*}(\vec{k}) = n\langle \frac{1}{\sqrt{n}}\pi_0 \mid \hat{h} \mid BO_{\pi^*}(\vec{k})\rangle$$
$$= i\beta \sin(2\pi k_a)$$

Finally, representing the interaction between $\frac{1}{\sqrt{n}}(\pi^*)_0$ and $BO_{\pi^*}(\vec{k})$ on Fig. 14, we obtain:

$$h_{\pi^*,\pi^*}(\vec{k}) = n\langle \frac{1}{\sqrt{n}}(\pi^*)_0 \mid \hat{h} \mid BO_{\pi^*}(\vec{k})\rangle$$
$$= \alpha - \beta(1 + \cos(2\pi k_a))$$
$$= \alpha - 2\beta \cos^2(\pi k_a)$$

(5.2) Band structure of the $(BNH_2)_n$ chain.
  (1) There are two mesomeric structures, as shown in Fig. 15.

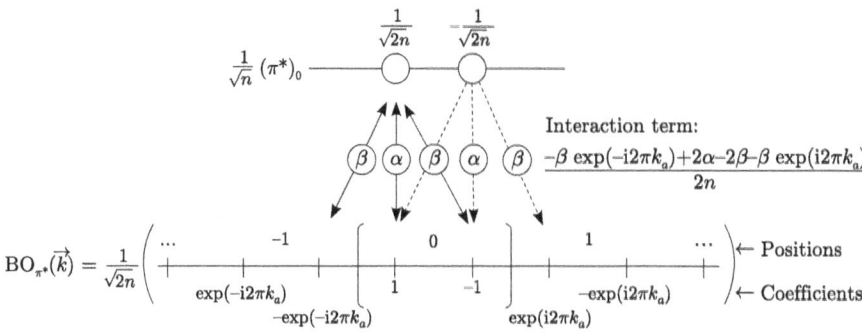

Fig. 14
Schematic evaluation of the interaction term between $\frac{1}{\sqrt{n}}(\pi^*)_0$ and $BO_{\pi^*}(\vec{k})$.

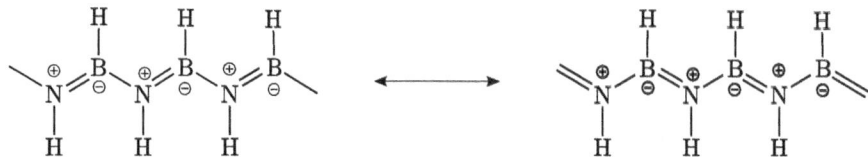

Fig. 15
Two Lewis structures for the $(BNH_2)_n$ chain.

(2) For exactly the same arguments put forward for *trans*-polyacetylene in the text, we can conclude that the Fermi level of this chain must be located within the π bands manifold built from the $2p_y$ orbitals of the nitrogen and boron atoms.

(3) (a) It is important to understand that this system is equivalent to the $(H_2)_n$ chain considered in the text. The only important difference is that the energies of the nitrogen and boron orbitals are not the same. Their values are $\alpha + \beta$ and $\alpha - \beta$, respectively (β is the interaction term between two adjacent $2p_y$ orbitals and α is the energy of a $2p_y$ orbital of the carbon atom). Thus, eqn (3.32) becomes:

$$h_{NN}(\vec{k}) = \alpha + \beta$$
$$h_{BB}(\vec{k}) = \alpha - \beta$$
$$h_{NB}(\vec{k}) = \beta(1 + \exp(-i2\pi k_a)) = (h_{BN}(\vec{k}))^*$$

if we generate the system from a unit cell with formula N–B (but not B–N).

Calculating the corresponding secular determinant, it is possible to obtain the following equation:

$$(h_{NN}(\vec{k}) - E(\vec{k}))(h_{BB}(\vec{k}) - E(\vec{k})) - |h_{NB}(\vec{k})|^2 = 0$$

the solutions of which are:

$$E_\pm(\vec{k}) = \alpha \pm \beta\sqrt{3 + 2\cos(2\pi k_a)}$$

(b) Writing the two crystal orbitals associated with a given value of the $\vec{k}$ wave vector as:

$$CO^\pm(\vec{k}) = c_N^\pm(\vec{k})(BO)_N(\vec{k}) + c_B^\pm(\vec{k})(BO)_B(\vec{k})$$

the secular equations are then:

$$\begin{cases} (\alpha - E^\pm(k_a) + \beta)c_N^\pm(\vec{k}) + (1 + \exp(-i2\pi k_a))\beta c_B^\pm(\vec{k}) = 0 \\ (1 + \exp(i2\pi k_a))\beta c_N^\pm(\vec{k}) + (\alpha - E^\pm(k_a) - \beta)c_B^\pm(\vec{k}) = 0 \end{cases}$$

At the X point, $(1 + \exp(i2\pi k_a)) = 0$ so the interaction term is nil and the crystal orbitals are simply the Bloch orbitals.

At the Γ point, one of the secular equations allows us to obtain the ratio between the $c_N^\pm(\Gamma)$ and $c_B^\pm(\Gamma)$ coefficients:

$$\frac{c_N^\pm(\Gamma)}{c_B^\pm(\Gamma)} = \frac{2}{\pm\sqrt{5} - 1}$$

i.e. approximately 1.6 and –0.6 for the two solutions, respectively.

(c) With these results in mind we can draw the band structure of Fig. 16.

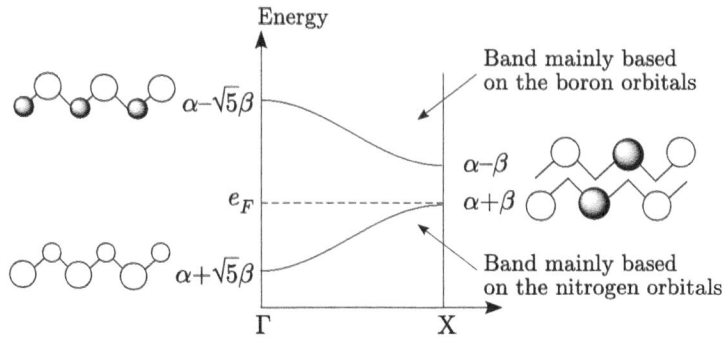

Fig. 16
Band structure for the $(BNH_2)_n$ chain.

(d) Since every cell contributes two electrons, the Fermi level lies on top of the lower band.

(e) The filled band is more heavily based on the nitrogen atoms whereas the empty band is more heavily based on the boron atoms. This result sounds a warning concerning the Lewis structures, on which the boron atoms hold a –1 charge while the nitrogen atoms hold a +1 charge. Since the filled band is more heavily based on the nitrogen atoms, the charges that may be reasonably be attributed to boron and nitrogen are actually considerably weaker in absolute value.

(4) Qualitative determination of the band structure

(a) We can now obtain the same band structure in a more qualitative way. First we will draw the band structure assuming that the Bloch orbitals $BO_N(k_a)$ and $BO_B(k_a)$ do not interact. Since the different $2p$ orbitals of nitrogen do not interact, the energy associated with $BO_N(k_a)$ will be a flat line with energy $\alpha + \beta$, which is the energy associated with an isolated $2p$ orbital of nitrogen. For the same reason, the energy associated with $BO_B(k_a)$ will be a flat line with energy $\alpha - \beta$. This non-interacting band structure is represented as broken lines in Fig. 17.

(b) Following the calculation developed in part 3(a) of this exercise, the interaction between the two Bloch orbitals is given by $h_{NB}(\vec{k}) = \beta\,(1 + \exp(-i2\pi k_a))$, which is a complex number with modulus $|\,2\beta\cos(\pi k_a)\,|$. This is maximum at Γ and minimum at X. Consequently, the Bloch orbitals $BO_N(k_a)$ and $BO_B(k_a)$ will interact in such a way that the interaction will be stronger as we approach the Γ point.

(c) Fig. 17 represents the energy of the Bloch orbitals as broken lines and those of the crystal orbitals resulting from their interaction as continuous lines. The strength of the interaction is sketched through the length of the arrows at each zone.

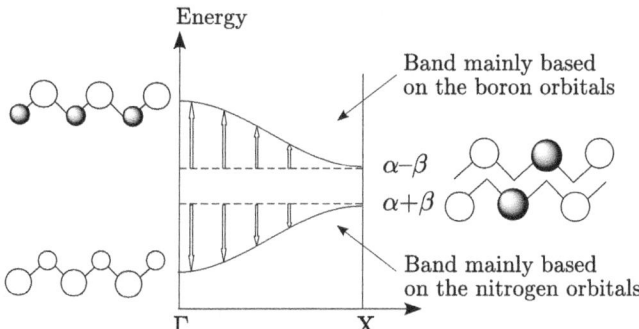

Fig. 17
Qualitative band structure for the $(BNH_2)_n$ chain.

The shape of the crystal orbitals at Γ may be obtained through mixing of the two Bloch orbitals $BO_N(k_a)$ and $BO_B(k_a)$, which leads to a bonding crystal orbital mainly based on the nitrogen orbitals and an antibonding crystal orbital mainly centred on the boron orbitals.

Chapter 6

(6.1) $t_{\vec{a}}(BO(k_a)) = \exp(-2i\pi k_a)BO(k_a)$
A crystal orbital is a linear combination of Bloch orbitals with the same exponential factor. Consequently a crystal orbital behaves as a Bloch orbital.

(6.2) It has been shown in the text that $t_{\vec{a}} \otimes (D_{4h})_0$ is equivalent to $(D_{4h})_{0/1}$. Consequently, this group leaves the chain invariant and generates the same space group as $T_n \otimes (D_{4h})_0$.

(6.3) The most complete point group P leaving one point of the reference cell invariant, as well as the chain, is the $D_{\infty h}$ group, which leaves

Solutions for exercises

invariant either the sodium atom or the chlorine atom of the reference cell. Consequently, this group is not unique.

(6.4) In fact there is a quite weak overlap between the $3s$ orbitals of the sodium atoms and the $3p_z$ orbitals of the chlorines of the Bloch orbitals $BO_{(3s)_{Na}}(\vec{k})$ and $BO_{(3p_z)_{Cl}}(\vec{k})$. In Fig 18 we represent the energies of these Bloch orbitals as they would be obtained from an extended Hückel calculation. Obviously, the weak bonding or antibonding character of these Bloch orbitals will not play any important role in the determination of the crystal orbitals that involve overlaps between adjacent atoms (i.e. between chlorine and sodium atoms) that are considerably stronger. This is the reason why we usually neglect such overlaps in the text, which in practise means that we work within the Hückel approximation, since we only consider overlaps between nearest neighbours.

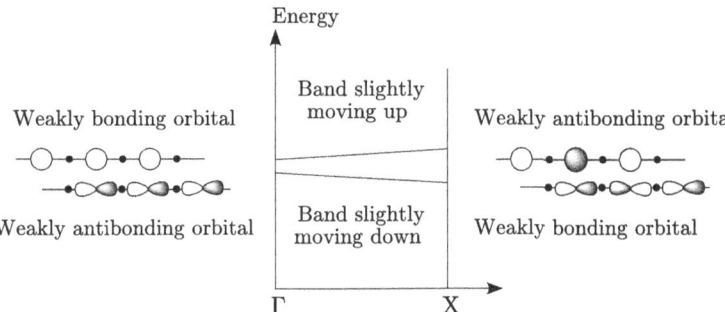

Fig. 18
Energies of the two Bloch orbitals of the NaCl chain.

(6.5) It is possible to consider the C_{2v} group leaving one carbon atom invariant and containing the symmetry operations E, C_{2x}, σ_{xz}, and σ_{xy}. This group contains as many symmetry operations as the group P considered in the text (C_{2h} group), and leaves the midpoint of the C–C bond invariant in such a way that its use will allow us to obtain the same information.

(6.6) Since $g_{C_{2z}}$ is equivalent to the product $\sigma_{xz} \cdot g_\sigma$, if we take into account the symmetry properties with respect to σ_{xz} and g_σ we also indirectly take into account those with respect to $g_{C_{2z}}$.

Chapter 7

(7.1) Different Lewis structures are possible, as shown Fig. 19

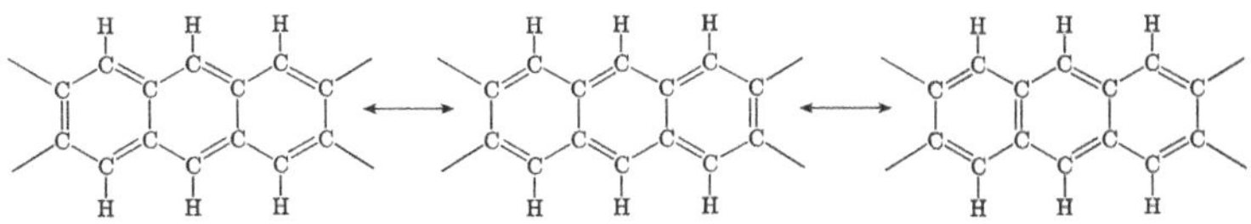

Fig. 19
Different Lewis structures for the polyacene.

(7.2) The π-type molecular orbitals of *cis*-butadiene and their occupation for the lowest energy state are shown in Fig. 20

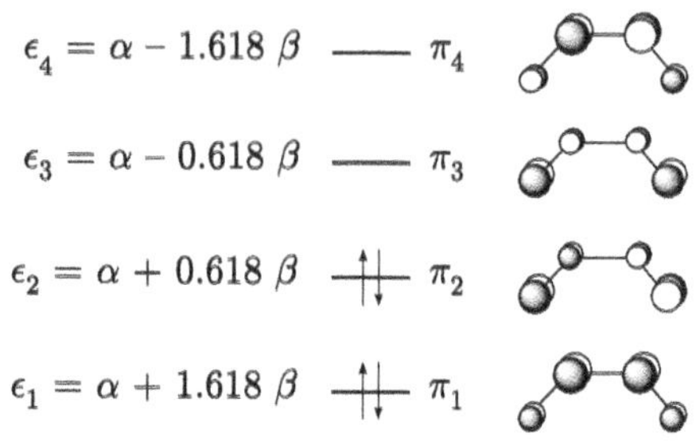

$$\epsilon_4 = \alpha - 1.618\,\beta \quad \underline{\hspace{2em}} \quad \pi_4$$

$$\epsilon_3 = \alpha - 0.618\,\beta \quad \underline{\hspace{2em}} \quad \pi_3$$

$$\epsilon_2 = \alpha + 0.618\,\beta \quad \underline{\uparrow\downarrow} \quad \pi_2$$

$$\epsilon_1 = \alpha + 1.618\,\beta \quad \underline{\uparrow\downarrow} \quad \pi_1$$

Fig. 20
π-type molecular orbitals of *cis*-butadiene.

(7.3) It would have been possible to use the D_{2h} group, which leaves invariant the centre of a hexagon.

(7.4) The energy diagram of the C_4H_2 fragment orbitals for the polyacene is shown in Fig. 21

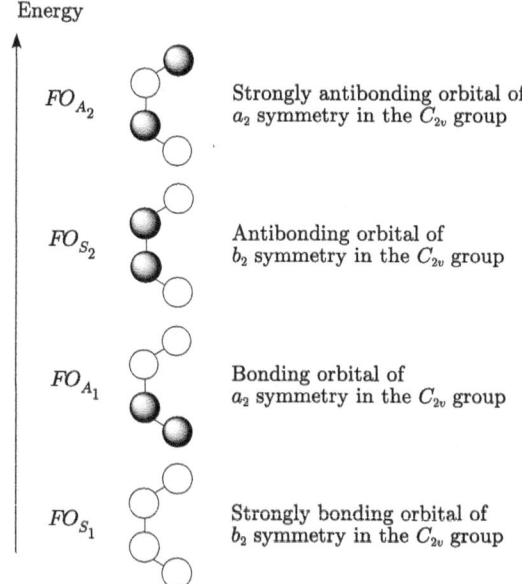

Fig. 21
Energy diagram of the C_4H_2 fragment orbitals for polyacene.

(7.5) The fragment orbital coefficients are the same in absolute terms whereas the molecular orbital coefficients of *cis*-butadiene are not.

(7.6) It would also have been possible to use the set of orbitals of Fig. 22, which are well adapted to the symmetry of polyacene:

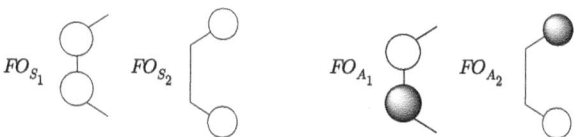

Fig. 22
A different set of fragment orbitals for C_4H_2.

(7.7) The interaction diagram between two Bloch orbitals symmetric with respect to the horizontal plane is shown in Fig. 23

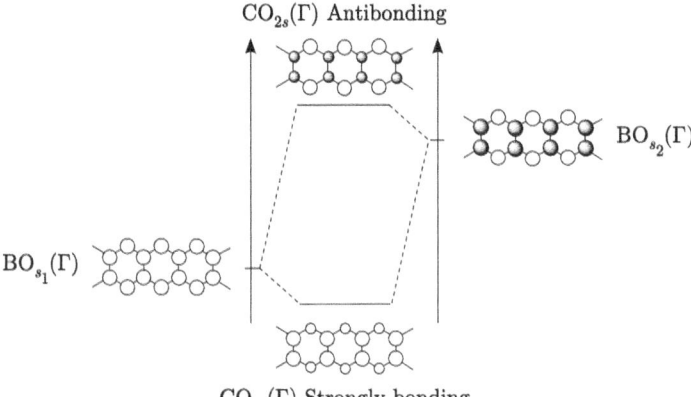

Fig. 23
Interaction diagram between two Bloch orbitals symmetric with respect to the polyacene horizontal plane.

and the interaction diagram for the Bloch orbitals that are antisymmetric with respect to the horizontal plane is shown Fig 24

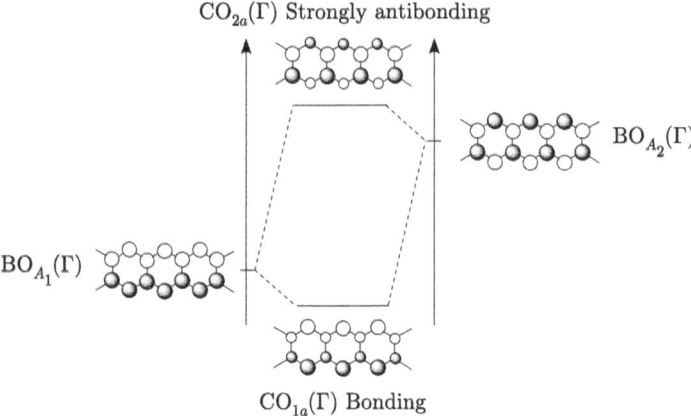

Fig. 24
Interaction diagram between two Bloch orbitals antisymmetric with respect to the polyacene horizontal plane.

The relative position between the four crystal orbitals is easy to find and is given in Fig 25.

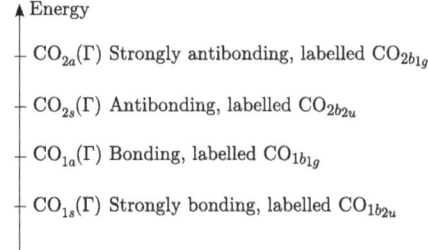

Fig. 25
Energy of the different crystal orbitals of polyacene at Γ.

(7.8) Having assessed the bonding, non-bonding or antibonding character of the crystal orbitals for the Γ and X points, Fig. 26 may be derived.

(7.9) If the group D_{2h} leaving the centre of a hexagon invariant had been used, the same crystal orbitals would had been obtained; only certain

Fig. 26
Energy of the different crystal orbitals of polyacene at Γ and X.

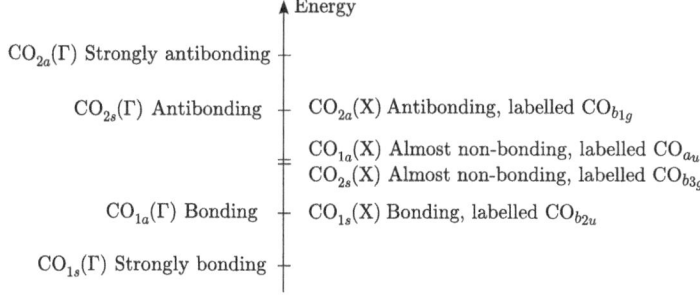

symmetry labels would be different. For instance, the orbitals labelled $CO_{b_{2u}}(X)$ and $CO_{b_{3g}}(X)$ in the text become orbitals of symmetry b_{3g} and b_{2u}, respectively, in the group D_{2h} leaving the centre of a hexagon invariant.

(7.10) Study of the electronic structure of distorted polyacene (see. Fig. 27).

Fig. 27
Distorted polyacene.

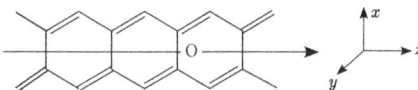

(a) The symmetry operations destroyed by the distortion are C_{2x}, C_{2z}, σ_{xy}, and σ_{yz}. The symmetry group appropriate for points Γ and X is C_{2h} containing the E, i, C_{2y}, and σ_{xz}, whereas the group appropriate for the other points is C_s, containing the E and σ_{xz} operations.

(b) Since all crystal orbitals associated with the π system of polyacene are antisymmetric with respect to the xz plane, they are of A'' symmetry in the C_s group. Consequently, no band crossing is allowed for k points different from Γ and X.

(c) The crystal orbitals $CO_{1b_{2u}}(\Gamma)$, $CO_{2b_{2u}}(\Gamma)$, $CO_{b_{2u}}(X)$, and $CO_{a_u}(X)$ of the non-distorted system are of a_u symmetry when using the C_{2h} symmetry, whereas the crystal orbitals $CO_{1b_{1g}}(\Gamma)$, $CO_{2b_{1g}}(\Gamma)$, $CO_{b_{3g}}(X)$, and $CO_{b_{1g}}(X)$ are of b_g symmetry. Consequently, when the system distorts, the crystal orbitals of the non-distorted system may interact and lead to new crystal orbitals. For instance, in Fig. 28 we represent how the $CO_{b_{2u}}(X)$ and $CO_{a_u}(X)$ crystal orbitals of the non-distorted system interact to lead to the $CO_{1a_u}(X)$ and $CO_{2a_u}(X)$ of the distorted system.

(d) In Fig. 29 we have schematically drawn the band structure of the distorted system, taking into account that band crossings are not allowed.

The crystal orbitals at Γ are practically identical to those of the non-distorted system because the energy difference between the orbitals that may interact under the distortion is quite large. Since there are four electrons to place in these bands, the Fermi level is located at the top of the $2a''$ band. The appropriate crystal orbital is $1b_g(X)$, which is slightly bonding as a result of the distortion.

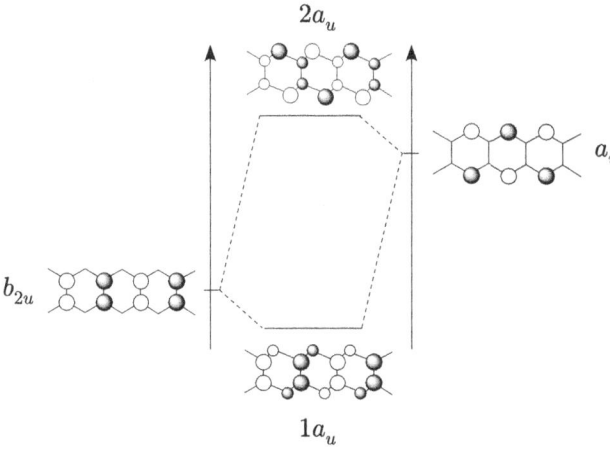

Fig. 28
$1a_u$ and $2a_u$ crystal orbitals for the regular polyacene at X.

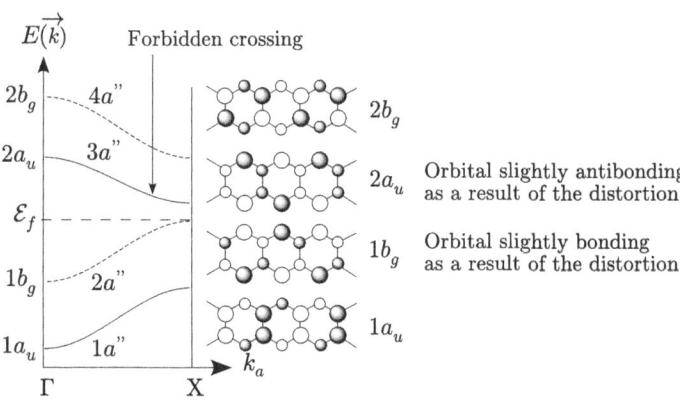

Fig. 29
Schematic band structure for the distorted polyacene.

Chapter 8

(8.1) Let us label the interaction term between the p_z and d_{z^2} orbitals in adjacent centres as β, as shown in Fig. 30. Then we may schematically draw the interaction shown Fig. 31 to calculate $\langle \mathrm{BO}_{6p_z}(\vec{k}) \mid \hat{h} \mid \mathrm{BO}_{5d_{z^2}}(\vec{k}) \rangle$. In this way we may write:

$$\langle \mathrm{BO}_{6p_z}(\vec{k}) \mid \hat{h} \mid \mathrm{BO}_{5d_{z^2}}(\vec{k}) \rangle = n \langle \frac{1}{\sqrt{n}} (p_z)_0 \mid \hat{h} \mid \mathrm{BO}_{5d_{z^2}}(\vec{k}) \rangle$$

$$= 2i\beta \sin(2\pi k_a)$$

This interaction term is nil at both points Γ and X, and maximum for $k_a = 1/4$.

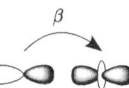

Fig. 30
Interaction between the $6p_z$ and $5d_{z^2}$ orbitals of the $[\mathrm{Pt(CN)}_4]_n$ chain.

Fig. 31
Schematic evaluation of $\langle BO_{6p_z}(\vec{k}) | \hat{h} | BO_{5d_{z^2}}(\vec{k}) \rangle$.

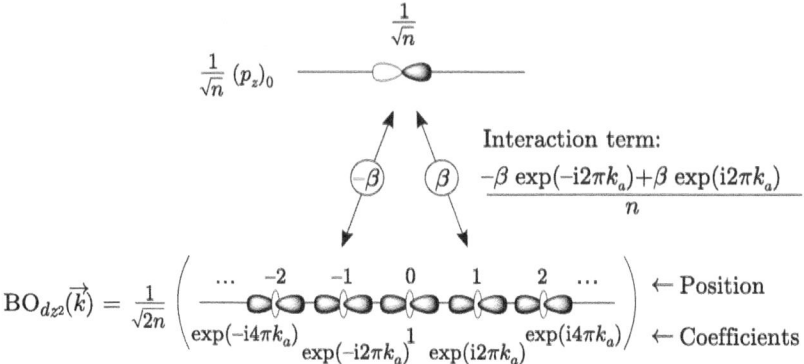

(8.2) Let us assume that the system contains *eclipsed* chains. Of course, the band structure of this system is practically identical to that of eclipsed KCP discussed in the text, especially regarding the d_{z^2} band. The formal oxidation state of platinum here is +2.33. This means that the d_{z^2} band is filled with $1.67 = 5/3$ electrons, i.e. it is 5/6 filled. Under such circumstances a hexamerisation should be expected. Thus, a structure such as that shown below, in which $d_1 = 2.72$ Å, $d_2 = 2.83$ Å, and $d_3 = 3.02$ Å, which corresponds to experimental observations, can be expected.

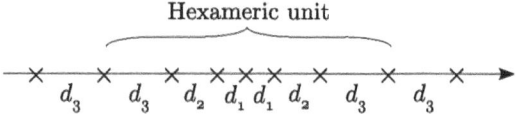

Fig. 32
Position of the platinum atoms after the distortion.

(8.3) The oxidation state of platinum in the hypothetically eclipsed compound $K_2Pt(CN)_4Li_{0.2}$ is +1.8. Consequently, there are 0.2 electrons in the $2a_1$ band. This means that the Fermi level occurs for a k_a value of $0.5 - 0.2/4 = 0.5 - 0.05 = 0.45$. It then follows that the compound must be metallic and tends undergo a 10-merisation at relatively low-temperature because $2k_a = 9/10$. Since doping with lithium results in a transfer of 0.2 electrons to the bottom of the $2a_1$ band, which contains metal–metal bonding levels, the metal–metal bond lengths in the metallic state will decrease with respect to those in the undoped system.

(8.4) Study of the electronic state of staggered KCP.

(a) The case of the eclipsed chain was considered in detail in Section 6.1.2, where it was shown that $T_n \otimes D_{4h}$ is the appropriate space group for this chain. The only difference between the eclipsed and staggered forms is the appearance of the different symmetry operations related to the screw rotation symmetry in the staggered chain. Thus, to verify that the space group of staggered KCP may be written as a direct product $T_n \otimes D_{4h} \otimes \{E, gC_{8z}\}$, it is only necessary to look at the product of the symmetry operations of the D_{4h} group with $g_{C_{8z}}$, where the latter is the screw axis along the main axis of the system.

(b–d) To find the symmetry properties of the Bloch orbitals generated by the d_{z^2} and p_z orbitals it is useful to remind ourselves of the main conclusions of Section 6.4.4 (see Fig. 6.23). In Fig. 33 we have shown the main results of the symmetry analysis. Bands 1 and 2 are those associated with the d_{z^2} orbitals and bands 3 and 4 are those associated with the p_z orbitals. The symmetry properties with respect to the $g_{C_{8z}}$ symmetry operation are also given.

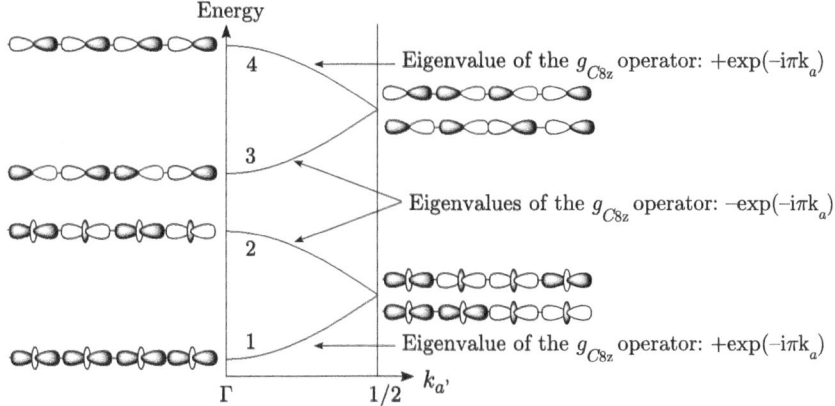

Fig. 33
Schematic band structure for staggered KCP.

According to this figure it is clear that bands 2 and 3 have the same symmetry properties with respect to $g_{C_{8z}}$. These bands also have the same symmetry behaviour with respect to the operations of the C_{4v} group. Consequently, bands 2 and 3 cannot cross inside the Brillouin zone. In contrast, at the Γ point, the Bloch orbitals of these bands have different symmetry properties in the D_{4h} group (a_{1g} and a_{1u}), which is the appropriate group for this point, so that the two orbitals cannot mix. In conclusion, whatever the pressure exerted the two bands cannot cross; they can only touch at the Γ point.

(8.5) In the linear chain ReCl$_4$N, the chlorine and nitrogen atoms must be considered to be Cl$^-$ and N^{3-} so that the formal oxidation state of rhenium is +7 (i.e. is a Re d^0). In the linear chain (Pt(NH$_3$)$_4$Cl^{2+})$_n$ the NH$_3$ ligands are formally neutral so that the oxidation state of platinum is +3 (i.e. a Pt d^7). In other words, rhenium does not formally provide any d electron to the first system, whereas platinum provides seven to the second.

(8.6) (a) See Fig. 34.

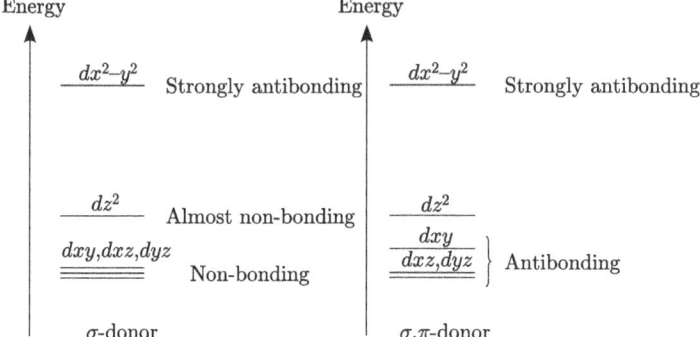

Fig. 34
Schematic orbitals for the ML$_4$ fragment when the ligand L is a σ donor (left) or a σ, π donor (right).

(b) Another possibility would have been to take the orbitals of an ML$_4$L$'$ fragment. The fragmentation used in the text has the advantage of clearly distinguishing those ligands (L) playing a secondary role in the electronic structure of the chain from those (L$'$) that play a primary role.

(c) The $6p_z$ orbitals of the metal are considerably higher in energy than those of the ligands so that, as a first approximation, it is legitimate to assume that they will play a minor role in the metal d-based bands of the system.

(d) These orbitals may be neglected in a first approximation for opposite reasons to those discussed in the previous exercise. The valence ns orbitals of the L$'$ ligands (L$'$ = Cl$^-$ and N^{3-}) are too low-lying in energy with respect to the d orbitals of the metal atom.

(e) Instead of choosing the metal atom M of the reference cell as the chain origin, the bridging ligand L$'$ could had also been chosen. Of course, as a consequence, there would be differences in the symmetry labels of the Bloch orbitals. For instance, those of the a_{1g} orbitals at the X point would become of a_{2u} symmetry.

(f) The following aspects of the relative band positioning are uncertain:
- how the top of the $2e$ band is positioned with respect to the $1b_2$ band
- where the top of the $2e$ band lies with respect to the bottom part of the $2a_1$ band
- where the top of the $2a_1$ band lies with respect to the $1b_1$ band.

(8.7) With respect to the orbitals of the metal in a square-planar environment there are three d orbitals which have changed shape. Because of the symmetry lowering, the (d_{xz}, d_{yz}) orbitals can interact and mix with the (p_x, p_y) orbitals and the d_{z^2} can interact and mix with the s orbital. Since these orbitals are metal–ligand antibonding, the mixing is such that this antibonding character is reduced. This is why the three orbitals hybridise in such a way that they concentrate outside the ligand region. The two orbitals remaining cannot mix with any of the s- and p-type orbitals under the distortion and remain unaltered.

(8.8) (a) When comparing the splitting of the d orbitals for a transition metal in octahedral and square-planar environments it is easy to see that in the first case there are two antibonding (e_g) levels whereas in the second there is just one (b_{1g}). Consequently, a platinum atom in an octahedral environment will tend to be in a +4 oxidation state. However, when the same atom is in a square-planar environment it will have a tendency to be in a +2 oxidation state. The outcome of the distortion observed in the chain leads exactly to this situation.

(b) One could imagine a distortion such that every other chlorine atom would be tightly bound to two platinum atoms, the other chlorines being almost isolated, i.e. –Pt-Cl-Pt–Cl–Pt-Cl-Pt–Cl–. However, from the chemical viewpoint there is nothing that could justify this view other than alternating platinum atoms in oxidation states +2 and +4 (i.e. one platinum atom tightly bound to two axial chlorine atoms and the other almost left alone, with only their basal ligands).

(8.9) (a) See Fig. 35

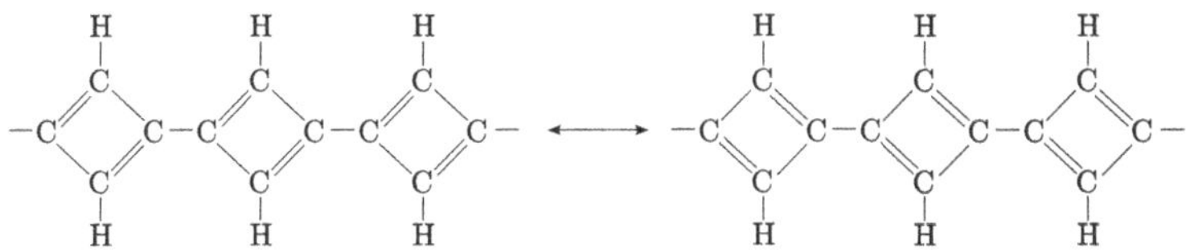

Fig. 35
Lewis structures for the $(C_4H_2)_n$ model compound.

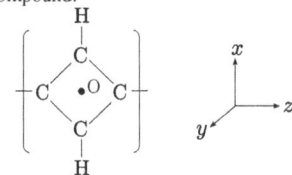

Fig. 36
Unit cell of the $(C_4H_2)_n$ system.

(b) The bands that will be located in the vicinity of the Fermi level are those of the π system. Thus, we will consider the $2p_z$ orbitals of the carbon atoms.

(c) See Fig. 35.

(d) The appropriate symmetry group for the Γ and X point is D_{2h}, which contains the operations E, i, σ_{xy}, σ_{xz}, σ_{yz}, C_{2x}, C_{2y}, and C_{2z}, assuming that the origin is taken at the centre of the ring in the reference cell. For the other k points the appropriate symmetry group is C_{2v}, containing the operations E, σ_{xz}, σ_{yz}, and C_{2z}.

(e) The simplest set of orbitals is that of the π system of cyclobutadiene which, taking into account the results for the H_4 system studied in Exercise 3.6, may be schematically represented as in Fig. 37.
Note: As usual, to simplify the scheme we have adopted a top view of the $2p_y$ orbitals (i.e. we have only depicted the lobe pointing towards the positive values of the y-axis, which consequently looks like a circle).

(f) Starting from the cyclobutadyene orbitals FO_1, FO_2, FO_3, and FO_4 it is possible to build the corresponding Bloch orbitals for any value of k_a, which will be denoted $BO_{FO_1}(k_a)$, $BO_{FO_2}(k_a)$, $BO_{FO_3}(k_a)$, and $BO_{FO_4}(k_a)$. The symmetry of these Bloch orbitals (obtained from the fragment orbitals (FOs)) for different points of the Brillouin zone are given in the table below.

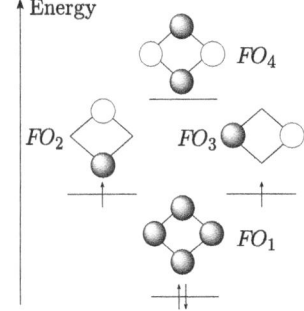

Fig. 37
Fragment orbitals used to generate the Bloch orbitals of the chain.

BO_{FO}	Γ point (D_{2h})	X point (D_{2h})	Other points (C_{2v})
BO_{F_1}	B_{2u}	B_{2u}	B_2
BO_{F_2}	B_{1g}	B_{1g}	A_2
BO_{F_3}	B_{3g}	B_{3g}	B_2
BO_{F_4}	B_{2u}	B_{2u}	B_2

According to this symmetry analysis, the Bloch orbitals $BO_{FO_1}(k_a)$ and $BO_{FO_4}(k_a)$ can interact and mix at Γ and X. However, this interaction is weak because the two orbitals are quite well separated in energy. For all other k points only one of the Bloch orbitals, $BO_{FO_2}(k_a)$, is also a crystal orbital because its symmetry is different from all others. The remaining three Bloch orbitals can interact for all points except Γ and X.

With this information in mind we can now draw the band structure below, in which the $1a_2$ and $2b_2$ bands cross for $k_a = 1/4$, at which point the energy is that of an isolated $2p_z$ orbital (see Fig. 38).

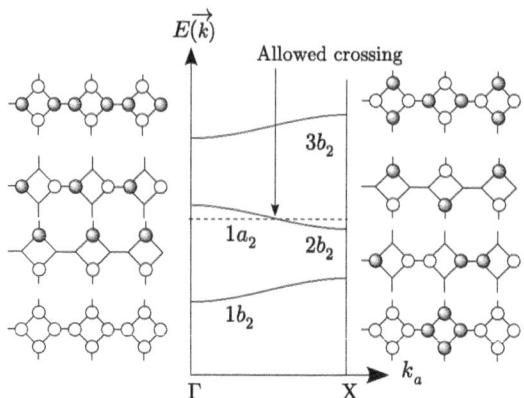

Fig. 38
Qualitative band structure for the $(C_4H_2)_n$ chain.

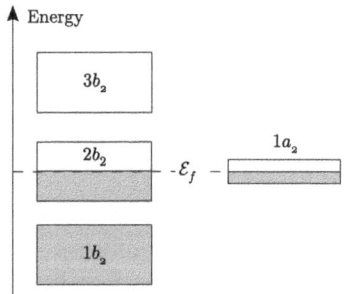

Fig. 39
Schematic band structure for the $(C_4H_2)_n$ chain.

Since there are four π-type electrons per unit cell, the equivalent of two bands must be filled. Consequently, the low-lying $1b_2$ will be completely filled and the $2b_2$ and $1a_1$ bands will be half-filled as schematically shown in Fig. 39. The Fermi level occurs at the energy of an isolated $2p_z$ orbital.

(g) The compound should exhibit metallic behaviour.

(8.10) Let us now consider the $(N_2B_2H_2)_n$ system, and the associated $(N_2B_2H_2Li)_n$ and $(N_2B_2H_2Br)_n$ systems. The study proceeds through exactly the same steps as before. The only difference lies in the fact that the $2p_z$ orbitals of boron and nitrogen are different. Band $1a_2$, which is exclusively centred on the nitrogen orbitals, will be lowered in energy whereas band $2b_2$, which is mostly based on the boron orbitals (but not exclusively because of the possible mixing with the other bands of the

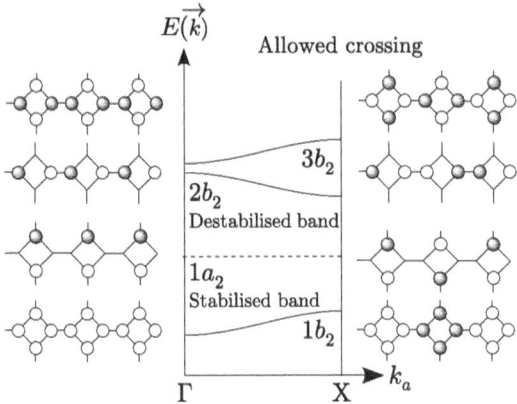

Fig. 40
Schematic band structure and crystal orbitals for the $(N_2B_2H_2)_n$ chain.

same symmetry), will be raised in energy. The band structure is shown in Fig. 40.

Since three of the four bands have the same b_2 symmetry, only the crossing of the $1a_2$ and $1b_2$ bands is likely. Introduction of Li or Br dopants allows a tuning of the number of electrons in the system. Note that the number of electrons per unit cell in the $(N_2B_2H_2)_n$, $(N_2B_2H_2Li)_n$, and $(N_2B_2H_2Br)_n$ systems is 4, 5, and 3, respectively. Consequently, we can predict the following band fillings and conductivity behaviour for these compounds:

- $(N_2B_2H_2)_n$: Bands $1b_2$ and $1a_2$ are completely filled and consequently semiconducting behaviour is predicted.
- $(N_2B_2H_2Br)_n$: Band $1b_2$ is completely filled and band $1a_2$ is half-filled. In a one-electron approach the compound is predicted to be metallic. However, as shown in Chapter 12, a half-filled, very flat band, as occurs here, does not usually lead to a metallic state but to a semiconductor, as a consequence of electronic repulsions. Localisation is induced by the electronic repulsion.
- $(N_2B_2H_2Li)_n$: Bands $1b_2$ and $1a_2$ are completely filled and band $2b_2$ is half-filled. Metallic-type conductivity may be anticipated.

(8.11) (a) The study of this chain is of course closely related to the discussion in Section 8.2, since the present system is a ML_5 chain. Consequently, we will generate the chain using the cell VO_5^{6-}.

(b-d) By following a reasoning similar to that used in Section 8.2 and taking into account that now the L ligand is a σ and π donor, we end up with a schematic band structure as shown Fig. 41.

(e) The oxidation state of vanadium in $(VO_5^{6-})_n$ is +4, so every cell brings one electron to the d bands of the system. Consequently, the three bands mainly centred on the ligands are completely filled, as should the equivalent of half a d band. If the crossing of the $1b_2$ and $2e$ bands occurs after the point with $k_a = 1/8$, the Fermi level cuts the $2e$ bands and is thus associated with $k_a = 1/8$.

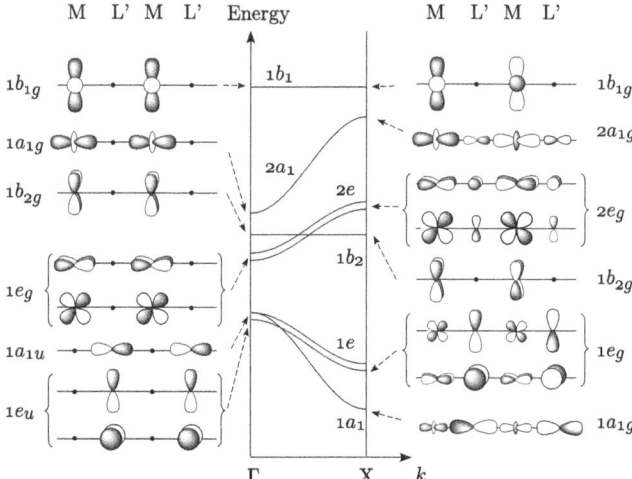

Fig. 41
Schematic band structure for the VO_5^{6-} chain.

If the crossing occurs before the point with $k_a = 1/8$, the Fermi level cuts both the $1b_2$ and $2e$ bands. If we denote as k_i the point at which the two bands cross, the Fermi level is associated with a k_f value such that:

$$2(4k_i) + 4k_f = 1$$

where $2(4k_i)$ is the number of electrons filling the $2e$ band and $4k_f$ is the number of electrons filling the $1b_2$ band.

(8.12) (a) Preamble (concerning the regular chain).
    (i) The oxidation state of niobium is +4, i.e. Nb is a $5d^1$ metal.
    (ii) Since niobium is formally a d^1 system, only the lowest part of the d-block bands will be filled. Because of the octahedral environment of the metal, the e_g bands will be well separated in energy from the t_{2g} bands. Consequently, we only need to consider in details the bands based on the d_{xz}, d_{yz}, and $d_{x^2-y^2}$ orbitals.
(b) In building the band structure of this chain, we will use the same approach as for the $ML_4'L$ chain; in other words we will fragment the NbX_4 chain into two units, NbX_2 and X_2. Such fragmentation distinguishes, in a natural way, the bridging ligands from those that are bonded to just one metal atom.
    (i) Every metal atom is octahedrally coordinated and brings just one electron to the d bands of the system. The d_{z^2} and d_{xy} orbitals point directly towards the six ligands and will participate either on the strongly bonding and mainly ligand-centred crystal orbitals or the strongly antibonding and mainly metal-centred crystal orbitals. Consequently the Fermi level will occur within the t_{2g}-block bands. This is why in the following we will not consider at all the crystal orbitals based on the d_{z^2} and d_{xy} orbitals.

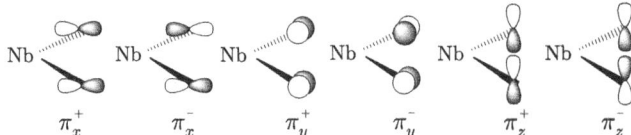

Fig. 42
Symmetry labels for the fragment orbitals of the X_2 ligand.

(ii) If we take a niobium atom as the origin O of the chain, the group appropriate for the Γ and X points is D_{2h}, whereas for all other points it is C_{2v}.

(iii) Let us start by working out the symmetry labels for the Bloch orbitals generated by either the transition metal atomic orbitals d_{xz}, d_{yz}, and $d_{x^2-y^2}$, or the six ligand orbitals of the X_2 fragment, which we will refer to as π_x^+, π_x^-, π_y^+, π_y^-, and π_z^+, π_z^-. These are represented in Fig. 42.

The symmetry labels for the different crystal orbitals are collected in the table below:

BO(FO)	Γ point (D_{2h})	X point (D_{2h})	Other points (C_{2v})
d_{yz}	B_{3g}	B_{3g}	A_2
d_{xz}	B_{2g}	B_{2g}	B_2
$d_{x^2-y^2}$	A_g	A_g	A_1
π_x^+	B_{3u}	A_g	A_1
π_x^-	B_{1g}	B_{2u}	B_1
π_y^+	B_{2u}	B_{1g}	B_1
π_y^-	A_g	B_{3u}	A_1
π_z^+	B_{1u}	B_{2g}	B_2
π_z^-	B_{3g}	A_u	A_2

On the basis of this symmetry analysis, we conclude that at Γ the Bloch orbitals $BO_{d_{yz}}(\Gamma)$ and $BO_{\pi_z^-}(\Gamma)$ can interact, as can the $BO_{d_{x^2-y^2}}(\Gamma)$ and $BO_{\pi_y^-}(\Gamma)$ pair. The transition-metal-based Bloch orbital, $BO_{d_{xz}}(\Gamma)$, cannot interact with ligand-based or other transition-metal-based Bloch orbitals and is thus already a crystal orbital of the system. At the X point the interactions allowed by symmetry are those between the Bloch orbitals $BO_{d_{xz}}(X)$ and $BO_{\pi_z^+}(X)$ as well as $BO_{d_{x^2-y^2}}(X)$ and $BO_{\pi_x^+}(X)$. The transition-metal-based Bloch orbital $BO_{d_{yz}}(X)$ cannot interact with ligand-based or other transition-metal-based Bloch orbitals and is thus already a crystal orbital of the system.

We thus can draw the energy-dependence of the mainly metal-centred Bloch orbitals $BO_{d_{x^2-y^2}}$, $BO_{d_{xz}}$, and $BO_{d_{yz}}$ (broken lines) as well as those of the crystal orbitals (continuous lines) resulting from their interaction with ligand-centred

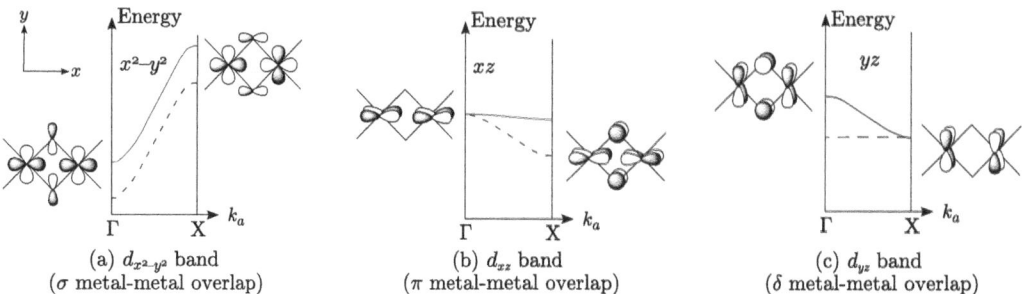

Fig. 43
Qualitative band structures for the t_{2g}-block orbitals.

(a) $d_{x^2-y^2}$ band (σ metal-metal overlap)
(b) d_{xz} band (π metal-metal overlap)
(c) d_{yz} band (δ metal-metal overlap)

Bloch orbitals (Fig. 43). Note that the interactions between the transition metal and bridging orbitals in these crystal orbitals are antibonding because the ligand-centred Bloch orbitals are σ and π donors. Since the corresponding ligand-based crystal orbitals are even lower lying and will thus all be completely filled, they are not represented in the figure.

It is important to note that in this chain the σ and π overlap between neighbouring niobium orbitals cannot be neglected. This is the reason for the energy dispersion in the metal-based Bloch orbitals, which of course follow the relative strength of the σ-, π-, and δ-type interactions. The t_{2g}-based band structure of the chain is simply the superposition of the three energy dispersion curves because the three crystal orbitals are of different symmetry for any k point.

(c) Points (a) and (b) above provide a basis to clearly understand the computational results for the NbCl$_4$ chain shown in Fig. 44. Note that the d_{xy} and d_{z^2} bands, also represented in this figure, are considerably higher in energy.

(i) For $k_a = \pm 1/4$ the three crystal orbitals cross. How can we understand this feature? As discussed in Chapter 3 (see Fig. 3.11), at this point it is possible to consider the nature of the Bloch orbitals on the basis of the transition-metal-based set of orbitals:

$$\{\sigma_{x^2-y^2}(\pm 1/4), \sigma_{xz}(\pm 1/4), \sigma_{yz}(\pm 1/4),$$
$$\delta_{x^2-y^2}(\pm 1/4), \delta_{xz}(\pm 1/4), \delta_{yz}(\pm 1/4)\}$$

and those of the bridging ligands. The advantage of using the basis:

$$\{\sigma_{x^2-y^2}(\pm 1/4), \sigma_{xz}(\pm 1/4), \sigma_{yz}(\pm 1/4),$$
$$\delta_{x^2-y^2}(\pm 1/4), \delta_{xz}(\pm 1/4), \delta_{yz}(\pm 1/4)\}$$

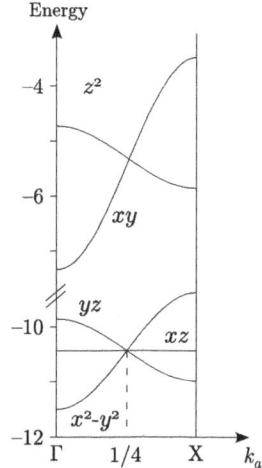

Fig. 44
Calculated band structure for the regular NbCl$_4$ chain.

is that every function is centred in every other niobium atom. Since the interactions between second transition metal neighbours is practically nil, there is no metal–metal interaction in any of these functions. In addition, as far as the interaction of the transition metal and ligands is concerned, the d_{xz}, d_{yz}, and $d_{x^2-y^2}$ orbitals interact in exactly the same way with the ligands because of (i) the octahedral nature of the coordination and (ii) the fact that in these functions the different octahedra are *effectively* isolated. Consequently, at $k_a = \pm 1/4$ the energy of the crystal orbitals is the same as that of the d_{xz}, d_{yz}, and $d_{x^2-y^2}$ orbitals of an octahedral NbX_6. This is why the three bands based on the d_{xz}, d_{yz}, and $d_{x^2-y^2}$ orbitals cross at this point.

(ii) Since every cell brings one electron to the t_{2g} bands, the equivalent of half a band must be filled. The occupation of these bands can be schematically represented as in the block diagram of Fig. 45 and consequently it is predicted that the chain will be metallic.

(iii) To build the band structure of the dimerised system, i.e. generated by a double cell, we can use the *folding* technique. The result is shown in Fig. 46.

(iv) When the dimerisation occurs gaps will open at the new border of the Brillouin zone (i.e. for $k'_a = 1/2$, which is labelled X' in Fig. 46). The gap will be larger as the interactions reinforced/diminished by the dimerisation become stronger. Consequently, since the band involving the $d_{x^2-y^2}$ orbitals is associated with σ metal–metal interactions, it will be this band that will be associated with the larger band gap. In contrast, since the band involving the d_{xz} orbitals is associated with δ metal–metal interactions, it will be this band that will be associated with the smaller band gap.

The band structures for a weak and strong dimerisation are shown in Fig. 47.

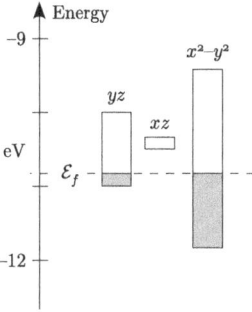

Fig. 45
Qualitative band structure for the regular NbCl$_4$ chain.

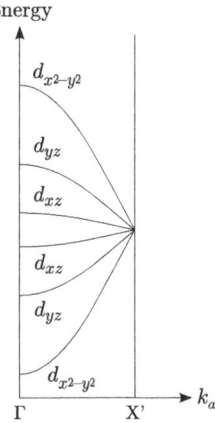

Fig. 46
Folded band structure for the dimerised system.

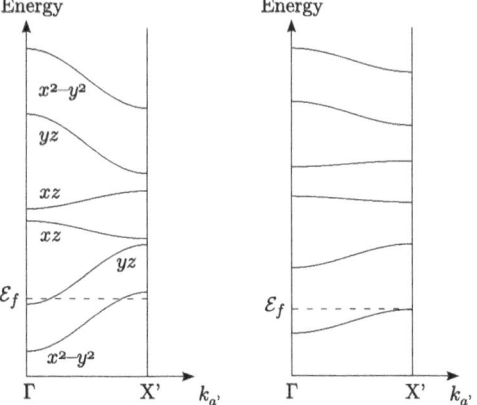

Fig. 47
Calculated band structure for a weak (left) and strong (right) dimerisation.

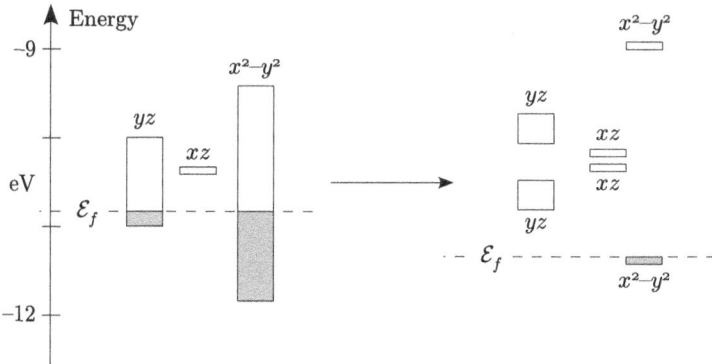

Fig. 48
Schematic band structure for a regular (left) and strongly dimerised (right) system.

These band structures clearly show that when weakly distorted the system retains its metallic behaviour. However, when the dimerisation becomes stronger the system behaves as a semiconductor. The band filling for the case of weak and strong dimerisations is schematically shown in Fig. 48.

(v) NbI$_4$ in its room-temperature and ambient-pressure form can be described by the right-hand part of Fig. 48, and exhibits a semiconducting character. When pressure is applied, the dimerisation is hindered and the description in the left-hand part of Fig. 48 is appropriate, thus predicting metallic behaviour.

Chapter 9

(9.1) The lattice vectors of a body-centred cubic lattice are:

$$\vec{a} = \frac{a}{2}(-\vec{u} + \vec{v} + \vec{w})$$

$$\vec{b} = \frac{a}{2}(\vec{u} - \vec{v} + \vec{w})$$

$$\vec{c} = \frac{a}{2}(\vec{u} + \vec{v} - \vec{w})$$

where a is the side of the conventional cube in the associated simple cubic lattice. The lattice vectors of the reciprocal lattice may be obtained by simple application of eqn (9.4), which leads to:

$$\vec{a}^* = \frac{2\pi}{a}(\vec{v} + \vec{w})$$

$$\vec{b}^* = \frac{2\pi}{a}(\vec{u} + \vec{w})$$

$$\vec{c}^* = \frac{2\pi}{a}(\vec{u} + \vec{v})$$

Thus, as shown in Figs 49a and b, the reciprocal lattice of a body-centred cubic lattice is a face-centred cubic lattice.

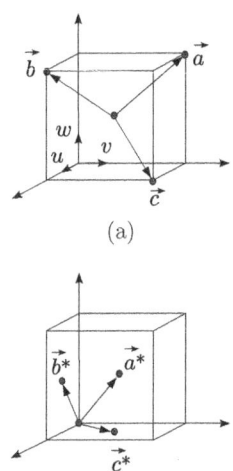

Fig. 49
Body-centred cubic lattice (a) and its reciprocal lattice (b).

(9.2) The lattice vectors of the face-centred cubic lattice that is the reciprocal lattice of a body-centred cubic lattice are given by:

$$\vec{a}^* = \frac{2\pi}{a}(\vec{v} + \vec{w})$$

$$\vec{b}^* = \frac{2\pi}{a}(\vec{u} + \vec{w})$$

$$\vec{c}^* = \frac{2\pi}{a}(\vec{u} + \vec{v})$$

The Brillouin zone may be obtained by drawing planes perpendicular to the 12 vectors to points near the origin: $\pi/a(\pm\vec{v} \pm \vec{w})$, $\pi/a(\pm\vec{u} \pm \vec{w})$, and $\pi/a(\pm\vec{u} \pm \vec{v})$. This procedure generates the rhombic dodecahedron shown in Fig. 50.

(9.3) Because of the orthogonal nature of the principal axes, the reciprocal lattice associated with a real-space tetragonal lattice is also tetragonal. Thus, the BZ is a parallelepiped with a square base. Looking at a general point of the BZ, it is clear that the fourth-order axis allows the reduction of the BZ to one-quarter of the full volume and the inversion symmetry associated with time reversal leads to a further reduction to one-eighth. In addition, the diagonal symmetry planes containing the fourth-order axis allow a further reduction to a volume of one-sixteenth of the initial BZ (see Fig. 51). Use of other symmetry elements does not allow us to further reduce the size of the zone and the triangular prism shown in Fig. 51 is then the irreducible part of the BZ.

(9.4) The hexagonal lattice possesses a six-fold symmetry axis, but the CH_4 molecule with a C–H bond pointing to the surface has only a three-fold symmetry axis in a direction perpendicular to the layer. Thus depending on the geometry of interaction, only certain symmetry elements of the hexagonal lattice will be kept. In case (a) the three-fold axis and the inversion associated with time reversal allow the reduction of the 2D BZ to one-sixth. In addition, the symmetry planes perpendicular to the hexagonal lattice containing the a, b and $-(a + b)$ axes allow a further reduction to a one-twelfth of the original BZ (see Fig. 52a). In case (b) no three-fold symmetry axis is kept. Only a plane containing the a axis is kept. Consequently, use of this plane and the inversion symmetry associated with time reversal allow the reduction of the Brillouin zone to one-quarter of the original area (see Fig. 52b). In case (c) one three-fold axis perpendicular to the hexagonal lattice as well as three symmetry planes

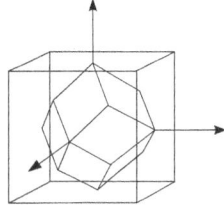

Fig. 50
Brillouin zone of the body-centred cubic lattice.

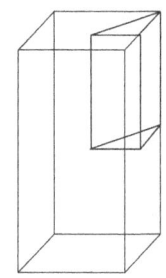

Fig. 51
Irreducible Brillouin zone for a tetragonal lattice.

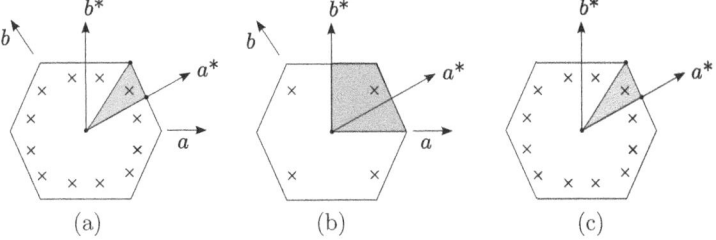

Fig. 52
Irreducible BZ for the three different models of interaction of a CH_4 molecule with a hexagonal lattice.

perpendicular to the lattice and containing the a^*, b^*, and $-(a^* + b^*)$ axes are also kept. Thus, use of these symmetry elements plus the inversion symmetry due to time reversal allows reduction of the full 2D BZ to one-twelfth (see Fig. 52c).

(9.5) The matrix elements for the 3×3 determinant needed to calculate the $E(\vec{k})$ vs $\vec{k}$ dependence of the three s, p_x, and p_y bands of a square lattice taking into account the s–p interaction can be easily calculated using eqn (9.14). Let us retain only the first nearest neighbours. Denoting the different β integrals as β_s and β_{s-p} for the σ-type interactions between two s orbitals and between one s and one p orbital in adjacent sites, and β_σ and β_π for the σ and π-type interactions between p orbitals in adjacent sites, it is possible to obtain:

$$h_{s,s}(\vec{k}) = \alpha_s + 2\beta_s(\cos(2\pi k_a) + \cos(2\pi k_b))$$

$$h_{p_x,p_x}(\vec{k}) = \alpha_p + 2\beta_\sigma \cos(2\pi k_a) + 2\beta_\pi \cos(2\pi k_b)$$

$$h_{p_y,p_y}(\vec{k}) = \alpha_p + 2\beta_\pi \cos(2\pi k_a) + 2\beta_\sigma \cos(2\pi k_b)$$

$$h_{s,p_x}(\vec{k}) = 2\beta_{s-p} \cos(2\pi k_a) = \left(h_{p_x,s}(\vec{k})\right)^*$$

$$h_{s,p_y}(\vec{k}) = 2\beta_{s-p} \cos(2\pi k_b) = \left(h_{p_y,s}(\vec{k})\right)^*$$

$$h_{p_x,p_y}(\vec{k}) = 0 = \left(h_{p_y,p_x}(\vec{k})\right)^*$$

Let us consider the additional terms needed to account for second-nearest neighbour interactions. Denoting as β'_s and β'_{s-p} the interactions between two s orbitals and between one s and one p orbital in second-nearest neighbour sites, β'_p the interaction between two p_x or two p_y orbitals in second-nearest neighbour sites, and β''_p the interaction between one p_x and one p_y orbitals in second-nearest neighbour sites. It can be easily deduced that the additional terms are:

$$4\beta'_s[\cos(2\pi k_a) \cdot \cos(2\pi k_b)] \quad \text{for } h_{s,s}(\vec{k})$$

$$4\beta'_p[\cos(2\pi k_a) \cdot \cos(2\pi k_b)] \quad \text{for } h_{p_x,p_x}(\vec{k}) \text{ and } h_{p_y,p_y}(\vec{k})$$

$$4\beta'_{s-p}[\cos(2\pi k_a) \cdot \cos(2\pi k_b)] \text{ for } h_{s,p_x} \text{ and } h_{s,p_y}$$

$$4\beta''_p[\cos(2\pi k_a) \cdot \cos(2\pi k_b)] \quad \text{for } h_{p_x,p_y}$$

(9.6) In a body-centred cubic lattice (see Fig. 49a) there are eight nearest neighbours located, with respect to the reference atom, at $(\pm a/2, \pm a/2, \pm a/2)$, where a is the side of the conventional cube. Consequently, the allowed energies are given by:

$$E(\vec{k}) = \alpha + \beta \left[(\exp(i\pi k_a) + \exp(-i\pi k_a)) \cdot \right.$$

$$\left. (\exp(i\pi k_b) + \exp(-i\pi k_b)) \cdot (\exp(i\pi k_c) + \exp(-i\pi k_c)) \right]$$

$$= \alpha + \beta[2\cos(\pi k_a) \cdot 2\cos(\pi k_b) \cdot 2\cos(\pi k_c)]$$

$$= \alpha + 8\beta[\cos(\pi k_a) \cdot \cos(\pi k_b) \cdot \cos(\pi k_c)]$$

In a face-centred cubic lattice (see Fig. 49b) there are twelve nearest neighbours located, with respect to the reference atom, at ($\pm a/2$, $\pm a/2$, 0), ($\pm a/2$, 0, $\pm a/2$), and (0, $\pm a/2$, $\pm a/2$), where a is the side of the conventional cube. Consequently, the allowed energies are given by:

$$E(\vec{k}) = \alpha +$$

$$\beta[(\exp(i\pi k_a) + \exp(-i\pi k_a)) \cdot (\exp(i\pi k_b) + \exp(-i\pi k_b))$$

$$+ (\exp(i\pi k_a) + \exp(-i\pi k_a)) \cdot (\exp(i\pi k_c) + (\exp(-i\pi k_c))$$

$$+ (\exp(i\pi k_b) + \exp(-i\pi k_b)) \cdot (\exp(i\pi k_c) + (\exp(-i\pi k_c))]$$

$$= \alpha + 4\beta[\cos(\pi k_a) \cdot \cos(\pi k_b)$$

$$+ \cos(\pi k_a) \cdot \cos(\pi k_c) + \cos(\pi k_b) \cdot \cos(\pi k_c)]$$

(9.7) Because of the equivalence of the three main directions, in a simple cubic lattice the energy equations giving the energy as a function of the $\vec{k}$ vector for the p_x, p_y, and p_z orbitals can be obtained from each other by cyclic permutation. Let us just consider the p_x orbital. Every site of the lattice has six first-nearest neighbours and twelve second-nearest neighbours. If we consider the first-nearest neighbour, two of them, those at ($\pm a$, 0, 0), undergo σ-type interactions whereas the remaining four, those at (0, $\pm b$, 0) and (0, 0, $\pm c$), undergo π-type interactions. We will denote these interactions β_σ and β_π. The contribution of first-nearest neighbours to $E(\vec{k})$ is thus:

$$E(\vec{k})_{fn} = \beta_\sigma (\exp(i2\pi k_a) + \exp(-i2\pi k_a))$$

$$+ \beta_\pi [(\exp(i2\pi k_b) + \exp(-i2\pi k_b))$$

$$+ (\exp(i2\pi k_c) + \exp(-i2\pi k_c))]$$

$$= 2\beta_\sigma \cos(2\pi k_a) + 2\beta_\pi \{(\cos(2\pi k_b)$$

$$+ \cos(2\pi k_c)\}$$

We can now consider the interactions with the twelve second-nearest neighbours. There are two different types of interactions: those associated with the p_x orbitals at ($\pm a$, $\pm b$, 0) and ($\pm a$, 0, $\pm c$) and those with the p_x orbitals at (0, $\pm b$, $\pm c$). We will denote these interactions as β_{sn} and β'_{sn}, respectively. The contribution of second-nearest neighbours to $E(\vec{k})$ is thus:

$$E(\vec{k})_{sn} = \beta_{sn}[(\exp(2i\pi k_a) + \exp(-2i\pi k_a)) \cdot (\exp(2i\pi k_b) + \exp(-2i\pi k_b))$$

$$+ (\exp(2i\pi k_a) + \exp(-2i\pi k_a)) \cdot (\exp(2i\pi k_c) + \exp(-2i\pi k_c))]$$

$$+ \beta'_{sn}((\exp(2i\pi k_b) + \exp(-2i\pi k_b)) \cdot (\exp(2i\pi k_c) + \exp(-2i\pi k_c)))$$

$$E(\vec{k})_{sn} = 4\beta_{sn}[\cos(2\pi k_a) \cdot \cos(2\pi k_b) + \cos(2\pi k_a) \cdot \cos(2\pi k_c)]$$

$$+ 4\beta'_{sn}[\cos(2\pi k_b) \cdot \cos(2\pi k_c)]$$

The final expression is thus:

$$E(\vec{k}) = \alpha + E(\vec{k})_{fn} + E(\vec{k})_{sn}$$

Solutions for exercises

Chapter 10

(10.1) If we write a molecular orbital for a polyatomic molecule as a linear combination of N_0 atomic orbitals $\{\chi_j, j = 1, \ldots, N_0\}$:

$$\phi_\mu = \sum_{j=1}^{N_0} c_{j\mu} \chi_j$$

the normalisation condition is such that:

$$1 = \sum_{j=1}^{N_0} |c_{j\mu}|^2 + \sum_{j=1}^{N_0} \sum_{j' \neq j}^{N_0} \text{Re}(c_{j\mu}^* c_{j'\mu}) S_{jj'}$$

Every contribution to the first term of this equation is the probability that the electron described by ϕ_μ is found on the atomic orbital χ_j, and is thus the *orbital population* of χ_j in molecular orbital ϕ_μ. The different contributions to the second term can be interpreted as being the probability that the same electron is shared by the atomic orbitals χ_j and χ'_j and is thus an *overlap population*. If the contributions for every pair of atomic orbitals are assumed to be equally shared by the two atomic orbitals (Mulliken population analysis), the previous equation can be written as:

$$1 = \sum_{j=1}^{N_0}(|c_{j\mu}|^2 + \sum_{j' \neq j}^{N_0} \text{Re}(c_{j\mu}^* c_{j'\mu} S_{jj'})) = \sum_{j=1}^{N_0} P_{j\mu}$$

and $P_{j\mu}$ is the *gross population* of atomic orbital χ_j in molecular orbital ϕ_μ. If these quantities are multiplied by a number giving the occupation of the molecular orbital ($n_\mu = 0, 1$, or 2), the gross population of atom A (P^A) is simply the addition of these quantities over all molecular orbitals and all atomic orbitals of atom A:

$$P^A = \sum_j^A \sum_\mu n_\mu P_{j\mu}$$

The same analysis can be developed for a periodic system. In that case, the crystal orbitals may be written as a linear combination of the Bloch orbitals associated with every atomic orbital of the repeat unit of the solid, in our case a 1D system:

$$\text{CO}_\mu(\vec{k}) = \sum_j c_{j\mu}(\vec{k}) \text{BO}_j(\vec{k}) = \frac{1}{\sqrt{n}} \sum_j \sum_\ell c_{j\mu}(\vec{k}) \exp(i 2\pi \ell k_a)(\chi_j)_\ell$$

where $(\chi_j)_\ell$ is the atomic orbital χ_j in cell ℓ. The normalisation condition is such that:

$$1 = \frac{1}{n} \sum_j \sum_\ell \sum_{j'} \sum_{\ell'} c_{j\mu}^*(\vec{k}) c_{j'\mu}(\vec{k}) \exp(i 2\pi k_a(\ell' - \ell)) S_{j\ell j'\ell'}$$

and the same population analysis leads to:

$$1 = \frac{1}{n} \sum_{\ell} \sum_{j} \Big\{ |c_{j\mu}(\vec{k})|^2$$

$$+ \sum_{\substack{j',\ell' \\ (j',\ell') \neq (j,\ell)}} \mathrm{Re}(c^*_{j\mu}(\vec{k}) c_{j'\mu}(\vec{k})\exp(i 2\pi k_a (\ell' - \ell))S_{j\ell j'\ell'} \Big\}$$

$$= \frac{1}{n} \sum_{\ell} \sum_{j} (P_{j\ell})_\mu(\vec{k})$$

where $(P_{j\ell})_\mu(\vec{k})$ is the gross population of atomic orbital χ_j in cell ℓ in crystal orbital CO_μ. If these quantities are multiplied by a number giving the occupation of the crystal orbital n_μ, the gross population of atom A, $P^A(\vec{k})$, is simply the addition of these quantities over all cells, crystal orbitals and atomic orbitals of atom A. However, it must be remembered that these gross populations are associated with a k point of the BZ. Thus to obtain the gross population for atom A, an integration over all the BZ must be carried out. In practice we can use a fine grid of k points, evaluate the associated $P^A(\vec{k})$ values, and multiply them by the normalised weight of the k point.

(10.2) The band structure for a uniform NbX_4 chain is discussed in Exercise 8.12. Since for X = Cl, Br, I, which are the cases of interest, the electron counting for the transition metal is d^1, only the lower lying t_{2g} bands need to be considered. The band structure is made up of a $x^2 - y^2$ band, which rises in energy from Γ to X, a yz band with the opposite slope, and a xz band which is essentially flat. The three bands cross at the $k_a = \pm 1/4$ point. Since both the $x^2 - y^2$ and yz bands have transition-metal–ligand antibonding interactions and the first contains metal–metal σ interactions whereas the second contains metal–metal δ interactions, the $x^2 - y^2$ is wider in energy. The details have been discussed in Exercise (8.12). With all these features in mind we can sketch the schematic DOS shown in Fig. 53. Since the transition metal atom is d^1 there is only one electron to fill the t_{2g} levels. Consequently, if the lower-lying $x^2 - y^2$ band was able to keep this electron it would be half-filled or, in other words, the contribution of this band to the DOS would be filled up to the middle. However, this is not possible because before this can be accomplished, the bottom part of the yz contribution starts to be filled. This means that before all levels containing σ-bonding metal–metal interactions, i.e. the lower half of the $x^2 - y^2$ band, can be filled, levels with neither metal–metal nor metal–bridging ligand interactions, i.e. the bottom part of the yz band (but remember that there are antibonding metal–non-bridging-ligand interactions), are filled. Consequently this transfer from the $x^2 - y^2$ to the yz levels results in a decrease in the metal–metal bonding interactions but a decrease in the metal–bridging-ligand antibonding interactions. Under a dimerisation, half of the metal–metal distances will shorten while the

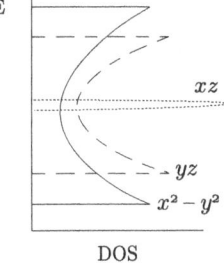

Fig. 53
Schematic DOS diagram showing the partial contributions of the transition metal t_{2g} orbitals for the uniform NbX_4 chain in the region of the t_{2g}-block bands.

metal–bridging ligand distances will be kept approximately constant. Thus, the bottom of the $x^2 - y^2$ band will be lowered while the top will be raised, i.e. the band will thus become wider and the $x^2 - y^2$ levels will progressively disappear from the central region of the DOS diagram and concentrate at the bottom and top of the diagram. By doing this, the direct metal–metal bonding interactions will become more stabilising at the price of adding some metal–bridging-ligand antibonding interaction. However, the increase in the metal–metal interaction will usually dominate, since it is associated with a noticeable decrease of the direct metal–metal distance whereas the metal–ligand distances vary comparatively less. When the metal–metal direct interaction strongly dominates over the metal-bridging-ligand ones, the system will evolve so as to optimise the metal–metal interactions, the lower $x^2 - y^2$ contribution will be considerably lowered and a band gap at the top of this band will appear. The system will then become a semiconductor. If the metal–metal interactions do not clearly dominate over the metal–bridging-ligand interactions, the driving force for the dimerisation will be smaller and most probably, although the lower $x^2 - y^2$ contribution will be lowered with respect to the yz contribution, some overlap still will be kept so that metallic behaviour will be retained.

(10.3) The electronic structure of many of these interesting phases has been discussed by R. Hoffmann and co-workers. A discussion of the electronic basis for the X–X distance variation can be found in ref. [16] of Chapter 10, and the reader is encouraged to read this publication to acquire some practice in the use of the DOS to gain a qualitative understanding of structural features in solids. However, it is always useful to try to anticipate the results of the calculations. Taking $BaMn_2P_2$ as a representative example of this family of phases, we can try to elaborate some kind of qualitative reasoning, which we can check with a simple calculation.

The $Mn_2P_2^{2-}$ layers in this phase can be seen as being built from the condensation of a series of square-planar pyramids with Mn atoms in the base and P atoms at the top. The important structural features to concentrate on are the following: (i) the Mn atoms are in a tetrahedral coordination so that the typical splitting of two orbitals below three is expected, (ii) the Mn–Mn distances are relatively short meaning there must be extended Mn–Mn interactions at work, and (iii) the P atoms being at the top of the pyramid, it is expected that the P p orbital perpendicular to the layer will mostly remain as a kind of non-bonding lone pair whereas the remaining orbitals will lead to the Mn–P bonding interactions. Consequently, we can sketch the DOS of this material as containing, in increasing order of energy: (i) a mostly phosphorous s contribution, (ii) a contribution dominated by the phosphorous p orbitals containing the metal–phosphorous bonding levels, at the top of which there will be the above mentioned mostly lone pair levels, (iii) a mostly transition-metal d contribution, which will contain the e_g below the t_{2g} contributions. Since there are relatively short metal–metal distances, these two contributions will be considerably spread

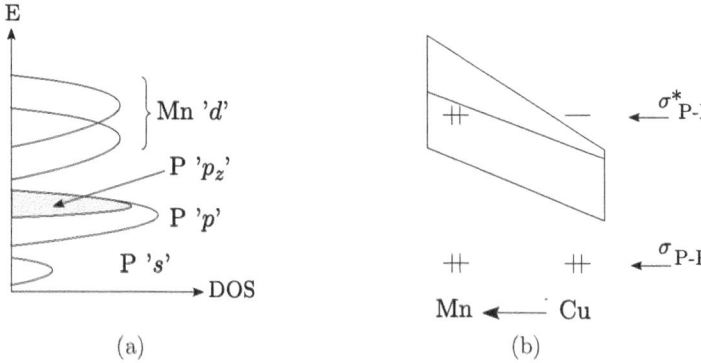

Fig. 54
Schematic diagrams for (a) the DOS of an $Mn_2P_2^{2-}$ layer and (b) the relative position of the broad transition-metal and narrow σ P–P and σ^* P–P contributions as a function of the nature of the transition metal atom in the 3D structure.

out and thus they will most likely appear as a single broad contribution. The Fermi level for all phases of this family will be found within this contribution. Such a schematic diagram is shown in Fig. 54a. At the highest energy will be the Mn–P antibonding levels.

We can now try to understand the question of the existence or otherwise of interlayer P–P bonds. Let us assume that the two layers are put together at a typical distance for a *moderate* P–P single bond (i.e. something like 2.45 Å). In that case the so-called lone-pair phosphorous levels of the two layers will interact and split into σ P–P bonding and σ^* P–P antibonding levels. These two types of level will stay relatively narrow in the solid because they are too far from each other. When the P–P interlayer distance increases, the splitting between the σ and σ^* levels will decrease and eventually, for large separations, they will be found as a single contribution at the top of the phosphorous contribution. Fig. 55 shows the results for a model calculation using the geometry of a typical AB_2X_2 phase, such as $BaMn_2P_2$, using short (2.45 Å) and long (3.79 Å) interlayer P–P distances. When projecting out of the DOS (dotted curve, the phosphorous p orbitals perpendicular to the layers (continuous line), the σ and σ^* P–P levels behave as expected and for long P–P distances appear below the transition-metal

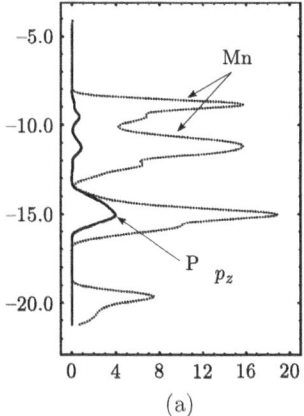
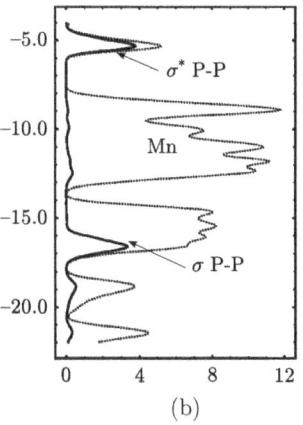

Fig. 55
Total DOS (dotted curve) and contribution of the phosphorus orbital perpendicular to the layers (i.e. p_z, continuous curve) for $Mn_4P_4^{4-}$ using the experimental structure for $BaMn_2P_2$ and interlayer P–P distances of 3.79 Å (experimental value) (a) and 2.45 Å (a typical P–P single bond distance) (b).

d contribution. A schematic diagram including the σ and σ^* P–P levels and some representation of the broad transition-metal contribution to the DOS can now be drawn. Essentially, as we move from left to right in the transition metal series of the periodic table, because of the increase of electronegativity and contraction of the orbitals we can expect that this broad transition-metal contribution will become lower in energy and less wide. Now we can superpose the two contributions, and we will get something like the schematic diagram of Fig. 54b. At the right of the diagram, for the more electronegative transition metal atoms, the antibonding σ^* levels will most certainly be empty, whereas for considerably less electronegative transition metal atoms these levels will be filled, so that there will not be a net P–P bond. Here the structure will tend to evolve to increase the interlayer distance and this will decrease the splitting between the σ and σ^* levels, keeping the occupation of the antibonding P–P levels. In contrast, for the transition metal atoms at the right, only the σ P–P bonding levels will be filled, the interlayer spacing will tend to decrease, but this will only additionally stabilise these bonding levels and destabilise the antibonding ones. Of course, in the solid there is a gradual change because of the more delocalised actual interactions, which modulate the filling of the σ^* P–P antibonding levels. Actual calculations (DOS and COOP curves) will certainly give detailed information. However, simplistic as it is, this argument clearly provides a rationale for the experimental observation that X–X bonding is increasingly favoured as we move towards the right of the transition metal series.

(10.4) The repeat unit of this 1D system contains two hydrogen atoms, so the analytical equations giving the energy as a function of $\vec{k}$ for the two bands are the solutions of the 2×2 secular determinant:

$$\begin{vmatrix} h_{11}(\vec{k}) - E(\vec{k}) & h_{12}(\vec{k}) \\ h_{21}(\vec{k}) & h_{22}(\vec{k}) - E(\vec{k}) \end{vmatrix} = 0$$

where the different matrix elements are given by:

$$h_{11}(\vec{k}) = \alpha + 2\beta \cos(2\pi k_a) = h_{22}(\vec{k})$$

$$h_{12}(\vec{k}) = k\beta = \left(h_{21}(\vec{k})\right)^*$$

Solution of the associated second-order equation leads to the equation:

$$E(\vec{k}) = \alpha + 2\beta \cos(2\pi k_a) \pm k\beta$$

Another way to deduce this simple equation would be to consider an infinite chain of hydrogen dimers. In this case the matrix elements would be:

$$h_{11}(\vec{k}) = (\alpha + k\beta) + 2\beta \cos(2\pi k_a)$$

$$h_{22}(\vec{k}) = (\alpha - k\beta) + 2\beta \cos(2\pi k_a)$$

$$h_{12}(\vec{k}) = 0 = \left(h_{21}(\vec{k})\right)^*$$

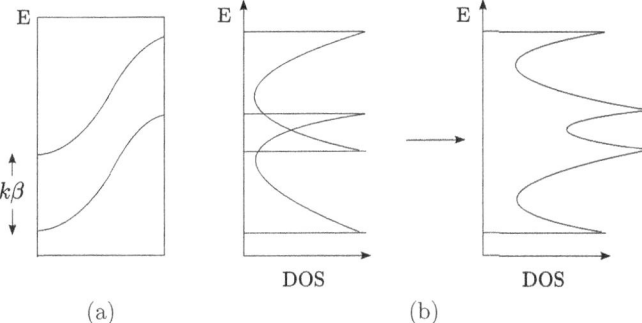

Fig. 56
Schematic band structure (a) and density of states (b) for a ladder of hydrogen atoms.

where the labels 1 and 2 refer to the bonding and antibonding levels of the dimer, obviously leading to the same solution. What this equation is telling us is that, at this level of approximation, the band structure for the ladder is simply made up of two typical 1D bands separated by an energy $k\beta$, i.e. for every value of $\vec{k}$, a bonding and antibonding combination of the two crystal orbitals separated by an energy difference equal to the interaction energy is created (see Fig. 56a). Consequently the DOS is just the superposition of typical 1D contributions (Fig. 56b).

Chapter 11

(11.1) The allowed energies for a square lattice system with only nearest-neighbour interactions are given by (see eqn (9.22))

$$E(\vec{k}) = \alpha + 2\beta(\cos(2\pi k_a) + \cos(2\pi k_b))$$

so that the upper and lower energy values of the band are given by $\alpha \pm 4\beta$. The Fermi level for a half-filled band is associated with the energy value $E(\vec{k}) = \alpha$. According to the previous equation, this will occur when $2\beta(\cos(2\pi k_a) + \cos(2\pi k_b)) = 0$ and consequently, when $\cos(2\pi k_a) = -\cos(2\pi k_b)$ which is the case when $k_b = 1/2 - k_a$. Consequently, as shown in Fig. 57, the Fermi surface for such system is a square with vertices at the X and equivalent points. When non-nearest-neighbour interactions are included, the Fermi surface becomes more rounded.

(11.2) When the hump octahedra are added to the Mo_4O_{18} chain in a zigzag way, the Mo_6O_{24} chain of Fig. 58a is generated. When these chains condense through edge-sharing in the way discussed in the text for $Tl_{0.33}MoO_3$, the MoO_3 layer of Fig. 58b is created (compare with the MoO_3 layer of Fig. 11.25d). In contrast with that of Fig. 11.25b, the chain of Fig. 58a possesses a twofold screw rotation axis along the chain direction. Simply on symmetry grounds (reminding ourselves of the discussion in Section 6.4) it can thus be predicted that for a hypothetical $A_{0.33}MoO_3$ (A = alkali metal atom) all bands must pair up at the Y = $(b^*/2, 0)$ point of the 2D Brillouin zone. Since such a system

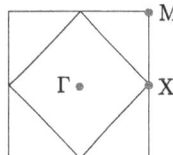

Fig. 57
Fermi surface for a square lattice with only nearest-neighbour interactions and one electron per site.

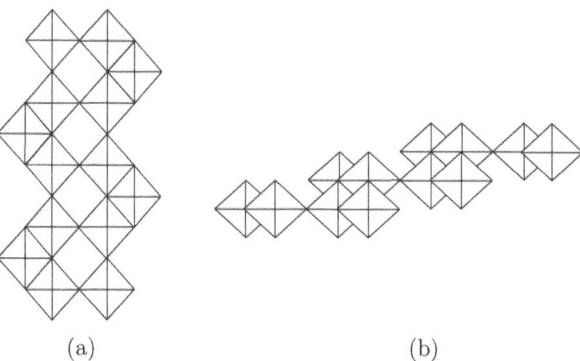

Fig. 58
Schematic representation of (a) the Mo_6O_{24} chains with zigzag hump octahedra and (b) the MoO_3 layer generated from these chains through edge sharing.

has only two electrons to fill the t_{2g}-block bands and consequently only the lowest band can be filled, we are led to the conclusion that in contrast with the situation for the red bronzes discussed in the text, where the hump octahedra were added in a parallel way, the system should now be metallic.

Note that the same conclusion could have been reached simply by considering the electronic structure of the ideal Mo_4O_{18} chain (see Figs 11.28 and 11.29). A consequence of the zigzag way in which the hump octahedra are found is that bands a and b at Y (see Fig. 11.29b and c, respectively) are completely equivalent and thus degenerate.

(11.3) Before looking at the qualitative band structure of the $Mo_{10}O_{30}^{3-}$ layer it is useful to notice that the pseudo-hexagonal Brillouin zone associated with the non-orthogonal vectors $\vec{b}'$ (i.e. $\vec{b}$) and $\vec{d}'$ can be rewritten in a rectangular representation as shown in Fig. 59. Note moreover that X′ (=$(d'^*/2, 0)$) and X (=$(d^*/2, 0)$) coincide but Y′ (=$(0, b'^*/2)$) and Y (=$(0, b^*/2)$) do not, i.e. $\vec{b}^* = \vec{b}'^* \sin\theta$, where θ is the angle between $\vec{b}'$ and $\vec{d}'$. Thus, although computationally it is better to use the cell defined by $\vec{b}'$ and $\vec{d}'$, for a qualitative treatment it is better to focus on the Γ, Y, and X points. Otherwise, when looking at the Γ → Y′ direction we also pick up part of the interchain interactions.

As shown in Fig. 11.37, the hump octahedra have two strong O...Mo–O alternations, so the t_{2g} levels will be high in energy and will not contribute to the bottom of the d-block bands. All other octahedra have one strong O...Mo–O alternation perpendicular to the b-direction. Consequently, only the xz orbital of the non-hump octahedra (i.e. those of the Mo_8O_{34} chain) will remain low in energy and will lead to the low-lying t_{2g}-block bands of the $Mo_{10}O_{30}$ layer. The cluster orbitals needed to build the xz bands of an ideal Mo_8O_{34} chain are those of Fig. 60. Just as in the case of the 2D red bronzes, their energy ordering can be obtained by counting the number of dots (i.e. (N)-type interactions). The result is shown in Fig. 61. The crystal orbitals at Γ and Y generated by cluster orbital a have the nodal patterns shown in Fig. 62a and b, respectively. The numbers of (N)-type interactions per unit cell in these crystal orbitals are 14 and 10, respectively. In other words, the number of (N)-type interactions in the crystal orbitals is the

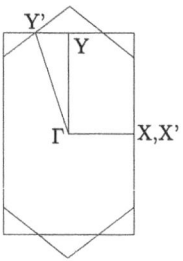

Fig. 59
Schematic representation showing that the pseudo-hexagonal Brillouin zone appropriate for the non-orthogonal vectors $\vec{b}'$ (i.e. $\vec{b}$) and $\vec{d}'$ can be rewritten in a rectangular representation that is more useful for a qualitative analysis.

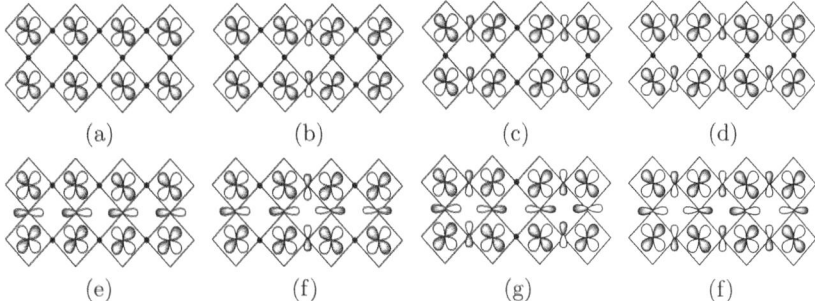

Fig. 60
(a)–(h) Cluster orbitals needed to construct the xz bands of the Mo_8O_{34} ideal chain.

same as in the initial cluster orbital at Y but increases by four at Γ. This stems from the antisymmetric character of the cluster orbital a with respect to the horizontal symmetry plane. Consequently, exactly the same is expected for crystal orbitals obtained from the cluster orbitals (b)–(d). The cluster orbitals (e)–(h) are symmetric with respect to the horizontal symmetry plane, so in the crystal orbitals derived from them, the number of (N)-type interactions is the same as that of the initial cluster orbital at Γ but increases by four at Y. Simple application of these counting rules leads to the qualitative band structure of Fig. 63.

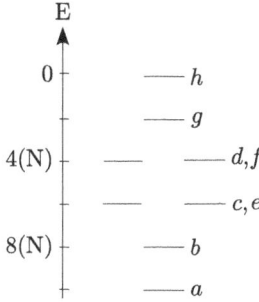

Fig. 61
Energy ordering of the different cluster orbitals.

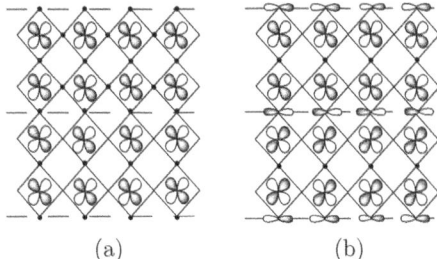

Fig. 62
Crystal orbitals generated by cluster orbital a at Γ (a) and Y (b).

As shown for the 2D red bronzes, the effect of the interchain interactions can be predicted by examining the phase relation between the xz orbitals of adjacent chains at the hump level in the crystal orbitals at Γ. If the orbital patterns around the hump octahedra are those of Fig. 64a and b, the effects of the interchain interactions are described by Fig. 11.34a and b, respectively. When viewed along the chain, the crystal orbitals a, b, and c at Γ in Fig. 63 can be drawn as in Fig. 65a, b, and c, respectively. Consequently, band a will behave as in Fig. 11.34a, but bands b and c will behave as in Fig. 11.34b. Thus, the lower part of the qualitative t_{2g} band structure of the ideal $Mo_{10}O_{30}$ layer (solid lines) is related to that of the ideal Mo_8O_{34} chain as schematically shown in Fig. 66. The agreement between the estimated (Fig. 66) and calculated (Fig. 11.38) band structures is excellent once the avoided nature of some crossings in the real calculation is taken into account, and one is reminded that the intrachain direction in the qualitative and

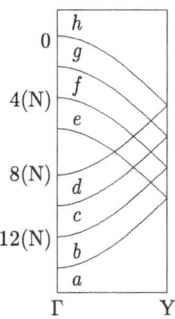

Fig. 63
xz bands of the ideal Mo_8O_{34} chain.

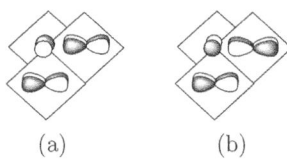

Fig. 64
(a)–(b) Orbital patterns around the hump octahedra in the crystal orbitals of the $Mo_{10}O_{30}$ layer.

calculated band structures are not exactly the same. For instance, note that the calculated crystal orbitals are not degenerate at Y′ because the screw axis symmetry is only a valid symmetry element along the $\Gamma \to Y$ direction.

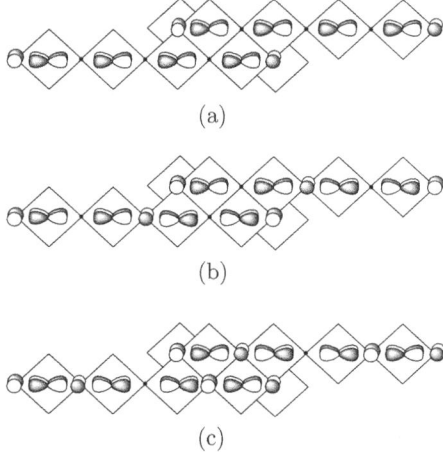

Fig. 65
(a)-(c) Top view of the crystal orbitals a, b, and c of the $Mo_{10}O_{30}$ layer at Γ.

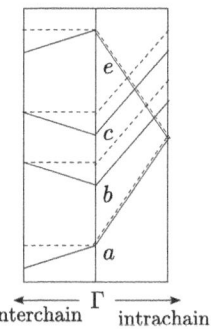

Fig. 66
Comparison of the bottom portion (bands a, b, c, and e) of the band structure for the ideal $Mo_{10}O_{30}$ layer (solid lines) with those of the Mo_8O_{34} chain (dashed lines).

Fig. 67
xy-type crystal orbitals at the Γ (a), X (b), and M (c) points for the VO_4 layer.

Fig. 68
xz-type crystal orbitals at the Γ (a), X (b), and M (c) points for the VO_4 layer.

(11.4) Using the oxidation states O^{2-} and Sr^{2+} it is clear that the vanadium atoms in this system are formally d^1. Consequently, we must only consider the t_{2g}-block bands. Let us first consider the band structure for a single octahedral VO_4 layer. The crystal orbitals for the xy and xz based bands at the Γ, X and M points are easy to draw and are shown in Fig. 67 and 68. Note that all octahedra have two axial oxygen contributions not drawn for clarity in Fig. 68 (the orbitals for the yz orbital are completely equivalent). Using the counting rules developed in Section 11.4.2 it is possible to draw the qualitative band structure of Fig. 69.

Construction of the qualitative t_{2g}-block band structure for the double octahedral V_2O_7 layers is easy. Fig. 70 shows the orbitals of every V–O_{ax}–V linkage (here the orbitals of the non-bridging oxygen orbitals

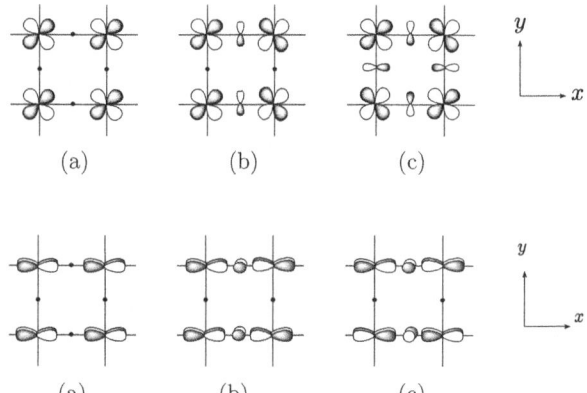

are not shown for simplicity). The two combinations based on the xy orbitals cannot interact with the p orbitals of the O_{ax} atom because of the local δ symmetry along the V–O_{ax}–V axis and, consequently, the two combinations are practically degenerate. The number of oxygen p-orbital contributions per unit cell of the associated double octahedral layer bands are thus identical, and just twice those of the single octahedral layer. The xz_+ (and yz_+) combination of Fig. 70 is lower in energy than the xz_- (and yz_-) because it does not have p-orbital contributions at the bridging axial position. The number of oxygen p-orbital contributions per unit cell of the double octahedral layer to the xz_+ and yz_+ based bands is twice that of the single octahedral layer minus two axial contributions (i.e. 2y = Y/2). Those of the xz_- and yz_- based bands are those of the xz_+ and yz_+ based bands plus one bridging contribution (Y) coming from the V–O_{ax}–V bridging position. Consequently, the qualitative band structure of Fig. 71 can be constructed.

(11.5) The analytical expression giving energy as a function of $\vec{k}$ for the two HOMO bands of the (TMTSF)$_2$X salts is:

$$E(\vec{k}) = \alpha + 2[t_{I_3}\cos(2\pi k_b) + t_{I_4}\cos(2\pi(k_a - k_b))]$$
$$\pm \mid t_{S_2} + t_{S_1}\exp(2i\pi k_a) + t_{I_2}\exp(2i\pi k_b)$$
$$+ t_{I_1}\exp(2i\pi(k_a - k_b)) \mid$$

where the different transfer integrals are those defined in Fig. 11.50c. Taking into account the numerical values for these integrals, which are those appropriate for the PF$_6$ salt at room temperature, it is clear that: (i) the intrastack S_1 and S_2 interactions are both approximately one order of magnitude larger than all others, and (ii) the interstack interactions I_1 and I_4 are considerably smaller than the other interstack interactions. Consequently, a possible approximation to the above equation could be obtained by defining average intra- and interstack transfer integrals as $t_S = 1/2(t_{S_1} + t_{S_2})$ and $t_I = t_{I_3} + \frac{t_{I_2}}{2}$ (reminding ourselves that every TMTSF molecule is involved in two I_3-type interactions but one I_2-type interaction). Thus, the above equation may be approximated by:

$$E(\vec{k}) = \alpha + 2[t_I\cos(2\pi k_b) \pm t_S\cos\frac{1}{2}(2\pi k_a)]$$

Using this equation the band structure can thus be obtained and analysed (it is convenient to take α as the energy zero). Shown in Fig. 72a and b are schematic band structures according to the approximate and full analytical equations. The dependency of several features of these band structures on the different transfer integrals are also shown. Note for instance that use of the approximate equation does not lead to the opening of a gap at X because use of this equation assumes that the stacks are uniform. However, when the full equation is used a gap is created and is given by $|2(t_{S_2} - t_{S_1}) + 2(t_{I_2} - t_{I_1})|$. Consequently, a gap opens up even without the interstack interactions – solely on the basis of the existence of two different intrastack interactions (i.e.

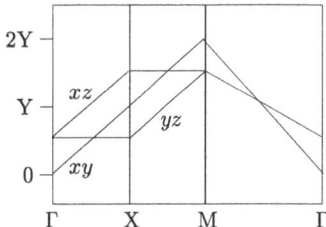

Fig. 69
Qualitative t_{2g}-block band structure for the VO$_4$ layer.

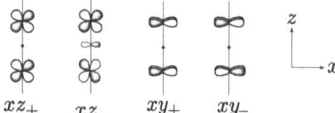

Fig. 70
xy and xz orbital patterns for the V–O_{ax}–V linkage of the double octahedral layer.

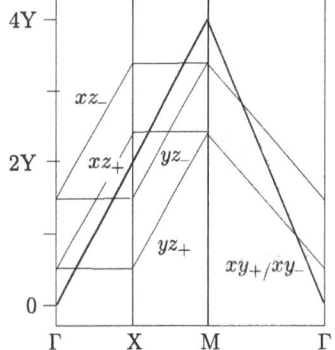

Fig. 71
Qualitative t_{2g}-block band structure for the V$_2$O$_7$ layer.

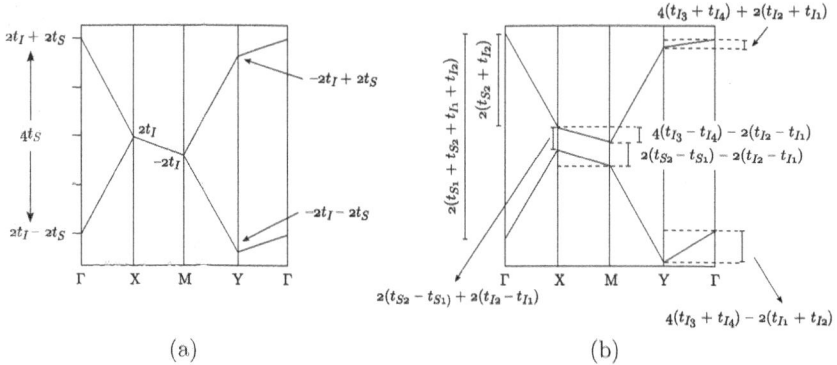

Fig. 72
Schematic band structure for the (TMTSF)$_2$X according to the approximate (a) and full (b) analytical equations.

a dimerisation gap) – but the interstack interactions also contribute to the gap. However, not all interchain interactions contribute. Depending on the values of the transfer integrals, the inter- and intrastack contributions could even be comparable. In addition, the total band width along a^* is given simply by $4t_S$ according to the simple treatment, but by $2(t_{S_2} + t_{S_1} + t_{I_2} + t_{I_1})$ according to the full treatment, i.e. both inter- and intrastack interactions participate, although the value is strongly dominated by the intrastack interactions. Note also that the band width of the upper of the two bands is given by $2(t_{S_2} + t_{I_2})$, which means that the width of upper HOMO band, which is half-filled and consequently is the band determining the transport properties, is dominated by the smaller of the intrastack interactions. The simplest way to look at the anisotropy of the system is to consider the ratio between the previously defined average t_S and t_I transfer integrals. Using the values for the PF$_6$ salt at room temperature, which are representative of the whole family of salts, it is clear that t_S is one order of magnitude larger. Since t_I is small but non-negligible it is clear that these salts are pseudo-1D systems. We can look in more detail at this question by using the analytical expressions to evaluate the energy differences associated with the X (1/2,0) and M (1/2,1/2) (or the Γ (0,0) and Y (0,1/2)) points. According to the approximate equation, at both the X and M points the two bands are degenerate and the corresponding energies are $+2t_I$ and $-2t_I$. Consequently, the dispersion along the X → M direction, which is a measure of the interstack interactions, is $4t_I$, which according to the values for the PF$_6$ salt is one order of magnitude smaller than the corresponding value for the Γ → X direction. If we are interested in more precise values for the dispersion along the interstack direction we can use the full equation. In that case the energy difference between the energy of the upper band at the X and M points is $4(t_{I_3} - t_{I_4}) - 2(t_{I_2} - t_{I_1})$. Using the values for PF$_6$ it is clear that the band goes down from X and M. The energy difference between the two bands at the M point is $2(t_{S_2} - t_{S_1}) - 2(t_{I_2} - t_{I_1})$. Since at X this energy difference was $2(t_{S_2} - t_{S_1}) + 2(t_{I_2} - t_{I_1})$, due to the negative value of the t_{I_1} and t_{I_2} transfer integrals and the larger strength of the latter, the gap decreases from X to M, i.e. the two bands have different dispersions. According to these results, the existence of a definite band gap separating the two bands depends essentially of the difference in energy of the upper band at M and the lower band at X. If $2(t_{S_2} - t_{S_1}) + 2(t_{I_2} - t_{I_1})$ is larger than $4(t_{I_3} - t_{I_4}) - 2(t_{I_2} - t_{I_1})$ there is a band gap. This is very important in discussing the physics of these salts since it indicates if the system should be described as a half-filled dimerised system or as a three-quarters filled system. In view of the typical values of the transfer integrals for this family, the existence or otherwise of this band gap is thus very sensitive to the value of the different interactions.

(11.6) The donor layers of the θ phases usually contain two donors per repeat unit, so the analytical expressions giving the energy as a function of $\vec{k}$ for the two HOMO bands are the solutions of the 2×2 secular determinant:

$$\begin{vmatrix} h_{11}(\vec{k}) - E(\vec{k}) & h_{12}(\vec{k}) \\ h_{21}(\vec{k}) & h_{22}(\vec{k}) - E(\vec{k}) \end{vmatrix}$$

where the different matrix elements are given by:

$$h_{11}(\vec{k}) = \alpha + 2t_c \cos(2\pi k_c) = h_{22}(\vec{k})$$

$$h_{12}(\vec{k}) = t_p(1 + \exp(i2\pi k_c) + \exp(i2\pi(k_c + k_a)) + \exp(i2\pi k_a))$$

$$= \left(h_{21}(\vec{k})\right)^*$$

Solution of the associated second-order equation leads to the equation:

$$E(\vec{k}) = \alpha + 2t_c \cos(2\pi k_c) \pm 4t_p \cos(\pi k_a) \cos(\pi k_c)$$

With this equation we can now draw a qualitative band structure. Taking α as the energy zero, the energies of the two bands at the Γ, X, Z, and M points are given by $E(\Gamma) = 2t_c \pm 4t_p$, $E(X) = 2t_c$, $E(Z) = -2t_c$ and $E(M) = -2t_c$. Typical values of the transfer integrals are 100 meV for t_p and from 0 to 40 meV for t_c, so that the qualitative band structure can be drawn (Fig. 73). Note that the total band width of these phases, as far as the t_c transfer integrals are not very large, is determined by the energy difference between the two bands at Γ and only depends of the t_p transfer integral.

(11.7) The donor layer of Fig. 11.64 is the same as that of Fig. 11.55, for which eqn (11.6) applies, depending on three different transfer integrals (t_{sc}, $t_{\pi c}$, and $t_{\pi c}$). The new cell is four times larger, which means that the BZ is four times smaller. Since the cell in real space doubles along the two main directions, the different $\vec{k}'$ of the new BZ are given by $\vec{k}' = k_{a'}\vec{a}'^* + k_{b'}\vec{b}'^*$, where $\vec{a}'^* = 1/2\vec{a}^*$ and $\vec{b}'^* = 1/2\vec{b}^*$ so that $k_{a'} = 2k_a$ and $k_{b'} = 2k_b$. The boundary conditions, $1/2 \geq k_{a'}, k_{b'} \geq -1/2$, lead to a translation of states with $|k_a| \geq 1/4$, $|k_b| \geq 1/4$ and $|k_a, k_b| \geq$

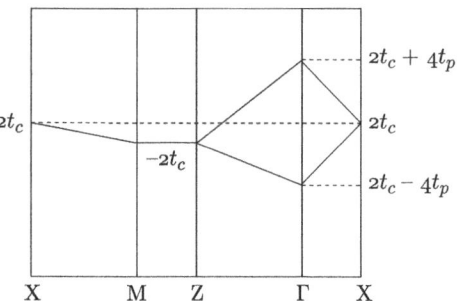

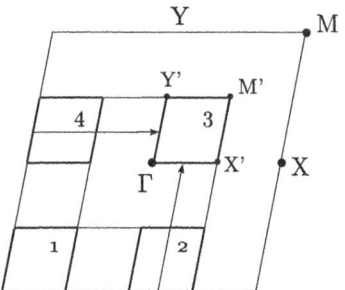

Fig. 73
Schematic band structure for the donor layers of the θ phases with two donors in the repeat unit according to the simple analytical equation.

Fig. 74
Schematic diagram illustrating the folding process.

1/4, to the inside of the folded BZ. The vectors needed for the folding process are those shown in Fig. 74. It is clear that if we label Γ, X, Y, and M the (0, 0), (1/2, 0), (0, 1/2), and (1/2, 1/2) points of the large BZ and Γ', X', Y', and M' those of the new, smaller BZ, the points Γ and Γ' coincide and the points X, Y, and M are projected into Γ'. Thus, the energies of the four bands at the centre of the new BZ will be given by eqn (11.6) appropriately modified to take into account the change in the axes and using the ($k_a = 0$, $k_b = 0$), ($k_a = 1/2$, $k_b = 0$), ($k_a = 0$, $k_b = 1/2$), and ($k_a = 1/2$, $k_b = 1/2$) values. If we want to plot the band structure along the $\Gamma' \to X' \to M' \to Y' \to \Gamma'$ path for instance, following the same procedure we find that the four bands will be associated with the energy values of the single molecule case for the four paths illustrated with bold lines in Fig. 74.

As mentioned above, $(a_1/a_2) = (c/-a)$ for β''-(BEDT–TTF)$_4$(guest)$_n$· [Re$_6$Q$_6$Cl$_8$] and $(a_1/a_2) = (b/-a)$ for (BEDO–TTF)$_2$· ReO$_4$· H$_2$O. Thus, direction a_2 is associated in both cases with the π-stacks, direction a_1 with the step-chains and direction (a_1+a_2) with the π-chains, so that the appropriate equation to use is:

$$E(\vec{k}) = \alpha + 2[t_{sc} \cos(2\pi k_{a_1}) + t_{\pi s} \cos(2\pi k_{a_2})$$
$$+ t_{\pi c} \cos(2\pi (k_{a_1} + k_{a_2}))]$$

Table 1 shows the energy values for these Γ', X', Y', M', and S' points, and these allow the construction of a generalised band structure for systems of this type. The labels 1, 2, 3, and 4 in this table refer to the four different paths of the large BZ projected onto the same path in the new, smaller BZ. When constructing the qualitative band structure, the energy values associated with the same label are those that must be correlated. To facilitate the comparison with the calculated band structures for the β''-(BEDT–TTF)$_4$(guest)$_n$ [Re$_6$Q$_6$Cl$_8$] and (BEDO–TTF)$_2$· ReO$_4$· H$_2$O salts where the π-stacks run along the first crystallographic direction, the labels X' and Y' should be relabelled Z and X for β''-(BEDT–TTF)$_4$(guest)$_n$· [Re$_6$Q$_6$Cl$_8$] and as Y and X for (BEDO–TTF)$_2$· ReO$_4$·H$_2$O. In this way, the qualitative band structure of Fig. 11.65 is generated for β''-(BEDT–TTF)$_4$(guest)$_n$· [Re$_6$Q$_6$Cl$_8$].

Table 1 Summary of the results for the band levels of the β'' lattice with a (2×2) repeat unit assuming that all HOMO levels are identical and equal to 0.

k-point	Band	Energy
Γ'	1	$2(-t_{sc} - t_{\pi s} + t_{\pi c})$
	2	$2(+t_{sc} - t_{\pi s} - t_{\pi c})$
	3	$2(+t_{sc} + t_{\pi s} + t_{\pi c})$
	4	$2(-t_{sc} + t_{\pi s} - t_{\pi c})$
X'	1	$-2t_{\pi s}$
	2	$-2t_{\pi s}$
	3	$+2t_{\pi s}$
	4	$+2t_{\pi s}$
Y'	1	$-2t_{sc}$
	2	$+2t_{sc}$
	3	$+2t_{sc}$
	4	$-2t_{sc}$
M'	1	$-2t_{\pi c}$
	2	$+2t_{\pi c}$
	3	$-2t_{\pi c}$
	4	$+2t_{\pi c}$
S'	1	$+2t_{\pi c}$
	2	$-2t_{\pi c}$
	3	$+2t_{\pi c}$
	4	$-2t_{\pi c}$

Chapter 12

Along this chapter, the notations $|\ |$ and $|\ \rangle$ stand for determinants

(12.1) The ground-state wavefunction of H_2 in the Hückel model is written in the localised basis set formed by the four distributions of Fig. 12.3:

$$\Psi_{GS}(t) = \cos\theta \left(\frac{|a\bar{b}\rangle - |\bar{a}b\rangle}{\sqrt{2}} \right) + \sin\theta \left(\frac{|a\bar{a}\rangle + |b\bar{b}\rangle}{\sqrt{2}} \right)$$

with $\theta = \pi/4$. In Fig. 12.1, σ_g and σ_u are the delocalised molecular orbitals deduced from the bonding and antibonding linear combinations of ϕ_a and ϕ_b:

$$\sigma_g = \frac{1}{\sqrt{2}}(\phi_a + \phi_b)$$

$$\sigma_u = \frac{1}{\sqrt{2}}(\phi_a - \phi_b)$$

Using the notations:

$$|\sigma_g \bar{\sigma}_g| = \frac{1}{2}|(a+b)(\bar{a}+\bar{b})| = \frac{1}{2}(|a\bar{a}\rangle + |b\bar{b}\rangle + |a\bar{b}\rangle + |b\bar{a}\rangle)$$

$$|\sigma_u \bar{\sigma}_u| = \frac{1}{2}|(a-b)(\bar{a}-\bar{b})| = \frac{1}{2}(|a\bar{a}\rangle + |b\bar{b}\rangle - |a\bar{b}\rangle - |b\bar{a}\rangle)$$

we show that:

$$\Psi_{GS}(t) = \left(\frac{\cos\theta + \sin\theta}{\sqrt{2}}\right)|\sigma_g\bar{\sigma}_g| + \left(\frac{\sin\theta - \cos\theta}{\sqrt{2}}\right)|\sigma_u\bar{\sigma}_u|$$

In the Hückel approximation, $\cos\left(\frac{\pi}{4}\right) = \sin\left(\frac{\pi}{4}\right)$, so that:

$$\Psi_{GS}(t) = \frac{2}{\sqrt{2}}\cos\left(\frac{\pi}{4}\right)|\sigma_g\bar{\sigma}_g| = |\sigma_g\bar{\sigma}_g|$$

(12.2) The ground-state wavefunction of $\hat{H}(t, U)$ is obtained by solving the linear equations:

$$\begin{pmatrix} -\epsilon_{GS} & 0 & +t & +t \\ 0 & -\epsilon_{GS} & -t & -t \\ +t & -t & U-\epsilon_{GS} & 0 \\ +t & -t & 0 & U-\epsilon_{GS} \end{pmatrix} \begin{pmatrix} c_1 \\ c_2 \\ c_3 \\ c_4 \end{pmatrix} = 0$$

with:

$$\epsilon_{ES} = \frac{U - \sqrt{U^2 + 16t^2}}{2} = \frac{U - X}{2}$$

This leads to:

$$c_1 = -c_2$$

$$c_3 = +c_4$$

$$\frac{c_1}{c_3} = -\frac{1}{4t}(U + X)$$

$$\frac{c_2}{c_4} = +\frac{1}{4t}(U + X)$$

and then to:

$$c_1 = -c_2 = \frac{\cos\theta}{\sqrt{2}}$$

$$c_3 = c_4 = \frac{\sin\theta}{\sqrt{2}}$$

$$\frac{c_3}{c_1} = \tan\theta = -\frac{4t}{U + X} = \frac{4|t|}{U + \sqrt{U^2 + 16t^2}}$$

(12.3) Within the Hubbard model, $\tan\theta$ gives the relative contribution of the covalent and ionic states to the ground-state wavefunction of H_2. Its asymptotes in the regimes of weak and strong repulsion can be deduced from a first-order Taylor expansion.

- In the limit of weak repulsion, $U \ll 4|t|$ so that:

$$\epsilon = \frac{U}{4|t|} \to 0$$

We can then write:

$$\lim_{\epsilon \to 0} \tan\theta = \lim_{\epsilon \to 0} \left(\frac{-4t}{U + \sqrt{U^2 + 16t^2}} \right)$$

$$= \lim_{\epsilon \to 0} \left(\frac{4|t|}{U + \sqrt{16t^2 \left(\frac{U^2}{16t^2} + 1\right)}} \right)$$

$$= \lim_{\epsilon \to 0} \left(\frac{4|t|}{U + 4|t|\sqrt{\epsilon^2 + 1}} \right)$$

$$= \lim_{\epsilon \to 0} \left(\frac{1}{\epsilon + \sqrt{\epsilon^2 + 1}} \right)$$

$$\approx \lim_{\epsilon \to 0} \left(\frac{1}{\epsilon + 1} \right) = (1+\epsilon)^{-1} \approx 1 - \epsilon = 1 - \frac{U}{4|t|}$$

- In the limit of strong repulsion, $U \gg 4|t|$ so that:

$$\epsilon' = \frac{4|t|}{U} \to 0$$

and

$$\lim_{\epsilon' \to 0} \tan\theta = \lim_{\epsilon' \to 0} \left(\frac{-4t}{U + \sqrt{U^2 + 16t^2}} \right)$$

$$= \lim_{\epsilon' \to 0} \left(\frac{4|t|}{U + |U|\sqrt{1 + \frac{16t^2}{U^2}}} \right)$$

$$= \lim_{\epsilon' \to 0} \left(\frac{\epsilon'}{1 + \sqrt{1 + \epsilon'^2}} \right) \approx \frac{\epsilon'}{2} = \frac{1}{2}\left(\frac{4|t|}{U}\right) = \frac{2|t|}{U}$$

NB: In these developments we have assumed that ϵ^2 and ϵ'^2 are negligible over ϵ and ϵ'.

(12.4) (a) We saw in Exercise (12.1) that $\Psi_{GS}(t, U)$ is written as:

$$\Psi_{GS}(t, U) = \left(\frac{\cos\theta + \sin\theta}{\sqrt{2}}\right) |\sigma_g \bar\sigma_g| + \left(\frac{\sin\theta - \cos\theta}{\sqrt{2}}\right) |\sigma_u \bar\sigma_u|$$

In the limit of strong repulsion, $\theta \to 0$ since $\tan\theta \to 0$. Hence:

$$\cos\theta + \sin\theta = -(\sin\theta - \cos\theta) = 1$$

which leads to:

$$\Psi_{GS}(t, U_{+\infty}) = \frac{1}{\sqrt{2}} \left(|\sigma_g \bar\sigma_g| - |\sigma_u \bar\sigma_u| \right)$$

which corresponds in the basis set to formed by the ϕ_a and ϕ_b local orbitals:

$$\Psi_{GS}(t, U_{+\infty}) = \frac{1}{\sqrt{2}} \left(|a\bar b\rangle - |\bar a b\rangle \right)$$

Using the notations:

$$|a\bar{b}\rangle = \frac{1}{\sqrt{2}}\{a(1)\bar{b}(2) - a(2)\bar{b}(1)\}$$

$$= \frac{1}{\sqrt{2}}\{\phi_a(r_1)\alpha(\sigma_1)\phi_b(r_2)\beta(\sigma_2) - \phi_a(r_2)\alpha(\sigma_2)\phi_b(r_1)\beta(\sigma_1)\}$$

$$|\bar{a}b\rangle = \frac{1}{\sqrt{2}}\{\bar{a}(1)b(2) - \bar{a}(2)b(1)\}$$

$$= \frac{1}{\sqrt{2}}\{\phi_a(r_1)\beta(\sigma_1)\phi_b(r_2)\alpha(\sigma_2) - \phi_a(r_2)\beta(\sigma_2)\phi_b(r_1)\alpha(\sigma_1)\}$$

we can show that $\Psi_{GS}(t, U_{+\infty})$ can be written as:

$$\Psi_{GS}(t, U_{+\infty}) = \frac{1}{\sqrt{2}}\{|a\bar{b}\rangle - |\bar{a}b\rangle\}$$

$$= \frac{1}{2}[\{\phi_a(r_1)\alpha(\sigma_1)\phi_b(r_2)\beta(\sigma_2) - \phi_a(r_2)\alpha(\sigma_2)\phi_b(r_1)\beta(\sigma_2)\}$$

$$= -\{\phi_a(r_1)\beta(\sigma_1)\phi_b(r_2)\alpha(\sigma_2) - \phi_a(r_2)\beta(\sigma_2)\phi_b(r_1)\alpha(\sigma_2)\}]$$

$$= \frac{1}{2}[\phi_a(r_1)\phi_b(r_2)\{\alpha(\sigma_1)\beta(\sigma_2) - \beta(\sigma_1)\alpha(\sigma_2)\}$$

$$= +\phi_a(r_2)\phi_b(r_1)\{\beta(\sigma_2)\alpha(\sigma_1) - \alpha(\sigma_2)\beta(\sigma_1)\}]$$

$$= \frac{1}{2}\{\phi_a(r_1)\phi_b(r_2) + \phi_a(r_2)\phi_b(r_1)\} \cdot \{\alpha(\sigma_1)\beta(\sigma_2) - \beta(\sigma_1)\alpha(\sigma_2)\}$$

showing that $\Psi_{GS}(t, U_{+\infty})$ is a antisymmetrised product of an antisymmetric spatial part and a symmetric spin part, i.e. a spin singlet. It is an eigenfunction of the $\hat{S}^2$ operator, while $|a\bar{b}|$ and $|\bar{a}b|$ are eigenfunctions of the $\hat{S}_z$ operator.

(b) The first-excited state of H_2 in the Hubbard model corresponds to the wavefunction associated with the energy $\epsilon_2 = 0$. This eigenvalue is equivalent to the ground-state energy $\epsilon_{GS}(t, U_{+\infty})$. Solving the linear equations:

$$\begin{pmatrix} 0 & 0 & +t & +t \\ 0 & 0 & -t & -t \\ +t & -t & U & 0 \\ +t & -t & 0 & U \end{pmatrix} \begin{pmatrix} c_1 \\ c_2 \\ c_3 \\ c_4 \end{pmatrix} = 0$$

for $\epsilon_{ES}(t, U)$ leads to:

$$\Psi_{ES}(t, U) = \frac{1}{\sqrt{2}}(|a\bar{b}\rangle + |\bar{a}b\rangle)$$

which can be written as:

$$\Psi_{ES}(t, U) = \frac{1}{\sqrt{2}}(|\sigma_g \bar{\sigma}_u| + |\bar{\sigma}_g \sigma_u|)$$

with:

$$|\sigma_g \bar{\sigma}_u| = \frac{1}{2} |(a+b)(\bar{a}-\bar{b})|$$

$$= \frac{1}{2}(|a\bar{a}\rangle - |a\bar{b}\rangle + |b\bar{a}\rangle - |b\bar{b}\rangle)$$

$$= \frac{1}{2}(|a\bar{a}\rangle - |a\bar{b}\rangle - |\bar{a}b\rangle - |b\bar{b}\rangle)$$

$$|\bar{\sigma}_g \sigma_u| = \frac{1}{2} |(\bar{a}+\bar{b})(a-b)|$$

$$= \frac{1}{2}(|\bar{a}a\rangle - |\bar{a}b\rangle + |\bar{b}a\rangle - |\bar{b}b\rangle)$$

$$= \frac{1}{2}(-|a\bar{a}\rangle - |\bar{a}b\rangle - |a\bar{b}\rangle + |b\bar{b}\rangle)$$

Following the same approach as for the spin singlet, this wavefunction can be written as:

$$\Psi_{ES}(t, U) = \frac{1}{\sqrt{2}}(|a\bar{b}\rangle + |\bar{a}b\rangle)$$

$$= \frac{1}{2}\{\phi_a(r_1)\phi_b(r_2) - \phi_b(r_1)\phi_a(r_2)\} \cdot \{\alpha(\sigma_1)\beta(\sigma_2) + \beta(\sigma_1)\alpha(\sigma_2)\}$$

showing that $\Psi_{ES}(t, U)$ is an antisymmetrised product of a symmetric spatial part and an antisymmetric spin part, i.e. a spin triplet. It is an eigenfunction of the $\hat{S}^2$ operator with $S = 1$ and $m_s = 0$, while $|a\bar{b}\rangle$ and $|\bar{a}b\rangle$ are eigenfunctions of the $\hat{S}_z$ operator.

(12.6) In the Hartree–Fock approximation, the energy of the n-interacting-particle system is given by:

$$E^{HF} = \langle \Psi_n | \sum_{i=1}^{n} \hat{F}_i | \Psi_n \rangle$$

$$= \sum_{i=1}^{n} \left\{ e_i + \frac{1}{2} \sum_{j \neq i} (J_{ij} - K_{ij}) \right\}$$

where $\Psi_n = \Psi(x_1, x_2)$ correspond to the two-electron wavefunctions of the H$_2$ molecule. Considering the Slater determinants:

$$\Psi_1 = |\phi_g \bar{\phi}_g|$$

$$= \frac{1}{\sqrt{2}}\{\phi_g(r_1)\alpha(\sigma_1)\phi_g(r_2)\beta(\sigma_2) - \phi_g(r_2)\alpha(\sigma_2)\phi_g(r_1)\beta(\sigma_1)\}$$

$$\Psi_2 = |\phi_g \phi_u|$$

$$= \frac{1}{\sqrt{2}}\{\phi_g(r_1)\alpha(\sigma_1)\phi_u(r_2)\alpha(\sigma_2) - \phi_g(r_2)\alpha(\sigma_2)\phi_u(r_1)\alpha(\sigma_1)\}$$

$$\Psi_3 = |\phi_g \bar{\phi}_u|$$
$$= \frac{1}{\sqrt{2}} \{\phi_g(r_1)\alpha(\sigma_1)\phi_u(r_2)\beta(\sigma_2) - \phi_g(r_2)\alpha(\sigma_2)\phi_u(r_1)\beta(\sigma_1)\}$$
$$\Psi_4 = |\phi_u \bar{\phi}_u|$$
$$= \frac{1}{\sqrt{2}} \{\phi_u(r_1)\alpha(\sigma_1)\phi_u(r_2)\beta(\sigma_2) - \phi_u(r_2)\alpha(\sigma_2)\phi_u(r_1)\beta(\sigma_1)\}$$

then the Hartree–Fock energy for Ψ_ℓ is:

$$E_\ell^{HF} = e_i^{\phi_k} + \frac{1}{2}\left(J_{ij}^{\phi_k\phi_l} - K_{ij}^{\phi_k\phi_l}\right) + e_j^{\phi_l} + \frac{1}{2}\left(J_{ji}^{\phi_l\phi_k} - K_{ji}^{\phi_l\phi_k}\right)$$

where e_i^k is the energy of electron i in spin-orbital k, and where J_{ij}^{kl} and K_{ij}^{kl} are the Coulomb and exchange energies between two electrons i and j lying in spin-orbitals k and l, respectively. Hence when the two electrons i and j have opposite spins, $K_{ij}^{\phi_k\phi_l} = K_{ji}^{\phi_k\phi_l} = 0$, independently of k and j. Moreover, when the two electrons lie in the same orbital, the Coulomb potential exerted by electron i on j is equivalent to the one exerted by electron j on i so that $e_i^{\phi_k} = e_j^{\phi_k} = e_k$ and $J_{ij}^{\phi_k\phi_k} = J_{ji}^{\phi_k\phi_k} = J_{kk} \neq 0$. We then get:

$$E_1^{HF} = e_1^{\phi_g} + e_2^{\phi_g} + J_{12}^{\phi_g\phi_g}$$
$$= 2e_g + J_{gg}$$
$$E_2^{HF} = e_1^{\phi_g} + e_2^{\phi_u} + J_{12}^{\phi_g\phi_u}$$
$$= e_g + e_u + J_{gu} - K_{gu}$$
$$E_3^{HF} = e_1^{\phi_g} + e_2^{\phi_u} + J_{12}^{\phi_g\phi_u}$$
$$= e_g + e_u + J_{gu}$$
$$E_4^{HF} = e_1^{\phi_u} + e_2^{\phi_u} + J_{12}^{\phi_u\phi_u}$$
$$= 2e_u + J_{uu}$$

Through this example, it is interesting to see that the Hartree–Fock energy can be written in different ways depending on whether the n-electron system is closed-shell (CS) or open-shell (OS). Indeed, if one writes the closed- and open-shell determinants as:

$$\Psi_n^{OS} = |\phi_i(x_1), \phi_j(x_2)\ldots\phi_{n-1}(x_{n-1})\phi_n(x_n)|$$
$$\Psi_n^{CS} = |\phi_i(x_1), \bar{\phi}_i(x_2)\ldots\phi_{n/2}(x_{n/2})\bar{\phi}_{n/2}(x_{n/2})|$$

then:

$$E^{OS} = \sum_{i=1}^{n}\left\{e_i + \frac{1}{2}\sum_{j\neq i}^{n}(J_{ij} - K_{ij})\right\}$$

$$E^{CS} = \sum_{i=1}^{n/2}\left\{2e_i + \frac{1}{2}\sum_{j\neq i}^{n/2}(2J_{ij} - K_{ij})\right\}$$

Appendix

Character tables

C_s	E	σ_{xy}	
A'	1	1	x, y, x^2, y^2, z^2, xy
A''	1	−1	z, xz, yz

C_{2v}	E	C_{2z}	σ_{xz}	σ_{yz}	
A_1	1	1	1	1	z, x^2, y^2, z^2
A_2	1	1	−1	−1	xy
B_1	1	−1	1	−1	x, xz
B_2	1	−1	−1	1	y, xz

C_{4v}	E	$2C_4$	C_2	$2\sigma_v$	σ_d	
A_1	1	1	1	1	1	z, x^2+y^2, z^2
A_2	1	1	1	−1	−1	
B_1	1	−1	1	1	−1	x^2-y^2
B_2	1	−1	1	−1	1	xy
E	2	0	−2	0	0	$(x,y), (xz, yz)$

C_{2h}	E	$2C_z$	i	σ_{xy}	
A_g	1	1	1	1	x^2, y^2, z^2, xy
B_g	1	−1	1	−1	xz, yz
A_u	1	1	−1	−1	z
B_u	1	−1	−1	1	x, y

D_{2h}	E	C_{2z}	C_{2y}	C_{2x}	i	σ_{xy}	σ_{xz}	σ_{yz}	
A_g	1	1	1	1	1	1	1	1	x^2, y^2, z^2
B_{1g}	1	1	−1	−1	1	1	−1	−1	xy
B_{2g}	1	−1	1	−1	1	−1	1	−1	xz
B_{3g}	1	−1	−1	1	1	−1	−1	1	yz
A_u	1	1	1	1	−1	−1	−1	−1	
B_{1u}	1	1	−1	−1	−1	−1	1	1	z
B_{2u}	1	−1	1	−1	−1	1	−1	1	y
B_{3u}	1	−1	−1	1	−1	1	1	−1	x

D_{4h}	E	$2C_4$	C_2	$2C_2'$	$2C_2''$	i	$2S_4$	σ_h	$2\sigma_v$	$2\sigma_d$	
A_{1g}	1	1	1	1	1	1	1	1	1	1	x^2+y^2, z^2
A_{2g}	1	1	1	−1	−1	1	1	1	−1	−1	
B_{1g}	1	−1	1	1	−1	1	−1	1	1	−1	x^2-y^2
B_{2g}	1	−1	1	−1	1	1	−1	1	−1	1	xy
E_g	2	0	−2	0	0	2	0	−2	0	0	(xz, yz)
A_{1u}	1	1	1	1	1	−1	−1	−1	−1	−1	
A_{2u}	1	1	1	−1	−1	−1	−1	−1	1	1	z
B_{1u}	1	−1	1	1	−1	−1	1	−1	−1	1	
B_{2u}	1	−1	1	−1	1	−1	1	−1	1	−1	
E_u	2	0	−2	0	0	−2	0	2	0	0	(x, y)

$C_{\infty v}$	E	$2C_\infty^\varphi$	...	$\infty \sigma_v$	
Σ^+	1	1	...	1	z, x^2+y^2, z^2
Σ^-	1	1	...	−1	
Π	2	$2\cos(\phi)$	...	0	$(x, y), (yz, xz)$
...	...	...	...	...	

$D_{\infty h}$	E	$2C_\infty^\varphi$	...	$\infty \sigma_v$	i	$2S_\infty^\varphi$	...	∞C_2	
Σ_g^+	1	1	...	1	1	1	...	1	x^2+y^2, z^2
Σ_g^-	1	1	...	−1	1	1	...	−1	
Π_g	2	$2\cos(\phi)$	...	0	2	$-2\cos(\phi)$	...	0	(yz, xz)
...	...	...	...	...	...	...	...	...	
Σ_u^+	1	1	...	1	−1	−1	...	−1	z
Σ_u^-	1	1	...	−1	−1	−1	...	1	
Π_u	2	$2\cos(\phi)$	...	0	−2	$2\cos(\phi)$	...	0	(x, y)
...	...	...	...	...	...	...	...	...	

Index

1D metal, 206, 232

A_n cyclic system, 96
A_n linear system, 101, 104
A_n square lattice system, 169
Acceptor stacks, 232, 238
Activated conductivity, 236, 238
AlB_2 3D systems, 196
Analytical approach
   to band structure, 241
Antibonding
   character, 30
   interaction, 39, 68
   orbital, 18, 51, 61, 86
Antiferromagnetism, 264
Antisymmetry
   product, 15, 277
   property, 277
Approximation
   Born–Oppenheimer, 15, 40
   DFT+U, 285
   Dimer splitting, 241
   GGA, 284
   Hartree, 276
   Hartree–Fock, 277
   LCAO, 15, 256
   LDA, 284
   mean-field, 273
   one-electron, 15
   rigid band, 197
   tight-binding, 256
Atomic orbital, 16
Avoided band crossing, 107, 114, 126, 165, 227, 231, 252

Back-donation (donation), 31
Band
   crossing, 107, 131, 165
   dispersion, 142, 172, 182, 195, 218, 222, 233, 234, 240, 247
   filling, 60, 72, 81, 197, 235, 236
   folding, 67, 75, 247, 248
   gap, 57, 60, 194, 209, 236
   structure, 5
   theory, 44, 70, 208, 236
   width (W), 54, 60, 233, 267
Band structure

α-$(BEDT–TTF)_2MHg(XCN)_4$, 243
β-$(BEDO–TTF)_{2.4}I_3$, 241
β''-$(BEDT–TTF)_4(1\text{-}4\text{-dioxane})_n \cdot [Re_6Q_6Cl_8]$, 251
$^2_\infty[H_n]$ square lattice, 168
$(H_2)_n$ dimerized chain, 58
$(ML_4L')_n$ linear chain, 146
$(NaCl)_n$ linear chain, 112
$(Pt(NH_2Et)_4Cl^{2+})_n$ linear chain, 149
$(ReCl_4N)_n$ linear chain, 147
$(TMTSF)_2PF_6$, 234
$(TMTSF)_3[Ta_2F_{11}]$, 237
$(TMTSF)_3[Ti_2F_8(C_2O_4)]$, 237
A_n linear chain, 108
BN hexagonal layers, 179
eclipsed KCP, 137
graphene layers, 173
H_n linear chain, 51, 54
MB_2 system, 199
polyacene, 123, 126
regular trans-polyacetylene, 87, 88
staggered KCP, 139
TaS_3 monoclinic phase, 216
topology, 237, 238, 244
Basis for a representation, 22, 46, 102
Basis set, 16, 22, 159, 260, 275
Bechgaard salts, 210
BEDO–TTF donor, 239
BEDT–TTF donor, 235, 250
Bloch
   energy, 52, 55
   function, 47, 55
   orbital, 45, 47, 49, 160
BN (boronitride) hexagonal system, 179
Boltzmann constant, 5
Bonding
   character, 30
   interaction, 68
   orbital, 18, 51, 61, 86
Born–Oppenheimer approximation, 15, 40, 273
Born–von Karman
   boundary conditions, 2, 44, 102
Boron–boron bonds, 196
Bridging ligand, 217
Brillouin zone, 48, 49, 72, 160
   construction, 161
   folding, 248, 250

Brillouin zone (*cont.*)
  of the graphene layers, 174
  of the hexagonal lattice, 199
  of the MB_2 system, 201
  of the square lattice, 166
  symmetry, 54, 162, 164, 194
Bronzes, 210
  blue $A_{0.3}MoO_3$, 215
  purple $A_{0.9}Mo_6O_{17}$, 215
  purple KMo_6O_{17}, 184, 227
  red $A_{0.33}MoO_3$, 215, 219

C_5H_5M system, 193
C_n cyclic group, 45, 46, 102
Character table (see Appendix), 45
Charge density wave (CDW), 73, 77, 209, 226, 231, 242
Charge transfer salts, 234
Cluster orbitals, 220
Commensurate distortion, 81
Conduction band, 12, 147, 239, 249, 267
Conductivity vs Resistivity, 13
Conductor (see Metal), 10
Configuration interaction (CI), 277
Correlation energy, 280, 282
Coulomb
  energy, 261, 276, 282, 285
  interaction, 260
  potential, 260, 273, 276
Coulomb hole, 283
Covalent character, 31
Covalent bonds, 184
Crystal orbital, 45, 49, 52, 165
Crystal orbital Hamilton population (COHP), 185
Crystal orbital overlap population (COOP), 185
  trans-polyacetylene, 186
  AlB_2 diborides, 197
  $FePS_3$ layered phase, 188
Crystal packing, 238
Current density, 8
Cyclobutadiene, 18, 25
Cylindrical symmetry, 139

Defects, 11, 268
Density
  matrix, 282
  probability, 15
Density functional theory (DFT), 274, 281
Density matrix renormalisation group (DMRG), 270
Density of states (DOS), 181

KMo_6O_{17}, 184
cyclobutadiene, 181
$FePS_3$ layered phase, 188
Diffuse X-ray scattering, 227
Dimer splitting approximation, 241
Dimerisation, 130
Donation (back-donation), 31
Donor stacks, 232, 236, 238, 250
Donor–acceptor salts, 232
Doping, 133, 141

Effective
  Hamiltonian, 265
  interaction, 265
  model, 265
Eigen-
  function, 165, 208, 264
  state, 9
  value, 264
  vector, 2
Elastic energy, 209
Electrical conductivity, 13, 224, 249
Electron
  –phonon coupling, 11, 72, 209
  collision, 7
  correlation, 277
  count, 147, 149, 192, 211, 219, 226, 238, 242, 288
  density, 208, 281
  diffraction, 227
  hopping, 233
  repulsion, 209, 267
  transfer, 258
Electron affinity (EA), 267
Electron repulsion
  intersite, 260
  on-site, 233, 260
Electronic
  distribution, 208
  instability, 207, 231, 233
  localisation, 233
  state
    ionic, 258
    neutral, 258
  wavefunction, 15
Electrostatic interaction, 184
Elementary building block, 193
Energy
  barrier, 40, 69
  dispersion, 142, 195, 222
  gap, 57, 60, 73
Equilibrium geometry, 69
Exchange
  energy, 277, 282, 285
  integral, 263, 277

Index

Exchange-correlation
  energy, 284
  functional, 283
  hole, 283

Fermi level, 5, 10
Fermi hole, 283
Fermi surface, 204, 207, 208, 227, 231, 248, 249
  α-(BEDT–TTF)$_2$MHg(XCN)$_4$, 243
  β-(BEDO–TTF)$_{2.4}$I$_3$, 241
  β''-(BEDT–TTF)$_4$(1-4-dioxane)$_n$·[Re$_6$Q$_6$Cl$_8$], 251
  (BEDO–TTF)$_2$ Cl(H$_2$O)$_x$, 247
  (BEDO–TTF)$_2$ ReO$_4$·H$_2$O, 247
  (BEDO–TTF)$_5$ [CsHg(SCN)$_4$]$_2$, 247
  (TMTSF)$_2$PF$_6$, 235
  1D (open), 243
  2D (closed), 243
  blue bronzes, 226
  hybridisation, 231
  superimposition, 252
  TaS$_3$ monoclinic phase, 214
  warping, 243
Fermi–Dirac, 5
Fermions, 257
First-principles methods, 196, 273
Flat band, 137, 218
Fock operator, 274
Folding, 64, 247
Fragment orbital, 16, 89, 123, 134, 143
Free electron model, 1, 10, 44
Functional, 275
  DFT+U, 285
  hybrid B3LYP, 285

Gap, 57, 194, 209, 270
Generalised gradient approximation (GGA), 284
Generator set (see Symmetry), 19
Glide plane, 115, 120, 130
Gross population, 183
Ground-state configuration, 15
Group theory, 14

H$_n$ square lattice system, 166
Hückel, 256
  extended Hückel method, 17, 196
  simple Hückel method, 17
Hamiltonian, 15
  $t - J$, 265
  Hückel, 17, 264
  Heisenberg, 265
  Hubbard, 260, 264
  matrix, 50, 161, 242, 258
Hartree
  energy, 276
  product, 274
Hartree–Fock (HF), 274
  energy, 277
Heisenberg model, 264
Hidden nesing, 227
Hidden nesting vector, 231
Highest occupied molecular orbital (HOMO), 234
Hohenberg–Kohn
  universal function, 281
Hopping integral, 233, 258, 270
Hubbard
  Hubbard model, 260
  model, 264

Incommensurate distortion, 83, 215
Insulator, 12
Interacting particle system, 256
Interaction
  δ type, 189
  π-type, 189
  σ-type, 189
  antisymmetric, 18
  long-range, 269
  short-range, 269
  symmetric, 18
Interaction diagram
  C$_4$H$_4$ π system, 30
  cyclobutadiene, 30
  ethylene C$_2$H$_4$, 86, 87
  H$_2$, 256
  ML$_6$ octahedral complex, 34
  ML$_6$, L π-acceptor, 37
  ML$_6$, L π-donor, 37
Interaction term β, 52
Interchain interactions, 222, 226
Intermolecular interactions, 233–235, 248
Intrachain interaction, 223
Intramolecular interactions, 234
Ionisation
  potential (IP), 267
Irreducible
  Brillouin zone, 163, 164
  representation, 22, 46
Isomorphism, 102

Jahn–Teller
  distortion, 40, 69
  effect, 40
    first-order, 41
    second-order, 42

k point, 161
　group, 104
　mesh, 182
KCP linear system, 133
Kohn–Sham orbitals, 281

Lagrange multipliers, 275
Lattice
　direct, 48, 157, 161
　reciprocal, 48, 157, 161
　vector, 159
　vibrations, 11, 68, 72
Lewis
　acid, 31
　base, 31
　structure, 85
Ligand, 31
　π-acceptor, 36, 143
　π-donor, 36, 143
　σ-donor, 33, 143
Linear combination
　of atomic orbitals (LCAO), 15, 256, 275
　of Bloch orbitals, 161
Local density approximation (LDA), 284
Local distortion, 248
Localised picture, 233, 260
Low-dimensional systems, 207, 224, 232, 233, 269
Lowest unoccupied molecular orbital (LUMO), 238

Magnetic interactions
　antiferromagnetic, 264, 288
　ferromagnetic, 264
Magnetic moment, 209
Magnetoresistance, 247
Many-body problem, 273
Many-body system, 256
Matrix product, 22
Mean-field approach, 273
Metal, 10, 11
Metal–insulator transition, 69, 81, 130, 208, 209, 233, 267
Metal-insulator transition, 224
Metal–metal bonding, 187
Mixed valence state, 193
ML_4L' linear system, 143
Molecular conductors, 232, 238
Molecular orbital, 15
Molecular orbital theory, 14
Momentum vector, 7
Mott–Hubbard transition, 267
MPS_3 layered system, 186
Mulliken
　overlap population, 185
　population analysis, 183
Multi-band systems, 182, 273

n-merisation, 77, 140, 153
NaCl linear system, 109
$NbSe_3$ system, 210
Nesting vector, 81, 207, 209, 226, 227, 231
Ni(dmit)$_2$ acceptor, 237
Nodal surface, 18, 73, 82, 83, 193
Non-bonding
　interaction, 34
　orbital, 36
Non-interacting particle system, 256
Normalisation
　condition, 259
　factor, 28, 266

Octahedral symmetry, 135
Ohm's law, 7
Operator
　annihilation, 265
　Coulomb, 275
　creation, 265
　exchange, 277
　Fock, 275
　kinetic, 273
　one-eletron, 273
　projection, 24, 46
　two-electron, 273
Orbital splitting, 187, 189
One-dimensional systems, 44
Organic conductors, 249
　α-(BEDT–TTF)$_2$MHg(XCN)$_4$, 242
　β-(BEDO–TTF)$_{2.4}$I$_3$, 239
　NR$_4$[Ni(dmit)$_2$]$_2$, 238
　NR$_4$[Ni(dmit)$_2$]$_2$, 269
　TTF-TCNQ, 232
Organic superconductor (TMTSF)$_2$PF$_6$, 232
Orthogonality condition, 29
Overlap
　δ type, 18
　π type, 18
　σ type, 18
　σ-type, 107
　density, 183
　matrix, 50, 161, 242
　mode, 238

Pauli
　exclusion principle, 15, 257, 277
Peierls
　distortion, 69, 142, 193

first-order, 72, 130, 153
second-order, 128, 131, 149
instability, 269
transition, 267, 271
Periodic system, 45
Perturbation theory, 263
Polyacene linear system, 122
Primitive cell, 157, 225
Projected DOS, 183
[(C_5H_5)Fe] chain, 195
[PtL_4Cl]$^{2+}$ (L=$EtNH_2$), 191
AlB_2 diborides, 197
$FePS_3$ layered phase, 188
KMo_6O_{17}, 184
Projector, 24, 46, 265
Pseudo-1D metal, 205, 224, 232, 235
$Pt(CN)_4^{2-}$ linear system, 102
$Pt(NH_3)_4Cl$ system, 188
Pyramidal symmetry, 149

Qualitative band structure
blue bronzes, 226
purple bronzes, 230
red bronzes, 217

Reducible representation, 23
Representation
of a group, 22
Resistivity, 11
anomaly, 184, 210, 227, 242
Resonance integral, 270
Rigid band approach, 197
Rotational symmetry, 46

SCF iterative procedure, 276
Schrödinger equation, 2, 49
Screening effect, 267, 285
Screw axis, 116, 120, 130, 225
Second quantisation, 264
Secular
determinant, 16, 50, 91, 161, 175, 259, 275
equations, 16, 91, 160
Self-consistent field (SCF), 274
Self-interaction
correction, 285
error, 285
Semi-empirical methods, 256
Semiconductor, 12
Semimetal, 11
Shubnikov–de Haas oscillations, 242, 250
Slater
determinant, 15, 257, 277, 278
orbitals, 17
Space group, 102, 115
Spin
contamination, 280
orbitals, 275
state, 262
singlet, 264
triplet, 264
Spin density wave (SDW), 209
Spin Hamiltonian, 262
Spin operator, 265
Spin sensity wave (SDW), 235
Square-planar symmetry, 135, 149
Stacking π interactions, 233
Stoichiometry, 238
Strongly correlated
orbitals, 286
systems, 256, 285, 286
Structural distortion, 207, 209, 227
dimerisation, 67
tetramerisation, 73, 216
Superconductor, 210
Symmetry
classes, 21
generator set, 19
lowering, 25, 68, 252, 288
operation, 19, 164
point group, 19, 113
properties, 18, 96, 168, 174
Symmorphism, 102, 114

T_n translation group, 45, 102, 116, 123
TaS_3 system, 210
Thermal excitation energy, 40, 270
Tight-binding approximation, 256, 259
Time reversal, 54, 162
TMTSF
donor, 234
HOMO, 234
Trans-polyacetylene, 85, 113
Transfer integral, 170, 235, 237, 239, 242, 250
Transition metal
complex, 31, 135
$Cu(NH_3)_6^{2+}$, 39
ML_6, 33
$ReNCl_4$, 32
diborides
ReB_2, 196
TiB_2, 197
oxidation state, 32, 147, 149, 212, 226
oxides
β-SrV_6O_{15}, 287
bronzes, 215
CoO, 286

Translational
  properties, 130, 131
  symmetry, 14, 44, 46, 55, 102, 160
  vector, 44
Transport properties, 236
Trigonal prismatic symmetry, 211

Uniform electron gas, 283
Unit cell, 44, 123, 157

Valence band, 12, 147, 267
van der Waals interactions, 211
van Hove singularity, 182
Variational principle, 274, 280

Wave vector, 7, 48, 208
Wave-particle dualism, 265
Wavefunction, 2
Wigner–Seitz cell, 161
Wolfram red salt, 188

The manufacturer's authorised representative in the EU for product safety is Oxford University Press España S.A. of El Parque Empresarial San Fernando de Henares, Avenida de Castilla, 2 - 28830 Madrid (www.oup.es/en or product.safety@oup.com). OUP España S.A. also acts as importer into Spain of products made by the manufacturer.
Printed and bound by CPI Group (UK) Ltd, Croydon, CR0 4YY

20/03/2026

02075336-0020